Core Level Spectroscopy of Solids

Advances in Condensed Matter Science

Edited by D.D. Sarma, G. Kotliar and Y. Tokura

Core Level Spectroscopy of Solids

Frank de Groot

Akio Kotani

CRC Press
Taylor & Francis Group
Boca Raton   London   New York

CRC Press is an imprint of the
Taylor & Francis Group, an **informa** business

CRC Press
Taylor & Francis Group
6000 Broken Sound Parkway NW, Suite 300
Boca Raton, FL 33487-2742

© 2008 by Taylor & Francis Group, LLC
CRC Press is an imprint of Taylor & Francis Group, an Informa business

---

Library of Congress Cataloging-in-Publication Data

Groot, Frank de, 1964-
    Core level spectroscopy of solids / Frank de Groot and Akio Kotani.
        p. cm. -- (Advances in condensed matter science ; v. 6)
    Includes bibliographical references and index.
    ISBN 978-0-8493-9071-5 (alk. paper)
    1. Solids--Spectra. 2. Spectrum analysis. 3. Electron spectroscopy. 4.
Energy-band theory of solids. I. Kotani, A. (Akio), 1941- II. Title. III. Series.

QC176.8.O6G76 2008
530.4'16--dc22   2007029414

Visit the Taylor & Francis Web site at
http://www.taylorandfrancis.com

and the CRC Press Web site at
http://www.crcpress.com

*This book is dedicated to our partners
Hedwig te Molder and Machiko.*

Contents

Preface

Core level spectroscopy is a powerful tool for the study of electronic states in solids where the precise information of valence electron states can be detected through excitation of core electrons with a local and sensitive probe. Traditionally, the study of the electronic and optical properties of solids was performed with photons in the visible and infrared region of the spectrum. Since the development of second generation synchrotron radiation sources in the 1970s, the use of x-rays has expanded widely. The most characteristic feature of an x-ray is that it excites a core electron; in other words, x-ray spectroscopy has opened a new door on core level spectroscopy. A direct implication is that the information on core level spectroscopy is local and element specific. The experimental study of core level spectroscopy has made remarkable progress using high-brilliance third-generation synchrotron radiation sources. In order to interpret new experimental data, theoretical study of core level spectroscopy has made continuous progress over the last 30 years and has made a major contribution to the physical understanding of spectroscopy.

This book examines fundamental and theoretical aspects of core level spectroscopy including x-ray photoemission spectroscopy, x-ray absorption spectroscopy, and resonant x-ray emission spectroscopy. With an emphasis on interpreting core level spectroscopy for transition metal and rare earth systems from a model based on charge transfer multiplet theory, this book presents typical experimental results and discusses various theoretical approaches such as atomic multiplet theory, ligand field theory, and charge transfer multiplet theory, making it an ideal reference for the research community.

The two authors, Frank de Groot and Akio Kotani, have been studying core level spectroscopy of solids for many years. Since the end of the 1980s, Frank de Groot has been involved in both experimental and theoretical studies on a wide range of materials. Akio Kotani has contributed to the theoretical development of this field continuously since his pioneering work in the 1970s. This book concentrates on the charge transfer multiplet theory for f and d electron systems. A wide range of other theoretical approaches are also touched upon briefly. All chapters were written jointly, although Chapters 2, 4, and 6 were mainly written by Frank de Groot and Chapters 3, 5, and 8 by Akio Kotani. It is the authors' hope that this book is useful for all researchers involved in x-ray and electron spectroscopy. This includes the users of x-ray spectroscopy and microscopy beamlines, plus all students and researchers interested in the study of core level spectroscopy.

Frank de Groot
Akio Kotani

Acknowledgments

This book is the outcome of twenty years of work in core level spectroscopy. I was thoroughly introduced to this field by an energetic John Fuggle. Theo Thole and George Sawatzky guided me in the theoretical analysis of core level spectra. I would like to thank them very much for having had the opportunity to work with them. Special thanks to Jeroen Goedkoop and Marco Grioni, who introduced me into the synchrotron world.

In addition, I would like to thank all the colleagues with whom I had the pleasure to discuss the various aspects of core level spectroscopy, to perform experiments at synchrotron radiation sources and to perform the calculations. I would like to mention John Inglesfield, Jan Vogel, Miguel Abbate, Leonardo Soriano, Francesca Lopez, Zhiwei Hu, Pieter Kuiper, Jan van Elp, Hao Tjeng, Michel van Veenendaal, Eric Pellegrin, Gerrit van der Laan, Jan Zaanen, Olle Gunnarsson, Lars Hedin, Teijo Åberg, Massimo Altarelli, Paolo Carra, Francesco Sette, CT Chen, Marie-Anne Arrio, Philippe Sainctavit, Christian Brouder, Delphine Cabaret, Alain Fontaine, Stephania Pizzini, Helio Tolentino, Cinthia Piamonteze, Steve Cramer, Pieter Glatzel, Uwe Bergmann, Xin Wang, Weiwei Gu, Simon George, Chi-chang Kao, Keijo Hämäläinen, Michael Krisch, Gyorgy Vanko, Ed Solomon, Eric Wasinger, Rosalie Hocking, Serena George de Beer, Axel Knop-Gericke, Michael Havecker, Claudia Dallera, Sergei Butorin, Jinghua Guo, Odile Stephan, Christian Colliex, John Rehr, Atsushi Fujimori, Takeo Jo, Arata Tanaka, Harohiko Ogasawara, Diek Koningsberger, Moniek Tromp, Jeroen van Bokhoven, Willem Heijboer, Ingmar Swart, and Bert Weckhuysen.

Frank de Groot

The manuscript of this book was written mainly at RIKEN/Spring-8 and PF/KEK, and I would like to thank Professor S. Shin (RIKEN) and Professors K. Nasu, T. Koide, and T. Iwazumi (PF) for their kind support and helpful discussions on core level spectroscopy. Parts of the manuscript were also prepared at ALS (Berkeley), IPCMS (Strasbourg) and LMCP (Paris), and thanks are due to Dr. Z. Hussain and Dr. J. H. Guo (ALS), Dr. J. C. Parlebas (IPCMS), and Professor A. Shukla (LMCP) for their hospitality and collaboration. In the preparation of the manuscript, two issues from a collection of papers entitled *Core Level Spectroscopy in Solids* (edited by K. Okada, S. Tanaka, T. Uozumi, and H. Ogasawara) and *Progress in Core-Level Spectroscopy of Condensed Systems* (edited by S. Shin, K. Okada, and T. Ohta) were very helpful. The two issues were edited on the occasion of my retirement from the University of Tokyo. The former is a collection of review papers written by myself and coauthors that was edited in 2003 (not for sale), and the latter is a collection of invited papers published in a special issue of *Journal of Electron Spectroscopy and Related Phenomena*, vol. 136, 2004. Special thanks are due to the editors of the two issues for their efforts. In addition to those colleagues just mentioned, I would also like to thank Professors I. Harada, T. Jo, Y. Kayanuma, Y. Maruyama, T. Miyahara, G. A. Sawatzky, L. Braicovich, C. Hague, J. Nordgren, C.-C. Kao, D. Chandesris, A. Bianconi, and Drs M. Nakazawa, M. Taguchi, T. Idé, M. Matsubara, K. Fukui, K. Asakura, Y. Harada, T. Nakamura, B. T. Thole, L. C. Duda, S. Butorin, P. Le Fèvre, M. Calandra, and many others not mentioned here, for their help and for discussions on a variety of subjects. It is my great pleasure to express particular gratitude to the late Professors T. Nagamiya and K. Motizuki, and Professors J. Kanamori and M. Tachiki for leading me to the theoretical study of condensed matter physics. Finally, my thanks go to Professor Y. Toyozawa for leading me to one of the most significant and enjoyable research fields, namely, core level spectroscopy.

Akio Kotani

We are both grateful to D.D. Sarma (series editor), Jill Jurgensen, and many others at CRC Press who have contributed to the production of this book.

Authors

Frank de Groot is an associate professor in the department of chemistry at Utrecht University. His work reflects a concern with both the theoretical and experimental aspects of x-ray spectroscopy. His current interest is in the use of x-ray spectroscopies for the study of the electronic and magnetic structure of condensed matter, in particular for transition metal oxides, (magnetic) nanoparticles and heterogeneous catalysts under working conditions. He has taught international workshops and courses on x-ray spectroscopy at universities in Berkeley, Davis, Trieste, Grenoble, Berlin, and Zurich. Synchrotron experiments have been performed in Orsay, Grenoble, Berlin, Hamburg, Karlsruhe, Lund, Zurich, Daresbury, Brookhaven, Stanford, Argonne, and Berkeley. He has given over 130 lectures and published more than 160 articles.

Frank de Groot studied chemistry at the University of Nijmegen where, in 1991, he received his PhD under the supervision of John Fuggle. The title of his thesis was "X-Ray Absorption Spectroscopy of Transition Metal Oxides." From 1992, he was a postdoctoral fellow at the LURE Synchrotron in Orsay, France. In 1995, he became a Royal Netherlands Academy of Arts fellow at the University of Groningen, under the supervision of George Sawatzky. He obtained a personal *vidi* grant from the Netherlands Organization for Scientific Research. In 2006, he was awarded the prestigious *vici* grant allowing him to further extend his line of research at Utrecht University.

Akio Kotani is a professor emeritus at the University of Tokyo. He received his doctorate degree from Osaka University in 1969. He worked as a research associate for three years at Osaka University and then for five years at the Institute for Solid State Physics of the University of Tokyo. As an associate professor, he spent four years at the Institute of Material Science of Tohoku University followed by six years at Osaka University. In 1987, he was appointed a full professor at Tohoku University, and moved again in 1991 to the Institute for Solid State Physics of the University of Tokyo. He retired from the University of Tokyo in 2003, and is now a senior visiting scientist at RIKEN/Spring-8 and a collaborative scientist at KEK-Photon Factory.

Professor Kotani is well known for his many distinguished contributions to the vast area of theoretical condensed matter physics. His early works involved the magnetic properties of the spin-density-wave in chromium metal, the electromagnetic properties in magnetic superconductors, the optical response from localized electrons coupled with a phonon field, etc. Among these extensive theoretical studies was his pioneering work on core-level spectroscopy, which was proposed in 1973 and 1974 to interpret peculiar features in core-level XAS and XPS spectra of strongly correlated electron systems. Since then, for three decades, he has worked to develop the theoretical aspect of core-level spectroscopy. He has encouraged many young researchers and experimentalists in this field, and his scientific predictions have often been guiding principles for experimentalists working with synchrotron light sources.

1 Introduction

When atoms condense into solids, the electronic states of outer electrons, the valence electron states (VES), are strongly affected depending on the atomic species and their arrangement in solids. The properties of solids are determined by the characteristic features of VES. The VES contain the partially filled conduction band states in metals and the filled valence and empty conduction band states in insulators, including the 4f electron states of rare earth (RE) elements and the 3d electron states of transition metal (TM) elements. Therefore, the study of VES is the main object of solid state physics. On the other hand, inner electrons (core electrons) of atoms are deeply bound inside the atoms and are almost unchanged even when atoms condense into solids. Core level spectroscopy is a powerful tool to study VES, where a core electron is excited by incident x-rays (or an incident electron beam) and the VES are detected with the core electron state (or core hole state) as a probe.

Let us consider a TM compound MnO as an example. The electron energy levels of Mn and O atoms are shown in Figure 1.1. An Mn atom has 3d, 4s, and 4p electrons (seven electrons, where the 4s and 4p levels are shown as continuum in Figure 1.1) as valence electrons and 1s, 2s, 2p, 3s, and 3p electrons (18 electrons) as core electrons. The O atom has four 2p valence electrons and 1s and 2s (4 electrons) as core electrons. If we regard MnO as a pure ionic solid, the Mn 4s and 4p electrons are transferred to the O 2p states to form the Mn^{2+} ion (with five 3d electrons) and the O^{2-} ion (with six 2p electrons), and MnO is stabilized by the ionic force between Mn^{2+} and the O^{2-}. Actually, there is a considerable covalency character in valence electrons in MnO, so that the Mn 3d and O 2p states are hybridized with each other. The 18 core electrons in Mn and 4 core electrons in O are almost the same as those in free atoms, so that their features are described by atomic physics. The solid state properties of MnO, that is, electric, magnetic, thermal, and other properties, are determined by VES (i.e. the degrees of the ionicity and covalency, the electron correlation, the existence of the energy gap, and so on). In core level spectroscopy, one of the 18 + 4 core electrons is excited by incident x-rays and the electronic states of Mn 3d and O 2p electrons are studied by taking advantage of the well-known character of the core electron states.

In order to excite core electrons, we need the incident photons with energies larger than the core level binding energies (up to about 30 keV), which belong mainly to the x-ray region: soft x-rays up to 2 keV and hard x-rays at higher energies. Until synchrotrons were utilized as the main photon sources, the study of core level spectroscopy was a long way behind the study of outer electron spectroscopy mainly using the visible and infrared photon-energy region. Recently, the development of core level spectroscopy has been remarkable, especially with high-brilliance third-generation synchrotron radiation sources.

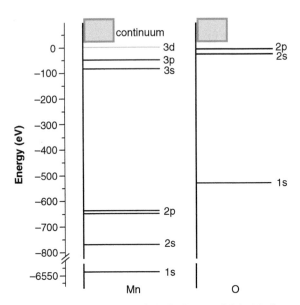

FIGURE 1.1   Energies of the core levels and VES of Mn and O in MnO.

There are various kinds of core level spectroscopy; three of the most important are x-ray photoemission spectroscopy (XPS), x-ray absorption spectroscopy (XAS), and x-ray emission spectroscopy (XES). In Figure 1.2, these spectroscopic processes are shown schematically for a simple case, as an example, where the VES consist of a filled valence band and an empty conduction band. In the case of XPS, a core electron is excited by the incident x-ray to the high-energy continuum states and detected as a photoelectron. In XAS, on the other hand, a core electron is excited near to the excitation threshold, which corresponds to the conduction band

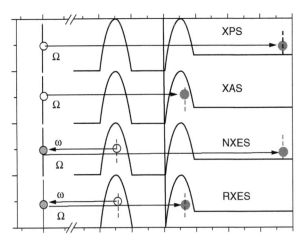

FIGURE 1.2   Schematic representation of XPS, XAS, NXES, and RXES (from top to bottom).

in this example. XPS and XAS are first-order optical processes that include only one photon, but XES is a second-order optical process where a core electron is excited by the incident x-ray and the excited state of the system decays radiatively by emitting x-rays. The XES is divided into two categories: when the core electron is excited by the incident x-ray to the high-energy continuum (as in the case of XPS), the XES is called normal XES (NXES), but when the core electron is excited near to the threshold (as in the case of XAS), it is called resonant XES (RXES). Each spectroscopy is discussed in more detail subsequently.

First we discuss XPS. In the historical development of the study of XPS, the one-electron character of XPS was first taken into account, and after that many-body effects were studied. Within the one-electron approximation, the kinetic energy ε of the photoelectron is determined by the energy conservation law as

$$\varepsilon = \hbar\Omega - (\varepsilon_F - \varepsilon_c) - \phi \tag{1.1}$$

where $\hbar\Omega$ is the incident photon energy, ε_c the core level, ε_F the Fermi energy, and ϕ is the work function. Theoretically, ε_c is obtained, for instance, by the Hartree-Fock approximation, and then Equation 1.1 represents the Koopmans' theorem. Experimentally, on the other hand, the value of ε_c can be estimated from the XPS spectrum if the other quantities in Equation 1.1 are known. Since the value of ε_c is not very much different from the corresponding value of the free atom, which is the characteristic of each element, XPS is useful for the analysis of elements in solids. A small deviation of ε_c from the free atom value, which is denoted by the chemical shift, provides us with the information on the chemical bonding of VES. The application of the chemical shift to the chemical analysis of molecules and solids was extensively made by Siegbahn et al. (1967), and their study played an important role in the development of XPS. In appreciation, the Nobel Prize in physics was awarded to Kai Siegbahn in 1982.

Koopmans' theorem does not generally hold because of the many-body effect beyond the Hartree-Fock approximation. Due to the progress in experimental techniques, the many-body charge-transfer effect in XPS can be observed as an asymmetry of the spectral shape and as satellite structures in various materials. When the core hole is created in the XPS process, VES are polarized by the core hole potential and screen it, as shown in Figure 1.3. The polarization of VES occurs mainly by the charge-transfer effect. Thus, the core hole plays a role of "test charge," which induces the charge transfer of VES, and the effect of charge transfer is reflected in the XPS spectrum as a spectral asymmetry and satellites.

A well-known example of the charge-transfer effect in XPS is the singularity in XPS spectra of simple metals. As shown by Anderson (1967) and Nozières and De Dominicis (1969), the polarization of conduction electrons around the core hole in simple metals gives rise to an asymmetric XPS line shape diverging at the threshold due to the "orthogonality catastrophe." A more drastic charge-transfer effect is exhibited in magnetic materials, which include the incompletely filled d or f states, for example, in materials including TM elements or RE elements. In these materials, the electrons in incomplete shells couple very strongly with the core hole, resulting in characteristic splitting of the XPS spectrum. The first theoretical interpretation for

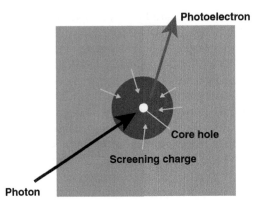

FIGURE 1.3　Excitation of a photoelectron by an x-ray photon creates a core hole that is screened by the surroundings.

the splitting of the 3d XPS spectrum in La metal was given by Kotani and Toyozawa (1974) with a simplified version of the single impurity Anderson model (SIAM). In the 3d XPS of La metal, two different types of final states occur: one is the final state where the core hole potential is screened by the charge transfer from the conduction band to the 4f state and the other is where the core hole potential is not screened by this type of charge transfer. The former is called "well-screened state" and the latter "poorly-screened state." Therefore, the 3d XPS spectrum of La metal is split into two peaks corresponding to the well-screened and poorly-screened final states. It should be mentioned that this mechanism of the charge-transfer effect succeeded in explaining the spectral splitting of XPS not only for La metal, but also for many materials including RE elements (and TM elements) with a ground state of $4f^n$ ($3d^n$) configuration, where $4f^{n+1}$ and $4f^n$ ($3d^{n+1}$ and $3d^n$) configurations in the final state correspond to the well-screened and poorly-screened states, respectively.

In the 1980s, the mixed valence effect (or fluctuating valence effect) in some intermetallic Ce compounds attracted much attention as an interesting many-body phenomenon. In trivalent Ce systems, for instance, γCe, $CeAl_2$, and $CeCu_2Si_2$, the ground state is in the $4f^1$ configuration, and the Ce 3d XPS spectrum splits into two peaks corresponding to the $4f^2$ well-screened and $4f^1$ poorly-screened final states. On the other hand, the ground state of the mixed valence Ce systems, for instance, αCe, $CePd_3$, and $CeRh_3$, is the quantum mechanical mixed state between the $4f^0$ and $4f^1$ configurations, and usually the Ce 3d XPS spectrum is split into three peaks. Gunnarsson and Schönhammer (1983a,b) extended the theory of Kotani and Toyozawa (1974), taking into account the spin and orbital degeneracy of the 4f state, to analyze the XPS spectra of mixed valence Ce systems. They formulated the XPS spectra by the $1/N_f$ expansion method, where N_f is the spin and orbital degeneracy of the 4f state. As a result of systematic analysis of XPS spectra (Fuggle et al., 1983) for various mixed valence Ce compounds, the mixing rate of the $4f^0$ and $4f^1$ configurations in the ground state is estimated quantitatively as one of the most powerful tools in the study of the mixed valence phenomenon.

The XPS study has also played an important role in the investigation of the mixed valence character in insulating systems including RE or TM elements. A typical example is the Ce 3d XPS of CeO_2. CeO_2 was traditionally considered to be a tetravalent system with pure $4f^0$ configuration in the ground state. However, the Ce 3d XPS spectrum of CeO_2 (Wuilloud et al., 1984) exhibited a three-peak structure similar to that in mixed valence Ce intermetallics, and from its theoretical analysis (Kotani et al., 1985) the ground state was found to be a strongly mixed state between $4f^0$ and $4f^1\underline{L}$ ($\underline{L}$ being a hole in the O 2p band) configurations due to the strong covalency hybridization between the Ce 4f and O 2p states. For insulating TM compounds, Zaanen et al. (1986) undertook a systematic analysis of XPS with the SIAM, and estimated the parameter values included in the model. Of those parameters, the charge-transfer energy Δ (defined by the energy cost to transfer an anion-valence-band electron to the TM 3d state) and the Coulomb interaction U_{dd} between 3d electrons were found to be two key parameters to characterize the system. Zaanen et al. (1985a) pointed out that when $U_{dd} < \Delta$, the insulating energy gap E_g is determined by U_{dd} (the well-known "Mott–Hubbard type insulator"), but when $U_{dd} > \Delta$, E_g is determined by Δ, and they called this new class of insulators "charge-transfer type insulators." After that, the mother materials (cuprates) of high T_c superconductors were revealed to be charge-transfer type insulators.

As mentioned previously, the charge-transfer effect induced by the core hole potential often plays an essential role in explaining the XPS spectral shape, so that the theoretical analysis of XPS spectra was made, in the initial stage, with SIAM (or cluster model) without taking into account the intra-atomic multiplet coupling effect due to the multipole components of Coulomb interaction. We denote this type of theory by charge transfer theory. More recently, the multiplet effect has also been incorporated in the charge transfer theory, for more detailed analysis of XPS, and such a theory is denoted by charge transfer multiplet (CTM) theory in the present book.

In XAS, a core electron is excited near to the threshold by the incident x-ray through the electric dipole transition (and sometimes the electric quadrupole transition in the hard x-ray region). Since the core electron state is well-known, the XAS spectrum provides us with important information on the symmetry-projected partial density of states of the excited states. The projected symmetry depends on the symmetry of the core electron states and the selection rule of the photo-excitation including the polarization of the incident x-ray.

The many-body charge-transfer effects in XAS have been studied for various systems as in the case of XPS. For XAS spectra of simple metals, the charge-transfer effect gives rise to the Fermi edge singularity (Nozières and De Dominicis, 1969), and for XAS spectra of mixed valence Ce compounds it causes satellite structures (Gunnarsson and Schönhammer, 1983b) in fair agreement with experimental results. In general, however, the satellite intensity in XAS spectra is much weaker than that in XPS spectra. Especially in 2p XAS of TM compounds, the satellite intensity is very weak, and the experimental XAS spectral shape is mainly determined by the intra-atomic multiplet coupling effects. The physical reason for the weaker charge-transfer effect in XAS, as against XPS, is the screening of the core hole potential by the photo-excited electron in XAS. Namely, in 2p XAS of TM compounds, the photo-excited electron stays in the 3d states and participates directly with the screening of the core hole potential, thus suppressing a further screening by the charge-transfer effect in going from initial to final states of XAS. On the other hand, in XPS, a core electron is excited to the high-energy continuum to be detected as a photoelectron and never participates in the screening of the core hole potential.

As mentioned previously, the charge transfer theory (without multiplet effect) is useful as a good starting approximation in the analysis of XPS spectra and the atomic multiplet effect gives some corrections to this first approximation. On the other hand, for 2p XAS of TM compounds, the atomic multiplet theory (with crystal field effect but without charge-transfer effect), which is often called "crystal field multiplet theory" or "ligand field multiplet (LFM) theory," serves as a good starting approximation, and the charge-transfer effect gives some corrections to this first approximation.

One of the most powerful applications of XAS to magnetic properties of materials is the circular polarization effect of XAS. In ferromagnetic or ferrimagnetic materials, the XAS intensity depends on the plus and minus (+ and −) helicity of the circularly polarized incident x-rays. The difference of XAS spectra for incident x-rays with plus and minus helicities is called x-ray magnetic circular dichroism (XMCD), and it is sensitive to the magnetic properties of VES. Above all, the sum rules of XMCD play an important role in the study of magnetic features of VES. Thole et al. (1992) and Carra et al. (1993) proposed XMCD sum rules, by which the orbital and spin magnetic moments of VES can be obtained, at least approximately, from energy-integrated intensities of XAS and XMCD spectra, and this method has been widely used in estimating orbital and spin magnetic moments of various materials.

For XES, both RXES and NXES are coherent second-order optical processes, where the excitation and de-excitation processes are coherently correlated by the second-order quantum formula, the so-called Kramers–Heisenberg formula (Kramers and Heisenberg, 1925). Since the intermediate states of RXES and NXES are, respectively the same as the final states of the first-order optical processes XAS and XPS, the information obtained from RXES and NXES is much greater than XAS and XPS. However, the intensity of the signal of RXES and NXES is much weaker than XAS and XPS, because the efficiency of the x ray emission is low, so that high-brilliance x-ray sources are required to obtain precise experimental data. The recent development of the RXES and NXES study is due to the implementation of undulator radiation from third-generation synchrotron radiation sources, as well as highly-efficient detectors. Since synchrotron radiation is tunable, it is useful for the measurements of RXES, by which we can obtain selected information connected directly with a specific intermediate state to which the incident x-ray energy is tuned (Rubensson, 2000; Kotani and Shin, 2001; Kotani, 2005).

On the XES treated in this book, we mainly concentrate on the theoretical and experimental study of RXES. RXES is one of the most important core-level spectroscopies, providing us with information that includes both x-ray absorption and emission processes and their correlation. Furthermore, RXES gives us bulk-sensitive, element-specific, and site-selective information. The RXES technique can be applied equally to metals and insulators, and can be performed in applied electric or magnetic fields, as well as under high pressure, since it is a photon-in and photon-out process.

RXES is divided into two categories depending on the electronic levels participating in the transition of x-ray emission. In the first category, the transition occurs from the valence state to the core state, and we have no core hole left in the final state of RXES. Typical examples are the 3d to 2p radiative decay of TM elements following the 2p to 3d excitation by the incident x-ray (denoted as 2p3d RXES) and the 3d4f RXES in RE elements. In this case, the difference of the incident and emitted x-ray energies (denoted as Raman shift), corresponds to the energy of electronic elementary excitations (i.e. the crystal field excitation, the charge transfer excitation, the correlation gap excitation, and so on). In this case, RXES is regarded as "resonant inelastic x-ray scattering (RIXS)," where the energy transfer corresponds to the energy of the elementary excitations of the valence electrons.

With this first category of RXES, we would like to point out the following facts: compared with the conventional (nonresonant) inelastic x-ray scattering, RXES has a larger intensity and depends on each intermediate state, which is convenient to identify the character of electronic excitations. Furthermore, compared with optical absorption spectroscopy, the selection rule of detecting electronic excitations in RXES is different, so that RXES and optical absorption give complementary information on elementary excitations. For example, the crystal field level excitation is forbidden in optical absorption, but allowed in RXES. If the experimental resolution of RXES is improved, electronic excitations across the Kondo gap, superconducting gap and so on will also be observed. In the hard x-ray region, the wavelength of the x-ray is comparable with the lattice spacing so that the momentum transfer in RXES corresponds to the wavenumber of the excitation mode and RXES also provides us with important information on the spatial dispersion of elementary excitations.

The second category of RXES is the case where the radiative decay occurs from a core state to another core state, so that a core hole is left in the final state of RXES. Typical examples are the 3p to 1s radiative decay following the 1s to 4p excitation in TM elements and the 3d to 2p radiative decay following the 2p to 5d excitation in RE elements. In general, the lifetime of a shallow core hole is longer than that of a deeper one. Further, the lifetime broadening of RXES is determined by the core hole in the final state, instead of the intermediate state. Taking advantage of these facts, we can use RXES measurements to detect a weak signal of core electron excitations that cannot be detected by using conventional XAS measurements because of the large lifetime broadening of a deep core hole. We can also obtain information on the spin-dependence of the core electron excitation by RXES measurements, which cannot be obtained by conventional XAS measurements.

The polarization-dependence in RXES gives important information on the symmetry of electronic states. For linearly polarized incident x-rays, two different polarization directions are often used to compare the resulting RXES spectra. Another important polarization-dependence in RXES is the magnetic circular dichroism (MCD) in ferromagnetic samples. The difference in RXES for circular polarized incident x-rays with plus and minus helicities gives important information on the magnetic polarization of electronic states in ferromagnetic materials.

The purpose of this book is to describe various aspects of core level spectroscopy of solids with XPS, XAS, and RXES as central subjects. Interesting experimental data, theoretical calculations, and underlying physics on these subjects are given, starting from fundamental aspects and covering the most recent topics.

In Chapter 2, fundamental aspects of core level spectroscopy are described. Basic features of core electron excitations, interaction of x-rays with matter, and various resonance processes are given. In addition to XAS, XPS, and XES, many other kinds of core level spectroscopy, such as electron energy loss spectroscopy (EELS), Auger electron spectroscopy (AES), appearance potential spectroscopy (APS) and so on, are introduced, as well as various resonance effects in core level spectroscopy.

In Chapter 3, the study of many-body charge-transfer effects are discussed mainly for XPS in order to understand the most important physical mechanisms underlying the XPS process, following the historical developments in these studies. It is shown that the charge transfer theory with the SIAM, but without the multiplet coupling effects, describe the most important features of the many-body effects in XPS for various systems including RE or TM elements. It is also shown that the charge-transfer effect is weaker in XAS than XPS.

Chapter 4 is devoted to various methods of theoretical calculations of core level spectroscopy, mainly for XAS. For XAS, the atomic multiplet effect is usually more important than the charge-transfer effect, so that we start from the simple free atom model with multiplet coupling effect, which is denoted by atomic multiplet theory. Then we add, as a solid state effect, the ligand field effect, or alternatively crystal field effect, to the atomic multiplet theory. We denote this theory as LFM theory, which describes XAS spectra of TM compounds considerably well. In a more detailed analysis of XAS, we combine LFM and charge transfer theories and we denote the theory as CTM theory, which describes both atomic multiplet and charge-transfer effects. In the analysis of XPS spectra, the LFM theory often fails to describe the important spectral features; thus we need the charge transfer theory, or for a more detailed analysis, we need the CTM theory.

Chapters 2–4 provide the reader with fundamental information of core level spectroscopy necessary to understand more advanced features of core level spectroscopy (mainly XPS, XAS, and RXES). Chapters 5–8 are devoted to detailed descriptions of the present status and current topics of core level spectroscopy. In Chapter 5, various features of XPS spectra are discussed, including the experimental data, theoretical analysis by CTM theory, and physical outcome obtained from the analysis. Resonant photoemission, resonant Auger electron, and resonant inverse photoemission spectroscopies are also discussed in some detail. Hard x-ray XPS is one of the more recent topics, and some of the new experimental results are introduced and analyzed theoretically.

Chapter 6 is devoted to a detailed description of XAS spectra. Various experimental aspects related to XAS are given in some detail, and characteristic features of XAS in TM compounds from $3d^0$ to $3d^9$ systems are discussed in a systematic manner. XAS spectra for RE and actinide systems are also reviewed. Chapter 7 is devoted to discussions on XMCD, where fundamental aspects of XMCD, sum rules of XMCD and their application to some ferromagnetic systems, plus a detailed description of XMCD in L edges of RE systems are given.

Finally, in Chapter 8, we discuss RXES spectra in detail. RXES is one of the most important fields of core level spectroscopy and has recently made such remarkable progress, as mentioned before, that we devote the largest space in this book to it. The most essential physical pictures for various recent topics on RXES are given, such as the charge-transfer and dd excitations in TM compounds and RE systems, characteristic electronic excitations and their spatial dispersions in high T_c cuprates, an RXES technique to detect weak electric-quadrupole and nonlocal electric-dipole transitions in the XAS pre-edge region of various materials, the MCD in RXES of ferromagnetic systems and so on.

2 Fundamental Aspects of Core Level Spectroscopies

In this chapter, we introduce some fundamental aspects of core level spectroscopies. The characteristics of core holes and core level spectroscopies will be given, and x-ray absorption and x-ray emission will be discussed on the basis of the interaction of x-rays with matter. We mainly confine ourselves to the one-electron approximation (the effect of electron–electron interactions on the core level spectroscopy will be discussed in detail from Chapters 3 to 8). At the end of the chapter, the x-ray and electron sources that are used for the core level spectroscopy are given.

2.1 CORE HOLES

The central component of core level spectroscopy is the core hole. A core hole can be defined as the absence of a core electron in a core level. A core hole is created by x-ray absorption or by x-ray photoemission experiments. It lives for ~10^{-15} seconds before it decays via radiative or nonradiative decay channels.

[handwritten annotation: femtosecond, $\Delta E \Delta t > \frac{\hbar}{2}$, $\hbar \approx \text{leV} \cdot \text{fs}$]

2.1.1 CREATION OF CORE HOLES

A core electron can be excited in at least five different ways:

1. X-ray absorption
2. X-ray scattering
3. Electron scattering
4. Proton (or ion) scattering
5. Electron capture in some isotopes

A core electron can be excited by the absorption of an x-ray photon to an empty state below the ionization threshold (an excitonic state) or to an essentially free electron above the ionization threshold and with a certain kinetic energy. This process is essentially the photoelectric effect and forms the basis of x-ray absorption spectroscopy (XAS), x-ray photoemission spectroscopy (XPS), and x-ray emission spectroscopy (XES).

Instead of being absorbed, an x-ray can also scatter on the electron density and loose part of its energy. The energy lost in the scattering process can be used to excite a core electron. This process is known as x-ray Raman scattering (XRS), because it is similar to normal Raman scattering. Also, an electron beam can be used to study the inelastic scattering processes. This process is known as electron energy loss spectroscopy (EELS). EELS can be performed with low-energy electrons and also with high-energy electrons in an electron microscope.

The scattering of protons or ions is used in proton (or particle) induced x-ray emission (PIXE). A proton beam with an energy of a few MeV is directed to the sample. The protons cause the electrons within the atoms of the target to be excited from core shells. The PIXE technique then makes use of the x-rays emitted when electrons fall back into the core hole. The core hole decay processes will be discussed subsequently.

Electron capture is a decay mode for chemical elements that occurs when there are too many protons in the nucleus of an atom, and there is not enough energy to emit a positron. In this case, one of the orbital electrons is captured by a proton in the nucleus, forming a neutron and a neutrino. Since the proton is changed to a neutron, the number of protons decreases by one and the atomic mass remains unchanged. By changing the number of protons, electron capture transforms the nuclide into a new element. Because the core electron neutralizes a proton, no extra valence electron is created. The charge of the nucleus plus core electrons together stays constant, which implies that there will be essentially no effect on the valence electrons. The difference with the excitation of a core electron out of the solid is that the screening processes will be different. The consequences of the differences in screening will be discussed for x-ray emission in Chapter 8.

2.1.2 DECAY OF CORE HOLES

An atom with a core hole is extremely unstable. The lifetime of the core hole is of the order of 10^{-15} s or 1 fs. The lifetime (τ) is linked to the uncertainty in the energy of the core hole (Γ) via the Heisenberg uncertainty relation:

$$\Gamma\tau \cong \hbar = 10^{-16}\,\text{eV s}. \tag{2.1}$$

A lifetime of 1 fs implies a lifetime broadening of 0.1 eV. There are two major decay processes: fluorescence and Auger. In fluorescence, another electron of the atom fills the core hole. This process can only occur if the energy of the electron is higher than the core hole. In other words, a shallow core electron can fill a deep core hole. Also a valence electron can fill a core hole. The energy difference between these two states is released as electromagnetic radiation. In most cases, this will be x-rays, but very shallow levels will release ultraviolet (UV) radiation or light. This is, for example, the case in x-ray excited optical luminescence (XEOL).

The second process is Auger decay. When an electron from the 2p shell drops to fill a vacancy formed by 1s shell ionization, the resulting x-ray photon with energy $\varepsilon_{2p} - \varepsilon_{1s}$ may excite a third electron, for example, another 2p electron. Such a process can be denoted as a 1s2p2p Auger process. Auger electron energies are usually described in the Barkla notation, see Table 2.1. The probability of Auger electron production increases as the difference between the energy states of the shells decreases. Light elements are more susceptible to the formation of Auger electrons by multiple ionizations. Thus the proportion of radiation emitted at characteristic wavelengths is lower than for heavier elements. Figure 2.1 shows that the

TABLE 2.1
Nomenclature for Core Level Spectra

Orbital*	Label[†]	E[‡] (Ni)	E[‡] (O)
1s	K	8333	543
2s	L_1	1008	42
$2p_{1/2}$	L_2	870	V[§]
$2p_{3/2}$	L_3	853	V[§]
3s	M_1	111	
$3p_{1/2}$	M_2	68	
$3p_{3/2}$	M_3	66	
$3d_{3/2}$	M_4	V	
$3d_{5/2}$	M_5	V	

* Orbital notation.
† Spectroscopic names (Barkla notation).
‡ Binding energies.
§ Valence state with a binding energy of a few eV.
Source: X-ray Data Booklet (2001) (LBNL, Berkeley).

proportion of Auger emission from the 1s core state is greater than 0.5 up to about $Z = 30$. For more details, see Figure 3.3 in Chapter 3.

The x-ray fluorescence or Auger decay will lead to other core holes, which subsequently decay until all core levels are filled again. This will give a cascade of fluorescence and Auger transitions. For example, a 1s core electron in Ni can be excited in a 1s XPS process. The 1s core hole is filled by a 2p electron in 1s2p XES or $K\alpha$ fluorescence. This creates a 2p core hole that can decay via 2p3p3p (LMM)

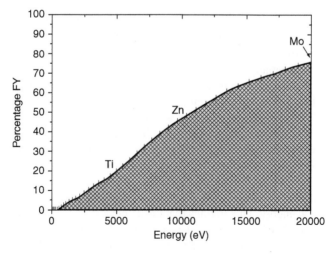

FIGURE 2.1　Percentage of fluorescence decay versus Auger decay as a function of the binding energy of the K edges. Indicated are Ti (4966 eV), Zn (9659 eV), and Mo (20,000 eV).

Auger transition, yielding an Auger electron. The two 3p core holes can decay via 3p3d3d Auger plus two further Auger electrons. The created 3d-holes decay via low-energy decay processes. The energy of the single x-ray photon that was absorbed is, by this process, is divided over the kinetic energy of the photoelectron, the photon energy of the $K\alpha$ fluorescence, and the kinetic energies of the three Auger electrons.

2.2 OVERVIEW OF CORE LEVEL SPECTROSCOPIES

We can now introduce various core level spectroscopies using a simple model. This model is not expected to give a correct interpretation of the spectral shapes and intensities observed, but it serves as a starting point on which more realistic models will be built. Two main approximations are:

1. *Density-of-states (DOS) approximation*: It is assumed that the ground state DOS, as calculated, for example, by density functional theory (DFT), can be used. This implies that it is also assumed that extra valence electrons have no effect.
2. *Core hole approximation*: It is assumed that the core hole created in x-ray absorption does not modify the DOS.

In addition, it is assumed for the moment that the transition matrix elements are constant and that the work function is zero.

2.2.1 CORE HOLE SPIN–ORBIT SPLITTING

A general ingredient of core level spectroscopies is the core hole spin–orbit splitting. Each core hole has an orbital angular momentum l. It is represented with the familiar notation s, p, d, and f for orbital angular momentum from 0 to 3 (in units of $\hbar$). The spin angular momentum s of each core hole is either +1/2 (spin-up) or −1/2 (spin-down) (in units of $\hbar$). The letter is preceded by the principal quantum number. In many spectroscopies, one uses a third symbol to indicate the total angular momentum J of the core hole. The total angular momentum is given by $l + s$ or $l - s$ (in units of $\hbar$). The coupling of the orbital and the spin angular momenta is given by the spin–orbit inter-action, which is essentially a relativistic effect. The spin–orbit interaction is very large for core holes and in general two peaks or structures will be visible in the spectrum, separated by the core hole spin–orbit splitting. If the orbital angular momentum is zero, the spin–orbit interaction is absent, whereas if the orbital angular momentum is one, the total angular momentum can be either 1/2 or 3/2 and two peaks will be observed. Similarly for a 3d state ($l = 2$), both the $3d_{3/2}$ and $3d_{5/2}$ states are present.

In spectroscopy, an alternative notation is often used. Instead of 1, 2, 3, and so on, for the principal quantum number, one uses K, L, M, and so on, and instead of the orbital angular momenta s, p, d, f, one uses 1, 2, 3, and so on, to label the peaks from high to low energy. This implies that the 2s-peak is called L_1. The next peak is the $2p_{1/2}$ peak, called L_2 and the last peak is the $2p_{3/2}$ peak, called L_3. They are collected in Table 2.1. If one speaks about the 2p states together, this is indicated as $L_{2,3}$. The relative intensity of the L_2 and L_3 peaks follows the simple rule that it is given by the degeneracy of the states. A $2p_{1/2}$ configuration consists of two states with the z component of total angular momentum j_z, which is also written as a magnetic quantum number m_j equal to −1/2 and +1/2 respectively. A $2p_{3/2}$ configuration consists of four states, with m_j equal to −3/2, −1/2, 1/2, and 3/2. Therefore, the intensity ratio of the $2p_{1/2}$ versus $2p_{3/2}$ peak is 1 : 2. The intensity ratio of the $3d_{3/2}$ to $3d_{5/2}$ peak is 2 : 3. This rule breaks down if there are other interactions, which are able to mix the spin–orbit split states. In Chapter 4, it will be shown that an important interaction is the overlap of the core hole wave function with the valence state wave functions, or in other words, the coupling of the core and valence angular momenta. That coupling will contrast sharply with the simple picture sketched here.

2.2.2 CORE HOLE EXCITATION SPECTROSCOPIES

An x-ray can be absorbed by a core electron that is excited to an empty state. In Figure 2.2, the schematic DOS of O 1s and 2p states of a typical TM compound is given. The energy of the O 1s core level measured from the Fermi energy is $\varepsilon_c =$ −530 eV (see Figure 1.1 of Chapter 1) and the O 2p states are split into a filled valence band (with energy ε_v) and an empty conduction band (with energy ε_w) due to the hybridization between the TM 3d, 4s, and 4p states with the O 2p states (for more details, see Chapter 6). In x-ray absorption, a 1s core electron (ε_c) is excited to an empty state with energy ε_w, where we use the subscript w to separate the empty states from the valence states that are given with subscript v. The XAS intensity as a function of the photon energy $\hbar\Omega$ identifies with the empty DOS, where the absorption

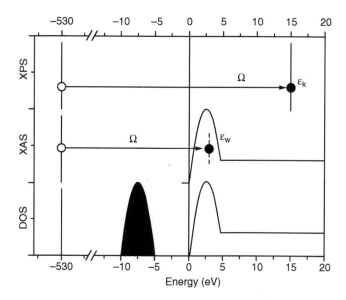

FIGURE 2.2 Schematic DOS of an oxide. The oxygen 1s core electron at energy −530 eV is excited by an x-ray. The bottom frame gives the oxygen p-projected density of states. The middle frame gives the oxygen 1s XAS spectrum. The top frame gives the oxygen 1s XPS spectrum. The arrow represents the electron transition by absorbing the incident photon with frequency Ω (energy $\hbar\Omega$).

energy is given by $\hbar\Omega = \varepsilon_w - \varepsilon_c$; matrix elements are neglected in this approach. As discussed previously, EELS and XRS measure (under certain conditions) the same spectrum.

XPS excites a core electron to a high-energy empty state with energy ε_k. The kinetic energy of the excited electron is measured, which yields a peak if the kinetic energy is equal to the x-ray excitation energy plus the core level energy ($\varepsilon_k = \hbar\Omega + \varepsilon_c$). As such, the core level energy ε_c (or the core level binding energy $|\varepsilon_c|$) is determined. Table 2.2 collects the descriptions of the various core level spectroscopies.

If a valence electron with energy ε_v is excited by the incident photon with energy $\hbar\Omega$ to become a photoelectron with energy ε_k, as shown at the top of Figure 2.3, the process is denoted as valence photoemission spectroscopy (VPES) or simply as

TABLE 2.2
Energy Effects of the Various Core Level Spectroscopies

Spectroscopy	Spectral Shape*	DOS	Chapter[†]
	Excitation		
XAS, EELS, XRS	$\hbar\Omega = \varepsilon_w - \varepsilon_c$	ε_w	6
	$\hbar\Omega = \varepsilon_k - \varepsilon_c$	—	
XPS, PES	$\varepsilon_k = \hbar\Omega + \varepsilon_c$	—	5
	$\varepsilon_k = \hbar\Omega + \varepsilon_v$	ε_v	
IPES, BIS	$\hbar\omega = \varepsilon_k - \varepsilon_w$	ε_w	—
	Decay		
XES	$\hbar\omega = \varepsilon_{c'} - \varepsilon_c$	—	8
	$\hbar\omega = \varepsilon_v - \varepsilon_c$	ε_v	
AES	$\varepsilon_k = \varepsilon_{c''} + \varepsilon_{c'} - \varepsilon_c$	—	5
	$\varepsilon_k = \varepsilon_v + \varepsilon_{c'} - \varepsilon_c$	ε_v	
	$\varepsilon_k = \varepsilon_v + \varepsilon_{v'} - \varepsilon_c$	$\varepsilon_v + \varepsilon_{v'}$	
	Resonance		
RPES	$\varepsilon_k - \hbar\Omega = \varepsilon_v$	ε_v	5
RXPS	$\varepsilon_k - \hbar\Omega = \varepsilon_{c'}$	—	
RAES	$\varepsilon_k - \hbar\Omega = \varepsilon_{c''} + \varepsilon_{c'} - \varepsilon_w$	—	5
	$\varepsilon_k - \hbar\Omega = \varepsilon_v + \varepsilon_{v'} - \varepsilon_w$	$\varepsilon_v + \varepsilon_{v'} - \varepsilon_w$	
	$\varepsilon_k - \hbar\Omega = \varepsilon_{c'} + \varepsilon_v - \varepsilon_w$	$\varepsilon_v - \varepsilon_w$	
RIPES	$\varepsilon_k - \hbar\omega = \varepsilon_w$	ε_w	5
	$\varepsilon_k - \hbar\omega = \varepsilon_w + \varepsilon_{w'} - \varepsilon_v$	$\varepsilon_w + \varepsilon_{w'} - \varepsilon_v$	
APS (= REELS)	$\varepsilon_k - \varepsilon_{k'} = \varepsilon_w - \varepsilon_v$	$\varepsilon_w - \varepsilon_v$	
RXES (= RIXS)	$\hbar\Omega - \hbar\omega = 0$	—	8
	$\hbar\Omega - \hbar\omega = \varepsilon_w - \varepsilon_v$	$\varepsilon_w - \varepsilon_v$	
	$\hbar\Omega - \hbar\omega = \varepsilon_{c'}$	—	
APECS	$\varepsilon_k - \varepsilon_{k'} = \varepsilon_{c''} + \varepsilon_{c'}$	—	5

Note: $\hbar\Omega$ and $\hbar\omega$ are used for the incident and emitted photon energies, respectively.
* The measured variable quantities are given before the equal sign and the constant values are given behind.
[†] The chapter that deals with this spectroscopy.

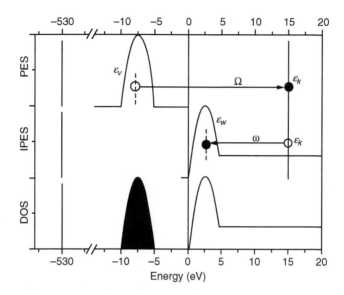

FIGURE 2.3 The bottom frame gives the oxygen p-projected density of states. The middle frame gives the IPES spectrum, and the top frame gives the valence band photoemission spectrum. The right-handed arrow with Ω and the left-handed arrow with ω represent the electron transition by absorbing a photon with frequency Ω and that by emitting a photon with frequency ω respectively.

photoemission spectroscopy (PES). Here ε_v is detected as $\varepsilon_v = -\hbar\Omega + \varepsilon_k$, and the valence band DOS is measured by PES.

The inverse photoemission (IPES) or Bremsstrahlung isochromat spectroscopy (BIS), which is the inverse process of PES, gives information related to XAS. The basic IPES experiment is where one shoots an electron towards the sample and this electron can be subsequently absorbed in the unoccupied DOS, where the energy loss is emitted as a photon with energy $\hbar\omega$. In this manner, the ε_w can be detected as $\varepsilon_w = \varepsilon_k - \hbar\omega$. There is no core electron involved, so BIS is not a core level spectroscopy. There are, in principle, two modes of measuring BIS: (*i*) one can scan the kinetic energy of the incident electron and detect the emitted photon at a fixed energy, or (*ii*) one can fix the kinetic energy of the incident energy and scan the energy of the emitted photon. Most BIS experiments are performed in combination with XPS, which makes it easier to use a fixed photon energy (i.e. to use a fixed x-ray monochromator).

The combination of PES and IPES experiments can be performed to measure the energy gap, $\varepsilon_w - \varepsilon_v$, between the empty conduction band and the filled valence band, as seen from Figure 2.3. If the gap is caused by the electron correlation, $\varepsilon_w - \varepsilon_v$ is exactly the definition of Hubbard correlation energy U_{dd}, as will be used in the later chapters. The group of Yves Baer pioneered this approach and measured the XPS and BIS spectra of rare earths (Cox et al., 1981a,b). In this chapter, we neglect correlation energies and implicitly assume that $U_{dd} = 0$.

2.2.3 CORE HOLE DECAY SPECTROSCOPIES

Once a core hole is created, it decays within a few femtoseconds. This is (within the model used) independent of the way the core hole is created, that is, by an x-ray, electron, proton, or electron capture. In x-ray emission, a deep core hole (ε_c) can be filled by a shallow electron ($\varepsilon_{c'}$) or an electron from the valence band (ε_v). The difference in energy of the two core states gives $\hbar\omega = \varepsilon_{c'} - \varepsilon_c$, as indicated in Figure 2.4. In this approximation, core–core XES thus just gives a peak at the binding energy difference; it is obvious that in reality, a much more complex situation occurs. If a valence electron fills the 1s core hole, one obtains the occupied DOS with the photon energy ($\hbar\omega$) equal to the valence band energy (ε_v) minus the core electron energy (ε_c).

In Auger electron spectroscopy (AES), a shallow core electron fills the deep core hole and another shallow core electron is excited out of the solid as a free electron. The final state contains two shallow core holes and the kinetic energy of the emitted electron equals $\varepsilon_k = \varepsilon_{c''} + \varepsilon_{c'} - \varepsilon_c$. If instead of two shallow core electrons, a core electron and a valence electron take part in the Auger process, one can detect again the occupied DOS, as indicated in Figure 2.5. It is also possible that two valence electrons take part in the Auger process. The summed energy of the two valence electrons is given by the difference between the kinetic energy and the binding energy, that is, $\varepsilon_k + \varepsilon_c (= \varepsilon_k - |\varepsilon_c|) = \varepsilon_v + \varepsilon_{v'}$. The occurrence of $\varepsilon_v + \varepsilon_{v'}$ in this equation implies that one measures the self-convolution of the occupied DOS. If one includes the electron correlation effects, the self-convolution of the two (3d) valence electrons includes the energy effect of U_{dd}. This is the basis of the Cini-Sawatzky method (Cini, 1977, 1978; Sawatzky, 1977). The assumption of this method is that the core-valence-valence Auger process is local, implying two localized valence holes that are affected by the correlation energy U_{dd}.

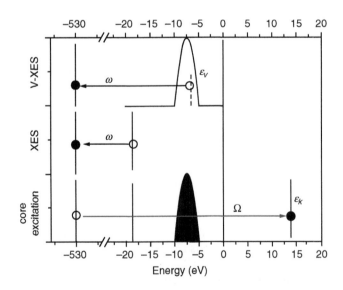

FIGURE 2.4 The bottom frame gives the core excitation process. The middle frame gives the core–core x-ray emission process. The top frame gives the valence band x-ray emission process.

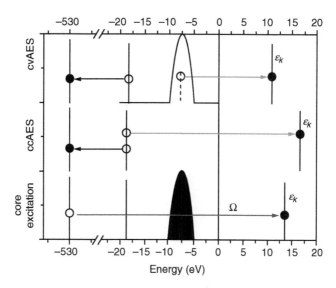

FIGURE 2.5 The bottom frame gives the core excitation process. The middle frame gives the core–core Auger process, where the kinetic energy of the emitted electron is only dependent on the core states involved. The top frame gives the core valence Auger process, where the kinetic energy of the emitted electron measures the valence band ε_v, similar to a PES experiment.

2.2.4 RESONANT PHOTOELECTRON PROCESSES

We have described core hole creation (XAS and XPS) separately from core hole decay (XES and AES). However, as soon as core hole creation takes place, decay occurs. This implies that close to the XAS absorption edge, the decay processes will be different from off-resonance excitations. A major effect is the coherence between excitation and decay as described subsequently. The coupling between excitation and decay implies that one obtains two-dimensional spectra for all resonance spectroscopies as a function of both the excitation and decay energies. The consequences are described in Chapter 8. Within the single particle model, the only effect is that the excited electron can take part in the decay process. This creates additional decay channels not present in off-resonant or normal AES and XES. The best-known resonant spectroscopy is resonant photoemission (RPES). Within the single particle model, this can be described as the two-step process, XAS followed by AES, where the electronic excitation energy in each step is given by

$$0 \xrightarrow{\text{XAS}} \varepsilon_w - \varepsilon_c \xrightarrow{\text{AES}} \varepsilon_k - \varepsilon_v. \tag{2.2}$$

In the final state, a hole exists in the valence band and one measures the occupied DOS, as can be seen in Figure 2.6. This is exactly the same as in a normal photoemission process. A difference is the matrix element of RPES versus PES. This will be discussed in Chapter 5. Another interesting feature of RPES is that the direct PES channel and the indirect XAS + AES channel have the same initial and final states, hence they interfere with each other.

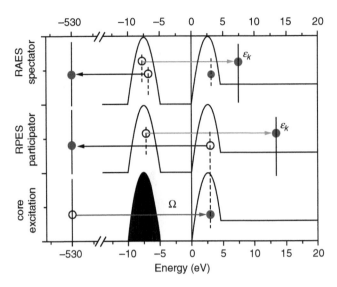

FIGURE 2.6   The bottom frame gives the resonant excitation to the empty DOS. The middle frame gives the RPES process where the excited core electron participates in the decay. This yields a final state that is identical to the final state of a normal valence band photoemission experiment. The top frame gives the resonant Auger electron process where the excited core electron does not participate in the decay.

Another RPES process includes a shallow core state instead of the valence band state, where first a deep core hole is created at resonance and subsequently it decays by Auger transition to a shallow core hole:

$$0 \xrightarrow{\text{XAS}} \varepsilon_w - \varepsilon_c \xrightarrow{\text{AES}} \varepsilon_k - \varepsilon_{c'}. \tag{2.3}$$

The final state in this participator channel is the same as the XPS final state, so that this process is denoted by resonant XPS (RXPS). The RXPS and direct XPS processes interfere with each other.

There is another possible Auger decay channel from the same core excitation:

$$0 \xrightarrow{\text{XAS}} \varepsilon_w - \varepsilon_c \xrightarrow{\text{AES}} -\varepsilon_v - \varepsilon_{v'} + \varepsilon_w + \varepsilon_k. \tag{2.4}$$

In this case, the excited electron of the intermediate state does not participate in the Auger process. This process is called the spectator channel, the final state of which contains two valence holes plus an extra electron in the conduction band ε_w in addition to the Auger electron. This can be viewed as a normal photoemission final state, plus a valence to conduction band excitation, and we denote it by resonant AES (RAES). The final state of RAES cannot be reached by direct PES, at least not within the single particle model that we use here. In fact, this final state is a crucial part of the many-body description of core level spectroscopy.

There are different types of RAES processes, where one or two valence holes in the above RAES process are replaced by shallow core hole(s):

$$0 \xrightarrow{\text{XAS}} \varepsilon_w - \varepsilon_c \xrightarrow{\text{AES}} -\varepsilon_v - \varepsilon_{c'} + \varepsilon_w + \varepsilon_k, \tag{2.5}$$

$$0 \xrightarrow{\text{XAS}} \varepsilon_w - \varepsilon_c \xrightarrow{\text{AES}} -\varepsilon_{c''} - \varepsilon_{c'} + \varepsilon_w + \varepsilon_k. \tag{2.6}$$

A related technique is Auger-photoemission coincidence spectroscopy (APECS), which measures a photoelectron in coincidence with an Auger electron. Effectively, this is also a resonance measurement. One energy axis is given by the excitation energy and its detected kinetic energy of the photoelectron. The second axis is given by the Auger decay. Some recent APECS experiments are discussed in Chapter 5.

$$0 \xrightarrow{\text{XPS}} \varepsilon_k - \varepsilon_c \xrightarrow{\text{AES}} -\varepsilon_{c'} - \varepsilon_{c''} + \varepsilon_k + \varepsilon_{k'}. \tag{2.7}$$

In a resonant IPES (RIPES) experiment, one excites with electrons that have a kinetic energy equal to the binding energy of a core state. An (inverse) Auger process can occur in which both the core electron and the impinging free electron are transferred to an electron in the conduction band. In a second step, a conduction electron can decay by x-ray emission to the core hole, which is the participator channel:

$$\varepsilon_k \xrightarrow{\text{AES}} -\varepsilon_c + \varepsilon_w + \varepsilon_{w'} \xrightarrow{\text{XES}} \varepsilon_w. \tag{2.8}$$

The RIPES spectator channel sees a valence electron decay to the core hole:

$$\varepsilon_k \xrightarrow{\text{AES}} -\varepsilon_c + \varepsilon_w + \varepsilon_{w'} \xrightarrow{\text{XES}} \varepsilon_w + \varepsilon_{w'} - \varepsilon_v. \tag{2.9}$$

A related method that makes use of this inverse Auger process is appearance potential spectroscopy (APS) (Hinkers et al., 1989). In APS, an electron beam with variable energy is impinging on a sample. When the electron energy is equal to the energy needed to excite a core electron, the AES transition as described in Equation 2.8 occurs. In APS spectroscopy, one distinguishes the decay of an electron or an x-ray. So-called x-ray-APS identifies with RIPES as described previously. If the excited state in an APS experiment is detected via Auger decay, one can observe "double Auger processes," that, for example, follow Equation 2.10 and decay to a final state with a valence hole plus a conduction electron.

$$\varepsilon_k \xrightarrow{\text{AES}} -\varepsilon_c + \varepsilon_w + \varepsilon_{w'} \xrightarrow{\text{AES}} \varepsilon_w - \varepsilon_v + \varepsilon_{k'}. \tag{2.10}$$

This experiment is also called Auger electron (AE)-APS (Hinkers et al., 1989). The AE-APS experiment is an electron-in electron-out experiment and as such can also be considered as a resonant EELS (REELS) experiment, where depending on angle and electron energy, one can change from the Auger matrix element to the XAS matrix element, as discussed in Section 2.5. This has been used by Gorschluter and Merz (1994, 1998) and Fromme et al. (1995) to measure dd excitations in NiO, which has been calculated by Jones et al. (2000). Fromme et al. (2001) used REELS to measure ff-transitions in rare earths. These REELS experiments are equivalent to the resonant x-ray emission spectroscopy (RXES) experiments discussed subsequently.

2.2.5 RESONANT X-RAY EMISSION CHANNELS

It is also possible to study x-ray emission processes at resonance. For example, Figure 2.7 shows that one can excite a core electron and detect participator RXES, in other words, the resonant elastic scattering path:

$$0 \xrightarrow{\text{XAS}} \varepsilon_w - \varepsilon_c \xrightarrow{\text{XES}} 0. \tag{2.11}$$

Spectator RXES leaves a hole in the valence band and an electron in the conduction band; in other words, it measures all valence band–conduction band excitations.

$$0 \xrightarrow{\text{XAS}} \varepsilon_w - \varepsilon_c \xrightarrow{\text{XES}} \varepsilon_w - \varepsilon_v. \tag{2.12}$$

Because of its similarity to Raman spectroscopy, RXES is also called resonant x-ray Raman spectroscopy (RXRS). Yet another name, often used to stress the relation to off-resonant inelastic x-ray scattering is resonant inelastic x-ray scattering (RIXS).

Another RXES channel is the excitation of a deep core state and the detection of the resonant decay of a shallow core state:

$$0 \xrightarrow{\text{XAS}} \varepsilon_w - \varepsilon_c \xrightarrow{\text{XES}} \varepsilon_w - \varepsilon_{c'}. \tag{2.13}$$

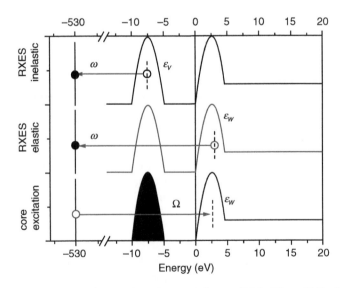

FIGURE 2.7 The bottom frame gives the ground state DOS. The middle frame gives the elastic RXES process where the excited electron decays again. The top frame gives the inelastic RXES process where an electron from another state decays, leaving behind a valence to conduction band excitation.

The final state consists of a shallow core hole and an electron in the unoccupied DOS. The energy differences between the incoming ($\hbar\Omega$) and emitted ($\hbar\omega$) x-rays equals the core (ε_c) to conduction band (ε_w) excitation energy. In other words, within the single particle model, core–core RXES identifies with a complicated detection technique of the XAS spectrum of the shallow core state. In Chapter 8, it will be explained that this identity is not correct due to different selection rules.

It is to be mentioned that the XES and AES treated in Section 2.2.3 are called "normal XES" (NXES) and "normal AES" (NAES), respectively, to distinguish them clearly from RXES and RAES. In NXES and NAES, the core hole creation process (the excitation of a core electron to a high-energy continuum) is independent of the XES and AES processes, at least in the one-electron approximation, so that they are first order quantum processes, while RXES and RAES are second order quantum processes. This is not the case for strongly correlated electron systems, where the creation of the core hole is correlated with XES and AES processes through the electronic relaxation around the core hole (see Chapter 8 for more details).

2.2.6   OVERVIEW OF THE RXES AND NXES TRANSITIONS

RXES and NXES occur for any allowed combination of core and valence states. In a 3d TM system, the 1s core state can be excited and the 1s core hole can be filled by 2p, 3p, or 4p (valence) electrons. For a 1s core hole, this yields 1s2p ($K\alpha$), 1s3p ($K\beta$), and 1s4p ($K\beta_{2,5}$) XES. We will use the core notation, where we will distinguish between 1s4p ($K\beta_{2,5}$) for the valence band and 1s4p$_{cross}$ ($K\beta''$) for the cross-over peak. The energies of these XES channels for Ti and Cu is given in Table 2.3. The nature of both transitions will be further explained subsequently.

Figure 2.8 shows the NXES spectra of a high-spin and low-spin divalent iron compound divided into their main components. The 1s2p ($K\alpha$) NXES is split due to the 2p spin–orbit coupling and the two constituents are denoted as $K\alpha_1$ and $K\alpha_2$, where $K\alpha_1$ relates to a $2p_{3/2}$ core hole final state and $K\alpha_2$ to a $2p_{1/2}$ core hole. The 1s3p ($K\beta$) NXES is split due to the 3p3d exchange interaction into $K\beta'$ and $K\beta_{1,3}$. Because the separation between these two peaks is mainly caused by the 3p3d

---

TABLE 2.3
Main NXES and RXES Spectra Measured for 3d TM Systems

Transition	Spectroscopic Name	Ti (eV)	Cu (eV)
1s2p	$K\alpha$	4510	8048
1s3p	$K\beta$	4932	8905
1s4p$_{cross}$	Cross-over	4946	8959
1s4p	Valence band	4966	8979
2p3s		395	810
2p3d	$L\alpha$	453	932

Note: It is assumed that a cross-over peak has an effective binding energy of ~20 eV.

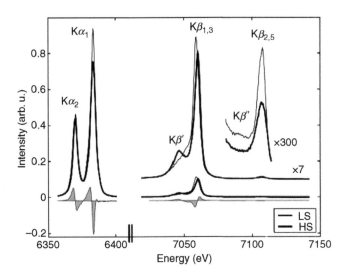

FIGURE 2.8   K shell NXES lines for a high-spin (HS, thick line) and low-spin (LS, thin line) Fe^{2+} compound. (Reprinted with permission from Vanko, G., et al., *J. Phys. Chem. B*, 110, 11647, 2006. Copyright 2006 by the American Chemical Society.)

exchange, high-spin Fe^{2+} compounds ($S = 2$) have two well-separated peaks, while low-spin Fe^{2+} compounds ($S = 0$) contain only a low-energy tail. The 1s4p valence band NXES is split into the cross-over peak ($K\beta''$) and the valence band NXES ($K\beta_{2,5}$). The valence band NXES is caused by the Fe 4p character that has hybridized with the valence states of the ligands (e.g. oxygen). In addition, there could be $K\beta_{2,5}$ intensity from quadrupole XES from occupied Fe 3d states. The $K\beta''$ peak is called cross-over because it can be considered as the XES channel from a ligand oxygen 2s "core" state to a Fe 1s state, which is a cross-over transition from oxygen to iron. It is better to view this transition as one from iron 4p to 1s, where the Fe 4p character has hybridized into the oxygen 2s core state. The intensity of the cross-over peak inversely correlates with the metal–ligand distance. This is caused by the fact that shorter distances imply stronger hybridization (Bergmann et al., 1999).

These spectra can be measured off-resonance after excitation with an x-ray energy well above the Fe K edge. In addition, one can use a high-energy electron or proton to excite the 1s core hole as, for example, used in energy dispersive x-ray emission (EDX) in an electron microscope. Another route to create a 1s core hole is by electron capture in isotopes of some elements. The same spectra can also be measured on-resonance in a RXES experiment. One can excite at the main dipole edge in a 1s4p transition, yielding 1s2p(4p) RXES, and so on. In addition, the 1s3d quadrupole pre-edge can be excited resonantly in 1s2p(3d) RXES. Instead of a 1s core hole, the 2p and the 3p core hole can also be used for XES experiments. For these soft x-ray energies, it is usual to perform resonant experiments. This includes 2p3d(3d) RXES, 2p3s(3d) RXES, and 3p3d(3d) RXES, where we will further omit the (3d) state and just write 2p3s RXES, and so on.

2.3   INTERACTION OF X-RAYS WITH MATTER

X-rays are defined as electromagnetic radiation with a wavelength smaller than UV radiation. The border between UV radiation and x-rays is not clearly defined, but is usually set at around 10 nm. The (angular) frequency ω is inversely proportional to the wavelength λ as $\omega = 2\pi c/\lambda$, and the x-ray energy is given by $\hbar\omega$. This defines the low-energy limit of ~100 eV. In principle, there is no lower limit of x-ray wavelengths although radiation below 1 pm is usually called gamma radiation. X-rays are subdivided into soft x-rays, between 100 eV and 3 keV and hard x-rays, above 3 keV. Soft x-rays are called "soft" because they do not penetrate air and, as such, are relatively safe to work with.

If an x-ray hits a solid material, the material starts to emit electrons. This is essentially the photoelectric effect that also occurs with UV radiation. The photons collide with the electrons in the metal. If the energy of a photon is less than the energy required to remove a core electron from the metal, then an electron will not be ejected, regardless of the intensity of the radiation. The energy required to remove an electron from the surface of a metal is called the work function of the metal and denoted by ϕ. However, if the energy of the photon is greater than ϕ, then an electron is ejected with a kinetic energy, $\varepsilon_k = \hbar^2 k^2/2m$, equal to the difference between the energy of the incoming photon and the work function. In the case of x-rays, the explanation is analogous, with the addition that a core electron is excited instead of a valence electron. The binding energy of the core electron must be added and the kinetic energy is given as:

$$\hbar^2 k^2/2m = \hbar\omega - E_B - \phi. \tag{2.14}$$

This equation defines the binding energy of a core level and x-ray photoemission experiments are used to measure core level binding energies. The binding energy E_B is closely related to the ionization energy that is defined as $E_B + \phi$.

In an XPS experiment, where NiO is excited with aluminum Kα radiation with an energy of 1487 eV, one observes peaks in the XPS spectrum at 607 eV and 637 eV, respectively. Assume that the work function of NiO is equal to 2 eV. Equation 2.14 yields the binding energies of the core levels associated with the observed peaks:

$$\hbar\omega - \varepsilon_k - \phi = E_B,$$
$$1487 - 607 - 2 = 878 \text{ eV},$$
$$1487 - 632 - 2 = 853 \text{ eV}.$$

These core level energies are both related to the 2p core level of the nickel ions in NiO. The 2p core level is split into two sub-peaks due to the 2p spin–orbit coupling.

The photoelectric effect is not the only interaction of x-rays with matter. Another important interaction is the elastic scattering of x-rays due to electrons, which is the origin of x-ray diffraction experiments. Elastic scattering is also known as Thomson scattering. In an x-ray diffraction experiment, a sample is hit with an x-ray and the diffracted x-rays are measured. The diffraction angle can be correlated with the interatomic distances using Bragg's law. With respect to spectroscopy, it is important to note that the amount of x-rays involved in x-ray diffraction is almost constant with respect to the photon energy. As such, it only gives a (nearly constant) background upon which the photoelectric effect is superimposed.

2.3.1 ELECTROMAGNETIC FIELD

Classically, the electromagnetic field can be represented by a complete set of plane waves. First, consider a single plane wave propagating in the x direction. Its wavevector $\mathbf{k}$ is along x, its electric field $\mathbf{E}$ is along y, perpendicular to x and its magnetic field $\mathbf{B}$ is along z, perpendicular to both x and y. The central object is the vector potential, which is given as:

$$\mathbf{A}(\mathbf{r}, t) = A_0 \mathbf{e}_y e^{i(kx-\omega t)} + A_0^* \mathbf{e}_y e^{-i(kx-\omega t)}. \tag{2.15}$$

The electric field $\mathbf{E}(\mathbf{r}, t)$ and the magnetic field $\mathbf{B}(\mathbf{r}, t)$ are given in terms of the vector potential. After rewriting the exponentials as cosine functions, this gives:

$$\mathbf{E}(\mathbf{r}, t) = -\frac{\partial}{\partial t} \mathbf{A}(\mathbf{r}, t) = 2i\omega A_0 \mathbf{e}_y \cos(kx - \omega t), \tag{2.16}$$

$$\mathbf{B}(\mathbf{r}, t) = \nabla \times \mathbf{A}(\mathbf{r}, t) = 2ik A_0 \mathbf{e}_z \cos(kx - \omega t). \tag{2.17}$$

2.3.2 TRANSITION TO QUANTUM MECHANICS

The transition from classical to quantum mechanics implies the change of the classical vector potential $\mathbf{A}(\mathbf{r}, t)$ to a quantum mechanical operator. Quantum mechanical problems are efficiently represented using so-called second quantization. The basic ingredients are the creation and annihilation operators. A creation operator $b_{\mathbf{k}\lambda}^\dagger$ creates a photon with wavevector $\mathbf{k}$ and polarization λ. Similarly, an annihilation operator is given by $b_{\mathbf{k}\lambda}$, and the creation and annihilation operators fulfill the commutation relation of Bose particles: $b_{\mathbf{k}\lambda}b_{\mathbf{k}'\lambda'}^\dagger - b_{\mathbf{k}'\lambda'}^\dagger b_{\mathbf{k}\lambda} = \delta_{\mathbf{k}\mathbf{k}'}\delta_{\lambda\lambda'}$. The number operator of the photon with $\mathbf{k}$ and λ is given by $n_{\mathbf{k}\lambda} = b_{\mathbf{k}\lambda}^\dagger b_{\mathbf{k}\lambda}$. An introduction to the use and properties of second quantization can be found in Chapter 12 of the book by Weissbluth (1978). The use of the creation and annihilation operators implies for the quantum-mechanical vector potential in the Schrödinger representation:

$$\mathbf{A}(\mathbf{r}) = \sum_{\mathbf{k},\lambda} A_0 \mathbf{e}_{\mathbf{k}\lambda} \left(b_{\mathbf{k}\lambda} e^{i\mathbf{k}\mathbf{r}} + b_{\mathbf{k}\lambda}^\dagger e^{-i\mathbf{k}\mathbf{r}} \right), \tag{2.18}$$

where $A_0 = \sqrt{2\pi\hbar c^2/V_s \omega_{\mathbf{k}}}$ with $\omega_{\mathbf{k}} = c|\mathbf{k}|$ and the system volume V_s in which the photon is normalized.

2.3.3 INTERACTION HAMILTONIAN

If x-rays interact with atomic electrons bound by a central potential $V(\mathbf{r})$, the overall interaction Hamiltonian can be written as:

$$H = H_{\text{RAD}} + H_{\text{ATOM}} + H_{\text{INT}}. \tag{2.19}$$

The radiation field Hamiltonian is written as a complete sum over $\mathbf{k}$ and λ degrees of freedom. The term of 1/2 is the zero point energy:

$$H_{\text{RAD}} = \sum_{\mathbf{k}\lambda} \hbar\omega_{\mathbf{k}} \left(n_{\mathbf{k}\lambda} + 1/2\right). \tag{2.20}$$

The Hamiltonian of the atomic electron consists of its kinetic term $\mathbf{p}^2/2m$ and the potential energy $V(\mathbf{r})$, which contains the Coulomb interaction with the nucleus

$$H_{\text{ATOM}} = \sum_{i} \left[\frac{\mathbf{p}_i^2}{2m} + V(\mathbf{r}_i) \right]. \tag{2.21}$$

Also, H_{ATOM} includes the Coulomb repulsion between electrons and the spin–orbit interaction, which are not described explicitly in Equation 2.21. The interaction Hamiltonian is treated as a small perturbation and the Coulomb gauge $\nabla\mathbf{A} = 0$ is used. Then the first-order terms of the interaction Hamiltonian are:

$$H_{\text{INT}(1)} = \frac{e}{mc}\sum_{i}\mathbf{p}_i \cdot \mathbf{A}(\mathbf{r}_i) + \frac{e}{2mc}\sum_{i}\sigma_i \cdot \nabla \times \mathbf{A}(\mathbf{r}_i). \tag{2.22}$$

The first term describes the interaction of the vector field $\mathbf{A}$ on the momentum operator $\mathbf{p}$ of an electron, or in other words the electric field $\mathbf{E}$ acting on the electron moments. The second term describes the magnetic field $\mathbf{B}$ ($= \nabla \times \mathbf{A}$) acting on the electron spin σ. In the second order of the perturbation, one finds the term:

$$H_{\text{INT}(2)} = \frac{e^2}{2mc^2}\sum_{i}\mathbf{p}_i \cdot \mathbf{A}(\mathbf{r}_i)^2. \tag{2.23}$$

For each $\mathbf{k}$ vector, the polarization is described by the two vectors $\mathbf{e}_{\mathbf{k}\lambda}$ ($\lambda = 1, 2$). They are perpendicular to the propagation direction $\mathbf{k}$ ($=x$) and perpendicular to each other. They are, for example, chosen in the y and z direction. Then the polarization vectors can be rewritten in the form of the irreducible tensor operator of rank one $\mathbf{e}_q$ with $\mathbf{e}_{-1} = -1/\sqrt{2}(\mathbf{e}_y - i\mathbf{e}_z)$ and $\mathbf{e}_{+1} = 1/\sqrt{2}(\mathbf{e}_y + i\mathbf{e}_z)$. The vectors $\mathbf{e}_y$ and $\mathbf{e}_z$ represent, respectively, the linear polarizations in the y and z directions, and $\mathbf{e}_{+1}$ and $\mathbf{e}_{-1}$ represent the circular polarizations with $+$ and $-$ helicities. The spin-vector of the photon can be oriented parallel ($q = +1$) or antiparallel ($q = -1$) to its momentum vector. Note that this polarization of synchrotron radiation is different from a conventional source.

2.3.4 GOLDEN RULE

The central role in the interaction of x-rays with matter is played by the Golden Rule. The Golden Rule states that the transition probability W between a system in its initial state Φ_i and final state Φ_f by absorbing the incident photon with energy $\hbar\Omega$ is given by:

$$W_{fi} = \frac{2\pi}{\hbar} |\langle \Phi_f |T|\Phi_i \rangle|^2 \delta(E_f - E_i - \hbar\Omega). \tag{2.24}$$

The initial and final state wave functions Φ_i are built from an electron part and a photon part, but in the following, we will not include the photon part explicitly in the discussion. The delta function takes care of the energy conservation and a transition takes place if the energy of the final state equals the energy of the initial state plus the x-ray energy. The squared matrix element gives the transition rate. The transition operator T is related to the interaction Hamiltonian H_{INT} with the Lippmann-Schwinger equation:

$$T = H_{INT} + H_{INT} \frac{1}{E_i - H + i\Gamma/2} T. \tag{2.25}$$

Γ is the lifetime broadening of an excited state and H is the Hamiltonian of the unperturbed system. The Lippmann–Schwinger equation is solved iteratively and in first order $T_1 = H_{INT(1)}$, which describes one-photon transitions (e.g. x-ray absorption, x-ray emission, and x-ray photoemission). The transition rates of these three processes are found by calculating the matrix elements of the transition operator T_1. Two-photon phenomena (e.g. x-ray scattering), are described with the transition operator in second order.

$$T_2 = H_{INT(2)} + H_{INT(1)} \frac{1}{E_i - H + i\Gamma/2} H_{INT(1)}. \tag{2.26}$$

Normal x-ray diffraction and small angle x-ray scattering (SAXS) experiments are described with $H_{INT(2)}$ only. In the case of resonant scattering experiments, the second term of T_2 is also included. This term describes the (virtual) absorption into an intermediate state and its subsequent decay into the final state.

2.4 OPTICAL TRANSITION OPERATORS AND X-RAY ABSORPTION SPECTRA

X-ray absorption, x-ray photoemission, and x-ray emission are all one-photon processes. They are described by the transition operator T_1, that is, by the first-order perturbation term of the interaction Hamiltonian $\mathbf{p} \cdot \mathbf{A}$. Omitting the summation over the index i of each electron, the interaction Hamiltonian is found by inserting the vector potential into $H_{INT(1)}$.

$$T_1 = \sum_{\mathbf{k},\lambda} \frac{e}{mc} A_0 \left[b_{\mathbf{k}\lambda} (\mathbf{e}_{\mathbf{k}\lambda} \cdot \mathbf{p}) e^{i\mathbf{k}\mathbf{r}} + \frac{\hbar}{2} b_{\mathbf{k}\lambda} (\mathbf{e}_{\mathbf{k}\lambda} \cdot \sigma \times \mathbf{k}) e^{i\mathbf{k}\mathbf{r}} \right]. \tag{2.27}$$

The first and second terms (in brackets) of Equation 2.27 represent, respectively, the electromagnetic interaction (nonrelativistic effect) and the spin interaction (relativistic effect). In the following, we disregard the spin interaction term, as it does not play an important role in the core level spectroscopy treated here. Then, T_1 can be rewritten using a Taylor expansion of $e^{i\mathbf{k}\mathbf{r}} = 1 + i\mathbf{k}\mathbf{r} + \cdots$. Limiting the equation to the first two terms, the transition operator is:

$$T_1 = \sum_{\mathbf{k},\lambda} b_{\mathbf{k}\lambda} \frac{e}{mc} A_0 [(\mathbf{e}_{\mathbf{k}\lambda} \cdot \mathbf{p}) + i(\mathbf{e}_{\mathbf{k}\lambda} \cdot \mathbf{p})(\mathbf{k} \cdot \mathbf{r})]. \tag{2.28}$$

Here the two terms in brackets represent, respectively, the electric dipole (ED) transition and the electric quadrupole (EQ) transition.

2.4.1 ELECTRIC DIPOLE TRANSITIONS

As seen from Equation 2.28, the EQ transition contains a small factor $\mathbf{k} \cdot \mathbf{r}$, compared with the ED transition. In the case of x-ray energies below some 10 keV, $\mathbf{k} \cdot \mathbf{r}$ is smaller than 10^{-2}. The transition probability is equal to the matrix element squared, hence the EQ term is smaller by 10^{-4} and can be neglected in the first approximation. This is the dipole approximation. The ED transition operator is given by:

$$T_1(ED) = \sum_{\mathbf{k},\lambda} b_{\mathbf{k}\lambda} \frac{e}{mc} A_0 (\mathbf{e}_{\mathbf{k}\lambda} \cdot \mathbf{p}). \tag{2.29}$$

2.4.2 ELECTRIC QUADRUPOLE TRANSITIONS

The next term in the Taylor expansion is $i(\mathbf{e}_{\mathbf{k}\lambda} \cdot \mathbf{p})(\mathbf{k} \cdot \mathbf{r})$. The squared matrix elements from this term are smaller by $(\mathbf{k} \cdot \mathbf{r})^2$. It is customary to rewrite this operator as:

$$i(\mathbf{e}_{\mathbf{k}\lambda} \cdot \mathbf{p})(\mathbf{k} \cdot \mathbf{r}) = \frac{1}{2} i[\mathbf{e}_{\mathbf{k}\lambda} \cdot (\mathbf{pr} + \mathbf{rp}) \cdot \mathbf{k}] + \frac{1}{2} i[\mathbf{e}_{\mathbf{k}\lambda} \cdot (\mathbf{pr} - \mathbf{rp}) \cdot \mathbf{k}]. \tag{2.30}$$

The symmetric combination gives rise to the EQ transitions. The antisymmetric combination is part of the magnetic dipole transition, which is disregarded. The wavevector $\mathbf{k}$ is rewritten as (ω/c) times the unit vector $\hat{k}$ and again the commutation law between the position operator $\mathbf{r}$ and the atomic Hamiltonian is used. One obtains:

$$T_1(EQ) \propto \mathbf{e}_{\mathbf{k}\lambda} \cdot \mathbf{rr} \cdot \hat{k} \tag{2.31}$$

The matrix element of $\mathbf{rr}$ must be calculated, but because $\mathbf{rr}$ is not an irreducible tensor, it is replaced by the quadrupole operator $\mathbf{Q} = \mathbf{rr} - 1/3r^2\delta_{ij}$. This gives for the EQ transition operator:

$$T_1(EQ) \propto \mathbf{e}_{\mathbf{k}\lambda} \cdot \mathbf{Q} \cdot \hat{k}. \tag{2.32}$$

2.4.3 DIPOLE SELECTION RULES

We found that in the the first approximation, x-ray absorption can be described with the dipole transition. Including this operator into the Fermi golden rule (Equation 2.24) gives for the x-ray absorption transition probability (W_{fi}) per unit time (Weissbluth, 1978):

$$W_{fi} = \frac{e^2}{\hbar c} \frac{4\Omega^3}{3c^2} n |\langle \Phi_f | \mathbf{r} | \Phi_i \rangle|^2 \delta(E_f - E_i - \hbar\Omega). \tag{2.33}$$

The first term $e^2/\hbar c$ is the fine structure constant α. The second term is proportional to the excitation frequency Ω to the third power. In practice, both the radiation field n (number of photons) and the excitation frequency Ω are not infinitely sharp but have a certain range of values.

In the case of an atom, the wave functions can be given J and M quantum numbers. The matrix element can be separated into a radial and an angular part according to the Wigner–Eckart theorem:

$$\langle\Phi_f(JM)|\mathbf{e}_{q\lambda}\cdot\mathbf{r}|\Phi_i(J'M')\rangle = (-1)^{J-M}\begin{bmatrix} J & 1 & J' \\ -M & q & M' \end{bmatrix}\langle\Phi_f(J)|\mathbf{e}_q\cdot\mathbf{r}|\Phi_i(J')\rangle. \quad (2.34)$$

The radial matrix element defines the line strength of a transition via:

$$S = e^2 \cdot |\langle\Phi_f(J)|\mathbf{e}_q\cdot\mathbf{r}|\Phi_i(J')\rangle|^2. \quad (2.35)$$

The triangular relations of the 3J-symbol determine the selection rules for x-ray absorption. The overall momentum quantum number J cannot be changed by more than 1; thus $\Delta J = +1, 0,$ or -1. The magnetic quantum number M is changed according to the polarization of the x-ray, that is, $\Delta M = q$. In addition, because the x-ray has an angular momentum of $l_{hv} = 1$, conservation of angular momentum gives $\Delta l_j = +1$ or -1; the angular momentum of the excited electron differs by 1 from the original core state. As x-rays do not carry spin, conservation of spin gives $\Delta s_j = 0$. For linearly polarized x-rays impinging on a sample under normal incidence, $q = \pm 1$, and for grazing incidence, $q = 0$. This gives a difference in the value of the 3J symbol, and hence (in potential) a polarization dependence. For circularly polarized x-rays, a dependence is found for a magnetic ground state. For an atomic nonmagnetic ground state, $M = \pm J$, and the 3J symbol is identical for $q = \pm 1$. The consequences for XMCD are discussed in Chapter 7. For extended final states (the Bloch-like wave functions in density functional methods), J is not a good quantum number and the only selection rules are $\Delta l_j \pm 1$ and $\Delta s_j = 0$. In the case of an excitation from a 1s core state, only p final states can be reached; from a p core state, s and d final states can be reached.

2.4.4 Transition Probabilities, Cross Sections, and Oscillator Strengths

The x-ray absorption transition probability per unit time was given in Equation 2.33. It can be rewritten with the line strength S as:

$$W_{fi} = \frac{1}{\hbar c}\frac{4\Omega^3}{3c^2}n\begin{bmatrix} J & 1 & J' \\ -M & q & M' \end{bmatrix}^2 S\delta(E_f - E_i - \hbar\Omega). \quad (2.36)$$

Omitting degeneracies and assuming the squared 3J-symbol to be unity, the transition probability yields:

$$W_{fi} = \frac{1}{\hbar c}\frac{4\Omega^3}{3c^2}nS\delta(E_f - E_i - \hbar\Omega). \quad (2.37)$$

The transition probability is proportional to the number of photons (n) or the photon flux (F_P). The x-ray absorption cross section (σ) is a property that is given by the transition probability divided by the photon flux as:

$$F_P = n \frac{\Omega^2}{\pi \hbar c^2}. \tag{2.38}$$

This yields for the x-ray absorption cross section (σ) given in m²:

$$\sigma = \frac{W_{fi}}{F_P} = \frac{4\pi^2 \Omega}{3c} S \delta(E_f - E_i - \hbar\Omega). \tag{2.39}$$

The cross section is directly proportional to the oscillator strength f as:

$$\sigma = \frac{2\pi^2 e^2}{mc} f. \tag{2.40}$$

The oscillator strength is a dimensionless quantity. We have omitted the role of the degeneracies of the initial and final states from the present discussion.

$$f = \frac{2m\Omega}{3\hbar} |\langle \Phi_f | \mathbf{r} | \Phi_i \rangle|^2 \delta(E_f - E_i - \hbar\Omega). \tag{2.41}$$

2.4.5 Cross Section, Penetration Depth, and Excitation Frequency

The delta-function implies that an infinitely sharp line is assumed. In experiment, this is not the case because of the lifetime broadening of the x-ray absorption lines, as discussed in Section 2.1.2. The lifetime broadening modifies the maximum cross section. We assume a monochromator that provides a constant energy within a 0.1 eV interval. If the lifetime broadening constant Γ is 0.2 eV, the lifetime broadening is 0.4 eV. With a lifetime broadening of 0.4 eV and a single δ-function resonance, there remains ~15% of the cross section at a 0.1 eV interval at the maximum. This defines the actual x-ray absorption cross section (σ_Γ or σ):

$$\sigma \approx 0.15 \cdot \frac{4\pi^2 \Omega}{3c} S. \tag{2.42}$$

The x-ray absorption cross section determines directly the penetration depth (λ_p) of the x-rays via $\lambda_p = 1/\rho\sigma$, where ρ is the density of the system. The cross section is typically given in Å². It is then convenient to define the inverse density, or in other words, the space used by a single atom as $V_{at} = 1/\rho$. In case of La metal, V_{at} is approximately 37 Å³ and the theoretical cross section $\sigma_\delta = 1$ Å², yielding a cross section σ of 0.15 Å². Together, this yields a penetration depth:

$$\Lambda_p = V_{at}/\sigma \approx 200 \text{ Å}. \tag{2.43}$$

This penetration depth is very short. In the case of 3d transition metals, the cross sections are smaller but the atomic volumes are also smaller. Typical values are

$\sigma = 0.05$ Å^2 and $V_{at} = 10$ Å^3, again yielding penetration depths of only 200 Å at the peak maxima of the L$_3$ edges. We can now also determine the average frequency with which each atom is excited. Assuming a beamline with a spot size of 20×20 μm, the volume where all (actually 65% of all) photons are absorbed is $(20 \times 10^4) \times (20 \times 10^4) \times 200 = 8 \times 10^{12}$ Å^3. In the case of harder x-rays, the attenuation length goes up to ~2 μm at 10 keV, increasing the absorbance volume to 8×10^{14} Å^3. Note that with soft x-ray microscopes, the volume will drastically decrease. For example, for beamline 11.0.2 at the ALS, the spot size is 20×20 nm, implying an absorbance volume of 8×10^6 Å^3. In the case of a 3d metal, a volume of 8×10^{12} Å^3 contains ~10^{12} atoms. Assuming a flux of 10^{12} photons per second, this implies that each atom is on average excited every second. At 10 keV, this number will go down to 0.01 excitations per second. Beamline 11.0.2 with a flux of 10^9 photons per second and ~10^6 atoms in the absorbance volume reaches a hit-frequency of 1000 excitations per second.

With a core hole lifetime of 10^{-15} s, it is clear that there is virtually no change of hitting an already excited atom with yet another photon. Off resonance, the L edge cross section drops by typically a factor of 100, decreasing the hit-frequency to less than once every 100 s. Each excitation process creates the possibility that the excited atom will be damaged; in particular, because of the ionization processes during core hole decay. If the excitation process leads to a significant increased chance for damage, it is obvious that after a few times, there will be serious damage.

2.4.6 X-Ray Attenuation Lengths

The attenuation length is defined as the penetration depth into the material where the intensity is lowered to $1/e$ of its value at the surface. $1/e$ is equal to 0.37 implying that 63% of the x-ray intensity is absorbed within the attenuation length. X-rays with energies less than 1 keV have an attenuation length of less than ~1 μm, implying that for soft x-rays to be transmitted through the sample, it has to be thin. Soft x-rays have a large absorption cross section with air; in particular, just above the oxygen K edge at 530 eV and up to 1 keV x-rays, the transmission through 10 cm of air is less than 10%. This implies that the experiments have to be performed either in a vacuum or with short optical paths. The x-ray transmissions and associated attenuation lengths can be calculated from the Web site of the center for x-ray optics (http://www-cxro.lbl.gov/).

2.5 INTERACTION OF ELECTRONS WITH MATTER

Core level spectra can also be measured with electrons. Depending on the energy (E) and momentum transfer (**q**) of the electron, one obtains a few different transition operators for the core transitions (Table 2.3). In short:

1. High E, low **q**: dipole operator
2. High E, not low **q**: dipole plus quadrupole operator
3. Low E: Coulomb operator

The most common route to measure core level spectra with electrons is using EELS in an electron microscope. Usually the momentum transfer is minimized and one obtains the dipole operator. Electron microscopes use energies of 100–300 keV.

With these energies, the electron can be approximated as a plane wave (e^{iqr}) and the cross section for inelastic scattering is given by:

$$\sigma \propto |\langle \Phi_f | e^{iqr} | \Phi_i \rangle|^2 \delta(E_f - E_i - \Delta E), \tag{2.44}$$

where ΔE is the electron energy loss. We expand the exponential factor as:

$$e^{iqr} = 1 + \mathbf{q} \cdot \mathbf{r} - \frac{1}{2}(\mathbf{q} \cdot \mathbf{r})^2 + \cdots. \tag{2.45}$$

For small scattering angles, $\mathbf{q} \cdot \mathbf{r}$ is small and all terms except $\mathbf{q} \cdot \mathbf{r}$ can be neglected. Within this dipole approximation, the EELS spectrum is exactly proportional to the x-ray absorption cross section:

$$\sigma_{EELS} \propto \sigma_{XAS} \propto |\langle \Phi_f | \mathbf{q} \cdot \mathbf{r} | \Phi_i \rangle|^2 \delta(E_f - E_i - \Delta E). \tag{2.46}$$

Because a high energy x-ray can also be approximated with a plane wave, exactly the same line of reasoning can be followed and x-ray energy loss (better known as XRS), can be measured on the same cross section as EELS and XAS:

$$\sigma_{XRS} \propto \sigma_{EELS} \propto \sigma_{XAS}. \tag{2.47}$$

In the case of larger scattering angles, the dipole approximation is not valid anymore and the quadrupole transitions become more important. In principle, this allows the EELS and XRS measurements to be tuned to, respectively, dipole and quadrupole transitions. The quadrupole transition operator also allows monopole transitions, for example, 1s2s transitions in Li (Krisch et al., 1997).

X-rays from synchrotron radiation sources are usually linearly polarized and for XMCD experiments, circular polarized x-rays are used. In the case of electrons, things are different. Recently, it has been shown that the relativistic effect on high-energy EELS effectively creates a linear polarization in the inelastic electron scattering within the dipole approximation. The direction of the linear polarization of the electrons ($\mathbf{Q}^*$) is given as (Schattschneider et al., 2005):

$$\mathbf{Q}^* = \frac{\mathbf{Q}_E + \theta_E \mathbf{k}(V^2/c^2)}{1 - (\hat{Q} \cdot \hat{k})(V^2/c^2)} \tag{2.48}$$

where $\mathbf{Q}_E$ is the wavevector, θ_E the scattering angle, $\mathbf{k}$ the incident electron wavevector, V the incident electron velocity, and c the speed of light. The effective wavevector $\mathbf{Q}^*$ determines the two polarizations, respectively, $\sigma \perp \mathbf{Q}^*$ and $\sigma \parallel \mathbf{Q}^*$. Because high-energy EELS is typically measured in a TEM with sub-nanometer resolution, one usually measures an oriented region, that is, single crystal region for crystals.

In the case of low-energy electrons, the transition operator cannot be approximated with a plane wave and the actual Coulomb operator $1/\mathbf{r}$ for electron scattering has to be used (Ogasawara and Kotani, 1996). It can be shown that the cross section can be written as:

$$\sigma_{LE\text{-}EELS} \propto |\langle \Phi_f | T_{eff} | \Phi_i \rangle|^2 \delta(E_f - E_i - \Delta E). \tag{2.49}$$

The final states that will be reached are the same as in XAS and high-energy EELS, but the matrix elements will be different. T_{eff} contains a summation over all direct Coulomb scattering and exchange scattering from the initial state (Φ_i) and electron (ε_i) to the final state (Φ_f) and scattered electron (ε_f):

$$T_{\text{eff}} = \sum_{i,j}\left[\langle \Phi_i \varepsilon_i \mid \frac{1}{\mathbf{r}} \mid \Phi_f \varepsilon_f \rangle - \langle \Phi_i \varepsilon_i \mid \frac{1}{\mathbf{r}} \mid \varepsilon_f \Phi_f \rangle \right] a_i^* a_f. \qquad (2.50)$$

The selection rules for LE-EELS are much more relaxed and many more final states are allowed. In particular, there will be significant intensities for nondipole transitions, such as the $\Delta J = 3$ transition. Examples are given in Chapter 6.

2.6   X-RAY SOURCES

Röntgen discovered x-rays whilst working with cathode-ray tubes (Röntgen, 1896). Using the principle of fast electrons hitting a metallic target, the first substantial gain in brilliance was not obtained until the introduction of the rotating anode sources in 1960. In an x-ray tube, electrons are accelerated between the anode and the cathode. When they hit the cathode with high energy, they are decelerated through inelastic collisions with atoms in the target material. If the energy of the electrons is high enough, this produces a continuous spectrum of x-rays. In addition, the impinging high-energy electrons excite core electrons from the cathode material. Core holes decay, in part, via radiative processes in which a shallow core electron fills a deep core hole, thereby emitting x-rays. These x-rays are characteristic for the cathode material and are confined to only a few energies.

X-ray tubes are used in many x-ray applications, for example, medical x-ray images and laboratory sources used in x-ray photoemission and x-ray diffraction. In x-ray spectroscopy, however, the role of x-ray tubes has been declining over the last few decades. The main reason is that x-ray tubes produce a maximum flux/brightness of 10^{10} photons per second, whereas x-rays produced with synchrotrons reach up to 10^{19} photons per second. The photon counts are usually given with the brightness. The brightness is normalized to an area of 1 mm^2 and further normalized to the x-ray beam divergence and the energy (range). It is clear that with 10^9 times more photons per second, much faster experiments can be carried out. In addition, it allows a range of new experiments that have low x-ray cross sections. It should be noted that the x-rays produced with synchrotrons have a very high intensity, but still they cannot be considered as x-ray lasers, because, in general, they have insufficient coherence. In addition to the flux/brightness, synchrotron x-rays have other important features that can be used [e.g. tunability, polarization properties, time structure, and (focal) spot size].

2.6.1   SYNCHROTRON RADIATION SOURCES

The name "synchrotron radiation" refers to the continuous band of electromagnetic spectrum that was accidentally discovered in a General Electric electron synchrotron in 1947. It includes infrared, visible, UV, and x-ray radiation. Around 1970, the first generation of synchrotron sources appeared. They used high-energy physics accelerators in parasitic mode. In these storage rings, the high-energy electrons were deflected by a magnetic field to keep them in the ring. The x-rays that were produced as such were considered a nuisance and were minimized. They could still be used for x-ray experiments. These experiments proved so promising that a number of dedicated second-generation sources were built around 1980. Third-generation synchrotron sources appeared in the 1990s, using a combination of focused electron beams and insertion devices. The resulting x-ray beams reach up to 10^{19} photons per second. An overview of all synchrotron radiation sources is to be found at http://www.lightsources.org/.

In a synchrotron, electrons are generated and accelerated to the desired energy before they are stored in a ring: the so-called storage ring. Without the presence of an external magnetic field, the electrons will travel in a straight line, but when a magnetic field is applied they will bend because of the Lorentz force, which is proportional to the electron velocity (v_e) and the magnetic field B (neglecting relativistic effects): $F \propto v_e \cdot B$. This force causes an acceleration of the particles and therefore radiation is formed. This radiation is emitted in the direction of motion of the particle. Within the bending magnet, the electron changes direction and a radiation beam in between the original and the new electron direction is emitted. The creation of radiation will cause the velocity of the electrons to reduce and therefore the radius of the circular orbit will decrease. After a short period the electrons will have lost their energy and, in order to prevent this, the energy loss is compensated by a radio frequency cavity. During the bending of the electron beam, a wide spectrum of radiation is formed.

Between the magnets, the electrons follow a straight path without the formation of radiation. Insertion devices can be placed in the electron beam in order to produce radiation. Examples of such devices are wigglers and undulators. Basically, these devices change the directions of the electrons continuously up and down and/or left and right along the straight path in order to produce more radiation than with a bending magnet. In an undulator, the electron beam is periodically deflected by weak magnetic fields. The spectral resolution of the radiation is proportional to the number of undulator periods and varying the magnetic field (i.e. the undulator gap), shifts the resonance wavelengths. Most radiation is emitted within a very small angle. Increasing the magnetic field creates relativistic effects and causes the motion of electrons in an undulator to become distorted. Many harmonics are generated that merge into a continuous spectrum from IR to hard x-rays. Compared to bending magnet radiation, wiggler radiation is enhanced by the number of magnet poles.

2.6.2   X-Ray Beamlines and Monochromators

An XAS spectrum originates from the fact that the probability of an electron to be ejected from a core level is dependent on the energy of the incoming beam. For this reason, the energy of the x-rays is varied during an experiment. Of course, this requires a monochromatic beam. However, the radiation generated by a synchrotron is polychromatic. Therefore, the desired wavelength has to be filtered from the polychromatic beam. For this purpose, a monochromator is installed in the beamline. In the case of hard x-rays, the monochromator is based on crystal optics. The incoming beam is diffracted on a crystal and the wavelength of the diffracted beam is given by Bragg's law $n\lambda = 2d \sin \theta$. By varying the angle θ between the x-ray propagation direction and the surface of the crystal, the energy of the x-ray beam can be varied. Typically one uses Si crystals, for example, Si(111) or Si(311). The diffraction plane determines the distance d and as such the angle θ needed for a particular energy. It also determines the energy resolution and because the life time broadening is usually above 1 eV, these crystals provide sufficient resolution. Higher resolution experiments need crystal planes with larger values of $h^2 + k^2 + l^2$, for example, Si(220). In principle, one crystal is enough to obtain the desired wavelength of the beam. In practice, a double crystal monochromator is used in order to obtain an outgoing beam that is parallel to the incoming beam. After the beam has passed the monochromator, its intensity (I_0) has to be measured. This can be done by an ionization chamber. In the next step, the x-ray interacts with the sample of interest. To measure the intensity after the sample a second ionization chamber can be used. This is the transmission mode of x-ray absorption.

Soft x-ray absorption experiments have much longer wavelengths (λ) and as such need crystal optics with large d-spacing. Between 800 eV and 1500 eV, beryl [$Be_3Al_2(SiO_3)_6$] crystals with a d-spacing of 15.95 Å are used. Below 800 eV, crystals with even longer d-spacing could be used but organic crystals, for example, were found to be quickly damaged by the x-ray beam. As a consequence, artificial gratings for the soft x-ray range were used. In the late 1980s, the Dragon (Chen and Sette, 1989) and SX700 monochromators (Petersen, 1982) were using artificial gratings, which dramatically improved the experimental resolution. An artificial grating has a much larger d-spacing. This implies that to reach the correct energy range, the gratings must be hit under grazing incidence. Recent improvements in the quality and line density (= d-spacing) of artificial grating means that they are now used for higher energies (i.e. ~1.5 keV).

2.6.3 OTHER X-RAY SOURCES

The x-ray anode, as used for the last 100 years for laboratory x-ray experiments and a number of other sources, still has promising uses for future applications. The principle is that a high-voltage difference is created and electrons are impacted on a metal anode thereby creating core holes. The x-ray emission of these core holes produces x-rays related to the various characteristic lines. In addition, very low intensity continuous x-rays are created (via Bremsstrahlung) that were used for x-ray absorption experiments until the development of synchrotron sources. The mechanism of Bremsstrahlung radiation involves the transition from the incoming electron into an unoccupied state. The energy lost in this process is emitted as x-rays, and this process is, in fact, used to measure the empty density of states via BIS or IPES. Effectively BIS yields, for a single electron energy, a white x-ray spectrum over a range of say 100 eV with all energies up to the original electron energy. These x-rays can be used in turn for an x-ray absorption experiment. The count rates of such experiments are probably over 10^{10} times lower than synchrotron experiments. If the electrons are created with a laser, x-ray anodes can be turned into pulsed x-ray sources.

Soft x-rays can be generated using high harmonics of optical range lasers. The procedure is to excite a rare gas (e.g. neon), which in turn will produce a pulse of high harmonics up to several 100 eV of x-ray energy (Froud et al., 2006). Harder x-rays can be produced from plasma sources. A metal is excited with a powerful laser, where the system can be optimized for characteristic x-rays or continuum x-rays. As far as the creation of the (harder) x-rays is concerned, it could essentially be viewed as a more intense variant of an x-ray anode, where the anode is turned into a plasma. Pulsed x-ray anodes, high harmonics and plasma sources are, in particular, useful for time-resolved experiments and plasma sources are strongly researched with respect to extreme UV (EUV) lithography (Bressler and Chergui, 2004).

A rather special x-ray source is formed naturally in space. They are used for x-ray absorption experiments in astronomy including the study of interstellar space by way of its x-ray absorption properties and by using pulsars as the x-ray source in combination with satellite x-ray detectors (de Vries et al., 2003a; Juett et al., 2004).

2.7 ELECTRON SOURCES

The principles of electron sources are relatively straightforward. The electrons are created by heating a cathode in a vacuum. The created electrons are then accelerated by an electric field. Instead of creating the electrons by heat, they can also be created by field emission. In field emission, a very large electric field extracts electrons out of the metal and, because of the presence of the field, this process takes place at a lower temperature then the usual thermionic emission. A lower temperature implies an electron beam that has a better-defined energy due to a smaller thermal broadening. An electron beam with high-energy resolution can also be obtained by using electric and/or magnetic lenses. Popular electron monochromators are hemispherical magnetic analyzers, as also used in PES, or the so-called Wien-filter, perpendicular electric and magnetic fields that are both perpendicular to the electron trajectory.

The energy of the electron source can be selected by the applied field in the electron accelerator and the energy resolution by the electron gun used plus any additional electron monochromators. High-resolution EELS experiments can, for example, be performed by using a hemispherical analyzer in the electron source stream and a second one as detector. This setup has been used by the high-resolution EELS setup of Fink (Fink et al., 1985; Fink, 1992) and it is also used in some high-resolution TEM-EELS microscopes. The most popular design of the EELS analyzer in TEM microscopes is at present, the GATAN electron analyzer coupled to a field emission gun, with or without monochromator. This yields energy resolutions for core levels below 0.2 eV (Krivanek et al., 1985; Krivanek and Patterson, 1990; Mitterbauer et al., 2003).

3 Many-Body Charge-Transfer Effects in XPS and XAS

3.1 INTRODUCTION

In this chapter, the many-body charge-transfer effects in x-ray photoemission spectroscopy (XPS) and x-ray absorption spectroscopy (XAS) are discussed. As the charge-transfer effect plays a more important role in XPS than it does in XAS, the major part of this chapter is devoted to XPS.

As already mentioned in Chapter 1, the development of XPS started from one-electron features and progressed to the analysis of elements in solids and the chemical analysis of molecules and solids. Subsequently, because of the progress that was made in experimental techniques, the many-body charge-transfer effect in XPS spectra became the main object in the study of XPS. A typical example is the anomalous XPS spectrum of simple metals resulting from an orthogonality catastrophe. More drastic charge-transfer effects were observed in f and d electron systems where charge transfer occurs from conduction (or valence) band states to the f or d electron states in order to screen the core-hole potential. This causes a splitting of the XPS spectra into well-screened and poorly-screened final states. By analysis of the charge-transfer effect in XPS, important information on valence electron states (VES) was found, including the f or d electron states. For mixed valence Ce compounds, in particular, the weight of the f electron occupancy in the ground state was estimated to be the most important (microscopic and local) information. Also, for various transition metal (TM) compounds, characteristic features of the insulating energy gap have been derived as well as the mixed valence features of the ground state.

In this chapter, a basic description of the XPS process is given together with general expressions of many-body charge-transfer effects in XPS. Then, the XPS features in simple metals, La metals, and mixed valence Ce intermetallics are described, followed by some discussions on the charge-transfer effect in XPS of insulation Ce compounds and insulating TM compounds. In this chapter, we confine ourselves to theoretical descriptions where the charge-transfer effect is mainly taken into account with the single impurity Anderson model (SIAM) (or cluster model), but the multiplet coupling effect is disregarded. The multiplet coupling effect usually plays a minor role in the XPS process compared to that of the charge-transfer effect. More complete descriptions of XPS spectra by charge transfer multiplet (CTM) theory (including both charge transfer and multiplet effects) are given in Chapter 5.

After treating the XPS spectra, similar descriptions of many-body charge-transfer effects in XAS are then given. However, in general, the charge-transfer effect in XAS proves to be much weaker than that in XPS. In the case of XAS, the core-hole potential is mainly screened by the photo-excited electron itself, and this suppresses the screening effect because of the charge-transfer effect. In XPS, the photo-excited electron does not screen the core-hole potential. The multiplet coupling effect and the crystal field effect play a more important role in XAS spectra than the charge-transfer effect. In Chapter 4, the analysis of various XAS spectra by ligand field multiplet (LFM) theory is given, where the effects of multiplet coupling and crystal field are taken into account but the charge-transfer effect is disregarded. CTM theory is covered in Chapter 6.

3.2 MANY-BODY CHARGE-TRANSFER EFFECTS IN XPS

3.2.1 BASIC DESCRIPTION OF THE XPS PROCESS

We divide our system into three subsystems: (*i*) a core electron system described by a single-electron wave function $|\phi_c\rangle$ with energy ε_c, (*ii*) a photoelectron system described by $|\phi_{nk}\rangle$ with energy ε_{nk} [which is a Bloch state specified by a band index n and a wavevector $\mathbf{k}$ inside a crystal, but connected to a plane wave with energy $\varepsilon (= \varepsilon_{nk})$ outside the crystal], and (*iii*) the VES. The subsystems (*i*) and (*ii*) are treated as independent single particles, except that a core hole left behind in the final state of XPS couples with the subsystem (*iii*) by a general interaction U. We disregard the interaction between (*ii*) and (*iii*). Then the interaction U is applied suddenly by the core electron excitation, and this treatment is denoted as the sudden approximation. The sudden approximation is valid in usual XPS experiments where the kinetic energy ε is sufficiently high. A more detailed description of the sudden approximation is given in Appendix A where the formulation of the XPS spectrum is made more precisely than in the present subsection. Note that care must be taken that the interaction of (*ii*) and (*iii*) does not give rise to a finite mean free path of the photoelectron. If the mean free path of the photoelectron is only a few times the lattice constant of the material system, the XPS spectrum is surface sensitive. We discuss this point at the end of this subsection (and also in Appendix A).

The subsystem (*iii*) is treated by the many-electron picture, and the Hamiltonian is generally written as H_0 and $H (= H_0 + U)$ in the initial and final states of XPS, respectively. The ground state $|0\rangle$ and each final state $|f\rangle$ of the VES (*iii*) satisfy the following equations:

$$H_0 |0\rangle = E_0 |0\rangle, \tag{3.1}$$

$$H |f\rangle = E_f |f\rangle. \tag{3.2}$$

If we disregard the interaction U, the XPS spectrum for incident photon energy $\hbar\Omega$ and the photoelectron kinetic energy ε is simply expressed, by the Golden Rule, as

$$F(\varepsilon, \Omega) = \sum_{n,\mathbf{k}} |\langle \phi_{nk} | \mathbf{p} \cdot \mathbf{e} | \phi_c \rangle|^2 \delta(\varepsilon_{nk} - \varepsilon_c - \hbar\Omega)\, \delta(\varepsilon - \varepsilon_{nk})$$

$$= f(\varepsilon)\delta(\varepsilon - \varepsilon_c - \hbar\Omega), \tag{3.3}$$

where

$$f(\varepsilon) = \sum_{n,\mathbf{k}} |\langle \phi_{n\mathbf{k}} | \mathbf{p} \cdot \mathbf{e} | \phi_c \rangle|^2 \delta(\varepsilon - \varepsilon_{n\mathbf{k}}). \tag{3.4}$$

Here, $\mathbf{p} \cdot \mathbf{e}$ represents the electric dipole transition operator (apart from a constant factor) for the incident photon polarization $\mathbf{e}$. Therefore, in this one-electron approximation, the XPS spectrum exhibits a single line at $\varepsilon - \varepsilon_c - \hbar\Omega = 0$. Here the origin of energies ε and ε_c is arbitrary, but in order to explicitly take the origin of ε at the vacuum level ε_{vac} (as usually done), we rewrite this equation as

$$(\varepsilon - \varepsilon_{\text{vac}}) + (\varepsilon_{\text{vac}} - \varepsilon_c) - \hbar\Omega = 0, \tag{3.5}$$

and replace $\varepsilon - \varepsilon_{\text{vac}}$ by ε. Further, we put the equation $\varepsilon_{\text{vac}} - \varepsilon_c = \phi + (\varepsilon_F - \varepsilon_c)$, then Equation 3.5 reduces to Equation 1.1 of Chapter 1.

Now we consider the effect of U. The initial and final states, respectively, are given by the direct products $|\phi_c\rangle|0\rangle$ and $|\phi_{n\mathbf{k}}\rangle|f\rangle$ whose energies are $\varepsilon_c + E_0$ and $\varepsilon_{n\mathbf{k}} + E_f$. Therefore, Equation 3.3 is modified to

$$F(\varepsilon, \Omega) = \sum_{n,\mathbf{k},f} |\langle \phi_{n\mathbf{k}} | \mathbf{p} \cdot \mathbf{e} | \phi_c \rangle|^2 |\langle f|0\rangle|^2 \delta(\varepsilon_{n\mathbf{k}} - \varepsilon_c + E_f - E_0 - \hbar\Omega)\delta(\varepsilon - \varepsilon_{n\mathbf{k}})$$

$$= f(\varepsilon)\sum_f |\langle f|0\rangle|^2 \delta(\varepsilon - \varepsilon_c + E_f - E_0 - \hbar\Omega). \tag{3.6}$$

It is to be noted that the overlap integral $\langle f|0\rangle$ has in general a finite value even if $|f\rangle \neq |0\rangle$, because the Hamiltonian for $|f\rangle$ is different to that for $|0\rangle$. In the usual situation of XPS, the function $f(\varepsilon)$ changes very slowly with ε so that we put it as a constant. Furthermore, we define the binding energy E_B by

$$E_B = \hbar\Omega - \varepsilon + \varepsilon_c, \tag{3.7}$$

taking its origin appropriately, and write the XPS spectrum $F(\varepsilon, \Omega)$ as $F(E_B)$. If we normalize the intensity of $F(E_B)$ by

$$\int dE_B F(E_B) = 1, \tag{3.8}$$

then $F(E_B)$ is given by

$$F(E_B) = \sum_f |\langle f|0\rangle|^2 \delta(E_B - E_f + E_0). \tag{3.9}$$

For more detailed derivation of the XPS formula, see Appendix A.

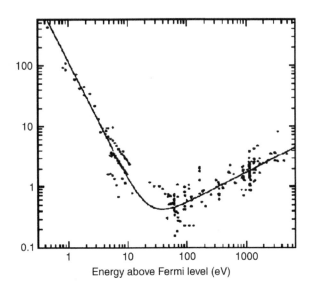

Energy above Fermi level (eV)

FIGURE 3.1 The electron mean free path in nanometres is given as a function of the energy of the escaping electron above the Fermi level. (From Seah, M.P., and Dench, W.A., *Surf. Interf. Anal.*, 1, 2, 1979. With permission.)

In the previous derivation of the XPS spectrum $F(E_B)$, the mean free path of the photoelectron is assumed to be infinite. However, the mean free path is finite and strongly depends on the kinetic energy of the photoelectron. Experimental data for the electron mean free path in various materials are shown in Figure 3.1 as a function of the electron kinetic energy. The behavior of all the data points is represented by the solid curve, which is called the "universal curve." If the electron energy is lower than 10 eV or higher than several keV, the mean free path is much larger than the lattice constant of the material systems; then our assumption of the infinite mean free path is acceptable and the expression of $F(E_B)$ given above is almost correct. On the other hand, for an electron energy of 10 eV to 500 eV, the mean free path is less than 10 Å, which means that photoelectrons excited within a depth of 10 Å from the surface (denoted by escape depth) can be detected and the XPS spectrum is surface-sensitive. Care has to be taken regarding whether the relevant XPS experimental data detects the bulk electronic states or the surface electronic states.

3.3 GENERAL EXPRESSIONS OF MANY-BODY EFFECTS

3.3.1 GENERAL DESCRIPTION

From a general viewpoint, how the response of VES to the core hole charge is reflected in the XPS spectrum (Kotani, 1987) is shown here. The multiplet coupling effect caused by the multipole Coulomb interaction within VES and between the VES and core states is disregarded. However, the charge-transfer effect is fully taken into account.

The XPS spectrum (Equation 3.9) can be rewritten in the following integral form:

$$F(E_B) = \langle 0| \sum_f \delta(E_B - E_f + E_0)|f\rangle\langle f|0\rangle$$

$$= \langle 0|\delta(E_B - H + E_0)|0\rangle$$

$$= \frac{1}{\pi\hbar}\mathrm{Re}\int_0^\infty dt \exp\left(i\frac{E_B}{\hbar}t\right)g(t), \tag{3.10}$$

where

$$g(t) = \langle 0(t)|f(t)\rangle, \tag{3.11}$$

$$|f(t)\rangle = \exp\left(-i\frac{H}{\hbar}t\right)|0\rangle, \tag{3.12}$$

$$|0(t)\rangle = \exp\left(-i\frac{E_0}{\hbar}t\right)|0\rangle, \tag{3.13}$$

and we have used Equations 3.1 and 3.2. For $t \geq 0$, $|f(t)\rangle$ represents the time variation of VES caused by the interaction with the core hole U, because $\exp[-i(H/\hbar)t]$ is the time development operator driven by the Hamiltonian H. The function $g(t)$ is the overlap integral between the state $|f(t)\rangle$ and the ground state $|0(t)\rangle$ without a core hole. Thus, $g(t)$ directly reflects the many-body response of VES to the core-hole potential, and the XPS spectrum $F(E_B)$ is given by the Fourier transform of $g(t)$. Note that if we extend $g(t)$ as an even function to the region $t < 0$, $F(E_B)$ is related to $g(t)$ by the usual Fourier transformation for $-\infty < t < \infty$. We denote $g(t)$ as the "generating function."

In the following, we put $\langle 0|U|0\rangle = 0$ by assuming that $\langle 0|U|0\rangle$ is already included in ε_c. Then, U describes the relaxation effect (or the redistribution effect) of VES caused by charge transfer because of the core-hole potential. If we neglect the effect of U by putting $U = 0$ in the above equations, the generating function $g(t)$ reduces to $g(t) = 1$, independent of time t. Then, the XPS spectrum $F(E_B)$ is given by a δ-function $\delta(E_B)$, namely the one-electron approximation for $F(E_B)$ given by Equation 3.3.

When $U \neq 0$, we can describe the behavior of $g(t)$ in a more explicit way. To this end, $g(t)$ is written as

$$g(t) = \langle 0|S(t)|0\rangle \tag{3.14}$$

where

$$S(t) = e^{i(H_0/\hbar)t}e^{-i(H/\hbar)t}. \tag{3.15}$$

By taking derivative of $S(t)$, we obtain the differential equation for $S(t)$:

$$\frac{dS(t)}{dt} = \frac{i}{\hbar}\left\{H_0 S(t) - e^{i(H_0/\hbar)t}(H_0 + U)e^{-i(H_0/\hbar)t}S(t)\right\} = -\frac{i}{\hbar}U(t)S(t), \quad (3.16)$$

where

$$U(t) = e^{i(H_0/\hbar)t}Ue^{-i(H_0/\hbar)t}. \quad (3.17)$$

By the iterative integration of the integrative equation, we have

$$S(t) = 1 - \frac{i}{\hbar}\int_0^t dt_1 U(t_1) + \left(\frac{i}{\hbar}\right)^2 \int_0^t dt_1 \int_0^{t_1} dt_2 U(t_1)U(t_2) + \cdots$$

$$= T_\tau \exp\left[-\frac{i}{\hbar}\int_0^t dt' \, U(t')\right], \quad (3.18)$$

where T_τ is the time-ordering operator. Therefore, $g(t)$ is represented by

$$g(t) = \langle 0 | T_\tau \exp\left[-\frac{i}{\hbar}\int_0^t dt' \, U(t')\right] | 0 \rangle. \quad (3.19)$$

3.3.2   Generating Function and Dielectric Response

By using the second-order cumulant approximation with respect to U, the generating function $g(t)$ is expressed as

$$g(t) = \exp\left[-\frac{1}{\hbar^2}\int_0^t dt_1 \int_0^{t_1} dt_2 C(t_1 - t_2)\right], \quad (3.20)$$

where $C(t_1 - t_2)$ is the time-correlation function of U defined by

$$C(t_1 - t_2) = \langle 0 | U(t_1) U(t_2) | 0 \rangle. \quad (3.21)$$

In many cases, the approximation is very good. Applying this formula to the dielectric response of VES and taking into account the Coulomb potential of the core hole charge, U is given by

$$U = \frac{1}{V_s}\sum_q \rho_q \phi_{-q}, \quad (3.22)$$

where V_s is the volume of the system, ρ_q is the $\mathbf{q}$ component of the polarization charge of VES and

$$\phi_q = 4\pi e/q^2, \quad (e > 0). \quad (3.23)$$

Hereafter, we choose the volume of the system as unity for simplicity. Substituting Equation 3.22 into Equation 3.21, we have

$$C(t_1 - t_2) = \sum_q \phi_q^2 \langle 0 | \rho_q(t_1) \rho_{-q}(t_2) | 0 \rangle. \tag{3.24}$$

In order to extend the theory to the finite temperature case, the ground state expectation value in this equation is replaced by the thermal average $\langle \rho_q(t_1) \rho_{-q}(t_2) \rangle$. We use the following relation between the charge correlation function and the dielectric function $\varepsilon(\mathbf{q}, \omega)$:

$$\langle \rho_q(t_1) \rho_{-q}(t_2) \rangle = -\frac{\hbar q^2}{4\pi^2} \int_{-\infty}^{\infty} d\omega \frac{e^{-i\omega(t_1 - t_2)}}{1 - e^{-\beta\omega}} \operatorname{Im} \left[\frac{1}{\varepsilon(\mathbf{q}, \omega)} \right], \tag{3.25}$$

where $\beta = 1/k_B T$. The generating function is finally expressed in the form

$$g(t) = \exp \left\{ \frac{1}{4\pi^2 \hbar} \sum_q q^2 \phi_q^2 \int_{-\infty}^{\infty} d\omega \frac{-i\omega + 1 - e^{-i\omega t}}{\omega^2(1 - e^{-\beta\hbar\omega})} \operatorname{Im} \left[\frac{1}{\varepsilon(\mathbf{q}, \omega)} \right] \right\}. \tag{3.26}$$

3.3.3 XPS SPECTRUM AND ITS LIMITING FORMS

Several quantities, which characterize the features of the XPS spectrum are now defined. First, a spectral function $J(\omega)$ is defined by the Fourier transform of the correlation function $C(t)$:

$$J(\omega) = \frac{1}{2\pi} \int_{-\infty}^{\infty} dt \, C(t) \, e^{i\omega t}. \tag{3.27}$$

Then, Equation 3.20 is rewritten as

$$g(t) = \exp \left[-\frac{1}{\hbar^2} \int_{-\infty}^{\infty} d\omega \, J(\omega) \frac{-i\omega t + 1 - e^{-i\omega t}}{\omega^2} \right], \tag{3.28}$$

so that the XPS spectrum is expressed as

$$F(E_B) = \frac{1}{\pi\hbar} \operatorname{Re} \int_0^{\infty} dt \exp \left[i \frac{E_B - \Delta_s}{\hbar} t - S + \int_{-\infty}^{\infty} d\omega \frac{J(\omega)}{(\hbar\omega)^2} e^{-i\omega t} \right]. \tag{3.29}$$

By expanding the exponential function in Equation 3.29 in the Taylor series, we obtain

$$F(E_B) = \sum_{n=0}^{\infty} F_n(E_B), \tag{3.30}$$

where

$$F_0(E_B) = e^{-S} \delta(E_B - \Delta_s) \tag{3.31}$$

and

$$F_n(E_B) = \frac{1}{n!} e^{-S} \int\limits_{-\infty}^{\infty} \cdots \int\limits_{-\infty}^{\infty} d\omega_1 \ldots d\omega_n \frac{J(\omega_1) \ldots J(\omega_n)}{(\hbar\omega_1)^2 \ldots (\hbar\omega_n)^2}$$

$$\times \delta(E_B - \Delta_s - \hbar\omega_1 - \cdots - \hbar\omega_n), \quad \text{for } n \geq 1. \tag{3.32}$$

Here, the quantities S and Δ_s are defined as

$$S = \int\limits_{-\infty}^{\infty} d\omega \frac{J(\omega)}{(\hbar\omega)^2}, \tag{3.33}$$

$$\Delta_s = \int\limits_{-\infty}^{\infty} d\omega \frac{J(\omega)}{\hbar\omega}. \tag{3.34}$$

We remark that $\exp(-S)$ is the intensity of the "zero line" and Δ_s represents the "relaxation energy" of the VES because of interaction U. The behavior of $F_n(E_B)$ is shown schematically in Figure 3.2. In the case of $U = 0$, we have $F_n(E_B) = \delta(E_B)$ as shown with the dashed line. For $U \neq 0$, the discrete line F_0 (the zero line) shifts by Δ_s with the intensity reduced to $\exp(-S)$, and the sideband structures, F_n, take place. The integrated intensity of F_n obeys the Poisson distribution:

$$\int\limits_{-\infty}^{\infty} dE_B F_n(E_B) = e^{-S} \frac{S^n}{n!}. \tag{3.35}$$

In the following, two limiting forms of $F(E_B)$ in the slow and rapid modulation limits, are given. To this end, two characteristic times are defined. One is the decay

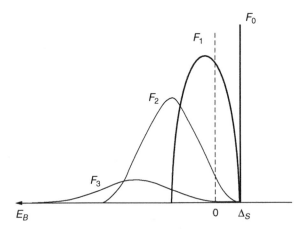

FIGURE 3.2 Schematic shapes of decomposed XPS spectra F_0–F_3 as a function of the binding energy. (Reprinted from Kotani, A., *Handbook on Synchrotron Radiation*, North-Holland, Amsterdam, 1987. With permission from Elsevier Ltd.)

time τ_c of the correlation function $C(t)$ and the other is the decay time τ_g of the generating function $g(t)$.

3.3.3.1 Slow Modulation Limit

When $\tau_c \gg \tau_g$, $C(t)$ scarcely changes in the time interval $0 \le t \le \tau_g$; thus $C(t) \cong C(0)$ can be put in the integrand of Equation 3.20. Then $g(t)$ can be approximated by the Gaussian form,

$$g(t) = \exp\left(-\frac{D^2}{2\hbar^2} t^2\right), \tag{3.36}$$

so that $F(E_B)$ is also given by the Gaussian function,

$$F(E_B) = \frac{1}{\sqrt{2\pi D^2}} \exp\left(-\frac{E_B^2}{2D^2}\right), \tag{3.37}$$

where

$$D^2 = C(0) = \int_{-\infty}^{\infty} d\omega\, J(\omega). \tag{3.38}$$

Since τ_g is estimated as $\tau_g \cong \hbar/D$, the condition $\tau_c \gg \tau_g$ is equivalent to

$$D\tau_c \gg 1. \tag{3.39}$$

3.3.3.2 Rapid Modulation Limit

When $\tau_c \ll \tau_g$, $C(t)$ decays so rapidly that we can approximate $J(\omega)$ as a constant

$$J(\omega) = \hbar\Gamma/\pi \tag{3.40}$$

with $\Gamma = D^2 \tau_c/\hbar$, then $g(t)$ decays exponentially as

$$g(t) = \exp\left(-\frac{\Gamma}{\hbar} t\right). \tag{3.41}$$

Therefore, the Fourier transform of it gives the XPS spectrum in the Lorentzian form:

$$F(E_B) = \frac{\Gamma/\pi}{E_B^2 + \Gamma^2}. \tag{3.42}$$

3.4 GENERAL EFFECTS IN XPS SPECTRA

3.4.1 SCREENING BY FREE-ELECTRON-LIKE CONDUCTION ELECTRONS

In the case where VES can be described by an electron gas system with the Coulomb interaction, the dielectric function is expressed in the random phase approximation as

$$\varepsilon(\mathbf{q},\omega) = 1 + \frac{4\pi e^2}{q^2} \sum_{\mathbf{k}} \frac{f(\varepsilon_{\mathbf{k}}) - f(\varepsilon_{\mathbf{k+q}})}{\varepsilon_{\mathbf{k+q}} - \varepsilon_{\mathbf{k}} - \hbar\omega}, \tag{3.43}$$

where $\varepsilon_{\mathbf{k}} = (\hbar k)^2/2m$, and $f(\varepsilon_{\mathbf{k}})$ is the Fermi distribution function. Now, consider the case where the core-hole potential is screened by the individual mode (i.e. by the low energy electron-hole pair excitations). By taking the low energy limit $\omega \to 0$ in $\varepsilon(\mathbf{q}, \omega)$, it is seen that

$$\mathrm{Im}\left[\frac{1}{\varepsilon(\mathbf{q},\omega)}\right] \propto \omega, \tag{3.44}$$

so that from Equation 3.26, $g(t)$ is found to behave for large t in the following asymptotic form

$$g(t) \propto t^{-\alpha} \tag{3.45}$$

with a positive constant α. Then, the XPS spectrum $F(E_B)$ diverges at the threshold in the inverse power form

$$F(E_B) \propto 1/E_B^{1-\alpha}, \tag{3.46}$$

if the value α is smaller than 1 (where the shift of the threshold energy Δ_s is disregarded). This spectral anomaly is called the "orthogonality catastrophe," since it originates from the fact that the ground state $|0\rangle$ of H_0 is orthogonal with the ground state (say $|\hat{0}\rangle$) of H:

$$\langle 0 | \hat{0} \rangle = \lim_{t\to\infty} \langle 0 | \exp\left(-i\frac{H}{\hbar}t\right) | 0 \rangle \propto \lim_{t\to\infty} g(t) = 0. \tag{3.47}$$

Next, consider the case where the core-hole potential is screened by the collective mode (the plasmon). By disregarding the dispersion of the plasmon, we obtain from Equation 3.43

$$\varepsilon(\mathbf{q},\omega) \cong \varepsilon(0,\omega) = 1 - \left(\frac{\omega_p}{\omega}\right)^2, \tag{3.48}$$

where ω_p is the plasma frequency given by

$$\omega_p = (4\pi n e^2/m)^{1/2} \tag{3.49}$$

with the electron density n. Therefore, we have

$$J(\omega) = -\frac{\hbar}{4\pi^2}\sum_{\mathbf{q}} q^2 \phi_{\mathbf{q}}^2 \frac{1}{1-e^{-\beta\hbar\omega}} \mathrm{Im}\left[\frac{1}{\varepsilon(\mathbf{q},\omega)}\right] \cong \frac{\hbar\omega_p}{8\pi}\sum_{\mathbf{q}} q^2 \phi_{\mathbf{q}}^2 \,\delta(\omega - \omega_p) \tag{3.50}$$

by assuming $\beta\hbar\omega_p \gg 1$, and from Equations 3.30 through 3.32, we obtain

$$F(E_B) = e^{-S} \sum_{n=0}^{\infty} \frac{S^n}{n!} \delta(E_B - \Delta_s - n\hbar\omega_p), \tag{3.51}$$

where

$$S = \frac{1}{8\pi\hbar\omega_p} \sum_{\mathbf{q}} q^2 \phi_{\mathbf{q}}^2, \tag{3.52}$$

$$\Delta_s = -\hbar\omega_p S. \tag{3.53}$$

Thus, the simultaneous excitation of plasmons is found to give rise to the satellites at $E_B = \Delta + n\hbar\omega_p$ in the XPS spectrum.

3.4.2 SCREENING BY LATTICE RELAXATION EFFECTS

In ionic crystals, for example, the core-hole potential is screened by the lattice polarization in the final state of XPS. In this case, the XPS spectrum can be obtained by simply replacing the dielectric function of VES with that of the lattice polarization. Assuming lattice polarization because of the optical phonon in a diatomic ionic crystal, the dielectric function is given by

$$\varepsilon(\omega) = \frac{\varepsilon_\infty(\omega_l^2 - \omega^2)}{\omega_t^2 - \omega^2}, \tag{3.54}$$

so that we obtain

$$\text{Im}\left[\frac{1}{\varepsilon(\mathbf{q},\omega)}\right] = -\frac{\pi}{2}\omega_l\left(\frac{1}{\varepsilon_\infty} - \frac{1}{\varepsilon_0}\right)[\delta(\omega - \omega_l) - \delta(\omega + \omega_l)]. \tag{3.55}$$

Here, ε_∞ and ε_0 are, respectively, the high frequency and the static dielectric constants, and ω_l and ω_t are, respectively, frequencies of longitudinal and transverse optical phonons whose dispersion is disregarded. In this system, the correlation time τ_c is estimated to be of the order of $1/\omega_l$, and by assuming $\beta\hbar\omega_l \ll 1$, the condition $\tau_c \gg \tau_g$ of the slow modulation limit is usually satisfied. Then, the XPS spectrum $F(E_B)$ is given by the Gaussian function of Equation 3.37, where D^2 is obtained as

$$D^2 = -\frac{1}{4\pi\beta}\left(\frac{1}{\varepsilon_\infty} - \frac{1}{\varepsilon_0}\right)\sum_{\mathbf{q}} q^2 \phi_{\mathbf{q}}^2. \tag{3.56}$$

Some remarks are given in the following:

1. The slow modulation limit corresponds to the classical limit, where the Gaussian spectrum is easily obtained by the Franck–Condon principle with the configuration coordinate model.
2. The above result is also obtained by taking the interaction U as the usual Fröhlich-type interaction.

3. When the condition $\tau_c \gg \tau_g$ is satisfied, the Gaussian broadening of the XPS spectrum occurs irrespective of the type of the relevant phonons (optical or acoustic).

3.4.3 SHAKE-UP SATELLITES

The electron-hole pair excitation in metals starts from $\omega = 0$, but in semiconductors and insulators, the energy of pair excitation modes (exciton and band-to-band excitation) is finite. Expressing the energy of the longitudinal exciton mode with wavevector $\mathbf{q}$ as $\varepsilon_{ex}(\mathbf{q})$, we have

$$\text{Im}\left[\frac{1}{\varepsilon(\mathbf{q}, \omega)}\right] \propto \delta(\omega - \varepsilon_{ex}(\mathbf{q})), \tag{3.57}$$

so that the simultaneous excitation of the longitudinal exciton or band-to-band electron excitation gives a satellite of the XPS spectrum at E_B separated by ε_{ex} from the main line. This is denoted by the shake-up satellite. The intra-atomic or intra-molecular electron excitation also gives the shake-up satellite, although the formalism with the dielectric function is not appropriate in this case.

3.4.4 LIFETIME EFFECTS

As shown in Chapter 2, the core hole has a finite lifetime because of the Auger transition and the radiative transition.

3.4.4.1 Auger Transition

In the Auger transition, two occupied electrons with energies ε_i and ε_j are scattered by the Coulomb interaction to the core level ε_c and a state with ε_k above the ionization threshold. The Auger transition is described by the interaction

$$U_A = \sum_{i,j,k} v_A(ij;ck) a_k^+ a_c^+ a_i a_j + \text{h.c.} \tag{3.58}$$

where a_l^+ and a_l ($l = i, j, c$, and k) are the creation and annihilation operators of the state l, respectively, and

$$v_A(ij;ck) = \int d\mathbf{r}\, d\mathbf{r}'\, \phi_k^*(\mathbf{r}) \phi_c^*(\mathbf{r}') \frac{e^2}{|\mathbf{r} - \mathbf{r}'|} \phi_j(\mathbf{r}') \phi_i(\mathbf{r})$$
$$- \int d\mathbf{r}\, d\mathbf{r}'\, \phi_k^*(\mathbf{r}) \phi_c^*(\mathbf{r}') \frac{e^2}{|\mathbf{r} - \mathbf{r}'|} \phi_i(\mathbf{r}') \phi_j(\mathbf{r}) \tag{3.59}$$

with the wave function ϕ_l of the states l. Since U_A explicitly includes the creation operator of the core state c, we apply our formalism to this case after including

$\varepsilon_c a_c^+ a_c$ in H_0 and H in Equations 3.1 and 3.2. Then the spectral function is obtained as

$$J(\omega) = \frac{1}{2\pi} \int_{-\infty}^{\infty} dt \langle 0 | a_c^+ U_A(t_1) U_A(t_2) a_c | 0 \rangle e^{i\omega(t_1 - t_2)}$$

$$= \sum_{i,j,k} | v_A(ij;ck) |^2 \delta\left(\omega - \frac{\varepsilon_k + \varepsilon_c - \varepsilon_i - \varepsilon_j}{\hbar} \right). \tag{3.60}$$

It can be remarked upon that the Auger transition usually corresponds to the rapid modulation case with a very small τ_c because the excited electron with ε_k has a large velocity and moves very rapidly away from the relevant core electron site. Therefore, the XPS spectrum is expressed by the Lorentzian function of Equation 3.42 with

$$\Gamma_A = \frac{\pi}{\hbar} J(0) = \pi \sum_{i,j,k} | v_A(ij;ck) |^2 \delta(\varepsilon_i + \varepsilon_j - \varepsilon_k - \varepsilon_c). \tag{3.61}$$

3.4.4.2 Radiative Transition

In the radiative decay, any occupied electrons i make a transition to the core state c. The radiative transition is expressed by the electric dipole approximation as

$$U_R = \sum_{i,q,\lambda} v_R(i; cq\lambda) a_c^+ a_i b_{q\lambda}^+ + \text{h.c.}, \tag{3.62}$$

where

$$v_R(i; cq\lambda) = \frac{e}{m} \sqrt{\frac{2\pi\hbar}{cqV_s}} \langle \phi_c | \mathbf{p} \cdot \mathbf{e}_{q\lambda} | \phi_i \rangle, \tag{3.63}$$

as given in Chapter 2. Here, $b_{q\lambda}^+$ is the creation operator of a photon with wavevector $\mathbf{q}$ and polarization λ, and V_s is the system volume in which the photon is normalized. The radiative transition is also a very rapid process, and from the calculation similar to that of the Auger transition, the XPS spectrum is given by the Lorentzian function with

$$\Gamma_R = \pi \sum_{i,q,\lambda} | v_R(i;cq\lambda) |^2 \delta(\varepsilon_i - \varepsilon_c - \hbar cq). \tag{3.64}$$

The radiative transition is less important in determining the core hole lifetime than Auger transition, except for deep core holes. For deep K-shell holes, the radiative process becomes more important when the atomic number Z becomes $Z \geq 31$. In Figure 3.3 (see also Figure 2.1 of Chapter 2), we show the energy width of the K-shell hole due to the Auger and radiative processes as a function of atomic number Z (Kotani and Toyozawa, 1979).

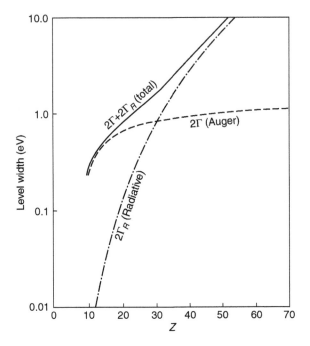

FIGURE 3.3 The line width of the K shell is shown as a function of the atomic number Z. The overall line width is divided into a radiative part ($2\Gamma_R$) and an Auger part (2Γ). (Reprinted from Kotani, A., and Toyozawa, Y., *Synchrotron Radiation*, Springer-Verlag, Berlin, 1979. With kind permission of Springer Science and Business Media.)

3.5 TYPICAL EXAMPLES OF XPS SPECTRA

3.5.1 SIMPLE METALS

The conduction band of simple metals can be described with the nearly free electron model. The orthogonality catastrophe and the plasmon satellite are actually observed in XPS of simple metals. In order to gain further insight into the orthogonality catastrophe, let us consider a simple metal where the conduction electrons are noninteracting spinless electrons described by the Hamiltonian

$$H_0 = \sum_{\mathbf{k}} \varepsilon_{\mathbf{k}} a_{\mathbf{k}}^+ a_{\mathbf{k}}, \qquad (3.65)$$

and the core-hole potential is a weak short-range potential

$$U = v \sum_{\mathbf{k},\mathbf{k}'} a_{\mathbf{k}}^+ a_{\mathbf{k}'}. \qquad (3.66)$$

Then the spectral function $J(\omega)$ is calculated as

$$J(\omega) = \frac{1}{2\pi} \int\limits_{-\infty}^{\infty} dt \, \langle 0|U(t)U|0\rangle \, e^{i\omega t}$$

$$= v^2 \sum_{k > k_F} \sum_{k' < k_F} \delta\left(\omega - \frac{\varepsilon_k - \varepsilon_{k'}}{\hbar}\right)$$

$$= (\rho v)^2 \int\limits_{0}^{D} d\varepsilon \int\limits_{-D}^{0} d\varepsilon' \, \delta\left(\omega - \frac{\varepsilon - \varepsilon'}{\hbar}\right)$$

$$= (\rho v)^2 \hbar^2 \omega, \quad \text{for } \hbar\omega \ll D, \tag{3.67}$$

where ρ is the density of states (at the Fermi level) of the conduction band, k_F is the Fermi wave number, and D is a cut-off energy of the order of the Fermi energy. We note that $J(\omega)$ is proportional to the excitation spectrum of an electron-hole pair in the conduction band, and the singularity described below comes from the fact that $J(\omega) \propto \omega$ for $\omega \to 0$. By substituting Equation 3.67 into Equation 3.28, the generating function is calculated as follows:

$$g(t) = \exp\left[-\frac{i\Delta_s t}{\hbar} - \int\limits_{-\infty}^{\infty} d\omega \frac{J(\omega)(1 - e^{-i\omega t})}{(\hbar\omega)^2}\right] \approx e^{-i(\Delta_s/\hbar)t} \left(\frac{iDt}{\hbar}\right)^{-(\rho v)^2}, \tag{3.68}$$

where we have used the integration

$$-\int\limits_{-\infty}^{\infty} d\omega \frac{J(\omega)(1 - e^{-i\omega t})}{(\hbar\omega)^2} = -\int\limits_{0}^{D/\hbar} d\omega \frac{(\rho v)^2 (1 - e^{-i\omega t})}{\omega}$$

$$= (\rho v)^2 \left[\int\limits_{0}^{Dt/\hbar} du \frac{\cos u - 1}{u} - i \int\limits_{0}^{Dt/\hbar} du \frac{\sin u}{u}\right]$$

$$\approx -(\rho v)^2 \log\left(\frac{iDt}{\hbar}\right), \tag{3.69}$$

which is valid asymptotically for large t. The Fourier transform of $g(t)$ is given by

$$F(E_B) = \frac{1}{\pi\hbar} \operatorname{Re} \int\limits_{0}^{\infty} dt \, e^{i[(E_B - \Delta_s)/\hbar]t} \left(\frac{iDt}{\hbar}\right)^{-(\rho v)^2}$$

$$= \operatorname{Re} \left[\frac{(iD)^{-g^2}}{\pi(E_B - \Delta_s)^{1 - g^2}} \int\limits_{0}^{\infty} dx \, e^{ix} x^{-g^2}\right]$$

$$= \begin{cases} \dfrac{1}{D\Gamma(g^2)\left(\dfrac{E_B - \Delta_s}{D}\right)^{1-g^2}} & \text{for } E_B \geq \Delta_s, \\[20pt] 0 & \text{for } E_B < \Delta_s, \end{cases} \tag{3.70}$$

where

$$g^2 = (\rho v)^2, \tag{3.71}$$

and $\Gamma(g^2)$ is the gamma function. Equation 3.68 is an asymptotic expression for $t \to \infty$, so that Equation 3.70 is also the asymptotic form for $E_B - \Delta_s \to +0$.

When the right-hand side of Equation 3.68 is expanded with respect to g^2, the nth-order term gives the contribution from the excitation of n electron-hole pairs from which $F_n(E_B)$ of Equation 3.32 results. Let us now consider how many electron-hole pairs are excited, on average, with the XPS process. Bearing in mind that the integrated intensity of $F_n(E_B)$ obeys the Poisson distribution, the average value of n is obtained as

$$\langle n \rangle = \sum_n n e^{-S} \frac{S^n}{n!} = S. \tag{3.72}$$

Further, from Equations 3.33 and 3.67, we obtain

$$S \propto \lim_{\omega \to 0}(-\log \omega) = \infty. \tag{3.73}$$

Thus, $\langle n \rangle$ diverges logarithmically, corresponding to an infinite number of electron-hole pair excitations with vanishingly small excitation energy.

What has been found is as follows: when a core electron is removed in the XPS process, the conduction electrons screen the core-hole potential. The screening process is equivalent to the excitation process of electron-hole pairs by the core-hole potential, as shown in Figure 3.4. The excitation of electron-hole pairs causes the relaxation of the conduction electron system around the core hole, the relaxation energy $|\Delta_s|$ being of the order of $(\rho v)^2 D$. At the same time, the excitation of electron-hole pairs manifests itself in the XPS spectrum as the sideband structure $F_n(E_B)$, where the excitation energy corresponds to $E_{ex} = E_B - \Delta_s$. Since the number of electron-hole pairs diverges logarithmically with vanishingly small excitation energy E_{ex}, the intensity of the XPS spectrum diverges as E_B tends to Δ_s. This is the mechanism of the orthogonality catastrophe. We have shown that $\langle n \rangle \to \infty$ comes from $S \to \infty$, which is equivalent to the orthogonality between $|0\rangle$ and $|\hat{0}\rangle$, because the intensity of the zero line is expressed as

$$|\langle 0|\hat{0}\rangle|^2 = e^{-S}. \tag{3.74}$$

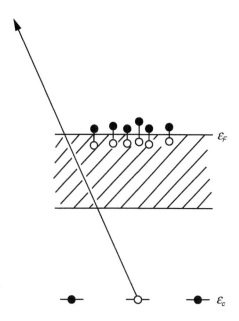

FIGURE 3.4 Schematic representation of XPS process in simple metals. (Reprinted from Kotani, A., *Handbook on Synchrotron Radiation, Vol. 2*, North-Holland, Amsterdam, 1987. With permission from Elsevier Ltd.)

When we extend the model by taking into account the electron spin and the core-hole potential with arbitrary strength and finite force range, $F(E_B)$ is still expressed as Equation 3.70, but g^2 in Equation 3.71 is replaced by

$$g^2 = 2\sum_l (2l+1)\left[\frac{\delta_l(\varepsilon_F)}{\pi}\right]^2, \tag{3.75}$$

where $\delta_l(\varepsilon_F)$ is the phase shift of the conduction electron (partial wave with angular momentum l) at the Fermi energy due to the scattering by the core-hole potential. The singular XPS spectrum, as well as the Fermi edge singularity in the XAS spectrum, were extensively studied from the theoretical aspect by Mahan (1967), Anderson (1967), Nozières and De Dominicis (1969), Hopfield (1969), Doniach and Šunjić (1970) and others, and had a significant impact on the theoretical and experimental study of core level spectroscopy. With this as a start, the study of the many-body response of VES to the core hole in the XPS process has been developed in various materials, as will be discussed in the following sections.

As an example of the experimental data of XPS in simple metals, the 2p XPS spectrum of Na is shown in Figure 3.5, which was measured at 300 K by Citrin et al. (1977). The spectral shape near the 2p threshold is analyzed by a convolution of the singular spectrum (Equation 3.70), the Gaussian spectrum (Equation 3.37) with D_{phonon} due to the lattice relaxation and the Lorentzian spectrum (Equation 3.42) with Γ_{hole} due to the core hole lifetime. By further taking into account the Gaussian broadening due to the instrumental resolution, as well as the effect of spin–orbit splitting

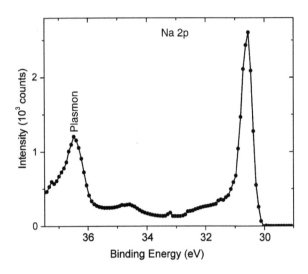

FIGURE 3.5 Experimental data of 2p XPS spectrum in Na metal. (Reprinted with permission from Citrin, P.H., Wertheim, G.K., and Baer, Y., *Phys. Rev. B*, 16, 4256, 1977. Copyright 1977 by the American Physical Society.)

of the 2p level, Citrin et al. (1977) obtained the following values: $g^2 = 0.198 \pm 0.015$, $2\Gamma_{hole} = 0.02 \pm 0.02$ eV, and $2.35D_{phonon}$ [= full-width at half-maximum (FWHM) value of Gaussian] $= 0.18 \pm 0.03$ eV.

In Figure 3.5, a plasmon satellite is also observed. When taking the plasmon energy dispersion quadratic into account with respect to the wave number, $J(\omega)$ has a ω-dependence of the form

$$J(\omega) \cong \frac{\omega_p^2}{\omega} \sqrt{\frac{\omega_p}{\omega - \omega_p}} \tag{3.76}$$

for $\omega > \omega_p$. Therefore, the one plasmon satellite is expressed as

$$F_1(E_B) = e^{-s} \frac{J(\omega)}{(\hbar\omega)^2}, \tag{3.77}$$

where

$$\omega \equiv \frac{E_B - \Delta_s}{\hbar}. \tag{3.78}$$

This expression explains the experimental asymmetric spectral shape with a tail in the high-energy side, as seen in Figure 3.5.

3.5.2 La Metal

In the La 3d XPS of La metal and La intermetallic compounds, such as $LaPd_3$, a weak satellite peak is observed on the lower binding energy side of a main peak.

The experimental data for La metal and LaPd$_3$ are shown in Figure 3.6a (Fuggle et al., 1983). In La metal, we see a weak satellite about 4 eV below the main peak, and for LaPd$_3$ the intensity of the satellite becomes considerably stronger. The occurrence of these satellite peaks cannot be explained by the exchange interaction between 4f and 3d states because La has no 4f electron in the ground state and thus no exchange splitting.

In order to interpret this satellite structure, Kotani and Toyozawa (1974) proposed a new mechanism, using the SIAM consisting of a well-localized 4f state and conduction electrons. The hybridization V is taken into account between 4f and conduction band states (Figure 3.7). In the ground state, the 4f level ε_f^0 is well above the Fermi level ε_F, so that the 4f level is empty. However, in the final state of the photoemission, as shown in Figure 3.7, the 4f level on the core hole site is assumed to be pulled down to

$$\varepsilon_f = \varepsilon_f^0 - U_{fc} \tag{3.79}$$

below ε_F because of the attractive core-hole potential $-U_{fc}$ acting on the 4f electron.

We can show that the XPS spectrum splits into two peaks, corresponding to the following two types of final states:

Type (A): A conduction electron near ε_F jumps into the 4f level ε_f through the hybridization V.

Type (B): The 4f level is still empty even after its energy has been lowered down to ε_f.

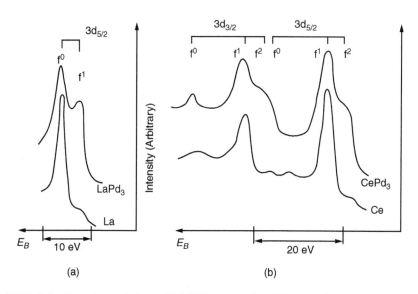

FIGURE 3.6 Experimental data of 3d XPS spectra for (a) La, LaPd$_3$ and (b) Ce, CePd$_3$. (Reprinted from Kotani, A., *Handbook on Synchrotron Radiation, Vol. 2*, North-Holland, Amsterdam, 1987. With permission from Elsevier Ltd.)

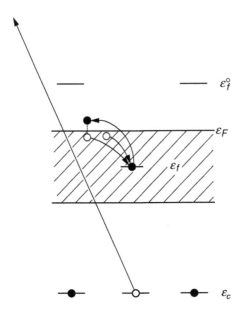

FIGURE 3.7 Model of the XPS process in La metal. (Reprinted from Kotani, A., *Handbook on Synchrotron Radiation, Vol. 2*, North-Holland, Amsterdam, 1987. With permission from Elsevier Ltd.)

The final state (A) is in the $4f^1$ configuration and it gives rise to the satellite peak of the XPS spectrum. On the other hand, the final state (B) is in the $4f^0$ configuration, and gives rise to the main peak.

The SIAM was originally proposed by Anderson (1961) in order to discuss the magnetic moment of an impurity atom of the 3d transition element in nonmagnetic host metals. In order to calculate the XPS spectrum of the La metal, Kotani and Toyozawa (1974) combined this model with a core electron state, which gives rise to a core-hole potential in the final state of XPS. Then, the Hamiltonian of the SIAM (including the core state) is given by

$$H = \sum_{\mathbf{k},\sigma} \varepsilon_{\mathbf{k}} a_{\mathbf{k}\sigma}^{+} a_{\mathbf{k}\sigma} + \varepsilon_f^0 \sum_{m,\sigma} a_{fm\sigma}^{+} a_{fm\sigma} + \sum_{\mathbf{k},m,\sigma} (V_{\mathbf{k}m} a_{\mathbf{k}\sigma}^{+} a_{fm\sigma} + V_{\mathbf{k}m}^{*} a_{fm\sigma}^{+} a_{\mathbf{k}\sigma})$$

$$+ \frac{U_{ff}}{2} \sum_{m,m',\sigma,\sigma'} a_{fm\sigma}^{+} a_{fm\sigma} a_{fm'\sigma'}^{+} a_{fm'\sigma'} + \varepsilon_c a_c^{+} a_c - U_{fc} \sum_{m,\sigma} a_{fm\sigma}^{+} a_{fm\sigma} (1 - a_c^{+} a_c). \quad (3.80)$$

The first term represents the conduction band, the second the 4f states (*m* represents the degeneracy of the 4f orbital), the third the hybridization between 4f and conduction band states, the fourth the Coulomb interaction between 4f electrons, the fifth the core state, and the last term is the attractive core-hole potential acting on the 4f states. The first four terms were originally introduced by Anderson and the last two terms were added by Kotani and Toyozawa (1973a,b, 1974). In this model, only the 4f states on a single atomic site are taken into account. Those on the other atomic sites are disregarded although they are also shown in Figure 3.7.

In order to extract some essential features of XPS in La metal, the Hamiltonian is simplified by disregarding the spin and orbital degeneracy and by assuming ε_f^0 is well above the Fermi level. Then the Hamiltonian is written as

$$H_0 = \sum_k \varepsilon_k a_k^+ a_k \tag{3.81}$$

for the initial state of XPS, and as

$$H = H_0 + \varepsilon_f a_f^+ a_f + V \sum_k (a_k^+ a_f + a_f^+ a_k) \tag{3.82}$$

for the final state.

In the following, $F(E_B)$ is given for the two types of final states, separately.

3.5.2.1   Final State of Type (A)

The 4f state is occupied in the final state of type (A). The generating function of this case is given by

$$g(t) = \frac{\rho V^2}{(\varepsilon_F - \tilde{\varepsilon}_f)^2} L_2(t) \exp\left[L_1(t)\right], \tag{3.83}$$

where ρ is the density of states of the conduction band at ε_F, $\tilde{\varepsilon}_f$ is the 4f level in the final state with an appropriate energy shift Δ_f by the hybridization V

$$\tilde{\varepsilon}_f = \varepsilon_f + \Delta_f \tag{3.84}$$

and $L_1(t)$ and $L_2(t)$ are, respectively, defined by

$$L_1(t) \cong -i\left(\frac{\Delta_f}{\hbar}\right) t - (\rho v_{\text{eff}})^2 \log\left(\frac{iDt}{\hbar}\right), \tag{3.85}$$

$$L_2(t) \cong \frac{\rho}{it} \exp\left[\frac{i}{\hbar}(\varepsilon_F - \varepsilon_f) t\right] \left(\frac{iDt}{\hbar}\right)^{-2\rho v_{\text{eff}}}, \tag{3.86}$$

with v_{eff} given by

$$v_{\text{eff}} \equiv -\frac{V^2}{\varepsilon_F - \varepsilon_f}. \tag{3.87}$$

Here we give an intuitive explanation of Equation 3.83, while a more detailed derivation is given in Appendix B. In the final state of type (A), a conduction electron just below ε_F jumps into the 4f level, and this process gives the prefactor $V^2/(\varepsilon_F - \tilde{\varepsilon})^2$

of Equation 3.83. Then we have (*i*) an extra electron in the 4f level and (*ii*) an extra hole near ε_F, and the time variation of the states (*i*) and (*ii*) give the contributions including $L_1(t)$ and $L_2(t)$ in Equation 3.83, respectively. The contribution of (*i*) is easily understood. As shown in Figure 3.7, the 4f electron jumps out above ε_F through V and another conduction electron below ε_F jumps into ε_f, giving rise to an electron-hole pair excitation near the Fermi level. This process is essentially the same as the electron-hole pair creation by the core-hole potential v in simple metals only if we replace the potential v by an effective scattering potential v_{eff} due to the second-order process of the hybridization V. Therefore, the factor $\exp[L_1(t)]$ is the same as the generating function (Equation 3.68) of simple metals but Δ_s and v are replaced by Δ_f and v_{eff}, respectively. The derivation of the factor $L_2(t)$ is more complicated. The hole near ε_F is scattered to another hole state through the effective potential v_{eff}, and the multiple scattering of the hole results in $L_2(t)$, as shown in Appendix B in more detail.

Substituting Equation 3.83 into Equation 3.10, we obtain

$$F(E_B) = \begin{cases} \dfrac{\rho V^2}{(\varepsilon_F - \tilde{\varepsilon}_f)^2 \Gamma(1 - 2g_{\mathrm{eff}} + g_{\mathrm{eff}}^2) \left(\dfrac{E_B + \varepsilon_F - \tilde{\varepsilon}_f}{D} \right)^{2g_{\mathrm{eff}} - g_{\mathrm{eff}}^2}} & \text{for } E_B \geq -(\varepsilon_F - \tilde{\varepsilon}_f), \\ \\ 0 & \text{for } E_B < -(\varepsilon_F - \tilde{\varepsilon}_f), \end{cases}$$

(3.88)

where

$$g_{\mathrm{eff}} = -\rho v_{\mathrm{eff}},$$

(3.89)

$$\tilde{\varepsilon}_f = \varepsilon_f + \Delta_f.$$

(3.90)

Therefore, $F(E_B)$ is found to diverge at the threshold.

3.5.2.2 Final State of Type (B)

In the final state (B), the 4f level is empty so that this final state can be obtained from $|0\rangle$ by taking into account the self-energy correction. The XPS spectrum is rewritten as

$$F(E_B) = -\frac{1}{\pi} \mathrm{Im} \langle 0 | \frac{1}{z - H} | 0 \rangle,$$

(3.91)

where

$$z = E_B + E_0 + i\eta, \quad \eta \to +0.$$

(3.92)

Thus, $F(E_B)$ is written as

$$F(E_B) = -\frac{1}{\pi} \mathrm{Im} \frac{1}{z - E_0 - \Sigma_0(z)}.$$

(3.93)

The self-energy $\Sigma_0(z)$ is expressed, up to the second-order in H' [defined by $H' = V \sum_k (a_k^+ a_f + a_f^+ a_k)$], as

$$\Sigma_0(z) = \langle 0|H' \frac{1}{z - H_{0f}} H'|0 \rangle$$

$$= V^2 \sum_{k < k_F} \frac{1}{z + \varepsilon_k - \varepsilon_f - E_0}$$

$$\cong \Delta_0 - i\pi\rho V^2, \tag{3.94}$$

where H_{0f} is given by $H_{0f} = H_0 + \varepsilon_f a_f^+ a_f$. So, the XPS spectrum is given by the Lorentzian function with a maximum at $E_B = \Delta_0$:

$$F(E_B) = \frac{\rho V^2}{(E_B - \Delta_0)^2 + (\pi\rho V^2)^2}. \tag{3.95}$$

The schematic shape of the XPS spectrum is shown in Figure 3.8. In the final state of type (A), which corresponds to the peak (A) in Figure 3.8, the core–hole charge is screened by the 4f electron because of the charge transfer from the conduction band to 4f state. Conversely, in the final state of type (B) [corresponding to the peak (B) in Figure 3.8], the core hole charge is not screened by the 4f electron. Spectrum (A) diverges because of the orthogonality catastrophe. Peak (B), on the other hand, has a Lorentzian broadening because of the finite lifetime $\tau = \hbar/(\pi\rho V^2)$ of the 4f hole (empty 4f state below ε_F) by its resonance transfer to the hole in the conduction band through V. The final states of (A) and (B), corresponding to $4f^1$ and $4f^0$ configurations, are denoted by "well-screened state" and "poorly-screened state," respectively.

In the La metal, the intensity of the well-screened peak is very small because of the small prefactor $\rho V^2/(\varepsilon_F - \tilde{\varepsilon}_f)^2$ in Equation 3.88. However, in LaPd$_3$ this factor is

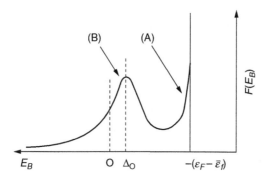

FIGURE 3.8 Schematic shape of the XPS spectrum obtained with the model of Figure 3.7. (Reprinted from Kotani, A., *Handbook on Synchrotron Radiation, Vol. 2*, North-Holland, Amsterdam, 1987. With permission from Elsevier Ltd.)

not very small because of the increase of ρ and V due to the existence of 4d states of Pd. Therefore, the intensity of the $4f^1$ well-screened peak in $LaPd_3$ is much stronger than that in La metal, as seen from Figure 3.6a.

A satellite structure, very similar to that of the La system, is also observed in systems containing trivalent Ce (i.e. $\gamma - Ce$, CePd, and $CeAl_2$). The experimental result of the Ce 3d XPS in $\gamma - Ce$ is shown in Figure 3.6b (Fuggle et al., 1983). This satellite can also be explained by essentially the same mechanism. In the ground state of the trivalent Ce, the 4f level ε_f^0 is well below ε_F, so that one 4f electron is already occupied. In the final state of the Ce 3d XPS, the 4f level is pulled down further by the core-hole potential and an electron transfer can occur from the conduction band to the 4f state. Therefore, the satellite arises from the $4f^2$ well-screened final state, while the main line corresponds to the $4f^1$ poorly screened final state. Furthermore, this mechanism can be generalized to any rare earth (RE) systems with a $4f^n$ ground state: The XPS spectrum of the system with the $4f^n$ ground state would be split into two peaks corresponding to the $4f^{n+1}$ well-screened final state and the $4f^n$ poorly screened final state.

In the rest of this subsection, we briefly discuss the effect of finite ε_f^0 on the XPS spectrum. In Figure 3.9, the XPS spectra calculated for various values of ε_f^0 is shown, taking into account the ε_f^0 level only on the core hole site (Kotani and Toyozawa, 1974). The calculation is made by exactly diagonalizing H_0 and H for a finite system where the conduction band consists of 50 discrete levels and the XPS spectrum of Equation 3.9 is computed numerically. In this figure, we fix $\rho V^2 = 0.05$ and $\varepsilon_f - \varepsilon_F = -0.5$, where the half-width of the conduction band is taken as energy units. It is found that when ε_f^0 is lowered, the intensity of the well-screened $4f^1$ peak increases. This is understood as follows: When ε_f^0 is finite, the $4f^1$ configuration is already mixed in the ground state $|0\rangle$ with a finite weight, although the weight is very small in the case of $\varepsilon_f^0 - \varepsilon_F > \rho V^2$. Since the photoemission intensity for the final state $|f\rangle$ is given by the overlap integral $|\langle f|0\rangle|^2$, the initial mixing of the $4f^1$ state enhances the intensity of the $4f^1$ final state. When $\varepsilon_f^0 - \varepsilon_F \cong \rho V^2$, the weight of the $4f^1$ configuration in the ground state becomes comparable with that of the $4f^0$ configuration, and this special state is denoted by the mixed valence state, which occurs in some Ce compounds and will be discussed in the next subsection.

3.5.3 Mixed Valence State in Ce Intermetallic Compounds

After the theory by Kotani and Toyozawa, the application of the SIAM to XPS spectra has been developed in two directions. One is the extension from La to Ce metals, especially to mixed valence Ce intermetallics. The other is from metallic systems to insulators.

As shown in the preceding subsection, the 4f level in La metal is well above ε_F and is well below ε_F in $\gamma - Ce$ metal. If the 4f level is close to ε_F (as in the case of $\alpha - Ce$), the 4f states hybridize strongly with the conduction band states, so that the ground state is a quantum-mechanically mixed state between the $4f^0$ and $4f^1$ configurations (denoted by a mixed valence state or a fluctuating valence state). Some intermetallic Ce compounds, such as $CePd_3$, $CeRh_3$, and $CeNi_3$, also behave as mixed valence compounds. In the Ce 3d XPS of mixed valence Ce compounds, three peaks were observed experimentally in each of the $3d_{5/2}$ and $3d_{3/2}$ core levels, as

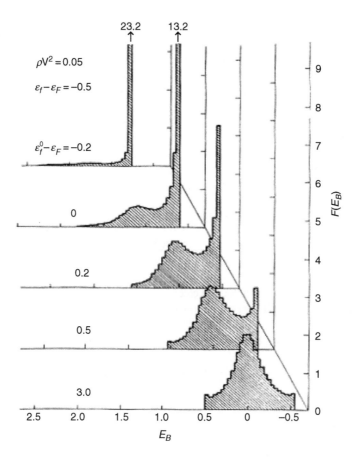

FIGURE 3.9 XPS spectra calculated for various values of the 4f level ε_f^0. (Reprinted from Kotani, A., *Handbook on Synchrotron Radiation, Vol. 2*, North-Holland, Amsterdam, 1987. With permission from Elsevier Ltd.)

shown in Figure 3.6b for $CePd_3$ (Fuggle et al., 1983). The mechanism of these three peaks can be understood, at least qualitatively, by applying the mechanism of the satellite in La metal and γ–Ce to the mixed valence systems: When we take into account the $4f^0$ component of the mixed valence ground state, we expect to have well-screened $4f^1$ and poorly screened $4f^0$ final states, as in the case of La. On the other hand, when we take into account the $4f^1$ component, we expect to have well-screened $4f^2$ and poorly screened $4f^1$ final states, as in the case of γ–Ce. Therefore, combining the two cases, we have three different configurations, $4f^0$, $4f^1$, and $4f^2$, in the final state, which give rise to the three peaks in XPS.

The first quantitative theoretical analysis of such a three-peak structure was performed by Gunnarsson and Schönhammer (1983a,b) with the SIAM. They took into account explicitly the Coulomb interaction U_{ff} between 4f electrons and the degeneracy N_f of the 4f state ($N_f = 14$), and calculated the 3d XPS by using the $1/N_f$ expansion method; the XPS intensity was expanded in the power series of $1/N_f$.

In the Hamiltonian of the SIAM, the conduction electron state was specified by the wavevector $\mathbf{k}$, but the basis states of the conduction electrons specified by energy ε and the spin and orbital degeneracy $v = (m, \sigma)$ were only taken approximately; the orbital angular momentum (l, m) was around that of the Ce site considered. The value of l can only be limited to $l = 3$, which couples with the 4f states. Then, the Hamiltonian in the initial state of XPS is written as

$$H_0 = \sum_v \int d\varepsilon \, \varepsilon \, a^+_{\varepsilon v} a_{\varepsilon v} + \varepsilon^0_f \sum_v a^+_{fv} a_{fv} + \sum_v \int d\varepsilon [V(\varepsilon) a^+_{\varepsilon v} a_{fv} + V(\varepsilon)^* a^+_{fv} a_{\varepsilon v}]$$
$$+ U_{ff} \sum_{v > v'} a^+_{fv} a_{fv} a^+_{fv'} a_{fv'}, \tag{3.96}$$

while the Hamiltonian in the final state is written as

$$H = H_0 - U_{fc} \sum_v a^+_{fv} a_{fv}. \tag{3.97}$$

This transformation of conduction electron states corresponds to the partial wave expansion and it is exact if the conduction electrons are regarded as free electron gas. For actual conduction electrons under the periodic potential, this transformation is approximately performed by

$$a^+_{\varepsilon v} = V(\varepsilon)^{-1} \sum_{\mathbf{k}} V^*_{\mathbf{k}m} \, \delta(\varepsilon - \varepsilon_{\mathbf{k}}) \, a^+_{\mathbf{k}\sigma} \tag{3.98}$$

with $V(\varepsilon)$ expressed as

$$\sum_{\mathbf{k}} V^*_{\mathbf{k}m} V_{\mathbf{k}m'} \, \delta(\varepsilon - \varepsilon_{\mathbf{k}}) = |V(\varepsilon)|^2 \, \delta_{m,m'}. \tag{3.99}$$

Gunnarsson and Schönhammer (1983a,b) proposed a systematic method to calculate the XPS spectrum in the power series of $1/N_f$. In Figure 3.10, we show schematically the basis states to diagonalize the Hamiltonian H_0 and H. For instance, the state A represents the Fermi vacuum $|A\rangle$, where all the conduction electrons below ε_F are occupied and the 4f level is empty. In the state B, $|B(\varepsilon)\rangle$, one conduction electron with energy ε is transferred to the 4f state:

$$|B(\varepsilon)\rangle = \frac{1}{\sqrt{N_f}} \sum_v a^+_{fv} a_{\varepsilon v} |A\rangle. \tag{3.100}$$

The states C and D are written as

$$|C(\varepsilon, \varepsilon')\rangle = \frac{1}{\sqrt{N_f(N_f - 1)}} \sum_{v,v'} a^+_{fv} a_{\varepsilon v} a^+_{fv'} a_{\varepsilon' v'} |A\rangle. \tag{3.101}$$

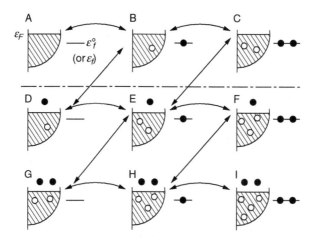

FIGURE 3.10 Schematic representation of basis states used to diagonalize the Hamiltonian of the single impurity Anderson model, where ε_f^0 (or ε_f) and ε_F denote the 4f level and the Fermi level, respectively, and the hatched areas denote occupied conduction band. (Reprinted from Kotani, A., *Handbook on Synchrotron Radiation, Vol. 2*, North-Holland, Amsterdam, 1987. With permission from Elsevier Ltd.)

$$|D(E, \varepsilon)\rangle = \frac{1}{\sqrt{N_f}} \sum_v a_{Ev}^+ a_{\varepsilon v} |A\rangle, \tag{3.102}$$

respectively.

It is important to note that the coupling among the states A, B, and C of Figure 3.10 occurs within the lowest order of $1/N_f$, but the coupling of these states with the other states, D, E, F and so on, occurs only as a higher-order correction with respect to $1/N_f$. Since the value N_f (= 14) is large, the lowest-order approximation provides us with sufficiently reliable results. Once we diagonalize H_0 and H, the XPS spectrum is calculated by

$$F(E_B) = \sum_f |\langle f|0\rangle|^2 \frac{\Gamma/\pi}{(E_B - E_f + E_0)^2 + \Gamma^2}, \tag{3.103}$$

where the ground state of H_0 is denoted by $|0\rangle$ (with energy E_0) and the eigenstates of H by $|f\rangle$ (with energy E_f) as in Equation 3.9. Furthermore, we take into account the spectral broadening Γ due to the lifetime of the core hole. In Figure 3.11, we show, as an example, a calculated spectrum (solid line) as well as an experimental one (dots), for the Ce 3d XPS of CeNi$_2$ (Gunnarsson and Schönhammer, 1983b). In this calculation, $|V(\varepsilon)|^2$ is assumed to be expressed as

$$|V(\varepsilon)|^2 = \frac{2V^2[D^2 - (\varepsilon - \varepsilon_0)^2]^{1/2}}{\pi D^2} \tag{3.104}$$

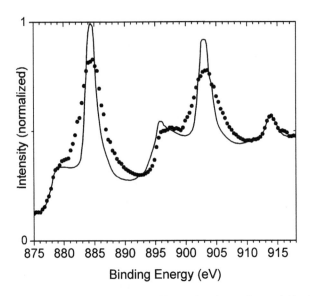

FIGURE 3.11 Comparison of theoretical (solid curve) and experimental (dots) results of the Ce 3d XPS in CeNi$_2$. (Reprinted with permission from Gunnarsson, O., and Schönhammer, K., *Phys. Rev. B*, 28, 4315, 1983. Copyright 1983 by the American Physical Society.)

with $\varepsilon_0 - \varepsilon_F = -1.995$ eV, $D = 2.005$ eV, and $2V^2/D = 0.13$ eV. The other parameter values are taken to be $\varepsilon_f^0 - \varepsilon_F = -1.3$ eV, $U_{fc} = 10.3$ eV, $U_{ff} = 5.5$ eV, and $\Gamma = 0.9$ eV. Two theoretical spectra $F(E_B)$ are superimposed with the weights 0.4 and 0.6 and with an appropriate energy separation to simulate the 3d spin–orbit splitting. The fractional intensity of the 4f^2 peak is found to be sensitive to the value of $2V^2/D$, whereas that of the 4f^0 peak is sensitive to the weight, $w(4f^0)$, of the 4f^0 configuration in the ground state. In this analysis, the value of $w(4f^0)$ is obtained to be $w(4f^0) = 0.19$, and the average 4f electron number in the ground state is estimated as $n_f \cong 1 - w(4f^0) = 0.81$ because $w(4f^2)$ is negligibly small.

From a systematic analysis of XPS spectra, the value of n_f was obtained for various Ce compounds (Gunnarsson and Schönhammer, 1983b; Fuggle et al., 1983). Before such an analysis had been made, the value of n_f was estimated from the data of lattice constant and susceptibility, and then the value of n_f distributed between 0 and 1 depending on the material. However, the value of n_f estimated from the XPS analysis is always found to be larger than about 0.7, ranging mainly between 0.8 and 1.0. Even for the materials, which were traditionally considered to have $n_f = 0$, such as the intermetallic Ce compounds with Ni, Ru, Rh and so on, the XPS analysis gives n_f around 0.8 (Table 3.1). As another example, in Figure 3.12 we show the comparison of theoretical and experimental results of the Ce 3d XPS for CePd$_3$. The calculation was made by Kotani and Jo (1986) with a method similar to that of Gunnarsson and Schönhammer (1983b) and the estimated value of n_f was 0.86.

In this way, XPS played an important role in the study of mixed valence Ce compounds. It is to be emphasized that XPS is a powerful technique of directly detecting the microscopic and local properties of the 4f state.

TABLE 3.1
Average 4f Electron Number n_f Estimated from the Analysis of XPS for Various Ce Compounds

	Metal		Insulator	
	n_f	Example	n_f	Example
Tetravalence	0	—	0	—
Nominal tetravalence	0.75–0.85	$CeRh_3$, $CeNi_3$	~0.5	CeO_2
Mixed valence	0.85–0.9	$CePd_3$, $CePt_3$	—	—
Trivalence	~1.0	$CeAl_2$, $CeCu_2Si_2$	~1.0	Ce_2O_3, CeF_3

3.5.4 INSULATING MIXED VALENCE Ce COMPOUNDS

Of many insulating RE compounds, the study of core level spectroscopy has been made most extensively for CeO_2. CeO_2 was traditionally considered to be tetravalent with the $4f^0$ ground state. However, according to experimental data of the Ce 3d XPS of CeO_2, a three-peak structure (somewhat similar to that of mixed valence Ce intermetallics), was observed for each of the $3d_{5/2}$ and $3d_{3/2}$ core levels (Burroughs et al., 1976; Wuilloud et al., 1984). Wuilloud et al. (1984) and Kotani et al. (1985) analyzed this structure with the SIAM, where the conduction band in the original model was replaced by a completely filled O 2p valence band. Between the Ce 4f and O 2p states, a covalency hybridization V was taken into account. They showed that if the

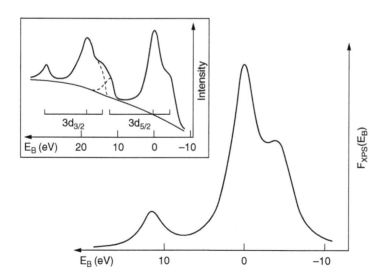

FIGURE 3.12 Calculated result of the Ce 3d XPS spectrum of $CePd_3$. Experimental result is shown in the inset for comparison. (Reprinted from Kotani, A., Jo, T., and Parlebas, J.C., *Adv. Phys.*, 37, 37, 1988. With permission.)

4f level was located near the valence band, it would be possible to have a mixed valence state between $4f^0$ and $4f^1\underline{L}$ configurations as a ground state, where $\underline{L}$ represents a hole in the valence band. Then, in the final state of XPS, an additional charge transfer can occur, and there are three different configurations ($4f^0$, $4f^1\underline{L}$, and $4f^2\underline{L}^2$), corresponding to the three peaks of the XPS.

The Hamiltonian is essentially the same as that of the original one except that the conduction band is replaced by a filled valence band. We now discretize (Kotani et al., 1985) the energy ε of the valence band as N levels of ε_k:

$$\varepsilon_k = \varepsilon_v - \frac{W}{2} + \frac{W}{N}\left(k - \frac{1}{2}\right), \tag{3.105}$$

where W is the width of the valence band and $k = 1, 2, \ldots, N$. Then the Hamiltonian H_0 is given by

$$H_0 = \sum_{k,v} \varepsilon_k a_{kv}^+ a_{kv} + \varepsilon_f^0 \sum_v a_{fv}^+ a_{fv} + \frac{V}{\sqrt{N}}\sum_{k,v}\left(a_{kv}^+ a_{fv} + a_{fv}^+ a_{kv}\right)$$
$$+ U_{ff}\sum_{v>v'} a_{fv}^+ a_{fv} a_{fv'}^+ a_{fv'}, \tag{3.106}$$

where the energy dependence of V is discarded for simplicity.

The calculated result is shown in Figure 3.13 (Kotani and Ogasawara, 1992). We have taken into account the three configurations $4f^0$, $4f^1\underline{L}$, and $4f^2\underline{L}^2$ for both ground and final states. The parameter values used are $\varepsilon_f^0 - \varepsilon_v = 1.6$ eV, $V = 0.76$ eV, $U_{ff} = 10.5$ eV, $U_{fc} = 12.5$ eV, $W = 3.0$ eV, and $\Gamma = 0.7$ eV (for more details on the parameter values, see Chapter 8). The value of N is taken sufficiently large so that the calculated spectrum converges well. The relative intensity and the energy separation of the three peaks are in good agreement with the experimental result (Wuilloud et al., 1984), as shown in the inset of Figure 3.13. With these parameter values, it is found that the ground state is a strong mixture of $4f^0$ and $4f^1\underline{L}$ configurations with the average 4f electron number $n_f = 0.52$.

In order to see the electronic structure more clearly, let us consider the limiting case of $N = 1$. Then, the Hamiltonian matrix is given by

$$\begin{pmatrix} 0 & \sqrt{N_f}\,V & 0 \\ \sqrt{N_f}\,V & \varepsilon_f^0 - \varepsilon_v(-U_{fc}) & \sqrt{2(N_f - 1)}\,V \\ 0 & \sqrt{2(N_f - 1)}\,V & 2(\varepsilon_f^0 - \varepsilon_v) + U_{ff}(-2U_{fc}) \end{pmatrix}, \tag{3.107}$$

where we take the three basis states:

$$|A\rangle = \prod_v a_{kv}^+ |vac\rangle, \quad (k = 1), \tag{3.108}$$

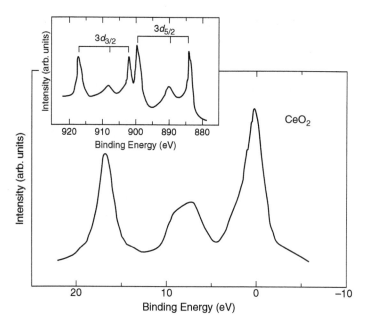

FIGURE 3.13 Calculated result of the Ce 3d XPS spectrum of CeO_2. Experimental result is shown in the inset for comparison. (Reprinted from Kotani, A., and Ogasawara, H., *J. Electron Spectrosc. Relat. Phenom.*, 60, 257, 1992. With permission from Elsevier Ltd.)

$$|B\rangle = \frac{1}{\sqrt{N_f}} \sum_v a_{fv}^+ a_{kv} |A\rangle, \tag{3.109}$$

$$|C\rangle = \sqrt{\frac{2}{N_f(N_f - 1)}} \sum_{v>v'} a_{fv}^+ a_{kv} a_{fv'}^+ a_{kv'} |A\rangle, \tag{3.110}$$

and $(-U_{fc})$ and $(-2U_{fc})$ in the Hamiltonian matrix are used only for the final state. Some remarks are given in the following:

1. The lowest-order approximation of the $1/N_f$ expansion gives the exact result for the insulating systems. This is because we have no empty (conduction) electron state above ε_F.
2. The effective hybridization between $4f^0$ and $4f^1\underline{L}$ configurations is $V_{eff} = \sqrt{N_f} V$ as a result of the degeneracy of the 4f state. If we take the basis states $|A\rangle$ and $|B(v)\rangle = a_{fv}^+ a_{kv} |A\rangle$, then the mixing matrix element is

$$\langle A | H_0 | B(v) \rangle = V, \tag{3.111}$$

but the Hamiltonian matrix is $(N_f + 1) \times (N_f + 1)$. By taking the linear combination

$$|B\rangle = \frac{1}{\sqrt{N_f}} \sum_v |B(v)\rangle, \tag{3.112}$$

the matrix is reduced to 2×2 with the effective mixing $V_{eff} = \sqrt{N_f} V$.

3. The $4f^0$ and $4f^1\underline{L}$ configurations are strongly mixed in the ground state of
   CeO_2, while the $4f^1\underline{L}$ and $4f^2\underline{L}^2$ configurations are strongly mixed in the
   final state of XPS.

A similar analysis has also been made for CeF_4, and the result is shown in
Figure 3.14 (Kotani et al., 1987b). The experimental results of the Ce 3d XPS of
CeF_4 (Kaindl et al., 1987) exhibit the three-peak structure, which is similar to that of
CeO_2 but the relative intensity of the third peak ($4f^0$ peak) is larger than that of CeO_2.
The parameter values that are obtained from this analysis are $\varepsilon_f^0 - \varepsilon_v = 4.0$ eV,
$V = 0.76$ eV, $U_{ff} = 8.5$ eV, $U_{fc} = 12.0$ eV, and the average 4f electron number in the
ground state is $n_f = 0.3$. It should be noted that the charge-transfer energy $\varepsilon_f^0 - \varepsilon_v$ is
larger than that of CeO_2, reflecting the larger electronegativity of the F ion than the
O ion. This is the main reason that n_f is smaller and the relative intensity of
the third peak is larger compared with CeO_2.

Of the insulating Ce compounds, CeO_2 and CeF_4 are in the mixed valence state,
and Ce_2O_3 and CeF_3 are in the trivalent state, as shown in Table 3.1.

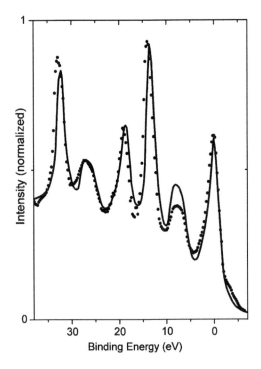

FIGURE 3.14 Comparison of theoretical (solid curve) and experimental (dots) results of the
Ce 3d XPS spectrum of CeF_4. (From Kotani, A., Jo, T., and Parlebas, J.C., *Adv. Phys.*, 37, 37,
1988. With permission.)

3.5.5 TRANSITION METAL COMPOUNDS

3.5.5.1 Model

Here we consider the 2p XPS of TM compounds. As a typical example, Figure 3.15 shows the 2p XPS of the TM difluoride series (from ZnF_2 to MnF_2). The dots are experimental data observed by Rosencwaig et al. (1971). There is no satellite peak in the Zn 2p XPS of ZnF_2 where the 3d shell is filled. However, for CuF_2, a strong satellite occurs on the higher binding energy side of the main peak (in each of the $2p_{3/2}$ and $2p_{1/2}$ components). The intensity of the satellite peak decreases in going from CuF_2 to MnF_2. This satellite was first considered to be caused by the shake-up transition between the metal 3d and 4s orbitals, but it is now well established that it originates from the charge transfer between the ligand 2p and metal 3d orbitals. The theoretical analysis of such a satellite structure has been successfully made using the SIAM or cluster model.

Here we consider a cluster consisting of a TM ion surrounded by F ions. The cluster is $(TM)F_6$ for TM = Ni–Mn and $(TM)F_4$ for TM = Cu, the symmetry

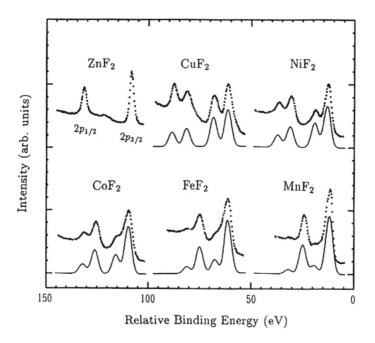

FIGURE 3.15 Experimental (dots) and theoretical (solid curves) results of 2p XPS spectra of transition metal dihalides. The theoretical spectra are obtained with the two-configuration model. (Reprinted from Kotani, A., and Okada, K., *Recent Advances in Magnetism of Transition Metal Compounds*, edited by A. Kotani and N. Suzuki, World Scientific, Singapore, 1993. Reprinted with permission from World Scientific Publishing Ltd.)

of which are taken approximately as O_h and D_{4h}, respectively. Since the 3d wave function of TM compounds is more extended in space than the 4f wave function of RE compounds, the 3d level and the hybridization strength between the 3d and ligand (F 2p) states depend more sensitively on the local arrangements of the cluster. Namely, they are specified by the irreducible representation of the O_h (or D_{4h}) symmetry group. The Hamiltonian of the system is written as

$$H_0 = \sum_{\Gamma,\sigma} \varepsilon_{\Gamma} a_{\Gamma\sigma}^+ a_{\Gamma\sigma} + \sum_{\Gamma,\sigma} \varepsilon_{d\Gamma} a_{d\Gamma\sigma}^+ a_{d\Gamma\sigma} + \sum_{\Gamma,\sigma} V(\Gamma)(a_{d\Gamma\sigma}^+ a_{\Gamma\sigma} + a_{\Gamma\sigma}^+ a_{d\Gamma\sigma})$$

$$+ U_{dd} \sum_{(\Gamma,\sigma) \neq (\Gamma',\sigma')} a_{d\Gamma\sigma}^+ a_{d\Gamma\sigma} a_{d\Gamma'\sigma'}^+ a_{d\Gamma'\sigma'}, \tag{3.113}$$

where $a_{\Gamma\sigma}^+$ and $a_{d\Gamma\sigma}^+$ are electron creation operators for a ligand molecular orbital Γ with spin σ and a 3d state (Γ, σ). The summation Γ runs over the irreducible representations t_{2g} and e_g for the (TM)F_6 cluster and over b_{1g}, a_{1g}, b_{2g}, and e_g for the (TM)F_4 cluster.

In the final state of XPS, the effect of core-hole potential is added to the Hamiltonian

$$H = H_0 - U_{dc} \sum_{\Gamma,\sigma} a_{d\Gamma\sigma}^+ a_{d\Gamma\sigma}. \tag{3.114}$$

3.5.5.2 Simplified Analysis

In order to understand the essential mechanism of the observed XPS spectra, let us consider a somewhat simplified situation: The basis states are limited to the two configurations $3d^n$ and $3d^{n+1}\underline{L}$, where $n = 9, 8, 7, 6,$ and 5 for TM = Cu, Ni, Co, Fe, and Mn, respectively. Furthermore, we disregard, for simplicity, the Γ-dependence in ε_{Γ} and $\varepsilon_{d\Gamma}$ (which are written as ε_{Γ} and ε_d, respectively). Then, the Hamiltonian H_0 is written in the following (2×2) forms (Kotani and Okada, 1993):

$$H_0 = E_0 |3d^n\rangle\langle 3d^n| + (E_0 + \Delta)|3d^{n+1}\underline{L}\rangle\langle 3d^{n+1}\underline{L}|$$

$$+ V_{eff}(|3d^n\rangle\langle 3d^{n+1}\underline{L}| + |3d^{n+1}\underline{L}\rangle\langle 3d^n|), \tag{3.115}$$

where

$$E_0 = n\varepsilon_d + \frac{n(n-1)}{2}U_{dd} + \sum_{\Gamma,\sigma} \varepsilon_{\Gamma}, \tag{3.116}$$

$$\Delta = \varepsilon_d - \varepsilon_{\Gamma} + nU_{dd}, \tag{3.117}$$

$$V_{\text{eff}} = \begin{cases} V(b_{1g}) & \text{for } MX_4 \text{ cluster,} \\ \sqrt{2V(e_g)^2 + (8-n)V(t_{2g})^2} & \text{for } MX_6 \text{ cluster.} \end{cases} \tag{3.118}$$

In the final state of XPS, the Hamiltonian is given from H_0 only by replacing Δ with

$$\Delta_f \equiv \Delta - U_{dc}, \tag{3.119}$$

apart from a constant shift of the total energy, which is not important in the following calculations.

The ground state $|g\rangle$ of the Hamiltonian H_0 is easily obtained in the form

$$|g\rangle = \cos\theta_0 |3d^n\rangle - \sin\theta_0 |3d^{n+1}\underline{L}\rangle, \tag{3.120}$$

where

$$\tan\theta_0 = \frac{1}{V_{\text{eff}}}\left[\sqrt{\left(\frac{\Delta}{2}\right)^2 + V_{\text{eff}}^2} - \frac{\Delta}{2}\right]. \tag{3.121}$$

Similarly, we obtain the eigenstates $|\pm\rangle$ in the final state of XPS, whose energies are given by

$$E_{\pm} = E_0 + \frac{\Delta_f}{2} \pm \sqrt{\left(\frac{\Delta_f}{2}\right)^2 + V_{\text{eff}}}. \tag{3.122}$$

Corresponding to the two states $|\pm\rangle$, we have two XPS peaks with the binding energy difference

$$\delta E_B = E_+ - E_- = \sqrt{\Delta_f^2 + 4V_{\text{eff}}^2} \tag{3.123}$$

and with intensity ratio

$$\frac{I_+}{I_-} = \frac{|\langle+|g\rangle|^2}{|\langle-|g\rangle|^2} = \tan^2(\theta - \theta_0), \tag{3.124}$$

where

$$\tan\theta = \frac{1}{V_{\text{eff}}}\left[\sqrt{\left(\frac{\Delta_f}{2}\right)^2 + V_{\text{eff}}^2} - \frac{\Delta_f}{2}\right]. \tag{3.125}$$

We discuss the behavior of I_+/I_- in the following two cases:

3.5.5.3   Case A: $\Delta_f > 0$ ($\Delta > U_{dc}$)

In this case, the final states $|+\rangle$ and $|-\rangle$ correspond mainly to $|3d^{n+1}\underline{L}\rangle$ and $|3d^n\rangle$ states, respectively. The splitting of the XPS spectrum is caused by an increase of the covalency mixing in going from the ground state to the XPS final state. In order to see this, we define the "covalency parameter" γ_0 (in the ground state) and γ (in the final state) as

$$\gamma_0 = \tan\theta_0, \tag{3.126}$$

$$\gamma = \tan\theta. \tag{3.127}$$

These quantities represent the strength of the covalency mixing, and they change between 0 (no covalency mixing) and 1 (maximum covalency mixing). Then I_+/I_- is written as

$$\frac{I_+}{I_-} = \left(\frac{\gamma - \gamma_0}{1 + \gamma\gamma_0}\right)^2. \tag{3.128}$$

Since $\Delta > U_{dc}$, the covalency mixing increases in the final state of XPS (i.e. $\gamma > \gamma_0$), and this effect is called "photo-induced covalency." The intensity ratio I_+/I_- is 0 in the limit of $\gamma = \gamma_0$ and takes the maximum value 1 in the limit of $\gamma = 1$ and $\gamma_0 = 0$. Therefore, the splitting of the XPS spectrum is caused by the photo-induced covalency.

3.5.5.4   Case B: $\Delta_f \leq 0$ ($\Delta \leq U_{dc}$)

In this case, the final states $|+\rangle$ and $|-\rangle$ correspond mainly to $|3d^n\rangle$ and $|3d^{n+1}\underline{L}\rangle$ states, respectively. The covalency parameter in the final state is now defined by

$$\gamma = \cot\theta, \tag{3.129}$$

so that γ takes the maximum value $\gamma = 1$ for $\Delta_f = 0$, and decreases with increasing $|\Delta_f|$, corresponding to the decrease of the covalency mixing. Then we have

$$\frac{I_+}{I_-} = \left(\frac{1 - \gamma\gamma_0}{\gamma + \gamma_0}\right)^2. \tag{3.130}$$

When both γ_0 and γ are much smaller than unity, the intensity I_+ is much larger than I_-; thus the final state $|+\rangle$ corresponds to the main line rather than the satellite. Conversely, when both γ_0 and γ are close to unity, the state $|+\rangle$ gives a satellite whose intensity is much smaller than that of $|-\rangle$.

On the basis of the above formulation, the analysis of the experimental data of 2p XPS for CuF_2-MnF_2 are made. From the data, two independent experimental

TABLE 3.2
Parameter Values Estimated from the Analysis Using the Two-Configuration Model

	Δ	V_{eff}	U_{dc}	n_d	I_+/I_-	δE_B	γ_0	γ	Case
CuF_2	6.3	2.85	10.4	9.13	0.80 (~0.8)	7.0 (~7)	0.38	0.51	B
NiF_2	8.4	2.83	10.4	8.08	0.60 (~0.6)	6.0 (~6)	0.31	0.71	B
CoF_2	9.8	3.00	10.4	7.07	0.40 (~0.4)	6.0 (~6)	0.28	0.90	B
FeF_2	11.2	3.16	10.4	6.06	0.26 (~0.2)	6.4 (~6.5)	0.26	0.88	A
MnF_2	13.0	3.32	10.4	5.05	0.15 (~0.1)	7.1 (~7)	0.24	0.68	A

Note: Δ, V_{eff}, U_{dc}, and δE_B are given in units of eV. The experimental values are given in parentheses.

values of I_+/I_- and δE_B are estimated and listed in parentheses in Table 3.2 (Kotani and Okada, 1993). With the present model, on the other hand, I_+/I_- and δE_B are expressed in terms of three independent parameters, Δ, V_{eff}, and U_{dc} $(= \Delta - \Delta_f)$. From a more sophisticated analysis, we know the value of $V(e_g)$ for NiF_2 is about 2.0 eV ($V_{eff} \cong 2.83$ eV) and that the value of U_{dc} is not very dependent on the metal ions Cu–Mn. Therefore, we first analyze the Ni 2p XPS of NiF_2 with the use of $V_{eff} \cong 2.83$ eV, and obtain $\Delta = 8.4$ eV and $U_{dc} = 10.4$ eV so as to reproduce the experimental results of I_+/I_- and δE_B. Then we analyze the 2p XPS of the other materials by assuming $U_{dc} = 10.4$ eV, and estimate the values of Δ and V_{eff}. The values of Δ, V_{eff} and U_{dc} estimated in this way are shown in Table 3.2 together with the calculated value of n_d (average 3d electron number in the ground state), I_+/I_-, δE_B, γ_0, and γ. The calculated results of the XPS spectra are also shown in Figure 3.15 with a Gaussian broadening of Γ (FWHM) = 4.0 eV.

It is interesting to see, from Table 3.2, the systematic variation of Δ and V_{eff} with the change of transition metal ions. Both Δ and V_{eff} increase almost monotonically from CuF_2 to MnF_2. The increase of Δ reflects the decrease of the electronegativity of the metal ion in going from Cu to Mn. As a result, we find $\Delta < U_{dc}$ (Case B) for CuF_2, NiF_2, and CoF_2, while $\Delta > U_{dc}$ (Case A) for FeF_2 and MnF_2. The crossover between Case A and Case B occurs around FeF_2. However, no drastic change of the spectral shape of the XPS is recognized around FeF_2 because the character of the final states $| \pm \rangle$ changes continuously and gradually around $\Delta \cong U_{dc}$ (with the change of Δ) because of the strong hybridization V_{eff}.

The increase of V_{eff} from NiF_2 to MnF_2 originates mainly from the increase of the unoccupied 3d state (2 for e_g states and $8 - n$ for t_{2g} states). If we assume $V(e_g)/V(t_{2g}) = -2.0$, we have $V(e_g) \cong 2.0$ eV for all $NiF_2 - MnF_2$.

From systematic analysis of TM 2p XPS for many TM compounds, we can estimate the value of the key parameters Δ, U_{dd}, $V(\Gamma)$, and so on. The estimated values are somewhat different, depending on the model of the analysis (the cluster model or the SIAM, the full multiplet calculation or the calculation without the multiplet effect), but the systematic trend of their variation is recognized with respect to the change of the TM element and also the anion element. This is also the basis of the Zaanen–Sawatzky–Allen model (Zaanen et al., 1985a) that will be introduced in Chapter 5.

3.6   MANY-BODY CHARGE-TRANSFER EFFECTS IN XAS

3.6.1   GENERAL EXPRESSIONS OF MANY-BODY EFFECTS

The expression of the XAS spectrum $\mu(\Omega)$ is given by

$$\mu(\Omega) = \sum_f |\langle f|M^+|0\rangle|^2 \delta(\hbar\Omega - E_f + E_0), \qquad (3.131)$$

where M^+ is the optical transition operator of the incident photon with energy $\hbar\Omega$. With the optical transition, a core electron is excited to the VES, so that M^+ describes the creation of an electron in VES. For instance, M^+ is given by

$$M^+ = \sum_{ki} M_k a_k^+, \qquad (3.132)$$

for our model of simple metals treated in Section 3.5.1, and

$$M^+ = M_c \sum_{ki} a_k^+ + M_f a_f^+ \qquad (3.133)$$

for our model of La metal in Section 3.5.2. The essential difference between XAS and XPS is the existence of an additional electron in VES in the final state of XAS. This photo-created electron also plays an important role in screening the core-hole potential. So far as the formal expressions given in Section 3.3.1 for XPS are concerned, the ones for XAS can be obtained from those for XPS by inserting the operator M^+ appropriately. For instance, the XAS spectrum $\mu(\Omega)$ is expressed as the Fourier transform of a generation function $g(t)$:

$$\mu(\Omega) = \frac{1}{\pi\hbar} \mathrm{Re} \int_0^\infty dt e^{i\Omega t} g(t), \qquad (3.134)$$

but now $g(t)$ is given by

$$g(t) = \langle 0| e^{i(H_0/\hbar)t} M e^{-i(H/\hbar)t} M^+ |0\rangle. \qquad (3.135)$$

In the following, we make explicit calculations of $\mu(\Omega)$ for several different systems.

3.6.2   XAS IN SIMPLE METALS

The Hamiltonian H_0 and the core-hole potential U are given by Equations 3.65 and 3.66, respectively. The generating function is expressed, using the linked-cluster expansion, as

$$g(t) = e^{i(E_0/\hbar)t} \langle 0| M e^{-i(H/\hbar)t} M^+ |0\rangle$$

$$= e^{i(E_0/\hbar)t} \left[\sum_{k,k'} M_{k'}^* M_k \langle 0| a_{k'} e^{-i(H/\hbar)t} a_k^+ |0\rangle \right]_c \langle 0| e^{-i(H/\hbar)t} |0\rangle, \qquad (3.136)$$

where $[...]_c$ means the contribution from the connected diagram, which was first calculated by Mahan (1967) with the most divergent approximation term in the form

$$[...]_c = \sum_{k>k_F} \frac{|M_k|^2}{it} (iDt)^{-2\rho v} e^{-i\varepsilon_k t}. \tag{3.137}$$

On the other hand, the last factor of Equation 3.136 is nothing more than the generating function of XPS and from Equation 3.68 we have

$$\langle 0|e^{-i(H/\hbar)t}|0\rangle = e^{-i(\Delta_s/\hbar)t} \left(\frac{iDt}{\hbar}\right)^{-(\rho v)^2}. \tag{3.138}$$

Therefore, the XAS spectrum is given by (Kotani and Toyozawa, 1979)

$$\mu(\Omega) = \begin{cases} \dfrac{|M_{k_F}|^2 \rho}{\Gamma(1-2g+g^2)\left(\dfrac{\hbar\Omega - \tilde{\varepsilon}_F}{D}\right)^{2g-g^2}} & \text{for } \hbar\Omega < \tilde{\varepsilon}_F, \\[20pt] 0 & \text{for } \hbar\Omega < \tilde{\varepsilon}_F, \end{cases} \tag{3.139}$$

where

$$g = -\rho v, \tag{3.140}$$

$$\tilde{\varepsilon}_F = \varepsilon_F + \Delta_s. \tag{3.141}$$

Therefore, the XAS spectrum diverges at the Fermi edge and is called the "Fermi edge singularity." The behavior of XAS is shown schematically in Figure 3.16b and compared with that of XPS (Figure 3.16a). The dashed lines represent the spectra

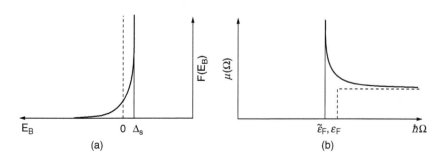

FIGURE 3.16 Schematic representations of (a) XPS and (b) XAS spectra of simple metals.

with no core-hole potential ($v = 0$). The exponent of the divergence is $-(2g - g^2)$ for XAS, while it is $-(1 - g^2)$ for XPS (see Equation 3.70). It is very interesting to see that the exponent of the singularity in XAS of simple metals is just the same as that of XPS in our model of La metal if we replace the core-hole potential v by v_{eff}, which was given in Section 3.5.2.

If we extend the model by taking into account the electron spin and the core-hole potential with arbitrary strength and finite force range, the XAS spectrum is expressed as (Nozières and De Dominicis, 1969)

$$\mu(\Omega) = W_{\ell-1}(\Omega)\left(\frac{D}{\hbar\Omega - \tilde{\varepsilon}_F}\right)^{\alpha_{\ell-1}} + W_{\ell+1}(\Omega)\left(\frac{D}{\hbar\Omega - \tilde{\varepsilon}_F}\right)^{\alpha_{\ell+1}}, \tag{3.142}$$

where we assume that the orbital angular momentum of the core state is ℓ (so that electric dipole transition is allowed to the conduction electron states with $\ell \pm 1$). In Equation 3.142, $W_{\ell \pm 1}(\Omega)$ is the XAS spectra in the limit of vanishing v and α_ℓ is given by

$$\alpha_\ell = \frac{2\delta_\ell(\varepsilon_F)}{\pi} - 2\sum_{l'}(2l'+1)\left[\frac{\delta_{l'}(\varepsilon_F)}{\pi}\right]^2 \tag{3.143}$$

with the phase shift of the conduction electron $\delta_\ell(\varepsilon_F)$.

According to experimental observations, the spectra of the L_{23} edges of Na, Mg, and Al metals are sharp and peaked, while those of the K edges of Li and Al metals are broad and rounded. If we assume that the screened potential of the core hole is of sufficiently short range, the s-wave scattering is predominant (so that $\delta_0 \gg \delta_1$, δ_2 and so on), then we obtain $\alpha_0 > 0$ and $\alpha_1 < 0$ from Equation 3.143, together with the Friedel sum rule. This means that the L edge should be peaked and the K edge should be rounded, in qualitative agreement with experimental results.

3.6.3 XAS IN La METAL

The Hamiltonians H_0 and H are given by Equations 3.81 and 3.82, respectively. The optical transition operator Equation 3.133 simulates the electric dipole transition from a La 3d core state to a La 4f state as well as that from the 3d core state to the f-symmetric partial wave states of conduction electrons (spin and orbital degeneracy is disregarded for simplicity) (Figure 3.17). In the limit of vanishing hybridization V, the XAS spectrum consists of a line spectrum at $\hbar\Omega = \varepsilon_f$ with the intensity $|M_f|^2$ and a continuous spectrum $|M_c|^2\rho$ above ε_F, as shown with dashed lines in Figure 3.18. The corresponding final states are given by

$$|f\rangle = a_f^+|0\rangle, \tag{3.144}$$

$$|k'\rangle = a_{k'}^+|0\rangle. \tag{3.145}$$

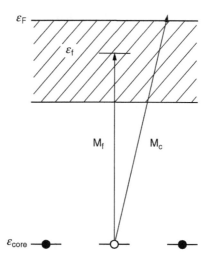

FIGURE 3.17 Model of XAS in La metals.

In the following, we show how this spectrum is affected by switching on the hybridization V for two different cases (Kotani and Toyozawa, 1973a).

3.6.3.1 Case A: $\varepsilon_f < \varepsilon_F$

(A-1) Threshold excitation ($\hbar\Omega \approx \varepsilon_f$)
When V is switched on, the state $|f\rangle$ is coupled with the states $a_k^+ a_{k'} |f\rangle$ by the effective potential v_{eff} (see Equation 3.87), which describes the second-order process ($f \rightarrow k$ and $k' \rightarrow f$). Therefore, we have an electron-hole pair excitation on the Fermi sea. The repetition of such a process results in the orthogonality catastrophe at the absorption edge as well as a shift of the edge Δ_f, which is exactly the same as the XPS spectrum of simple metals (see Section 3.5.1) if we replace v_{eff} by the direct

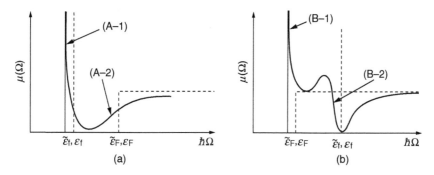

FIGURE 3.18 Schematic representation of XAS spectra calculated with the model shown in Figure 3.17. The figures (a) and (b) correspond to the cases of $\varepsilon_f < \varepsilon_F$ and $\varepsilon_f > \varepsilon_F$, respectively.

core-hole potential v. Namely, the XAS threshold diverges, as shown in Figure 3.18a, in the inverse power form:

$$\mu(\Omega) \approx (\hbar\Omega - \tilde{\varepsilon}_f)^{-(1-g_{eff}^2)}, \tag{3.146}$$

$$g_{eff} = -\rho v_{eff}, \tag{3.147}$$

$$\tilde{\varepsilon}_f = \varepsilon_f + \Delta_f. \tag{3.148}$$

(A-2)   Fermi edge ($\hbar\Omega \approx \varepsilon_f$)
The final state $|k'\rangle$, which gives a discontinuous Fermi edge at $\hbar\Omega = \varepsilon_F$, is mixed resonantly with $|f\underline{k}\,k'\rangle \equiv a_f^+ a_k |k'\rangle$ (with $\varepsilon_k \approx \varepsilon_f$) by the hybridization V. Then, by the second-order process $|k'\rangle \rightarrow |f\,\underline{k}\,k'\rangle \rightarrow |k'\rangle$, the state $|k'\rangle$ has the self-energy correction $\Sigma_0(z)$ (see Equation 3.94); that is, the state $|k'\rangle$ has the energy shift Δ_0 and the lifetime broadening $\Gamma \approx \pi\rho V^2$. Therefore, the Fermi edge is blurred out in the form

$$\mu(\Omega) \approx |M_c|^2 \rho \left\{ \frac{1}{2} + \frac{1}{\pi} \tan^{-1} \left[\frac{\hbar\Omega - \tilde{\varepsilon}_F}{\pi\rho V^2} \right] \right\} \tag{3.149}$$

with

$$\tilde{\varepsilon}_F = \varepsilon_F + \Delta_0. \tag{3.150}$$

3.6.3.2   Case B: $\varepsilon_f > \varepsilon_F$

(B-1)   Fermi edge ($\hbar\Omega \approx \varepsilon_f$)
In this case, the core electron is excited to the Fermi edge and thus the situation is exactly the same as the Fermi edge singularity in simple metals except for replacing the direct core-hole potential v with the effective potential v_{eff}. Thus, we have

$$\mu(\Omega) \approx (\hbar\Omega - \tilde{\varepsilon}_F)^{-(2g_{eff} - g_{eff}^2)}, \tag{3.151}$$

and the XAS spectrum diverges at the Fermi edge as shown in Figure 3.18b.

(B-2)   Excitation to the 4f level ($\hbar\Omega \approx \varepsilon_f$)
The excitations of the core electron to the f and c states interfere with each other through the hybridization V, resulting in the Fano resonance (Fano, 1961). The XAS spectrum exhibits an asymmetric peak and dip characteristic of the Fano resonance as shown schematically in Figure 3.18b. For more details of this XAS behavior, see the papers by Kotani and Toyozawa (1973a,b).

According to the experimental results of 4d XAS in La and Ce metals, no sharp Fermi edge is observed at the Fermi level, which was determined by XPS measurements, and furthermore, one or several sharp peaks corresponding to the 4d–4f transition have been recognized below the presumed Fermi level (Suzuki et al., 1972). This situation is qualitatively in agreement with the spectrum of Figure 3.18a.

3.6.4 Ce 3d XAS of Mixed Valence Ce Compounds

Gunnarsson and Schönhammer (1983b) calculated the Ce 3d XAS of mixed valence Ce intermetallic compounds. Figure 3.19 shows the result for $CeNi_2$ with the same parameter values as those used in the calculation of 3d XPS (Figure 3.11) compared with the experimental result. The XAS spectrum exhibits two peaks, which are interpreted to originate mainly from the $4f^2$ (main) and $4f^1$ (satellite) configurations in each of the $3d_{5/2}$ and $3d_{3/2}$ components.

As mentioned previously, the ground state is a linear combination of the states A, B, and C (Figure 3.10) in the lowest-order terms of the $1/N_f$ expansion method. In the final state of Ce 3d XAS, we have an additional 4f electron (say the state fv_0), excited from the core state by the incident photon. Therefore, the final states are linear combinations of the following basis states:

$$|A'\rangle = a^+_{fv_0}|A\rangle, \qquad (3.152)$$

$$|B'(\varepsilon)\rangle = \frac{1}{\sqrt{(N_f - 1)}} a^+_{fv_0} \sum_{v \neq v_0} a^+_{fv} a_{\varepsilon v} |A\rangle, \qquad (3.153)$$

$$|C'(\varepsilon, \varepsilon')\rangle = \frac{1}{\sqrt{(N_f - 1)(N_f - 2)}} a^+_{fv_0} \sum_{v,v'(\neq v_0)} a^+_{fv} a_{\varepsilon v} a^+_{fv'} a_{\varepsilon' v'} |A\rangle, \qquad (3.154)$$

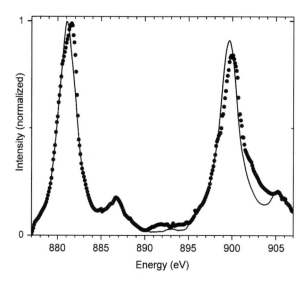

FIGURE 3.19 Experimental (dots) and theoretical (solid curve) results of the Ce 3d XAS spectrum of $CeNi_2$. (Reprinted with permission from Gunnarsson, O., and Schönhammer, K., *Phys. Rev. B*, 28, 4315, 1983. Copyright 1983 by the American Physical Society.)

corresponding to the $4f^1$, $4f^2$, and $4f^3$ configurations, respectively. The important point is that the occupations of the $4f^2$ and $4f^3$ configurations in the ground state are negligibly small because their energies $2\varepsilon_f^0 - \varepsilon + U_{ff}$ and $3\varepsilon_f^0 - \varepsilon - \varepsilon' + 3U_{ff}$ are much higher than those of $4f^0$ and $4f^1$ configurations. Therefore, we can disregard the contribution of the $4f^3$ configuration in the XAS spectrum. Compared with the three-peak structure of XPS, the many-body charge-transfer effect is weaker in XAS, resulting in the double-peak structure, and the essential reason for this is that in XAS the photo-excited 4f electron (fv_0) screens the core-hole potential to suppress the further screening by the charge-transfer effect.

The calculation of the Ce 3d XAS spectrum of CeO_2 was made by Kotani et al. (Jo and Kotani, 1988b; Kotani and Ogasawara, 1992), and the result is shown in Figure 3.20. We see a double-peak structure similar to that of $CeNi_2$, which are assigned as $4f^2\underline{L}$ (main) and $4f^1$ (satellite) configurations in the final state, in reasonable agreement with the experimental result shown in the inset. If we take into account only two configurations ($4f^0$ and $4f^1\underline{L}$ in the ground state and $4f^1$ and $4f^2\underline{L}$ in the final state), and consider the case of $N = 1$, the Hamiltonian matrix Equation 3.107 for XPS reduces, in the case of XAS, to

$$\begin{pmatrix} 0 & \sqrt{N_f}V \\ \sqrt{N_f}V & \varepsilon_f^0 - \varepsilon_v \end{pmatrix} \tag{3.155}$$

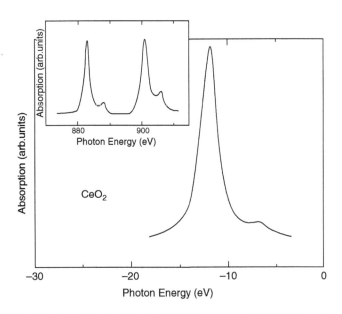

FIGURE 3.20 Calculated result of the Ce 3d XAS spectrum for CeO_2. Experimental result is shown in the inset. (Reprinted from Kotani, A., and Ogasawara, H., *J. Electron Spectrosc. Relat. Phenom.*, 60, 257, 1992. With permission from Elsevier Ltd.)

for the ground state and

$$
\begin{pmatrix}
\varepsilon_f^0 - U_{fc} & \sqrt{N_f - 1}\, V \\
\sqrt{N_f - 1}\, V & 2\varepsilon_f^0 - \varepsilon_v + U_{ff} - 2U_{fc}
\end{pmatrix}
\tag{3.156}
$$

for the final state. If we disregard the difference between $\sqrt{N_f}\, V$ and $\sqrt{N_f - 1}\, V$, and the difference between U_{ff} and U_{fc}, the structure of the two Hamilton matrices coincide with each other. Thus, no charge transfer occurs in going from the ground state to the final state.

3.6.5   Ce L$_3$ XAS

We have shown that in Ce 3d XAS, the photo-excited 4f electron directly screens the core-hole potential. In Ce L$_3$ XAS, on the other hand, a core electron (2p$_{3/2}$ state) is excited to the Ce 5d conduction band, instead of the Ce 4f state. Then the photo-excited 5d electron also partly screens the core hole charge and, as shown in this subsection, this screening effect due to the 5d electron plays an important role in determining the spectral shape of the L$_3$ XAS.

Let us consider the Ce L$_3$ XAS of CeO$_2$. The core-hole potential acting on the 4f state, $-U_{fc}$, because of the 2p core hole is known to be nearly the same as that due to the 3d core hole, which we have already treated in the discussion on Ce 3d XPS. So we take them to be the same. Then, if we disregard the role of the photo-excited 5d electron, the final state interaction caused by the core hole is almost the same for Ce 3d XPS and L$_3$ XAS. According to experimental data, however, the spectral features of the L$_3$ XAS are very different from those of the 3d XPS. As shown in the inset of Figure 3.21, the experimental L$_3$ XAS (Bianconi et al., 1987) has only two peaks whose energy separation is about 8 eV, whereas the 3d XPS has three peaks and the energy separation of the outermost two peaks is about 16 eV, as seen from Figure 3.13.

Jo and Kotani (1985) calculated the L$_3$ XAS of CeO$_2$ by taking into account the screening of the core-hole potential by the 5d electron, which is treated by introducing the core-hole potential acting on the 5d electron, $-U_{dc}$. At the same time, they introduced the repulsive Coulomb interaction U_{fd} between the 5d and 4f electrons, in order to avoid the overscreening by both 4f and 5d screening charges.

The Hamiltonian of the initial state is given by

$$
H_0 + \sum_k \varepsilon_{dk} a_{dk}^+ a_{dk}
\tag{3.157}
$$

and that of the final state by

$$
H_0 + \sum_k \varepsilon_{dk} a_{dk}^+ a_{dk} - U_{fc} \sum_v a_{fv}^+ a_{fv} - \frac{U_{dc}}{N} \sum_{k,k'} a_{dk}^+ a_{dk'} + \frac{U_{fd}}{N} \sum_{k,k',v} a_{fv}^+ a_{fv} a_{dk}^+ a_{dk'},
\tag{3.158}
$$

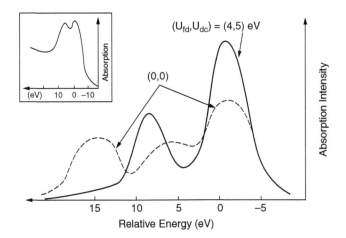

FIGURE 3.21 Calculated results of the Ce L_3 XAS spectra for CeO_2 compared with the experimental result in the inset. (Reprinted from Kotani, A., and Jo, T., *J. Physique*, C8/915, 1986. With permission.)

where H_0 is given by Equation 3.106 and ε_{dk} is the energy of the Ce 5d conduction band, which is expressed as

$$\varepsilon_{dk} = \varepsilon_d - \frac{W_d}{2} + \frac{W_d}{N}\left(k - \frac{1}{2}\right)$$
(3.159)

similar to Equation 3.105. The parameter values are taken to be the same as those determined from the analysis of 3d XPS (see Section 3.5.4), and the new parameters are taken as $U_{fd} = 4.0$ eV and $U_{dc} = 5.0$ eV. The calculated result is shown by the solid curve in Figure 3.21, which is in good agreement with the experimental result shown in the inset, if we take into account an appropriate background contribution. On the other hand, the result with $U_{fd} = 0$ and $U_{dc} = 0$ is nothing but a more broadened version (because of the Ce 5d bandwidth) of the Ce 3d XPS.

The essential reason for the two-peak structure of L_3 XAS is easily understood by considering the limiting case of $N = 1$ (both for the O 2p and Ce 5d bands). The Hamiltonian of the ground state is given by Equation 3.107 with $U_{fc} = 0$ and that of the final state by

$$\begin{pmatrix} \varepsilon_d - U_{dc} & \sqrt{N_f}\,V & 0 \\ \sqrt{N_f}\,V & \varepsilon_d - U_{dc} + \varepsilon_f^0 - \varepsilon_v - U_{fc} + U_{fd} & \sqrt{2(N_f - 1)}\,V \\ 0 & \sqrt{2(N_f - 1)}\,V & \varepsilon_d - U_{dc} + 2(\varepsilon_f^0 - \varepsilon_v) + U_{ff} - 2U_{fc} + 2U_{fd} \end{pmatrix}.$$
(3.160)

In Figure 3.22, three energy eigenvalues in the final state are shown as a function of U_{fd} where the dashed lines I, II, and III correspond to the energies of $4f^2\underline{L}^2$, $4f^1\underline{L}$, and $4f^0$ configurations without the hybridization and the solid curves are those after

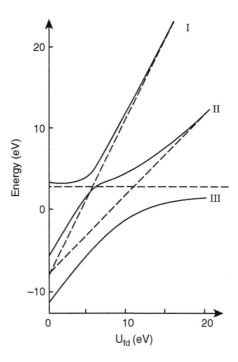

FIGURE 3.22 Final-state energies of the Ce L_3 XAS calculated in the limit of $N = 1$ and shown as a function of the 4f-5d Coulomb interaction U_{fd}. (Reprinted from Jo, T., and Kotani, A., *Solid State Commun.*, 54, 451, 1985. With permission from Elsevier Ltd.)

the hybridization (Jo and Kotani, 1985). In the limit of $U_{fd} = 0$, the three energies are essentially the same as those of 3d XPS, so that we have three-peak structure. However, with increasing U_{fd} the energies of $4f^2\underline{L}^2$ and $4f^1\underline{L}$ configurations increase by $2U_{fd}$ and U_{fd}, respectively. For $U_{fd} = 4.0$ eV, the energies of $4f^0$ and $4f^2\underline{L}^2$ configurations become close to each other and they mix only weakly through the mixing with the $4f^1\underline{L}$ configuration. Thus, they cannot be resolved into two XAS peaks if we take into account the spectral broadening by the core hole lifetime. This is the reason for the two XAS peaks, one for mainly the $4f^1\underline{L}$ configuration and the other for the $4f^0$ and $4f^2\underline{L}^2$ unresolved configurations.

In the case of $N = 1$, mentioned previously, the role of the core-hole potential $-U_{dc}$ is not important and results only in a uniform energy shift of the final state. This is because the Ce 5d state is already localized for $N = 1$, irrespective of the existence of the potential $-U_{dc}$. When we take into account a finite bandwidth of the 5d states, on the other hand, the 5d wave function is extended in space, so that it is not much affected by the local interaction U_{fd} without the potential $-U_{dc}$. The role of $-U_{dc}$ is to localize the 5d states around the core hole site (the 5d screening effect of the core-hole potential), and then the 5d electron couples strongly with the localized 4f electron through U_{fd}.

As another example, we show in Figure 3.23 (Kotani et al., 1987b) the calculated Ce L_3 XAS spectra for CeF_4 with $(U_{fd}, U_{dc}) = (4, 5)$ eV (solid curve) and $(0, 0)$ eV (dashed curve). The other parameter values are the same as those used in the analysis of the Ce 3d XPS (see Section 3.5.4). The result of the dashed curve is a more

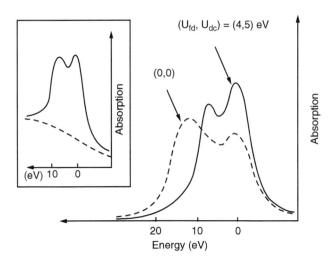

FIGURE 3.23   Calculated results of the Ce L$_3$ XAS spectra for CeF$_4$ compared with the experimental result in the inset. (Reprinted from Kotani, A., Jo, T., and Parlebas, J.C., *Adv. Phys.*, 37, 37, 1988. With permission.)

broadened version of the 3d XPS (Figure 3.14), and it is very different from the experimental result (Kaindl et al., 1987) shown in the inset. On the other hand, the solid curve is in good agreement with the experimental result, if we consider the background contribution (the dashed curve in the inset).

It should be mentioned that the effect of $-U_{dc}$ and U_{fd} is also important in explaining the La L$_3$ XAS spectral shape of insulating La compounds, LaF$_3$ and La$_2$O$_3$ (Kotani et al., 1987a; Kotani et al., 1988). The 3d XPS of these La compounds exhibits a two-peak structure, corresponding to the bonding and antibonding states of 3d^{9}4f^0 and 3d^{9}4f^1<u>L</u> configurations. An example is given for La$_2$O$_3$ in Chapter 5. Conversely, the La L$_3$ XAS spectra of these compounds exhibit a single peak with an asymmetric spectral shape. The analysis of these spectra was made in a similar manner to that for Ce compounds and the results for $(U_{fd}, U_{dc}) = (4, 5)$ eV were found to be in good agreement with experimental XAS spectra (not shown here).

For metallic La and Ce compounds, the effect of $-U_{dc}$ and U_{fd} is much weaker than that in insulating La and Ce compounds because these potentials are screened by conduction electrons. The typical value of U_{fd} is reduced to 1.0–2.0 eV from that in insulating systems (~4.0 eV), but its effect on the XAS spectral shape is still important. Here we show the result for LaPd$_3$ (Kotani et al., 1987c; Kotani et al., 1988). The inset of Figure 3.24 shows the experimental La L$_3$ XAS spectrum (solid curve) and its second derivative (dotted curve), while the upper and lower panels are the calculated results with $(U_{fd}, U_{dc}) = (2.0, 1.7)$ eV and (0, 0) eV, respectively. In the calculation, $-U_{dc}$ and U_{fd} are assumed to act only on the photo-excited 5d electron. As shown in Figure 3.5a, the La 3d XPS spectrum has a double-peak structure corresponding to 3d^{9}4f^0 and 3d^{9}4f^1 <u>L</u> configurations, and the result calculated with $(U_{fd}, U_{dc}) = (0, 0)$ eV is a more broadened version of the 3d XPS. Thus, the peak of the XAS spectrum is considerably rounded and its second derivative does not agree with

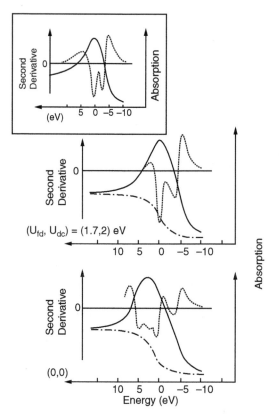

FIGURE 3.24 Calculated results of the La L_3 XAS spectrum (solid curve) and its second derivative (dotted curve) for LaPd$_3$ compared with the experimental result in the inset. (Reprinted from Kotani, A., Jo, T., and Parlebas, J.C., *Adv. Phys.*, 37, 37, 1988. With permission.)

the experimental result. On the other hand, the result with $(U_{fd}, U_{dc}) = (2.0, 1.7)$ eV is in good agreement with the experiment because the energy of the 3d^{9}4f^1 $\underline{L}$ configuration is pushed up towards the 3d^{9}4f^0 configuration by the effect of U_{fd}. Thus, the XAS peak becomes sharper, similar to the experimental result.

For metallic mixed valence Ce compounds, the importance of the effect of $-U_{dc}$ and U_{fd} will be discussed in detail in Chapters 5 and 8.

3.6.6 XAS IN TRANSITION METAL COMPOUNDS

We calculate the 2p XAS spectra of TM compounds with the same model as that in Section 3.5.5. Similar to the consideration in the preceding subsection, the Hamiltonian H in the final state of XAS is given by

$$H = E_1 |3d^{n+1}\rangle\langle 3d^{n+1}| + (E_1 + \Delta + U_{dd} - U_{dc})|3d^{n+2}\underline{L}\rangle\langle 3d^{n+2}\underline{L}|$$

$$+ V'_{\text{eff}}(|3d^{n+1}\rangle\langle 3d^{n+2}\underline{L}| + |3d^{n+2}\underline{L}\rangle\langle 3d^{n+1}|), \tag{3.161}$$

where $E_1 = E_0 + \varepsilon_d - U_{dc}$ and V'_{eff} is the effective hybridization strength (not explicitly given here) similar to Equation 3.118 in XPS.

The ground state is given by Equation 3.120 with Equation 3.121, and the final states are given by the same expression in XPS only by replacing Δ_f with

$$\Delta'_f = \Delta + U_{dd} - U_{dc} \tag{3.162}$$

and by replacing V_{eff} with V'_{eff}. Numerical calculations for I_+/I_- and $\delta E \equiv E_1 - E_-$ are easily made and here, for example, we give the results for NiF_2 and FeF_2. We use the parameter values from Table 3.2 and the value of $U_{dd} - U_{dc}$ is fixed at -0.2 eV (see Table 5.1 of Chapter 5). For NiF_2, we have $V'_{\text{eff}} = V(e_g) = V_{\text{eff}}/2 = 1.41$ eV and $\Delta'_f = 8.2$ eV, so that $I_+/I_- = 0.017$ and $\delta E = 8.67$ eV. For FeF_2, we have $V'_{\text{eff}} = 2.72$ eV, $\Delta'_f = 11.0$ eV, $I_+/I_- = 0.00073$, and $\delta E = 12.27$ eV. Compared with $I_+/I_- = 0.60$ (NiF_2) and 0.26 (FeF_2) in XPS, the value of I_+/I_- is found to be much smaller. This is the effect of the photo-excited 3d electron, which screens the core-hole potential directly and suppresses further screening because of the charge-transfer effect. This is essentially the same situation as mentioned in the Ce 3d XAS (Section 3.6.4). The effect is very clear, if we consider a limiting situation where $V'_{\text{eff}} = V_{\text{eff}}$ and $\Delta'_f = \Delta$. Then we have $I_+/I_- = 0$, namely the phase of the wave function in the ground state is exactly the same as that of the lower energy final state and orthogonal with the higher energy final state. Consequently, the intensity of the XAS satellite (higher energy final state) vanishes by the phase cancellation effect. It should also be noted that the value of δE is larger than δE_B in XPS because the value of Δ'_f is larger than Δ.

Experimental results of 2p XAS in various TM compounds will be given in the following chapters. In most cases, the main structures of the 2p XAS spectra are determined by atomic multiplet structures originating from multipole components of the electron–electron Coulomb interaction, the spin–orbit interaction and the crystal field effect, and the many-body charge-transfer effect plays only a minor role. This is in strong contrast with the 2p XPS of TM compounds where the many-body charge-transfer effect plays the dominant role as we have already seen and the atomic multiplet effect is minor. The essential difference of XAS and XPS spectra is that, in XAS, the photo-excited electron stays in the 3d state and participates directly in the screening of the core hole charge while, in XPS, it is excited to a high energy continuum that is detected as a photoelectron and it never participates in the screening.

For 3d XPS and 3d XAS of mixed valence Ce compounds and 2p XPS of transition metal compounds, the many-body charge transfer calculation (without multiplet effect) works as a good starting approximation and the atomic multiplet effect gives some corrections to this first approximation. On the other hand, for 2p XAS of transition metal compounds, the LFM theory (with crystal field effect but without charge-transfer effect) serves as a good starting approximation and the charge-transfer, effect gives some corrections to this first approximation. In any case, the most complete model is the CTM theory, which includes all of the crystal field, charge transfer, and multiplet effects. The estimation of I_+/I_- in XAS given in this section is too crude but, using the CTM theory, it can be somewhat improved upon.

3.7   COMPARISON OF XPS AND XAS

In order to see the large difference between XPS and XAS spectra, it is instructive to compare the XPS and XAS transitions on the ionic state energy diagram. In Figure 3.25, we show an ionic state energy diagram of XPS. Here we consider, for example, the Ni 2p XPS process of a divalent Ni compound, where we assume that $\Delta = 4$ eV, $U_{dd} = 7.5$ eV, and $U_{dc} = 8.5$ eV (close to those of NiO given in Chapter 8). The two curves represent the energies of ground state configurations and final state configurations. If the energy of the $3d^8$ configuration (3d count = 8) in the ground state is taken as the origin of the energy, then that of the $3d^9\underline{L}$ (3d count = 9) is Δ. If we take the energy of the $3d^8$ configuration in the final state as E_0, the energy of $3d^9\underline{L}$ is $E_0 + \Delta - U_{dc}$. In Figure 3.25, E_0 is taken to be 10 eV (the offset energy). The energy of the $3d^{10}\underline{L}^2$ configuration (3d count = 10) is $2\Delta + U_{dd}$ for the ground state and $E_0 + 2\Delta + U_{dd} - 2U_{dc}$ for the final state. Going in the opposite direction, the energy of $3d^7L$ is equal to $-\Delta + U_{dd}$ for the ground state and $E_0 - \Delta + U_{dd} + U_{dc}$ for the final state, where the energy difference of L (ligand electron) and $\underline{L}$ (ligand hole) is neglected for simplicity. This figure is based on the work of John Fuggle who introduced this figure for VPES spectra of RE systems (Fuggle et al., 1983). The arrows indicate the XPS transitions in which a 2p core electron is excited to an empty state. This implies the transition from $3d^8$ to $2p^53d^8$ and so on. The energy of $2p^53d^9\underline{L}$ is lower than that of $2p^53d^8$, which means that the ionic configuration at the lowest energy is dominated by the $2p^53d^9\underline{L}$ configuration (i.e. the well screened state).

A similar ionic state energy diagram of the Ni 2p XAS with the same parameter values is shown in Figure 3.26. The final state energies of 2p XAS are exactly the

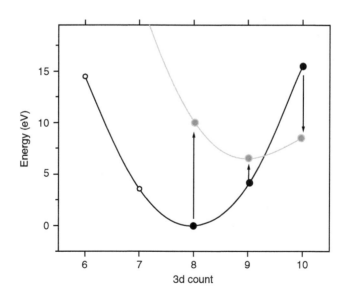

FIGURE 3.25   The energy effects in a 2p XPS experiment of a Ni^{2+} ion, using $\Delta = 4$ eV, $U_{dd} = 7.5$ eV and $U_{dc} = 8.5$ eV. The dark curve describes the ground state. The light curve is the 2p XPS spectrum offset by 10 eV. The arrows indicate the 2p XPS transitions.

same as for the 2p XPS. The difference is that the transitions are vertical for 2p XPS ($3d^8$ to $2p^53d^8$) and diagonal for 2p XAS ($3d^8$ to $2p^53d^9$). There is a small difference between the parabola-like curves of the ground and excited states and the excitation energies of the two transitions are $E_0 + \Delta - U_{dc}$ (from $3d^8$ to $2p^53d^9$) and $E_0 + \Delta + U_{dd} - 2U_{dc}$ (from $3d^9\underline{L}$ to $2p^53d^{10}\underline{L}$). Apart from the small shift, $U_{dd} - U_{dc} = -1$ eV, the energies are the same. This implies that the ground-state and final-state mixing are similar, which implies that the process is dominated by the transitions from the bonding combination of $3d^8$ and $3d^9\underline{L}$ to the bonding combinations of $2p^53d^9$ and $2p^53d^{10}\underline{L}$. This indicates clearly that the charge-transfer effect in XAS is much smaller than that in XPS. This effect is discussed in more detail in Section 4.3.

In this chapter, we have discussed the many-body charge-transfer effects that have more importance for XPS. As a consequence, the main part of this chapter has been devoted to XPS. In Chapters 5 and 8, respectively, we describe XPS and RXES along with the CTM theory. In Chapters 4, 6, and 7, we mainly focus on XAS and details of the LFM theory and CTM theory are presented.

Before closing this section, some comments are made on the screening of the core hole charge by the photo-excited electron in XAS. This effect suppresses the charge transfer when going from the ground state to the final state. However, this does not mean that the charge-transfer effect is suppressed in each of the ground and final states. For instance, the ground state of NiF_2 is a mixed-state configuration of 92% $3d^8$ and 8% $3d^9\underline{L}$ because of the charge-transfer effect (Table 3.2). With their more covalent nature, the mixing will be still larger for $NiCl_2$ and NiO. Therefore, even though the XAS spectra of these materials are very well described by the LFM theory, it cannot be assumed that the electronic states are also well described by this theory.

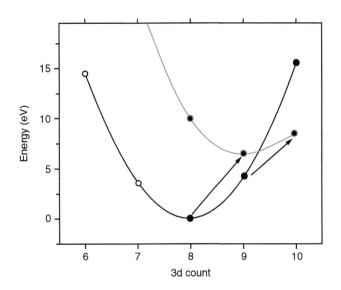

FIGURE 3.26 The energy effects in a 2p XAS experiment of a Ni^{2+} ion, using $\Delta = 4$ eV, $U_{dd} = 7.5$ eV and $U_{dc} = 8.5$ eV. The dark curve describes the ground state. The light curve is the 2p XPS spectrum offset by 10 eV. The arrows indicate the 2p XAS transitions.

As the LFM theory disregards the charge-transfer effect, the ground state of these materials is in the pure $3d^8$ configuration, which is incorrect. The essential point is that the charge-transfer effect can be renormalized by modifying the crystal field strength and the Slater integrals describing the multiplet coupling, in the calculation of XAS by the LFM theory. However, it is not the case in the calculation of XPS together with the calculation of the ground and excited electronic states.

4 Charge Transfer Multiplet Theory

4.1 INTRODUCTION

In Chapter 2, we used the single electron excitation description of x-ray absorption and the other core level spectroscopies. However, in the core level spectroscopies of strongly correlated electron systems, such as rare earth (RE) systems and transition metal (TM) compounds, the single electron excitation picture often breaks down. The many-body effects beyond the single electron excitation are mainly divided into the charge-transfer effect and the intra-atomic multiplet coupling effect.

In Chapter 3, the many-body charge-transfer effect in x-ray photoemission spectroscopy (XPS) and x-ray absorption spectroscopy (XAS) was discussed, and it was shown that the charge-transfer effect plays an essential role in RE 3d XPS and TM 2p XPS, but that the charge-transfer effect is much weaker in XAS than XPS. On the other hand, the multiplet coupling effect is of primary importance when describing XAS, especially TM 2p XAS, as is shown in this chapter and Chapter 6. The charge transfer multiplet (CTM) theory is where both the charge-transfer and multiplet effects are taken into account, which is applicable to the calculation of most XPS and XAS spectra in RE and TM systems.

In this chapter, we outline the basic aspects of the CTM theory and discuss the importance of the multiplet effect (as well as the charge-transfer effect) over the single electron excitation picture in XAS of TM compounds. Applying the single electron excitation description to TM oxides, good agreement is found for the oxygen K edges whereas for the metal $L_{2,3}$ edges, the agreement is poor. The reason for this discrepancy is not that the density-of-states (DOS) is wrongly calculated, but that the DOS is not observed in such x-ray absorption processes. The reason for the deviation from the DOS is the strong overlap of the core wave function with the valence wave functions. The potential overlap of core and valence wave functions is also present in the ground state but, because all core states are filled, it is not effective and one can approximate the core electrons with their charge. In the final state of an x-ray absorption process, a partly filled core state is found (e.g. a $2p^5$ configuration). In the case of a system with a partly filled 3d band, for example NiO, the final state will have an incompletely filled 3d band. For NiO, this can be approximated as a $3d^9$ configuration. The 2p hole and the 3d hole have radial wave functions that overlap significantly. This wave function overlap is an atomic effect that can be very large. It creates final states that are found after the vector coupling of the 2p and 3d wave functions. This effect is well known in atomic physics and actually plays a crucial role in the calculation of atomic spectra. Experimentally, it has been shown that while the direct core hole potential is largely screened, these so-called multiplet effects are hardly screened in the solid state. This implies that the atomic multiplet effects are of the same order of magnitude in atoms and in solids.

In the case of strongly correlated excited states (or localized states) in XAS, like the TM 3d states or the RE 4f states, the interaction between the electrons, as well as the interaction of the electrons with the core hole after the absorption process, has to be taken into account explicitly. To be more specific, localized 3d states imply

1. *Charge-transfer effects*: The energy differences between different occupations of the 3d states, yielding large reordering effects in the energy positions of configurations because of the core hole potential.
2. *Multiplet effects*: Large 2p3d two-electron integrals.

In order to describe the CTM theory in this chapter, we start with an atomic model where only the interactions within the absorbing atom are considered, without any influence from the surrounding atoms. Solid state effects are then introduced as a perturbation. This can be justified if the intra-atomic interactions are much larger than the one between the atoms. The first effect of the solid that is included is the effect of the crystal field in a solid. This takes into account the symmetry aspects of the solid. The second effect of the solid is the effect of bonding, which is described by the charge-transfer effect. This effect also takes care of the screening effects in the presence of a core hole.

In Appendix F, some discussion is presented on the theoretical methods for treating the single electron excitation picture, which is useful for the calculation of the O 1s XAS of TM oxides, as well as the XAS calculations in the systems where the f or d shells are empty or completely filled. We look at the Fermi Golden Rule for x-ray absorption to pinpoint where the multiplet and charge-transfer effects are removed in the single electron excitation description that identifies the final state in XAS with the DOS. In the dipole approximation, the Fermi Golden Rule was found in Chapter 2 to be:

$$W_{fi} \propto \sum_q |\langle \Phi_f | \mathbf{e}_q \cdot \mathbf{r} | \Phi_i \rangle|^2 \, \delta(\hbar\Omega - E_f + E_i). \quad (4.1)$$

This Fermi Golden Rule is a general expression and uses the initial state (Φ_i) and final state (Φ_f) wave functions. These wave functions are not exactly known and in practical calculations, one must make approximations to actually calculate the x-ray absorption cross-section. In the single electron excitation model, the initial state wave function is rewritten as a core state (c) and the final state wave function as an empty state (ε). Hence, it is implicitly assumed that all other electrons do not participate in the x-ray induced transition. The matrix element is then rewritten to a single electron matrix element, abbreviated with M, that is:

$$|\langle \Phi_f | \mathbf{e}_q \cdot \mathbf{r} | \Phi_i \rangle|^2 = |\langle \Phi_i \underline{c}\varepsilon | \mathbf{e}_q \cdot \mathbf{r} | \Phi_i \rangle|^2 \approx |\langle \varepsilon | \mathbf{e}_q \cdot \mathbf{r} | c \rangle|^2 \equiv M^2. \quad (4.2)$$

The delta function in Equation 4.1 implies that one observes the density of empty states (ρ) and using the one electron approximation this yields:

$$I_{XAS} \sim M^2 \rho. \tag{4.3}$$

The remaining problem in the single electron excitation model is then the determination of the DOS ρ where, in principle, ρ must be calculated in the presence of the core hole (ρ^*). The main methods that start from this approximation are based on the local (spin) density approximation (LDA or LSDA) of the density functional theory (DFT). For more details on the LSDA of DFT and some computer codes for XAS calculations with LSDA together with some methods that go beyond LSDA (see Appendix F).

4.2 ATOMIC MULTIPLET THEORY

Atomic multiplet theory is the description of the atomic structure with quantum mechanics. There exist many textbooks that discuss this issue at great length. The basic aspects only, which are needed for a general understanding of the concepts, are touched on here.

The starting point is the Schrödinger equation of a single electron in an atom:

$$H\Psi = E\Psi \tag{4.4}$$

with the Hamiltonian H is equal to:

$$H = \frac{-\hbar^2}{2m}\nabla^2 - \frac{Ze^2}{\mathbf{r}}. \tag{4.5}$$

The kinetic energy term contains the electron mass m and the derivatives to the three directions x, y, and z. The nuclear term describes the Coulomb attraction of the nucleus with the atomic number Z. The solutions of this equation are the atomic orbitals as defined by their quantum numbers n_p, l, and m_l. Table 4.1 gives the atomic orbitals, along with their spectroscopic names. The principal quantum number is

TABLE 4.1
Principal (n_p), Azimuthal (l), and Magnetic (m_l) Quantum Numbers and Their Spectroscopic Names

Level	Edge	$n_p - l$	m_l Values
1s	K	1–0	0
2s	L_1	2–0	0
2p	L_2 and L_3	2–1	–1, 0, +1
3s	M_1	3–0	0
3p	M_2 and M_3	3–1	–1, 0, +1
3d	M_4 and M_5	3–2	–2, –1, 0, +1, +2

usually written as n, but we use n_p in order to discriminate from the 3d (or 4f) electron number n in an atom.

The atomic wave functions (Ψ_{LM}) of the 3d electrons are:

$$\Psi(3d_{z^2}) = \frac{1}{81}\left(\frac{1}{6\pi a_0^7}\right)^{1/2} r^2 e^{-r/3a_0}(3\cos^2\theta - 1) = \Psi_{20},$$

$$\Psi(3d_{x^2-y}) = \frac{1}{81}\left(\frac{1}{6\pi a_0^7}\right)^{1/2} r^2 e^{-r/3a_0}\sqrt{3}\sin^2\theta\cos 2\phi,$$

$$\Psi(3d_{xy}) = \frac{1}{81}\left(\frac{1}{6\pi a_0^7}\right)^{1/2} r^2 e^{-r/3a_0}\sqrt{3}\sin^2\theta\sin 2\phi,$$ (4.6)

$$\Psi(3d_{yz}) = \frac{1}{81}\left(\frac{1}{6\pi a_0^7}\right)^{1/2} r^2 e^{-r/3a_0}2\sqrt{3}\sin\theta\cos\theta\sin\phi,$$

$$\Psi(3d_{yz}) = \frac{1}{81}\left(\frac{1}{6\pi a_0^7}\right)^{1/2} r^2 e^{-r/3a_0}2\sqrt{3}\sin\theta\cos\theta\cos\phi.$$

(handwritten margin note: different linear combinations of the Y_{LM}, $L=2$, $M=-L...+L$)

In atoms where more than one electron is present, there are two additional terms in the atomic Hamiltonian: the electron–electron repulsion (H_{ee}) and the spin–orbit coupling of each electron (H_{ls}). The total Hamiltonian is then given by:

$$H_{ATOM} = \sum_N \frac{\mathbf{p}_i^2}{2m} + \sum_N \frac{-Ze^2}{\mathbf{r}_i} + \sum_{pairs} \frac{e^2}{\mathbf{r}_{ij}} + \sum_N \zeta(\mathbf{r}_i)\, l_i \cdot s_i.$$ (4.7)

Here $\sum_N$ is the abbreviation of $\sum_{i=1}^N$. The kinetic energy and the interaction with the nucleus are the same for all electrons in a given atomic configuration. They define the average energy of the configuration (H_{av}). The electron–electron repulsion and the spin–orbit coupling define the relative energy of the different terms within this configuration. The main difficulty when solving this equation is that H_{ee} is too large to be treated as a perturbation. A solution to this problem is given by the central field approximation, in which the spherical average of the electron–electron interaction is separated from the nonspherical part. The spherical average $\langle H_{ee}\rangle$ is added to H_{av} to form the average energy of a configuration. In the modified electron–electron Hamiltonian H'_{ee}, the spherical average has been subtracted

$$H'_{ee} = H_{ee} - \langle H_{ee}\rangle = \sum_{pairs}\frac{e^2}{\mathbf{r}_{ij}} - \left\langle \sum_{pairs}\frac{e^2}{\mathbf{r}_{ij}}\right\rangle.$$ (4.8)

The two interactions H'_{ee} and H_{ls} therefore determine the energies of the different terms within the atomic configuration.

4.2.1 TERM SYMBOLS

The overview of quantum numbers of a single electron and two electrons in an atom, as well as their nomenclature, is given in Table 4.2. The principal quantum number

TABLE 4.2
Overview of the Quantum Numbers and Their Nomenclature

Name	Symbol	Values (Single Electron)	Values (Two Electrons)
Principal quantum number	n_p	$n_{p(max)} = \infty$ steps of 1 $n_{p(min)} = 1$	
Azimuthal quantum number or orbital angular momentum quantum number	L or l	$l_{(max)} = n - 1$ steps of 1 $l_{(min)} = 0$	$L_{(max)} = l_1 + l_2$ steps of 1 $L_{(min)} = |l_1 - l_2|$
Magnetic quantum number	M_L or m_L	$m_{l(max)} = 1$ steps of 1 $m_{l(min)} = -l$	$M_{L(max)} = L$ steps of 1 $M_{L(min)} = -L$
Spin angular momentum quantum number	S or s	$1/2$	$S_{(max)} = 1$ $S_{(min)} = 0$
Spin magnetic quantum number	M_S or m_s	$m_{s(max)} = 1/2$ $m_{s(min)} = -1/2$	$M_{S(max)} = S$ $M_{S(min)} = -S$
Total angular momentum quantum number	J or j	$j_{(max)} = l + 1/2$ $j_{(min)} = l - 1/2$	$J_{(max)} = L + S$ steps of 1 $J_{(min)} = |L - S|$
Total magnetic quantum number	M_J or m_J	$m_{j(max)} = j$ steps of 1 $m_{j(min)} = -j$	$M_{J(max)} = J$ steps of 1 $M_{J(min)} = -J$

n_p is not important for the angular symmetry of a state. For a single electron, the quantum numbers are indicated with the orbital angular momentum l, the spin angular momentum s of 1/2, the total angular momentum j with two values $l + 1/2$ and $l - 1/2$, the magnetic quantum number m_l, the spin magnetic quantum number m_s, and the total magnetic quantum number m_j.

For a two-electron configuration, the maximum orbital angular momentum L is equal to the addition of the two individual orbital angular momenta, l_1 and l_2. The same rule applies to the spin angular momentum, implying that the spin angular momentum of two electrons can be either 1 or 0, and the total angular momentum J takes the values from $|L - S|$ to $L + S$ by step 1.

For multi-electron configurations with quantum numbers L, S, and J, in general, a term symbol is written as $^{2S+1}L_J$, where the orbital angular momentum L is indicated with their familiar notation: S for $L = 0$, P for $L = 1$, and so on. In the absence of spin–orbit coupling, all terms with the same L and S have the same energy, giving an energy level that is $(2L + 1)(2S + 1)$-fold degenerate. When spin–orbit coupling is important, the terms are split in energy according to their J-value with a degeneracy of $2J + 1$. The quantity $2S + 1$ is called the spin multiplicity of the term and the terms are called singlet, doublet, triplet, quartet, and so on according to $S = 0$, 1/2, 1, 3/2, and so on.

A single s electron has an orbital angular momentum $l = 0$, a spin angular momentum $s = 1/2$ and a total angular momentum $j = 1/2$. There is only one term symbol $^2S_{1/2}$. For one p electron, $l = 1$, $s = 1/2$, and j can be 1/2 or 3/2, corresponding to term symbols $^2P_{1/2}$ and $^2P_{3/2}$. Similarly, a single d electron has term symbols $^2D_{3/2}$ and $^2D_{5/2}$ and a single f electron $^2F_{5/2}$ and $^2F_{7/2}$. The degeneracy of these states is given by $2j + 1$, which gives the well-known 2 : 1 ratio for $^2P_{1/2}$ and $^2P_{3/2}$ (L_2 and L_3 edges) and 3 : 2 for $^2D_{3/2}$ and $^2D_{5/2}$ (M_4 and M_5 edges).

The usual approach to solve the Hamiltonian is to distinguish two situations, LS-coupling and jj-coupling. In the case of valence electrons, one finds that the spin–orbit coupling is small, especially for light elements. In that case, one can neglect the spin–orbit coupling H_{LS} and use pure LS-coupling. The valence electrons of heavier atoms have a stronger spin–orbit coupling and one has to include the spin–orbit coupling into the description of the ground state. In the case of a single core hole, not interacting with any valence electrons, the core hole spin–orbit coupling (H_{LS_c}) is often larger than the electrostatic interactions and pure jj-coupling can be used. However, in many cases, it is necessary to use intermediate coupling. For example, in the case of a core hole interacting with valence holes, one usually finds that $H_{LS_c} > H'_{EE} > H_{LS'}$ and intermediate coupling is necessary.

Why is the coupling scheme important? An important answer is obtained when looking at the commutation relation of the respective Hamiltonian terms. For the electrostatic interactions, one finds that it commutes with the orbital angular momentum, the spin angular momentum, and the total angular momentum. This implies that the electrostatic interactions are diagonal in L and S (thus also in J). In contrast, the spin–orbit coupling commutes neither with L or S, only with J. This implies that if spin–orbit coupling is important, the overall Hamiltonian is only diagonal in J. Because this is the basic situation in core level spectroscopy, in most sections intermediate coupling is used as the basic rule and LS-coupling and jj-coupling is used only to give simpler examples.

4.2.2 Some Simple Coupling Schemes

In the case of a TM ion, the important configuration for the initial state of the absorption process is $3d^n$. In the final state with a 3s or a 3p core hole, the configurations are $3s^13d^{n+1}$ and $3p^53d^{n+1}$. As the principal quantum number has no influence on the coupling scheme, the same term symbols can be found for 4d and 5d systems or for 2p and 3p core holes.

In case of a $2p^2$ configuration, the first electron has six quantum states available, the second electron only five. This is because of the Pauli exclusion principle that forbids two electrons to have the same quantum numbers n_p, m_l, and m_s. Because the sequence of the two electrons is not important, dividing the number of combinations by two gives 15 possible combinations.

A 2p electron has quantum numbers $l = 1$ and $s = 1/2$. This gives the six individual combinations with $m_l = +1$, 0, or −1 and $m_s = +1/2$ or −1/2. We will use a shorthand notation and write $|1, +\rangle$ to indicate the quantum numbers. One can create a two-electron state by adding two of these $|m_{la}, m_{sa}\rangle$ combinations, for example, $|1, +\rangle + |1, -\rangle$. This yields a state with $|M_L, M_S\rangle$ quantum numbers equal to $|2, 0\rangle$. The 15 combinations of adding two 2p electrons are indicated in Table 4.3.

In Table 4.4, it can be seen that there are three states with $|M_L, M_S\rangle = |0, 0\rangle$, two states $|M_L, M_S\rangle = |1, 0\rangle$ and $|-1, 0\rangle$ and a number of other states. Working out the

TABLE 4.3
Fifteen Combinations of States $|m_{la}, m_{sa}\rangle$ and $|m_{lb}, m_{sb}\rangle$ of a $2p^2$ Configuration

$\lvert m_{la}, m_{sa}\rangle$	$\lvert m_{lb}, m_{sb}\rangle$	$\lvert M_L, M_S\rangle$	#	$\lvert m_{la}, m_{sa}\rangle$	$\lvert m_{lb}, m_{sb}\rangle$	$\lvert M_L, M_S\rangle$	#
$\lvert 1, +\rangle$	$\lvert 1, -\rangle$	$\lvert 2, 0\rangle$	1	$\lvert 1, -\rangle$	$\lvert -1, -\rangle$	$\lvert 0, -1\rangle$	1
$\lvert 1, +\rangle$	$\lvert 0, +\rangle$	$\lvert 1, 1\rangle$	1	$\lvert 0, +\rangle$	$\lvert 0, -\rangle$	$\lvert 0, 0\rangle$	3
$\lvert 1, +\rangle$	$\lvert 0, -\rangle$	$\lvert 1, 0\rangle$	1	$\lvert 0, +\rangle$	$\lvert -1, +\rangle$	$\lvert -1, 1\rangle$	1
$\lvert 1, +\rangle$	$\lvert -1, +\rangle$	$\lvert 0, 1\rangle$	1	$\lvert 0, +\rangle$	$\lvert -1, -\rangle$	$\lvert -1, 0\rangle$	1
$\lvert 1, +\rangle$	$\lvert -1, -\rangle$	$\lvert 0, 0\rangle$	1	$\lvert 0, -\rangle$	$\lvert -1, +\rangle$	$\lvert -1, 0\rangle$	2
$\lvert 1, -\rangle$	$\lvert 0, +\rangle$	$\lvert 1, 0\rangle$	2	$\lvert 0, -\rangle$	$\lvert -1, -\rangle$	$\lvert -1, -1\rangle$	1
$\lvert 1, -\rangle$	$\lvert 0, -\rangle$	$\lvert 1, -1\rangle$	1	$\lvert -1, +\rangle$	$\lvert -1, -\rangle$	$\lvert -2, 0\rangle$	1
$\lvert 1, -\rangle$	$\lvert -1, +\rangle$	$\lvert 0, 0\rangle$	2				

Note: The # column counts the degeneracy of the total symmetry $|M_L, M_S\rangle$ states. These 15 $|M_L, M_S\rangle$ states can be put into a table collecting their overall M_L and M_S quantum numbers. This yields the result as in Table 4.4.

symmetry properties of these states, a number of so-called irreducible representations (irrep) will be found. An irrep defines a single configuration with a defined L and S value. The energies of all the $|M_L, M_S\rangle$ states within an $|L, S\rangle$ irrep are the same. The rules on quantum numbers as outlined previously also apply for irreps. This gives a lead to derive the irreps directly from the number of $|M_L, M_S\rangle$ states as given in the table. The presence of a $|2, 0\rangle$ implies that this state is part of an irrep with L equal to, at least, 2. An irrep with $L = 2$ has five states with M_L values

TABLE 4.4
(Top) Number of States with a $|M_L, M_S\rangle$ Combination; (Bottom) All the Possible Term Symbols of a $3p^2$ Configuration

$2p^2$	1	$M_S = 0$	-1
$M_L = 2$	0	1	0
$M_L = 1$	1	2	1
$M_L = 0$	1	3	1
$M_L = -1$	1	2	1
$M_L = -2$	0	1	0
$M_L = 2$		^1D	
$M_L = 1$	^3P	^1D ^3P	^3P
$M_L = 0$	^3P	^1D ^3P ^1S	^3P
$M_L = -1$	^3P	^1D ^3P	^3P
$M_L = -2$		^1D	

between -2 and $+2$. The only associated M_S value is $M_S = 0$, which implies that there is an irrep with $|L, S\rangle = |2, 0\rangle$, which is a ^1D term symbol.

Removing these five states from the table will then leave ten states containing $M_L = \pm 1$ and $M_S = \pm 1$. The next term symbol that is found has $|L, S\rangle = |1, 1\rangle$, which is a ^3P term symbol. A ^3P term symbol has nine states and we are left with one additional state with $M_S = M_L = 0$. This state belongs to an $|L, S\rangle = |0, 0\rangle$ term symbol, or ^1S. We have found that the $2p^2$ configuration contains the terms ^3P, ^1D, and ^1S, with the respective degeneracies of $3 \times 3 = 9$, $1 \times 5 = 5$ and $1 \times 1 = 1$. Thus, it can be checked that total degeneracy adds up to 15. Including J in the discussion, we have the values $^1D_2$, $^1S_0$, and $^3P_2$ plus $^3P_1$ plus $^3P_0$. Focusing on the J-values, we have two $J = 0$, one $J = 1$, and two $J = 2$ values. Because x-ray absorption calculations are carried out in the intermediate coupling, the J-value is important; the total calculation is split into its various J-values.

The term symbols of a 2p3p configuration do not have to obey the Pauli principle. There will be 6×6 combinations of terms. The term symbols can be determined directly by multiplying the individual term symbols. Multiplication of terms A and B is written as $A \otimes B$. Since both L and S are vectors, the resulting terms have possible values of $|L_A - L_B| \le L \le L_A + L_B$ and $|S_A - S_B| \le S \le S_A + S_B$. For $^2P \otimes {}^2P$, this gives $L = 0, 1, 2$ and $S = 0$ or 1. This gives the term symbols ^1S, ^1P, ^1D and ^3S, ^3P, ^3D. The respective degeneracies are 1, 3, 5 and 3, 9, 15 adding up to 36, as indicated in Table 4.5. Adding the J-values, for the singlet states just a single J is found (i.e. $^1S_0$,

---

TABLE 4.5
Configurations of s and p Electrons

Configurations	J	Term Symbols	Degeneracy	$\Sigma(2J + 1)$
$1s^0$	0	$^1S_0$	1	1
$1s^1$	1/2	$^2S_{1/2}$	1	2
$1s^1 2s^1$	0	$^1S_0$	1	4
	1	$^3S_1$	1	
$2p^1 = 2p^5$	½	$^2P_{1/2}$	1	6
	3/2	$^2P_{3/2}$	1	
$2p^2 = 2p^4$	0	$^1S_0 \, ^3P_0$	2	
	1	$^3P_1$	1	15
	2	$^1D_2 \, ^3P_2$	2	
$2p^3$	½	$^2P_{1/2}$	1	
	3/2	$^4S_{3/2} \, ^2P_{3/2} \, ^2D_{3/2}$	3	20
	5/2	$^2D_{5/2}$	1	
$2p^1 3p^1$	0	$^1S_0 \, ^3P_0$	2	
	1	$^1P_1 \, ^3S_1 \, ^3P_1 \, ^3D_1$	4	36
	2	$^1D_2 \, ^3P_2 \, ^3D_2$	3	
	3	$^3D_3$	1	

Note: The term symbols are sorted for their J value. The third column gives the number of term symbols per J value. The last column gives the overall degeneracy of the configuration.

$^1P_1$, $^1D_2$). The triplet states each form three J-term symbols $^3P_2$ plus $^3P_1$ plus $^3P_0$ and $^3D_3$ plus $^3D_2$ plus $^3D_1$. The ^3S state has $L = 0$, hence also only one J state as $^3S_1$. We find two term symbols with $J = 0$ ($^1S_0$ and $^3P_0$), four with $J = 1$, three with $J = 2$, and one with $J = 3$. The degeneracies can be checked by adding the $2J + 1$ values, yielding $(2 \times 1) + (4 \times 3) + (3 \times 5) + (1 \times 7) = 36$.

4.2.3 TERM SYMBOLS OF d-ELECTRONS

The LS term symbols for a $3d^1 4d^1$ configuration can be similarly found by multiplying the term symbols for the configurations $3d^1$ and $4d^1$. For $^2D \otimes ^2D$, this gives $L = 0$, 1, 2, 3, or 4 and $S = 0$ or 1. The 10 LS term symbols of the $3d^1 4d^1$ configuration are ^1S, ^1P, ^1D, ^1F, ^1G plus ^3S, ^3P, ^3D, ^3F, ^3G. The total degeneracy of the $3d^1 4d^1$ configuration is 100. In the presence of the spin–orbit coupling, a total of 18 symbols are found.

Due to the Pauli exclusion principle, a $3d^2$ configuration does not have the same degeneracy as the $3d^1 4d^1$ configuration. In total, there are $10 \times 9/2 = 45$ possible states. Following the same procedure as for the $2p^2$ configuration, one can write out all 45 combinations of a $3d^2$ configuration and sort them by their M_L and M_S quantum numbers, as indicated in Table 4.6.

TABLE 4.6
(Top) Number of States with a $|M_L, M_S\rangle$ Combination; (Bottom) All the Possible Term Symbols of a $3d^2$ Configuration

M_L	$M_S = 1$	$M_S = 0$	$M_S = -1$
4	0	1	0
3	1	2	1
2	1	3	1
1	2	4	2
0	2	5	2
−1	2	4	2
−2	1	3	1
−3	1	2	1
−4	0	1	0

M_L	$M_S = 1$	$M_S = 0$	$M_S = -1$
4		^1G	
3	^3F	^1G ^3F	^3F
2	^3F	^1G ^3F ^1D	^3F
1	^3F ^3P	^1G ^3F ^1D ^3P	^3F ^3P
0	^3F ^3P	^1G ^3F ^1D ^3P ^1S	^3F ^3P
−1	^3F ^3P	^1G ^3F ^1D ^3P	^3F ^3P
−2	^3F	^1G ^3F ^1D	^3F
−3	^3F	^1G ^3F	^3F
−4		^1G	

Analysis of the combinations of the allowed M_L and M_S quantum numbers yields the term symbols ^1G, ^3F, ^1D, ^3P, and ^1S. This is a subset of the term symbols of a $3d^1 4d^1$ configuration. The term symbols can be divided into their J-quantum numbers as $^3F_2$, $^3F_3$, $^3F_4$, $^3P_0$, $^3P_1$, $^3P_2$, $^1G_4$, $^1D_2$, and $^1S_0$.

In the case of a $3d^3$ configuration, a similar approach shows that the possible spin-states are doublet and quartet. By adding the degeneracies, it can be checked that a $3d^3$ configuration has 120 different states, that is, $10 \times 9/2 \times 8/3$. The general formula to determine the degeneracy of a $3d^n$ configuration is:

$$\binom{10}{n} = \frac{10!}{(10-n)!n!}. \tag{4.9}$$

One can show that the term symbols of a configuration $3d^n$ do also exist in a configuration $3d^{n+2}$, for $n + 2 \leq 5$. Thus the term symbols of $3d^4$ contain all term symbols of $3d^2$ that contain the ^1S term symbol of $3d^0$. Similarly, the term symbols of $3d^5$ contain all term symbols of $3d^3$ that contain the ^2D term symbol of $3d^1$. In addition, there is a symmetry equivalence of holes and electrons; hence $3d^3$ and $3d^7$ have exactly the same term symbols.

A new result with respect to the s and p electrons is that two states with an identical term symbol are found for a $3d^3$ configuration. To distinguish both term symbols, the seniority number is introduced. The seniority number is the number n of the d^n configuration for which a term symbol occurs first. For example, as the ^2D term symbol occurs for a $3d^1$ configuration, it has seniority number 1. The term symbol could be rewritten as ^2_1D. The second ^2D term symbol of $3d^3$ takes its seniority number from the next lowest number n where this term symbol occurs and can be written as ^2_3D. If only one term symbol occurs, its seniority number will not be included in its notation.

The configurations of the $2p^5 3d^n$ final states are important for the 2p x-ray absorption edge. The term symbols of the $2p^5 3d^n$ states are found by multiplying the configurations of $3d^n$ with a ^2P term symbol. For example, $^2P \otimes {}^3P$ yields $^2S + {}^2P + {}^2D + {}^4S + {}^4P + {}^4D$. Referring to Tables 4.7 and 4.8, the last two columns give the number of term symbols for each J value and the degeneracy. Tables 4.7 and 4.8 with J-value degeneracies are also important for crystal field effects. The total degeneracy of a $2p^5 3d^n$ state is given in Equation 4.10

$$6 \times \binom{10}{n} = 6 \times \frac{10!}{(10-n)!n!}. \tag{4.10}$$

For example, a $2p^5 3d^5$ configuration has 1512 possible states. Analysis shows that these 1512 states are divided into 205 term symbols, implying in principle 205 possible final states. Whether or not all these final states have finite intensity depends on the selection rules. This will be further discussed in Chapter 6. Tables 4.7 and 4.8 are important to determine the number of allowed transitions in a 2p x-ray absorption spectrum. For example, a $3d^0$ to $2p^5 3d^1$ transition contains three peaks in its atomic spectral shapes. This result can be inferred from the dipole selection rules, that is, $\Delta J = -1, 0$, or $+1$ with the exception that $\Delta J \neq 0$ if $J = J' = 0$. From Table 4.7, it can be

TABLE 4.7
Symmetries of the 3d^n Systems

Configuration	J	Term Symbols	Degeneracy
3d^0	0	^{1}S	1
$\Sigma = 1$			$\Sigma = 1$
3d^2	0	^{1}S ^{3}P	2
3d^8	1	^{3}P	1
	2	^{1}D ^{3}P ^{3}F	3
	3	^{3}F	1
	4	^{1}G ^{3}F	2
$\Sigma = 45$			$\Sigma = 9$
3d^4	0	^{1}S ^{1}S ^{3}P ^{3}P ^{5}D	5
3d^6	1	^{3}P ^{3}P ^{3}D ^{5}D	4
	2	^{1}D ^{1}D ^{3}P ^{3}P ^{3}D ^{3}F ^{3}F ^{5}D	8
	3	^{1}F ^{3}D ^{3}F ^{3}F ^{3}G ^{5}D	6
	4	^{1}G ^{1}G ^{3}F ^{3}F ^{3}G ^{3}H ^{5}D	7
	5	^{3}G ^{3}H	2
	6	^{1}I ^{3}H	2
$\Sigma = 210$			$\Sigma = 34$
3d^1	3/2	^{2}D	1
3d^9	5/2	^{2}D	1
$\Sigma = 10$			$\Sigma = 2$
3d^3	1/2	^{2}P ^{4}P	2
3d^7	3/2	^{2}P ^{2}D ^{2}D ^{4}P ^{4}F	5
	5/2	^{2}D ^{2}D ^{2}F ^{4}P ^{4}F	5
	7/2	^{2}F ^{2}G ^{4}F	3
	9/2	^{2}G ^{2}H ^{4}F	3
	11/2	^{2}H	1
$\Sigma = 120$			$\Sigma = 19$
3d^5	1/2	^{2}S ^{2}P ^{4}P ^{4}D	4
	3/2	^{2}P ^{2}D ^{2}D ^{2}D ^{4}P ^{4}D ^{4}F	7
	5/2	^{2}D ^{2}D ^{2}D ^{2}F ^{2}F ^{4}P ^{4}D ^{4}F ^{4}G	10
	7/2	^{6}S	7
	9/2	^{2}F ^{2}F ^{2}G ^{2}G ^{4}D ^{4}F ^{4}G	5
	11/2	^{2}G ^{2}G ^{2}H ^{4}F ^{4}G	3
	13/2	^{2}H ^{2}I ^{4}G	1
$\Sigma = 252$		^{2}I	$\Sigma = 37$

Note: For each number of electrons, all term symbols are given, selected by their J value. The number of representations per J value is given, with the total number of representations in the bottom line. The total number of states is given in the first column.

found that 3d^0 has a ground state with $J = 0$ and from Table 4.8, it can be found that within the 2p^{5}3d^1 final state, there are three states with $J' = 1$. In the case of a 3d^6 ground state, Table 4.7 gives the ^{5}D$_4$ ground state and from Table 4.8, it can be found that within the 2p^{5}3d^7 final state, there are 24 states with $J' = 3$, 18 with $J' = 4$, and 11 with $J' = 5$. This implies a total of 53 peaks for the atomic 3d^6 to 2p^{5}3d^7 transition.

TABLE 4.8
Symmetries of the $2p^53d^n$ Systems

Configuration	J	Term Symbols	Degeneracy
$2p^53d^{10}$	1/2	^2P	1
	3/2	^2P	1
$\Sigma = 6$			$\Sigma = 2$
$2p^53d^2$	1/2	$^2S\ ^2P3\ ^4P\ ^4D2^*$	7
$(2p^53d^8)$	3/2	$^2P3\ ^2D3\ ^4S\ ^4P\ ^4D2\ ^4F$	11
	5/2	$^2D3\ ^2F3\ ^4P\ ^4D2\ ^4F\ ^4G$	11
	7/2	$^2F3\ ^2G2\ ^4D2\ ^4F\ ^4G$	9
	9/2	$^2G2\ ^2H\ ^4F\ ^4G$	5
	11/2	$^2H\ ^4G$	2
$\Sigma = 45$			$\Sigma = 45$
$2p^53d^4$	1/2	$^2S2\ ^2P7\ ^4P4\ ^4D6\ ^6D\ ^6F$	21
$(2p^53d^6)$	3/2	$^2P7\ ^2D8\ ^4S2\ ^4P4\ ^4D6\ ^4F5\ ^6P\ ^6D\ ^6F$	35
	5/2	$^2D8\ ^2F9\ ^4P4\ ^4D6\ ^4F5\ ^4G4\ ^6P\ ^6D\ ^6F$	39
	7/2	$^2F9\ ^2G7\ ^4D6\ ^4F5\ ^4G4\ ^4H2\ ^6P\ ^6D\ ^6F$	36
	9/2	$^2G7\ ^2H5\ ^4F5\ ^4G4\ ^4H2\ ^4I\ ^6D\ ^6F$	26
	11/2	$^2H5\ ^2I2\ ^4G4\ ^4H2\ ^4I\ ^6F$	15
	13/2	$^2I2\ ^2K\ ^4H2\ ^4I$	6
	15/2	$^2K\ ^4I$	2
$\Sigma = 1260$			$\Sigma = 180$
$2p^53d^1$	0	^3P	1
$(2p^53d^9)$	1	$^1P\ ^3P\ ^3D$	3
	2	$^1D\ ^3P\ ^3D\ ^3F$	4
	3	$^1F\ ^3D\ ^3F$	3
	4	^3F	1
$\Sigma = 60$			$\Sigma = 12$
$2p^53d^3$	0	$^1S\ ^3P4\ ^5D2$	7
$(2p^53d^7)$	1	$^1P3\ ^3S2\ ^3P4\ ^3D6\ ^5P\ ^5D2\ ^5F$	19
	2	$^1D4\ ^3P4\ ^3D6\ ^3F5\ ^5S\ ^5P\ ^5D2\ ^5F\ ^5G$	25
	3	$^1F4\ ^3D6\ ^3F5\ ^3G4\ ^5P\ ^5D2\ ^5F\ ^5G$	24
	4	$^1G3\ ^3F5\ ^3G4\ ^3H2\ ^5D2\ ^5F\ ^5G$	18
	5	$^1H2\ ^3G4\ ^3H2\ ^3I\ ^5F\ ^5G$	11
	6	$^1I\ ^3H2\ ^3I\ ^5G$	5
	7	^3I	1
$\Sigma = 270$			$\Sigma = 110$
$2p^53d^5$	0	$^1S\ ^3P7\ ^5D3$	11
	1	$^1P5\ ^3S2\ ^3P7\ ^3D9\ ^5P3\ ^5D3\ ^5F3$	32
	2	$^1D6\ ^3P7\ ^3D9\ ^3F10\ ^5S\ ^5P3\ ^5D3\ ^5F3\ ^5G2\ ^7P$	45
	3	$^1F7\ ^3D9\ ^3F10\ ^3G7\ ^5P3\ ^5D3\ ^5F3\ ^5G2\ ^5H\ ^7P$	46
	4	$^1G5\ ^3F10\ ^3G7\ ^3H5\ ^5D3\ ^5F3\ ^5G2\ ^5H\ ^7P$	37
	5	$^1H4\ ^3G7\ ^3H5\ ^3I2\ ^5F3\ ^5G2\ ^5H$	24

Continued

TABLE 4.8 (continued)
Symmetries of the 2p⁵3dⁿ Systems

Configuration	J	Term Symbols	Degeneracy
	6	¹I2 ³H5 ³I2 ³K ⁵G2 ⁵H	13
	7	¹K ³I2 ³K ⁵H	5
	8	³K	1
Σ = 1512			Σ = 205

Note: For each number of electrons, all term symbols are given, selected by their *J* value. The last column gives the number of representations per *J* value and the total number of representations and the total number of states.

*Notations such as ⁴D2 should be read as two ⁴D configurations.

4.2.4 MATRIX ELEMENTS

Previously, the number and symmetry of the states of a certain $3d^n$ configuration were found. The next task is to calculate the matrix elements of these states with the Hamiltonian H_{ATOM} given by Equation 4.7. As discussed previously, H_{ATOM} includes the electron–electron interaction H_{ee} and the spin–orbit coupling H_{ls}:

$$H_{ee} + H_{ls} = \sum_{pairs} \frac{e^2}{r_{ij}} + \sum_N \zeta(\mathbf{r}_i)\, l_i \cdot s_i. \tag{4.11}$$

We first discuss the matrix elements of the electron–electron interaction. Because this Hamiltonian commutes with L^2, S^2, L_z, and S_z, the off-diagonal elements are all zero. A simple example is a 1s2s configuration consisting of ^1S and ^3S terms. The respective energies can be shown as:

$$\langle {}^1S \,|\, \frac{e^2}{r_{12}} \,|\, {}^1S \rangle = F^0(1s2s) + G^0(1s2s), \tag{4.12}$$

$$\langle {}^3S \,|\, \frac{e^2}{r_{12}} \,|\, {}^3S \rangle = F^0(1s2s) - G^0(1s2s). \tag{4.13}$$

Note that the triplet state is threefold degenerate and the average energy of the 1s2s configuration equals $F^0(1s2s) - 1/2\, G^0(1s2s)$. F^0 and G^0 are the Slater–Condon parameters (or Slater parameters) for the direct Coulomb repulsion and the Coulomb exchange interaction, respectively. The main result can be stated as "the singlet and the triplet state are split by the exchange interaction." This energy difference is $2\, G^0(1s2s)$. An analogous result is found for a 1s2p state for which the singlet and triplet states are split by $2/3\, G^0(1s2p)$. The prefactor is determined by the degeneracy

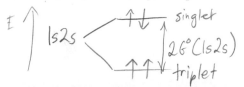

of the 2p state. The general formulation of the matrix elements of two-electron wave functions can be written as:

$$\langle ^{2S+1}L_J \,|\frac{e^2}{\mathbf{r}_{12}}|\, ^{2S+1}L_J \rangle = \sum_k f_k F^k + \sum_k g_k G^k. \tag{4.14}$$

To obtain this result, the radial parts F^k and G^k have been separated using the Wigner–Eckhart theorem and Hamiltonian $1/\mathbf{r}_{12}$ has been expanded. The angular parts f_k and g_k can be calculated using angular momentum coupling and the result is conveniently written in terms of $3j$- and $6J$-symbols:

$$f_k = (2l_1 + 1)(2l_2 + 1)(-1)^L \begin{pmatrix} l_1 & k & l_1 \\ 0 & 0 & 0 \end{pmatrix}\begin{pmatrix} l_2 & k & l_2 \\ 0 & 0 & 0 \end{pmatrix}\begin{Bmatrix} l_1 & l_2 & L \\ l_2 & l_1 & k \end{Bmatrix}, \tag{4.15}$$

$$g_k = (2l_1 + 1)(2l_2 + 1)(-1)^S \begin{pmatrix} l_1 & k & l_2 \\ 0 & 0 & 0 \end{pmatrix}\begin{pmatrix} l_1 & k & l_2 \\ 0 & 0 & 0 \end{pmatrix}\begin{Bmatrix} l_1 & l_2 & L \\ l_1 & l_2 & k \end{Bmatrix}. \tag{4.16}$$

We have chosen to write f_k and g_k in a similar way. The two $3j$-symbols for g_k are the same and could of course be rewritten as a single $3j$-symbol squared. For equivalent electrons, g_k is not present and f_k can be simplified by setting $l = l_1 = l_2$. This equation will be used to determine the energies for some examples. The values of k can be determined from the triangle conditions of the $3j$-symbols. The triangle conditions state that for $3j$-symbols with all zeros for m_j, the sum of the j-values must be even and that the maximum j-value is equal to the sum of the two others. Using the two $3j$-symbols of Equation 4.15, this implies for f_k that k must always be even. $k = 0$ is always present and the maximum value for k equals two times the lowest value of l; otherwise, one of the two $3j$-symbols equals zero. For g_k, it implies that k is even if $l_1 + l_2$ is even and k is odd if $l_1 + l_2$ is odd. The maximum value of k equals $l_1 + l_2$. Table 4.9 gives the possible k values for the simplest configurations.

In the case of the 1s2s configuration ($l_1 = l_2 = 0$), the electrons are in different orbitals and thus G_k is not equal to F_k. Using Equation 4.15, it is obtained for the ^1S state that both f_k and g_k are equal to one. Their $3j$-symbol and the $6J$-symbol consist of six zeros and are equal to one. In the case of the ^3S state, f_0 is again equal to 1. However, because S is equal to 1, g_0 is equal to –1. This reproduces the results as given above. The result that f_0 equals one is a general result for all two-electron states. It can be shown that f_0 is equal to the number of permutations $[n(n-1)/2]$ of n electrons.

TABLE 4.9
Possible Values of k for the Angular Coefficients f_k and g_k

Configuration	f_k	Configuration	f_k	g_k	Configuration	f_k	g_k
1s^2	0	1s2s	0	0	1s2p	0	1
2p^2	0 2	2p3p	0 2	0 2	2p3d	0 2	1 3
3d^2	0 2 4	3d4d	0 2 4	0 2 4	3d4f	0 2 4	1 3 5
4f^2	0 2 4 6						

A second example is the energy of the ^1S state of $2p^2$. Because the two electrons are positioned in the same shell, there are no exchange terms. We now have $l_1 = l_2 = 1$ and

$$f_2(^1S) = (3)(3) \begin{pmatrix} 1 & 2 & 1 \\ 0 & 0 & 0 \end{pmatrix} \begin{pmatrix} 1 & 2 & 1 \\ 0 & 0 & 0 \end{pmatrix} \begin{Bmatrix} 1 & 1 & 0 \\ 1 & 1 & 2 \end{Bmatrix} = 9 \cdot \frac{2}{15} \cdot \frac{1}{3} = \frac{2}{5}. \quad (4.17)$$

With f_0 again equal to one, the energy is equal to $F^0 + 2/5\, F^2$.

4.2.5 ENERGY LEVELS OF TWO d-ELECTRONS

The energies of the representations of the $3d^2$ configuration are found from the calculation of f_2 and f_4 for the five term symbols ^1S, ^3P, ^1D, ^3F, and ^1G. We have $l_1 = l_2 = 2$, which implies that the prefactor $(2l_1 + 1)(2l_2 + 1)$ is equal to 25. The $3j$-symbols are only dependent on l (equal to 2) and k (equal to 2 and 4). Both $3j$-symbols squared are equal to $2/35$. For all five states this gives a prefactor of $(25 \times 2)/35$ and f_k is equal to:

$$f_k = \frac{10}{7} - 1^L \begin{Bmatrix} 2 & 2 & L \\ 2 & 2 & k \end{Bmatrix}. \quad (4.18)$$

To a very good degree of approximation, the Slater–Condon parameters F^2 and F^4 have a constant ratio: $F^4 = 0.62F^2$. The last column gives the approximate energies of the five term symbols as listed in Table 4.10. In the case of the 3d TM ions, F^2 is approximately equal to 10 eV. For the five term symbols, the energies are, respectively: ^3F at -1.8 eV, ^1D at -0.1 eV, ^3P at $+0.2$ eV, ^1G at $+0.8$ eV, and ^1S at $+4.6$ eV. The ^3F-term symbol has the lowest energy and is the ground state of a $3d^2$ system. This

TABLE 4.10
Energies of the Five Term Symbols of a $3d^2$ Configuration

	f_2		f_4		Energy
^1S	$\dfrac{10}{7} \begin{Bmatrix} 2 & 2 & 0 \\ 2 & 2 & 2 \end{Bmatrix}$	2/7	$\dfrac{10}{7} \begin{Bmatrix} 2 & 2 & 0 \\ 2 & 2 & 4 \end{Bmatrix}$	2/7	$0.46F^2$
^3P	$-\dfrac{10}{7} \begin{Bmatrix} 2 & 2 & 1 \\ 2 & 2 & 2 \end{Bmatrix}$	3/21	$-\dfrac{10}{7} \begin{Bmatrix} 2 & 2 & 1 \\ 2 & 2 & 4 \end{Bmatrix}$	$-4/21$	$0.02F^2$
^1D	$\dfrac{10}{7} \begin{Bmatrix} 2 & 2 & 2 \\ 2 & 2 & 2 \end{Bmatrix}$	$-3/49$	$\dfrac{10}{7} \begin{Bmatrix} 2 & 2 & 2 \\ 2 & 2 & 4 \end{Bmatrix}$	4/49	$-0.01F^2$
^3F	$-\dfrac{10}{7} \begin{Bmatrix} 2 & 2 & 3 \\ 2 & 2 & 2 \end{Bmatrix}$	$-8/49$	$-\dfrac{10}{7} \begin{Bmatrix} 2 & 2 & 3 \\ 2 & 2 & 4 \end{Bmatrix}$	$-1/49$	$-0.18F^2$
^1G	$\dfrac{10}{7} \begin{Bmatrix} 2 & 2 & 4 \\ 2 & 2 & 2 \end{Bmatrix}$	4/49	$\dfrac{10}{7} \begin{Bmatrix} 2 & 2 & 4 \\ 2 & 2 & 4 \end{Bmatrix}$	1/441	$0.08F^2$

Note: The energy in the last column is calculated using the fact that the radial integrals F^2 and F^4 have a constant ratio of 0.62.

is in agreement with Hund's rules, which will be discussed in the next section. The three states, ^1D, ^3P, and ^1G, are close in energy at some 1.7 to 2.5 eV above the ground state. The ^1S state has a very high energy of 6.3 eV above the ground state—the reason being that two electrons in the same orbit strongly repel each other.

Three closely related notations are used to indicate the radial integrals: (*i*) the Slater–Condon parameters F^k, (*ii*) the normalized Slater–Condon parameters F_k, and (*iii*) the Racah parameters A, B, and C. The Slater–Condon parameters and Racah parameters are related through the relationships as indicated in Table 4.11. The bottom half of the table uses the relationship between F^2 and F^4 and it further uses a typical F^2 value of 10 eV and a F^0 value of 8 eV.

4.2.6 MORE THAN TWO ELECTRONS

For three or more electrons, the situation is considerably more complex. It is not straightforward to write down an antisymmetrized three-electron wave function. It can be shown that the three-electron wave function can be built from two-electron wave functions with the use of the so-called coefficients of fractional parentage. The coefficients of fractional parentage are indicated with $C_{L_1S_1}^{LS}$. The three-electron wave function with quantum numbers LS is generated from a series of two-electron wave functions with quantum numbers L_1S_1 by

$$|d^n[LS]\rangle = \sum_{L_1S_1} C_{L_1S_1}^{LS} |d^{n-1}[L_1S_1]d'\rangle. \tag{4.19}$$

In some cases, LS-degeneracies do occur and a seniority number must be added to the summation. An example of the coefficients of fractional parentage is given for the two quartet states of $3d^3$. They can be formed only from the triplet states of $3d^2$:

$$|d^3[^4P]\rangle = -\sqrt{\frac{8}{15}} |d^2[^3P]d'\rangle - \sqrt{\frac{7}{15}} |d^2[^3F]d'\rangle, \tag{4.20}$$

TABLE 4.11
Relations between the Slater Condon Parameters and Racah Parameters

Slater–Condon	Normalized	Racah
F^0	$F_0 = F^0$	$A = F_0 - 49F_4$
F^2	$F_2 = F^2/49$	$B = F_2 - 5F_4$
F^4	$F_4 = F^4/441$	$C = 35F_4$
$F^0 = 8.0$	$F_0 = 8.0$	$A = 7.3$
$F^2 = 10.0$	$F_2 = 0.41$	$B = 0.13$
$F^4 = 6.2$	$F_4 = 0.014$	$C = 0.49$

Note: The top half gives the relationships between the three
parameter sets. The bottom half uses $F^0 = 8.0$ eV,
$F^2 = 10.0$ eV, and $F^4 = 6.2$ eV.

$$|d^3[^4F]\rangle = -\sqrt{\frac{3}{15}}\ |d^2[^3P]d'\rangle + \sqrt{\frac{12}{15}}\ |d^2[^3F]d'\rangle. \qquad (4.21)$$

Tables with all of the coefficients of fractional parentage for the d^n states are presented in Cowan (1981). Previously, we have seen that in the case of two electrons in the same shell, the angular parameter f_k can be calculated as:

$$f_k = (2l+1)^2 \cdot (-1)^L \begin{pmatrix} l & k & l \\ 0 & 0 & 0 \end{pmatrix}^2 \begin{Bmatrix} l & l & L \\ l & l & k \end{Bmatrix}. \qquad (4.22)$$

For three or more electrons, the $6J$-symbol is replaced by normalized $U^{(k)}_{L'LS}$ matrix elements:

$$f_k = (2l+1)^2 \cdot (-1)^L \begin{pmatrix} l & k & l \\ 0 & 0 & 0 \end{pmatrix}^2 \left[\frac{1}{2L+1} \sum_{L'} \left(U^{(k)}_{L'LS} \right)^2 - \frac{n}{2l+1} \right], \qquad (4.23)$$

$$U^{(k)}_{L'LS} = \sum_{L_1 S_1} (-1)^{L_1 + L + k + l} \sqrt{(2L+1)(2L'+1)}\, C^{LS}_{L_1 S_1} C^{L'S}_{L_1 S_1} \begin{Bmatrix} L & K & L' \\ l & L_1 & l \end{Bmatrix}. \qquad (4.24)$$

Thus, the $U^{(k)}_{L'LS}$ matrix elements can be calculated as a summation over all coefficients of fractional parentage times the $6J$-symbol due to the recoupling of the LS and $L'S$ quantum numbers of the three electron states. The $U^{(k)}_{L'LS}$ matrix elements of the $3d^n$ configurations are also tabulated in Cowan (1981).

4.2.7 MATRIX ELEMENTS OF THE 2p³ CONFIGURATION

With the general formulas given previously, we will show the example of three-electron wave functions in the $2p^3$ configuration. The first task is again to determine the term symbols. When writing out the electron state combinations, it is found that only the $1^\uparrow 0^\uparrow 1^\uparrow$ combination has $M_S = 3/2$. Analysis of the M_l and M_s quantum numbers reveals that the term symbols are ^4S, ^2D, and ^2P. The next task is to find the energies of these three term symbols. As indicated previously, f_0 is equal to the number of permutations being three. The only other possible k-value is f_2. Using Equation 4.23 with $k = 2$ and $l = 1$, we obtain:

$$f_2 = \frac{3}{5} \left[\frac{1}{2L+1} \sum_{L'} \left(U^{(2)}_{L'LS} \right)^2 - 1 \right]. \qquad (4.25)$$

From the tabulated $U^{(k)}_{L'LS}$ values of $2p^3$, one finds that there is only one nonzero value, which is the $U^{(2)}_{1 2 \frac{1}{2}}$ matrix element coupling the ^2D and ^2P states. Its value is $\sqrt{3}$. This gives the energies of the term symbols as given in Table 4.12. It is found that the ^4S state has the lowest energy. The energy of the ^2D state is higher by $9/25\ F^2$ and the ^2P state has the highest energy, being $15/25\ F^2$ higher. The fact that the lowest energy state is the ^4S state is in agreement with the Hund's rules, which will be discussed in the next section.

TABLE 4.12
Values of f_0 and f_2

	f_0	f_2	Energy
^4S	$n(n-1)/2 = 3$	$3/5(0-1) = -3/5$	$3F^0 - 15/25\ F^2$
^2P	$n(n-1)/2 = 3$	$3/5(1-1) = 0$	$3F^0$
^2D	$n(n-1)/2 = 3$	$3/5(3/5-1) = -6/25$	$3F^0 - 6/25\ F^2$

4.2.8 HUND'S RULES

We now discuss the ground state symmetries of the TM compounds, which are characterized with a partly-filled 3d band. The term symbols with the lowest energy are found after calculating the matrix elements, following the rules as described previously. Finding the ^3F state as the ground state of a $3d^2$ configuration is an example of Hund's rules. On the basis of experimental information, Hund formulated three rules to determine the ground state of a $3d^n$ configuration (Hund, 1925, 1927). For $3d^n$ configurations, the rules are correct, as confirmed by the atomic multiplet calculations. The three rules are:

1. Term symbol with maximum S.
2. Term symbol with maximum L.
3. Term symbol with maximum J (if the shell is more than half-full).

The energy of a configuration is the lowest if the electrons are as far apart as possible. The first of Hund's rules, "maximum spin," can be understood from the Pauli principle: electrons with parallel spins must be in different orbitals, which implies larger separations and, hence, lower energies. This is, for example, evident for a $3d^5$ configuration where the ^6S state has its five electrons divided over the five spin-up orbitals, which minimizes their repulsion. In the case of $3d^2$, the first of Hund's rules implies that either the ^3P or the ^3F-term symbol must have lowest energy. The second of Hund's rules states that the ^3F-term symbol is lower than the ^3P-term symbol. Again the reason is that the ^3F wave function tends to minimize electron repulsion. In a ^3F configuration, the electrons are orbiting in the same direction. That implies that they can stay a larger distance apart on average since they could always be on the opposite side of the nucleus. For a ^3P configuration, some electrons must orbit in the opposite direction and therefore pass close to each other once per orbit. This leads to a smaller average separation of electrons and therefore a higher energy (Weissbluth, 1978). The effects of spin–orbit coupling are well known in the case of core states. A 2p core state has $^2P_{3/2}$ and $^2P_{1/2}$ states. The state with the lowest energy is $^2P_{3/2}$. The physical background of the third of Hund's rules is that the scalar product $S \cdot L$ is negative if the spin and orbital angular momentum are in opposite directions. Since the coefficient of $S \cdot L$ is positive, this implies that the lowest J gives the ground state for a single 2p electron. If the shell is more than half-full, the effect on the 2p hole is inverted. Consider, for example, the 2p XAS or XPS spectrum of nickel. The $^2P_{3/2}$ peak is positioned at approximately 850 eV and the $^2P_{1/2}$ at about 880 eV, and the $^2P_{1/2}$

state can decay to the $^2P_{3/2}$ state by the Coster–Kronig Auger process. Note that the state with the lowest binding energy is related to the lowest energy of the configuration. The case of the 2p core state is an example of Hund's third rule: the configuration $2p^5$, which is more than half-full, implies that the highest J-value has the lowest energy. The third rule implies that the ground state of a $3d^8$ configuration is $^3F_4$, while it is $^3F_2$ in case of a $3d^2$ configuration.

4.2.9 FINAL STATE EFFECTS OF ATOMIC MULTIPLETS

The 2p x-ray absorption process excites a 2p core electron into the empty 3d shell and the transition can be described as $2p^6 3d^0 \rightarrow 2p^5 3d^1$. The $2p^5 3d^1$ configuration contains two new terms in its Hamiltonian: the 2p spin–orbit coupling and 2p3d multiplet effects. The final state atomic energy matrix consists of terms related to the two-electron Slater integrals (H_{ELECTRO}) and the spin–orbit couplings of the 2p (H_{LS-2p}) and the 3d electrons (H_{LS-3d}):

$$H_{\text{eff}} = H_{\text{ELECTRO}} + H_{LS-2p} + H_{LS-3d},$$

$$H_{\text{ELECTRO}} = \langle 2p^5 3d^1 | \frac{e^2}{\mathbf{r}_{12}} | 2p^5 3d^1 \rangle, \tag{4.26}$$

$$H_{LS-2p} = \langle 2p | \varsigma_p l_p \cdot s_p | 2p \rangle,$$

$$H_{LS-3d} = \langle 3d | \varsigma_d l_d \cdot s_d | 3d \rangle.$$

To show the individual effects of these interactions, each will now be introduced separately. A series of five calculations are shown, in which:

(a) All final state interactions are set to zero: $H = 0$
(b) 2p spin–orbit coupling is included: $H = H_{LS-2p}$
(c) Slater–Condon parameters are included: $H = H_{\text{ELECTRO}}$
(d) 2p spin–orbit coupling and Slater–Condon parameters are included: $H = H_{\text{ELECTRO}} + H_{LS-2p}$
(e) 3d spin–orbit coupling is included: $H = H_{\text{ELECTRO}} + H_{LS-2p} + H_{LS-3d}$.

First, we analyze the symmetry properties of the calculation. The ground state has $^1S_0$ symmetry and the term symbols of the final state are $^2P \times {}^2D$, which gives the 12 states $^1P_1$, $^1D_2$, $^1F_3$, $^3P_{012}$, $^3D_{123}$, and $^3F_{234}$. The x-ray absorption transition matrix elements that need to be calculated are:

$$I_{\text{XAS}} \propto \langle 3d^0 | p | 2p^5 3d^1 \rangle^2. \tag{4.27}$$

The symmetry aspects are:

$$I_{\text{XAS}} \propto \langle [^1S_0] | [^1P_1] | [^{1,3}\text{PDF}] \rangle^2. \tag{4.28}$$

Using the full Hamiltonian, 12 states are obtained that are built from the 12 term symbols according to the interactions that have been included. The irreducible representations (i.e. the states with the same J-value), block out in the calculation, implying that all matrix terms that mix J-values are zero. The symmetry of the dipole transition is given as $^1P_1$, according to the dipole selection rules, which state that $\Delta J = +1, 0, -1$ but not $J' = J = 0$. Also, within LS coupling, $\Delta S = 0$ and $\Delta L = 1$. The dipole selection rule reduces the number of final states that can be reached from the ground state. The J-value in the ground state is zero. In this case, the dipole selection rule proclaims that the J-value in the final state must be one, thus only the three term symbols ($^1P_1$, $^3P_1$, and $^3D_1$) can obtain finite intensity. Only this 3×3 matrix needs to be considered in the calculation (de Groot et al., 1990a).

As shown in Table 4.13, we start by setting all final state interactions to zero. The energy levels are given subsequently. They are labeled from top to bottom (^3P, ^3D, and ^1P), indicating the approximate term symbol related to the state. The original term symbols (^3P, ^3D, and ^1P) are given in the first row, second row, and bottom row of the eigenvector matrix, respectively. The intensity of the states is indicated on the right. With all interactions at zero, the complete energy matrix is zero along with the 2p binding energy. The states are all pure states and because of the dipole selection rules, all the intensity goes to the $^1P_1$ state as indicated in Table 4.13.

Inclusion of the 2p spin–orbit coupling H_{LS-2p} of 3.776 eV creates nondiagonal elements in the energy matrix. In other words, the LS-character of the individual states is mixed. In this case, as only the 2p spin–orbit coupling is included, the result is rather simple with the triplet states at $-1/2\varsigma_p$ and the singlet state at $+3/2\varsigma_p$. The eigenvector matrix shows that the three states are mixtures of the three pure states (i.e. the first state is, in fact: $^3P = 0.5\ ^3P_1 - 0.866\ ^3D_1$). The intensities of the three states are directly given by the square of the percentage of the $^1P_1$ character. This gives the familiar result that the triplet states, or $2p_{3/2}$ states, have twice the intensity of the singlet, or $2p_{1/2}$ states (Table 4.14).

Next, only the pd Slater–Condon parameters are included, keeping the 2p spin–orbit coupling at zero. The Slater–Condon parameters are reduced to 80% of their atomic Hartree–Fock values and F^2, G^1, and G^3 are 5.042, 3.702, and 2.106 eV,

---

TABLE 4.13
Energy Matrix and Eigenvectors of the 3 × 3 Matrices of the $2p^53d^1$ Final States with $J = 1$

Energy Matrix	Eigenvectors
$\begin{vmatrix} 0 & 0 & 0 \\ 0 & 0 & 0 \\ 0 & 0 & 0 \end{vmatrix}$	$\begin{vmatrix} 1 & 0 & 0 \\ 0 & 1 & 0 \\ 0 & 0 & 1 \end{vmatrix}$

Energy Levels		Intensities
0.00	^3P	0.00
0.00	^3D	0.00
0.00	^1P	1.00

Note: The bottom half of the table gives the resulting energies and intensities. All final state interactions are set to zero.

TABLE 4.14
Energy Matrix and Eigenvectors of the
3 × 3 Matrices of the $2p^53d^1$ Final States
with $J = 1$, after Inclusion of the 2p
Spin–Orbit Coupling

Energy Matrix			Eigenvectors		
0.944	1.635	2.312	0.5	−0.5	−0.707
1.635	−0.944	1.335	−0.866	−0.288	−0.408
2.312	1.335	0.000	0.0	0.816	−0.577

Energy Levels		Intensities
−1.888	^3P	0.00
−1.888	^3D	0.666
+3.776	^1P	0.333

respectively. This gives the three states as −1.345, 0.671, and 3.591 eV, respectively. Only the $^1P_1$ state has a finite intensity and its energy is shifted to an energy of 3.591 eV above the centre of gravity. The two other states have zero intensity. It can be seen that the pd Slater–Condon parameters are diagonal in the LS-terms; therefore, the three states are pure in character (Table 4.15).

In part (b), the nondiagonal terms of the 2p spin–orbit coupling make all three states into mixtures of the individual term symbols. 2p spin–orbit coupling further creates the 2 : 1 intensity ratio, thereby shifting most of the ^1P character to a lower energy. In part (c), it was found that the Slater–Condon parameters shift the ^1P state to a higher energy and that the triplet states have a considerably lower energy. If one includes both the 2p spin–orbit coupling and the pd Slater–Condon parameters, the result will depend on their relative values. In the case of the 2p core hole of Ti^{4+}, the Slater–Condon parameters are relatively large and most intensity goes to the

TABLE 4.15
Energy Matrix and Eigenvectors of the
3 × 3 Matrices of the $2p^53d^1$ Final States
with $J = 1$, after Inclusion of the 2p3d
Slater–Condon Parameters

Energy Matrix			Eigenvectors		
−1.345	0	0	1	0	0
0	0.671	0	0	1	0
0	0	3.591	0	0	1

Energy Levels		Intensities
−1.345	^3P	0.00
+0.671	^3D	0.00
+3.591	^1P	1.00

TABLE 4.16

Energy Matrix and Eigenvectors of the 3 × 3 Matrices of the $2p^53d^1$ Final States with $J = 1$, after Inclusion of the 2p3d Slater–Condon Parameters and the 2p Spin–Orbit Coupling

Energy Matrix			Eigenvectors		
1.615	1.635	2.312	0.297	−0.776	0.557
1.635	−2.289	1.335	−0.951	−0.185	0.248
2.312	1.335	3.591	0.089	0.603	0.792

Energy Levels		Intensities
−2.925	^3P	0.008
+0.207	^3D	0.364
+5.634	^1P	0.628

$2p_{1/2}$ state. The triplet states are separated by 3 eV and the lowest ^3P energy state is extremely weak, gaining less than 1% of the total intensity (Table 4.16). In the next section, four similar spectra with different ratios of Slater–Condon parameters and core hole spin–orbit couplings are compared to show the variations in their spectral shapes.

For completeness, we also include the 3d spin–orbit coupling in the final calculation. Because the 3d spin–orbit coupling is only 32 meV, its influence on the spectral shape is negligible in the present case. The energy position of the ^1P state shifts by 40 meV and its intensity drops by 0.4% of the total intensity. The effects on the intensities and energies have been included as the extra number in Table 4.17. It should be noted

TABLE 4.17

Energy Matrix and Eigenvectors of the 3 × 3 Matrices of the $2p^53d^1$ Final States with $J = 1$, after Inclusion of the 2p3d Slater–Condon Parameters, the 2p Spin–Orbit Coupling and the 3d Spin–Orbit Coupling

Energy Matrix			Eigenvectors		
1.575	1.649	2.293	0.303	−0.774	0.555
1.649	−2.313	1.301	−0.949	−0.195	0.246
2.293	1.301	3.591	0.082	0.601	0.795

Energy Levels		Intensities
−2.954 − 0.029	^3P	0.007 − 0.001
+0.212 + 0.005	^3D	0.361 − 0.003
+5.594 − 0.040	^1P	0.632 + 0.004

Note: The second number indicates the change in energy level and intensity due to the 3d spin–orbit coupling.

TABLE 4.18
Relative Intensities, Energy, Core Hole Spin–Orbit Coupling, and F^2 Slater–Condon Parameters Are Compared for Four Different $^1S_0$ Systems

Edge	Ti 2p	Ti 3p	La 3d	La 4d
Average energy (eV)	464.00	37.00	841.00	103.00
Core spin–orbit (eV)	3.78	0.43	6.80	1.12
F^2 Slater–Condon (eV)	5.04	8.91	5.65	10.45
Intensities				
Prepeak	0.01	10^{-4}	0.01	10^{-3}
$p_{3/2}$ or $d_{5/2}$	0.72	10^{-3}	0.80	0.01
$p_{1/2}$ or $d_{3/2}$	1.26	1.99	1.19	1.99

that the 3d spin–orbit coupling can have very significant effects on the spectral shape of 3d compounds if the 3d-shell is partly filled in the ground state.

We compare a series of x-ray absorption spectra of tetravalent titanium 2p and 3p edges and the trivalent lanthanum 3d and 4d edges. The ground states of Ti^{4+} and La^{3+} are $3d^0$ and $4f^0$, respectively, and they share a ^1S ground state. The transitions at the four edges are respectively:

- Ti^{4+} $L_{2,3}$ edge: $3d^0 \rightarrow 2p^53d^1$
- Ti^{4+} $M_{2,3}$ edge: $3d^0 \rightarrow 3p^53d^1$
- La^{3+} $M_{4,5}$ edge: $4f^0 \rightarrow 3d^94f^1$
- La^{3+} $N_{4,5}$ edge: $4f^0 \rightarrow 4d^94f^1$

These four calculations are equivalent and all spectra consist of three peaks with $J = 1$. The changes are the values of the atomic Slater–Condon parameters and core hole spin–orbit coupling (see Table 4.18). The G^1 and G^3 Slater–Condon parameters have an approximately constant ratio with respect to the F^2 value. The important factor for the spectral shape is the ratio of the core spin–orbit coupling and the F^2 value. Finite values of both the core spin–orbit and the Slater–Condon parameters cause the presence of the prepeak. It can be seen in Table 4.18 that the 3p and 4d spectra have small core spin–orbit couplings, implying small $p_{3/2}$ ($d_{5/2}$) edges and extremely small prepeak intensities. The deeper 2p and 3d core levels have larger core spin–orbit splitting with the result of a $p_{3/2}$ ($d_{5/2}$) edge of almost the same intensity as the $p_{1/2}$ ($d_{3/2}$) edge and a larger prepeak. Note that none of these systems comes close to the single-particle result of a 2 : 1 ratio of the p edges or the 3 : 2 ratio of the d edges. Figure 4.1 shows the x-ray absorption spectral shapes. They are given on a logarithmic scale to make the pre-edges visible.

4.3 LIGAND FIELD MULTIPLET THEORY

The starting point of the ligand field multiplet (LFM) model is to approximate the TM as an isolated atom surrounded by a distribution of charges, which mimic the

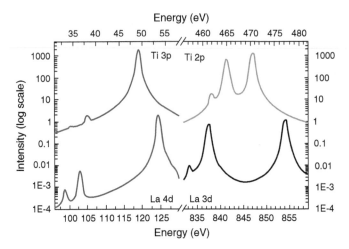

FIGURE 4.1 The La^{3+} 4d and 3d plus Ti^{4+} 3p and 2p x-ray absorption spectra as calculated for isolated ions. The intensity is given on a logarithmic scale to make the pre-edge peaks visible. The intensities of titanium have been multiplied by 1000. (From de Groot, F., *Coord. Chem. Rev.*, 249, 31, 2005. With permission from Elsevier Ltd.)

system, molecule, or solid, around the TM. At first sight, this seems to be a very simplistic model and one might doubt its usefulness to explain experimental data. However, it turned out that such a simple model was successful in explaining a large range of experiments [i.e. optical spectra, electron paramagnetic resonance (EPR) spectra, and magnetic moments].

A TM ion in the gas phase has five degenerate 3d orbitals. When the metal ion is placed in a crystal with six neighboring ions equidistant on the three axes, the crystal field (or ligand field) is octahedral and its symmetry properties belong to the cubic group O$_h$. The effect of this cubic crystal field is that the five 3d orbitals will loose their degeneracy and become split in energy. The main reason for the success of the LFM model is that the explained properties are strongly determined by symmetry considerations. With its simplicity in concept, the LFM model could make full use of the results of group theory. Group theory also makes possible a close link to atomic multiplet theory. Group (theoretically speaking, the only thing ligand field theory does is translate), or branch, the results are obtained in atomic symmetry to cubic symmetry and further to any other lower point groups. The mathematical concepts for these branchings are well developed. We will focus on these group theoretical results and their effects on the ground states as well as on the spectral shapes.

4.3.1 Ligand Field Multiplet Hamiltonian

The LFM Hamiltonian consists of the atomic Hamiltonian as outlined in Chapter 3, to which a crystal field is added:

$$H_{\text{LFM}} = H_{\text{ATOM}} + H_{\text{CF}}, \tag{4.29}$$

$$H_{\mathrm{ATOM}} = \sum_{N} \frac{\mathbf{p}_i^2}{2m} + \sum_{N} \frac{-Ze^2}{\mathbf{r}_i} + \sum_{\mathrm{pairs}} \frac{e^2}{\mathbf{r}_{ij}} + \sum_{N} \zeta(\mathbf{r}_i)\, l_i \cdot s_i, \qquad (4.30)$$

$$H_{\mathrm{CF}} = -e\phi(\mathbf{r}). \qquad (4.31)$$

The only term added to the atomic Hamiltonian is the crystal field, which consists of the electronic charge e times a potential that describes the surroundings $\phi(\mathbf{r})$. The potential $\phi(\mathbf{r})$ is written as a series expansion of spherical harmonics Y_{LM}s:

$$\phi(\mathbf{r}) = \sum_{L=0}^{\infty} \sum_{M=-L}^{L} \mathbf{r}^L A_{LM} Y_{LM}(\psi, \phi). \qquad (4.32)$$

The crystal field is regarded as a perturbation to the atomic result. This implies that it is necessary to determine the matrix elements of $\phi(\mathbf{r})$ with respect to the atomic 3d orbitals $\langle 3d | \phi(\mathbf{r}) | 3d \rangle$. The matrix elements can be separated into a spherical part and a radial part, as was also done for the atomic Hamiltonian. The radial part of the matrix elements yields the strength of the crystal field interaction. The spherical part of the matrix element can be written completely in Y_{LM} symmetry, where the two 3d electrons are written as Y_{2m}. This gives:

$$\langle Y_{2m_2} | Y_{LM} | Y_{2m_1} \rangle = (-1)^{m_2} \sqrt{15(2L+1)/4\pi} \begin{pmatrix} 2 & L & 2 \\ -m_2 & M & m_1 \end{pmatrix} \begin{pmatrix} 2 & L & 2 \\ 0 & 0 & 0 \end{pmatrix}. \qquad (4.33)$$

The second 3j-symbol is zero unless L is equal to 0, 2, or 4. This limits the crystal field potential for 3d electrons to:

$$\phi(\mathbf{r}) = A_{00} Y_{00} + \sum_{M=-2}^{2} \mathbf{r}^2 A_{2M} Y_{2M} + \sum_{M=-4}^{4} \mathbf{r}^4 A_{4M} Y_{4M}. \qquad (4.34)$$

The first term $A_{00} Y_{00}$ is a constant. It will only shift the atomic states and it is not necessary to include this term explicitly if the spectral shape is calculated.

4.3.2 CUBIC CRYSTAL FIELDS

A large range of systems possess a TM ion surrounded by six/eight neighbors. The six neighbors are positioned on the three Cartesian axes, that is, on the six faces of a cube surrounding the TM. They form the octahedral field. The eight neighbors are positioned on the eight corners of the cube and form the cubic field. Both these systems belong to the O_h point group. O_h symmetry is a subgroup of the atomic SO_3 group.

The calculation of the x-ray absorption spectral shape in atomic symmetry involved the calculation of the matrices of the initial state, the final state and the transition. The initial state is given by the matrix element $\langle 3d^n | H_{\mathrm{ATOM}} | 3d^n \rangle$, which for a particular J-value in the initial state gives $\sum_{J} \langle J | 0 | J \rangle$. The same applies for the final state matrix element $\langle 2p^5 3d^{n+1} | H_{\mathrm{ATOM}} | 2p^5 3d^{n+1} \rangle$, where $\sum_{J'} \langle J' | 0 | J' \rangle$ is calculated for the values of J' that fulfill the selection rule (i.e. $J' = J - 1$, J, and $J + 1$).

TABLE 4.19
Branching Rules for the Symmetry
Elements by Going from SO$_3$ Symmetry
to O$_h$ Symmetry

SO$_3$		O$_h$ (Butler)	O$_h$ (Mulliken)
S	0	0	A$_1$
P	1	1	T$_1$
D	2	$2 + \hat{1}$	E + T$_2$
F	3	$\hat{0} + 1 + \hat{1}$	A$_2$ + T$_1$ + T$_2$
G	4	$0 + 1 + 2 + \hat{1}$	A$_1$ + E + T$_1$ + T$_2$

The dipole matrix element $\langle 3d^n | p | 2p^5 3d^{n+1} \rangle$ implies the calculation of all matrices that couple J and J': $\sum_{J,J'} \langle J | 1 | J' \rangle$. To calculate x-ray absorption spectrum in a cubic crystal field, these atomic transition matrix elements must be branched to cubic symmetry. This is essentially the only task to fulfill.

Table 4.19 gives the branching from SO$_3$ to O$_h$ symmetry. This table can be determined from group theory (Butler, 1981). This table implies that an S symmetry state in atomic symmetry branches only to an A$_1$ symmetry state in octahedral symmetry. This is the case, because the symmetry elements of an s orbital in O$_h$ symmetry are determined by the character table of A$_1$ symmetry; that is, whatever symmetry operation is applied, an s orbital remains an s orbital. This is not the case for the other orbitals. For example, a p orbital can be described with the characters of the T$_1$ symmetry state in the O$_h$ symmetry (e.g. the class G$_3$), a two-fold rotation around x inverts the p orbital. A d orbital, or a D symmetry state in SO$_3$, branches to E plus T$_2$ symmetry states in octahedral symmetry. This can be related to the character table by adding the characters of E and T$_2$ symmetry, yielding the overall characters 5, −1, 1, −1, and 1, which describe the properties of d orbitals in O$_h$ symmetry; that is, the dimension of a d orbital is 5 and the class G$_4$ (a four-fold rotation around x) inverts the d orbitals. This is a well-known result: a 3d electron is separated into t$_{2g}$ and e$_g$ electrons in octahedral symmetry where the symmetries include the gerade-notation of the complete O$_h$ character table.

The following observations can be made: the dipole transition operator has p-symmetry and is branched to T$_1$ symmetry. Having a single symmetry in O$_h$ symmetry, there will be no dipolar angular dependence in x-ray absorption. The quadrupole transition operator has d-symmetry and is split into two operators in O$_h$ symmetry; in other words, there will be different quadrupole transitions in different directions. The Hamiltonian is given by the unity representation A$_1$ of the symmetry under consideration.

We can lower the symmetry from octahedral O$_h$ to tetragonal D$_{4h}$ and describe this symmetry lowering again with a branching table. Table 4.20 gives the branching table from O$_h$ to D$_{4h}$ symmetry.

An atomic s orbital is branched to D$_{4h}$ symmetry according to the branching series $S \rightarrow A_1 \rightarrow A_1$. In other words, it is still the unity element, and it will always be the unity element in all symmetries. An atomic p orbital is branched according to

TABLE 4.20
Branching Rules for the Symmetry Elements by Going from
O_h Symmetry to D_{4h} Symmetry

O_h (Butler)	O_h (Mulliken)	D_{4h} (Butler)	D_{4h} (Mulliken)
0	A_1	0	A_1
$\tilde{0}$	A_2	2	B_1
1	T_1	$1 + \hat{0}$	$E + A_2$
$\hat{1}$	T_2	$1 + \hat{2}$	$E + B_2$
2	E	$0 + 2$	$A_1 + B_1$

$P \rightarrow T_1 \rightarrow E + A_2$. Adding the characters of E and A_2 yields 3, −1, 1, −1, and −1, implying that a two-fold rotation around the z-axis inverts a p orbital, and so on. Similarly, an atomic d orbital is branched according to $D \rightarrow E + T_2 \rightarrow A_1 + B_1 + E + B_2$. Adding the characters of these four representations yields 5, 1, −1, 1, and 1. The dipole transition operator has p-symmetry and hence is branched to $E + A_2$ symmetry; in other words, the dipole operator is described with two operators in two different directions implying an angular dependence in the x-ray absorption intensity. The quadrupole transition operator has d-symmetry and is split into four operators in D_{4h} symmetry; in other words, there will be four different quadrupole transitions in different directions/symmetries. The Hamiltonian is given by the unity representation A_1. Similarly, as in O_h symmetry, the atomic G-symmetry state branches into the Hamiltonian in D_{4h} symmetry according to the series $G \rightarrow A_1 \rightarrow A_1$. In addition, it can be seen that the E symmetry state of O_h symmetry branches to the A_1 symmetry state in D_{4h} symmetry. The E symmetry state in O_h symmetry is found from the D and G atomic states. This implies that the series $G \rightarrow E \rightarrow A_1$ and $D \rightarrow E \rightarrow A_1$ also become part of the Hamiltonian in D_{4h} symmetry. The three branching series in D_{4h} symmetry are in Butler's notation given as $4 \rightarrow 0 \rightarrow 0$, $4 \rightarrow 2 \rightarrow 0$, and $2 \rightarrow 2 \rightarrow 0$ and the radial parameters related to these branches are indicated as X_{400}, X_{420}, and X_{220}. The X_{400} term is already important in the O_h symmetry. This term is closely related to the cubic crystal field term 10 Dq as will be discussed subsequently.

4.3.3 Definitions of the Crystal Field Parameters

In order to compare the X_{400}, X_{420}, and X_{220} definition of crystal field operators to other definitions like Dq, Ds, Dt, we compare their effects on the set of 3d functions. The most straightforward way to specify the strength of the crystal field parameters is to calculate the energy separations of the 3d functions. In O_h symmetry, there is only one crystal field parameter X_{40}. This parameter is normalized in a manner that creates unitary transformations in the calculations. The result is that it is equal to $1/18 \times \sqrt{30}$ times 10 Dq, or 0.304 times 10 Dq. In tetragonal symmetry (D_{4h}), the crystal field is given by three parameters, X_{400}, X_{420}, and X_{220}. An equivalent description is to use the parameters Dq, Ds, and Dt. Table 4.21 gives the action of the X_{400}, X_{420}, and X_{220} on the 3d orbitals and relates the respective symmetries to the linear combination of X parameters, the linear combination of the Dq, Ds, and Dt

TABLE 4.21
Energy of the 3d Orbitals Is Expressed in X_{400}, X_{420}, and X_{220} in the Second Column and in Dq, Ds, and Dt in the Third Column

Γ	Energy Expressed in X-Terms	In D-Terms	Orbitals
b_1	$30^{-\frac{1}{2}} \cdot X_{400} - 42^{-\frac{1}{2}} \cdot X_{420} - 2.70^{-\frac{1}{2}} \cdot X_{220}$	$6\,Dq + 2\,Ds - 1\,Dt$	$x^2 - y^2$
a_1	$30^{-\frac{1}{2}} \cdot X_{400} + 42^{-\frac{1}{2}} \cdot X_{420} + 2.70^{-\frac{1}{2}} \cdot X_{220}$	$6\,Dq - 2\,Ds - 6\,Dt$	z^2
b_2	$-2/3 \cdot 30^{-\frac{1}{2}} \cdot X_{400} + 4/3 \cdot 42^{-\frac{1}{2}} \cdot X_{420} - 2.70^{-\frac{1}{2}} \cdot X_{220}$	$-4\,Dq + 2\,Ds - 1\,Dt$	xy
e	$-2/3 \cdot 30^{-\frac{1}{2}} \cdot X_{400} - 2/3 \cdot 42^{-\frac{1}{2}} \cdot X_{420} + 70^{-\frac{1}{2}} \cdot X_{220}$	$-4\,Dq - 1\,Ds + 4\,Dt$	xz, yz

parameters and the specific 3d orbitals of that particular symmetry. From this table, we can relate these parameters and write X_{400}, X_{420}, and X_{220} as a function of Dq, Ds, and Dt.

$$X_{400} = 6 \cdot 30^{\frac{1}{2}} \cdot Dq - 7/2 \cdot 30^{\frac{1}{2}} \cdot Dt$$
$$X_{420} = -5/2 \cdot 42^{\frac{1}{2}} \cdot Dt$$
$$X_{220} = -70^{\frac{1}{2}} \cdot Ds.$$

The inverse relationships imply:

$$Dq = 1/6 \cdot 30^{-\frac{1}{2}} \cdot X_{400} - 7/30 \cdot 42^{-\frac{1}{2}} \cdot X_{420}$$
$$Ds = -70^{-\frac{1}{2}} \cdot X_{220}$$
$$Dt = -2/5 \cdot 42^{-\frac{1}{2}} \cdot X_{420}$$

These relations allow for the quick transfer of, for example, the values of Dq, Ds, and Dt, from optical spectroscopy to the X-values as used in x-ray absorption.

4.3.4 Energies of the 3d^n Configurations

Table 4.10 gives the energy levels of a 3d^8 configuration and Table 4.7 gives the ground states of the 3d^n configurations in atomic symmetry. The crystal field effect modifies these energy levels by the additional terms in the Hamiltonian. We will use the 3d^8 configuration as an example to show the effects of the O_h and D_{4h} symmetry (Table 4.22). Assuming for the moment that the 3d spin–orbit coupling is zero, in O_h symmetry the five term symbols in spherical symmetry split into 11 term symbols. Their respective energies can be calculated by adding the effect of the cubic crystal field 10 Dq to the atomic energies. The diagrams of the respective energies with respect to the cubic crystal field, normalized to the Racah parameter B, are known as the Tanabe–Sugano diagrams (Sugano et al., 1970).

Figure 4.2 gives the Tanabe–Sugano diagram for the 3d^8 configuration. The ground state of a 3d^8 configuration in O_h symmetry has a $^3A_2$ symmetry. If the crystal field energy is 0.0 eV, one has effectively the atomic multiplet states. From low energy to high energy, one can observe, respectively, the ^3F, ^1D, ^3P, ^1G, and ^1S states. Including a crystal field strength splits these states—for example, the ^3F state is split into

TABLE 4.22

Five Symmetry States of a 3d⁸ Configuration in SO₃ Symmetry

	Relative Energy (ev)	Symmetries in O_h	Symmetries in D_{4h}
^1S	4.6	$^1A_1$	$^1A_1$
^3P	0.2	$^3T_1$	$^3E + {}^3A_2$
^1D	−0.1	$^1E + {}^1T_2$	$^1A_1 + {}^1B_1 + {}^1E + {}^1B_2$
^3F	−1.8	$^3A_2 + {}^3T_1 + {}^3T_2$	$^3B_1 + {}^3E + {}^3A_2 + {}^3E + {}^3B_2$
^1G	0.8	$^1A_1 + {}^1T_1 + {}^1T_2 + {}^1E$	$^1A_1 + {}^1E + {}^1A_2 + {}^1E + {}^1B_2 + {}^1A_1 + {}^1B_1$

Note: Energies for Ni^{2+} are given in the first and second columns. The third column gives the respective symmetries of these states in O_h symmetry and column 4 in D_{4h} symmetry. In both cases, the spin–orbit coupling has not yet been included.

$^3A_2 + {}^3T_1 + {}^3T_2$. At higher crystal field strengths, states start to change their order and they cross. For states to actually cross each other or show noncrossing behavior, it depends on whether their symmetries allow them to form a linear combination of states. This also depends on the inclusion of the 3d spin–orbit coupling.

Figure 4.3 shows the effect of the reduction of the Slater–Condon parameters. Figure 4.3 is the same as Figure 4.2 up to a crystal field of 1.5 eV. Then, for this crystal field value, the Slater–Condon parameters have been reduced from their atomic value, indicated with 80% of their Hartree–Fock value, to 0%. The spectrum for 0% has all its Slater–Condon parameters reduced to zero. In other words, the 2p3d coupling has been turned off, and one essentially observes the energies of a 3d⁸ configuration (i.e. two 3d holes). This single particle limit has three configurations, respectively, the two holes in $e_g e_g$, $e_g t_{2g}$, and $t_{2g} t_{2g}$ states. The energy difference between $e_g e_g$ and $e_g t_{2g}$ is exactly the crystal field value of 1.5 eV. Figure 4.3 effectively shows

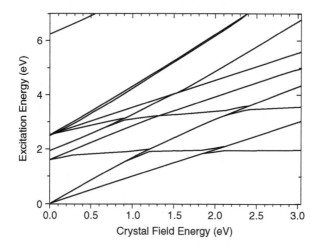

FIGURE 4.2 Tanabe–Sugano diagram for a 3d⁸ configuration in O_h symmetry.

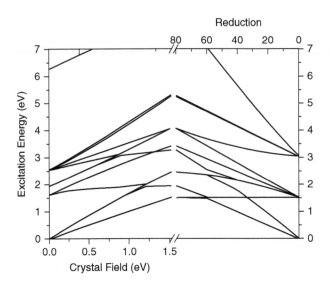

FIGURE 4.3   Tanabe–Sugano diagram for a 3d^8 configuration in O$_h$ symmetry including the effect of reduced Slater–Condon parameters.

the transition from the single particle picture to the multiplet picture for the 3d^8 ground state.

The ground state of a 3d^8 configuration in O$_h$ symmetry always remains ^{3}A$_2$. The reason is clear if one compares these configurations to the single particle description of a 3d^8 configuration. In a single particle description, a 3d^8 configuration is split by the cubic crystal field into the t$_{2g}$ and the e$_g$ configuration. Having found these configurations, one adds the eight 3d electrons one-by-one to these configurations. The t$_{2g}$ configuration has the lowest energy and can contain six 3d electrons. The remaining two electrons are placed in the e$_g$ configuration where both have a parallel alignment according to Hund's rules. The result is that the overall configuration is t$_{2g}^6$e$_g(\uparrow)^2$. This configuration identifies with the ^{3}A$_2$ configuration.

In general, a 3d configuration splits into an e$_g$ and a t$_{2g}$ configuration and both configurations are further split by the Stoner exchange splitting J. The Stoner exchange splitting J is given as a linear combination of the Slater–Condon parameters as $J = (F_2 + F_4)/14$. The Stoner exchange splitting is an approximation to the effects of the Slater–Condon parameters and, in fact, a second parameter C, the orbital polarization, can be used in combination with J. The orbital polarization C is given as $C = (9F_2 - 5F_4)/98$. Often, this orbital polarization is omitted from single particle descriptions. In that case, the multiplet configuration ^{3}A$_2$ is not exactly equal to the single particle configuration t$_{2g}^6$e$_g(\uparrow)^2$. We assume for the moment that the effect of the orbital polarization will not modify the ground states.

Table 4.23 shows that for the configurations 3d^4, 3d^5, 3d^6, and 3d^7, there are two possible ground state configurations in O$_h$ symmetry. A high-spin ground state that originates from the Hund's rule ground state and a low-spin ground state for which all t$_{2g}$ levels are filled at first. One can directly relate the symmetry of a configuration

TABLE 4.23

Configurations 3d⁰ to 3d⁹ Are Given in O_h Symmetry for All Possible HS and LS States

Conf.	Ground State in SO_3	HS Ground State in O_h	HS Ground State in Single Particle Models	LS Ground State in Single Particle Models	LS Ground State in O_h
$3d^0$	$^1S_0$	$^1A_1$	—	—	—
$3d^1$	$^2D_{3/2}$	$^2T_2$	t_{2g+}^{1}	—	—
$3d^2$	$^3F_2$	$^3T_1$	t_{2g+}^{2}	—	—
$3d^3$	$^4F_{3/2}$	$^4A_2$	t_{2g+}^{3}	—	—
$3d^4$	$^5D_0$	^5E	$t_{2g+}^{3} e_{g+}^{1}$	$t_{2g+}^{3} t_{2g-}^{1}$	$^3T_1$
$3d^5$	$^6S_{5/2}$	$^6A_1$	$t_{2g+}^{3} e_{g+}^{2}$	$t_{2g+}^{3} t_{2g-}^{2}$	$^2T_2$
$3d^6$	$^5D_2$	$^5T_2$	$t_{2g+}^{3} e_{g+}^{2} t_{2g-}^{1}$	$t_{2g+}^{3} t_{2g-}^{3}$	$^1A_1$
$3d^7$	$^4F_{9/2}$	$^4T_1$	$t_{2g+}^{3} e_{g+}^{2} t_{2g-}^{2}$	$t_{2g+}^{3} t_{2g-}^{3} e_{g+}^{1}$	^2E
$3d^8$	$^3F_4$	$^3A_2$	$t_{2g+}^{3} e_{g+}^{2} t_{2g-}^{3}$	—	—
$3d^9$	$^2D_{5/2}$	^2E	$t_{2g+}^{3} e_{g+}^{2} t_{2g-}^{3} e_{g-}^{1}$	—	—

Note: The third column gives the (high spin) HS term symbols and the last column the (low spin) LS term symbols. The fourth and fifth columns give the respective occupations of the t_{2g} and e_g orbitals.

to the partly-filled sub-shell in the single particle model. A single particle configuration with one t_{2g} electron has T_2 symmetry, two t_{2g} electrons imply T_1 symmetry and one e_g electron implies E symmetry. If the t_{2g} electrons are filled and the e_g electrons (of the same spin) are empty, then the symmetry is A_2. Finally, if both the t_{2g} and e_g states (of the same spin) are filled, the symmetry is A_1. The nature of the ground state is important—it will be shown subsequently that E symmetry states are susceptible to Jahn–Teller distortions and T_1 and T_2 symmetry states are susceptible to the effects of the 3d spin–orbit coupling.

The transition from high-spin to low-spin ground states is determined by the cubic crystal field 10 Dq and the exchange splitting J. The exchange splitting is present for every two parallel electrons. Table 4.23 gives the high-spin and low-spin occupations of the t_{2g} and e_g spin-up and spin-down orbitals (t_{2g+}, e_{g+}, t_{2g-}, and e_{g-}). The $3d^4$ and $3d^7$ configuration differs by one t_{2g} versus e_g electron, yielding an energy difference equal to the crystal field splitting 10 Dq. The $3d^5$ and $3d^6$ configurations differ by 2D. The exchange interaction J is slightly different for $e_g e_g$, $e_g t_{2g}$, and $t_{2g} t_{2g}$ interactions, as indicated in Table 4.24. The last column can be used to estimate the transition point. For this column, the exchange splittings were assumed to be equal, yielding the simple rules that, for $3d^4$ and $3d^5$ configurations, high-spin states are found if the crystal field splitting is less than $3J$. In the case of $3d^6$ and $3d^7$ configurations, the crystal field value should be less than $2J$ for a high-spin configuration. Because J can be estimated as 0.8 eV, the transition points are approximately 2.4 eV for $3d^4$ and $3d^5$, respectively, 1.6 eV for $3d^6$ and $3d^7$. In other words, $3d^6$ and $3d^7$ materials have a tendency to be low-spin compounds. This is particularly true for $3d^6$ compounds because of the additional stabilizing nature of the $3d^6$ $^1A_1$ low-spin ground state.

TABLE 4.24
High-Spin and Low-Spin Distribution of the 3d Electrons for the
Configurations $3d^4$ to $3d^7$

Configuration	High Spin	Low Spin	10 Dq (D)	Exchange (J)	J/D
$3d^4$	$t_{2g+}^{\,3}\,e_{g+}^{\,1}$	$t_{2g+}^{\,3}\,t_{2g-}^{\,1}$	1D	$3J_{te}$	3
$3d^5$	$t_{2g+}^{\,3}\,e_{g+}^{\,2}$	$t_{2g+}^{\,3}\,t_{2g-}^{\,2}$	2D	$6J_{te} + J_{ee} - J_{tt}$	~3
$3d^6$	$t_{2g+}^{\,3}\,e_{g+}^{\,2}\,t_{2g-}^{\,1}$	$t_{2g+}^{\,3}\,t_{2g-}^{\,3}$	2D	$6J_{te} + J_{ee} - 3J_{tt}$	~2
$3d^7$	$t_{2g+}^{\,3}\,e_{g+}^{\,2}\,t_{2g-}^{\,2}$	$t_{2g+}^{\,3}\,t_{2g-}^{\,3}\,e_{g+}^{\,1}$	1D	$3J_{te} + J_{ee} - 2J_{tt}$	2

Note: The fourth column gives the difference in crystal field energy; the fifth column gives the difference
in exchange energy. For the last column, we have assumed that $J_{te} \sim J_{ee} \sim J_{tt} = J$.

4.3.5 Symmetry Effects in D_{4h} Symmetry

In D_{4h} symmetry, the t_{2g} and e_g symmetry states split further into e_g and b_{2g}, respectively, a_{1g} and b_{1g}. Depending on the nature of the tetragonal distortion, either the e_g
or the b_{2g} state have the lowest energy. Table 4.25 shows that all configurations from
$3d^2$ to $3d^8$ have a low-spin possibility in D_{4h} symmetry. Only the $3d^2$ configuration
with the e_g state as a ground state does not possess a low-spin configuration. The $3d^1$
and $3d^9$ configurations contain only one unpaired spin, thus they have no possibility
of obtaining a low-spin ground state. It is important to notice that a $3d^8$ configuration

TABLE 4.25
Branching of the Spin-Symmetry States and Its Consequence on the States
That Are Found after the Inclusion of Spin–Orbit Coupling

Conf.	Ground State in SO_3	HS Ground State in O_h	Spin in O_h	Degeneracy	Overall Symmetry in O_h
$3d^0$	$^1S_0$	$^1A_1$	A_1	1	A_1
$3d^1$	$^2D_{3/2}$	$^2T_2$	U_1	2	$U_2 + G$
$3d^2$	$^3F_2$	$^3T_1$	T_1	4	$E + T_1 + T_2 + A_1$
$3d^3$	$^4F_{3/2}$	$^4A_2$	G	1	G
$3d^4$	$^5D_0$	^5E	$E + T_2$	5	$A_1 + A_2 + E + T_1 + T_2$
		$3T_1$	T_1	4	$E + T_1 + T_2 + A_1$
$3d^5$	$^6S_{5/2}$	$^6A_1$	$G + U_2$	2	$G + U_2$
		$2T_2$	U_1	2	$G + U_2$
$3d^6$	$^5D_2$	$^5T_2$	$E + T_2$	6	$A_1 + E + T_1 + T_1 + T_2 + T_2$
		$1A_1$	A_1	1	A_1
$3d^7$	$^4F_{9/2}$	$^4T_1$	G	4	$U_1 + U_2 + G + G$
		$2E$	U_1	1	G
$3d^8$	$^3F_4$	$^3A_2$	T_1	1	T_2
$3d^9$	$^2D_{5/2}$	^2E	U_1	1	G

Note: The fourth column gives the spin-projection and the fifth column its degeneracy. The last column
lists all the symmetry states after inclusion of spin–orbit coupling.

as, for example, found in Ni^{2+} and Cu^{3+} can yield a low-spin configuration. Actually, this low-spin configuration is found in the trivalent parent compounds of the high T_C superconducting oxides (Hu et al., 1998a,b). The D_{4h} symmetry ground states are particularly important for those cases where O_h symmetry yields a half-filled e_g state. This is the case for $3d^4$ and $3d^9$ plus low-spin $3d^7$. These ground states are unstable in octahedral symmetry and will relax to, for example, a D_{4h} ground state (the well-known Jahn–Teller distortion). This yields the Cu^{2+} ions with all states filled except the $^1A_{1g}$ hole.

4.3.6   EFFECT OF THE 3d SPIN–ORBIT COUPLING

As discussed previously, the inclusion of a 3d spin–orbit coupling will lead to the multiplication of the spin and orbital angular momenta to a total angular momentum. In this process, the familiar nomenclature for the ground states of the $3d^n$ configurations is lost. For example, the ground state of Ni^{2+} in octahedral symmetry is in total symmetry that is referred to as T_2 and not as $^3A_2$. Also in total symmetry, the spin angular momenta are branched to the same symmetry group as the orbital angular momenta, yielding for a $^3A_2$ ground state an overall ground state of $T_1 \otimes A_2 = T_2$. It turns out that in many cases, it is better to omit the 3d spin–orbit coupling because it is "quenched," for example, by solid state effects. This has been found to be the case for CrO_2. A different situation is found for CoO where the explicit inclusion of the 3d spin–orbit coupling is essential for a good description of the 2p x-ray absorption spectral shape. In other words, 2p x-ray absorption is able to determine the different role of the 3d spin–orbit coupling in CrO_2 (quenched) and CoO (not quenched), respectively (de Groot, 1994).

Table 4.25 gives the spin-projection to O_h symmetry. The ground states with an odd number of 3d electrons have a ground state spin angular momentum that is half-integer (Butler, 1981). Table 4.25 shows that the degeneracy of the overall symmetry states is often not exactly equal to the spin number as given in the third column. For example, the $^3T_1$ ground state is split into four configurations, not three as one would expect. If the 3d spin–orbit coupling is small (and if no other state is close in energy), two of these four states are quasi-degenerate and one finds essentially three states. This is, in general, the case for all situations. Note that the $^6A_1$ ground state of $3d^5$ is split into two configurations. These configurations are degenerate as far as the 3d spin–orbit coupling is concerned. However, because of differences in the mixing of excited term symbols, a small energy difference can be found. This is the origin of the small, but nonzero, zero field splitting in the EPR analysis of $3d^5$ compounds.

Figure 4.4 shows the Tanabe–Sugano diagram for a $3d^7$ configuration in O_h symmetry. Only the excitation energies from 0.0 eV to 0.4 eV are shown to highlight the high-spin low-spin transition at 2.25 eV, and also the important effect of the 3d spin–orbit coupling. It can be observed that the atomic multiplet spectrum of Co^{2+} has a large number of states at low energy. All these states are part of the $^4F_{9/2}$ configuration that is split by the 3d spin–orbit coupling. After applying a cubic crystal field, most of these multiplet states are shifted to higher energies and only four states remain at low energy. These are the four states of $^4T_1$ as indicated in Table 4.25. All these four states remain within 0.1 eV from the U_1 ground state.

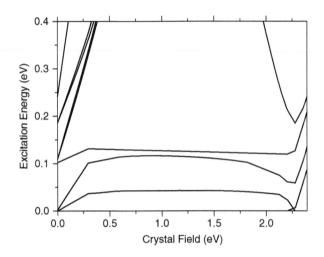

FIGURE 4.4   Tanabe–Sugano diagram for a 3d^7 configuration in O$_h$ symmetry.

This description is actually correct and has been shown in detail for the 2p x-ray absorption spectrum of CoO (de Groot, 1991, 1994), which has a cubic crystal field of 1.2 eV. At 2.25 eV, the high-spin low-spin transition is evident. A new state is coming from high energy and a G symmetry state replaces the U$_1$ symmetry state at the lowest energy. In fact, there is a very interesting complication: due to the 3d spin–orbit coupling, the G symmetry states of the ^{4}T$_1$ and ^{2}E configurations mix and form linear combinations. Just above the transition point, this linear combination will have a spin-state that is neither high-spin nor low-spin and, in fact, a mixed spin-state can be found.

4.3.7   CONSEQUENCES OF REDUCED SYMMETRY

Table 4.26 gives all matrix element calculations that have to be carried out for 3d^n → 2p^{5}3d^{n+1} transitions in SO$_3$ symmetry for the J-values up to 4. We will use the transitions 3d^0 → 2p^{5}3d^1 as examples. 3d^0 contains only $J = 0$ symmetry states. This limits the calculation for the ground-state spectrum to only one ground state, one transition, and one final state matrix element. We are now going to apply the SO$_3$ → O$_h$ and O$_h$ → D$_{4h}$ branching rules to these tables.

In octahedral symmetry, one has to calculate five matrices for the initial and final states and 13 transition matrices. Note that this is a general result for all even numbers of 3d electrons, as there are only these five symmetries in the O$_h$ symmetry. In the 3d^0 case, the ground state branches to A$_1$ and only three matrices are needed to generate the spectral shape: $\langle$A$_1$|A$_1$|A$_1$$\rangle$ for the 3d^0 ground state, $\langle$A$_1$|T$_1$|T$_1$$\rangle$ for the dipole transition, and $\langle$T$_1$|A$_1$|T$_1$$\rangle$ for the 2p^{5}3d^1 final state.

4.3.8   3d^0 SYSTEMS IN OCTAHEDRAL SYMMETRY

In this section, we focus on the discussion of the crystal field effects on the spectral shape of 3d^0 systems. The 3d^0 systems are rather special because they are

TABLE 4.26
Matrix Elements in SO_3 Symmetry Needed for the Calculation of 2p X-Ray Absorption (Up to a Value of $J = 4$)

General Calculation $3d^n \rightarrow 2p^53d^{n+1}$ in SO_3 Symmetry

Initial State	Transition	Final State
$\langle 0\,\vert\,0\,\vert\,0 \rangle$	$\langle 0\,\vert\,1\,\vert\,1 \rangle$	$\langle 0\,\vert\,0\,\vert\,0 \rangle$
$\langle 1\,\vert\,0\,\vert\,1 \rangle$	$\langle 1\,\vert\,1\,\vert\,0 \rangle$	$\langle 1\,\vert\,0\,\vert\,1 \rangle$
	$\langle 1\,\vert\,1\,\vert\,1 \rangle$	
	$\langle 1\,\vert\,1\,\vert\,2 \rangle$	
$\langle 2\,\vert\,0\,\vert\,2 \rangle$	$\langle 2\,\vert\,1\,\vert\,1 \rangle$	$\langle 2\,\vert\,0\,\vert\,2 \rangle$
	$\langle 2\,\vert\,1\,\vert\,3 \rangle$	
$\langle 3\,\vert\,0\,\vert\,3 \rangle$	$\langle 3\,\vert\,1\,\vert\,2 \rangle$	$\langle 3\,\vert\,0\,\vert\,3 \rangle$
	$\langle 3\,\vert\,1\,\vert\,3 \rangle$	
	$\langle 3\,\vert\,1\,\vert\,4 \rangle$	
$\langle 4\,\vert\,0\,\vert\,4 \rangle$	$\langle 4\,\vert\,1\,\vert\,3 \rangle$	$\langle 4\,\vert\,0\,\vert\,4 \rangle$
	$\langle 4\,\vert\,1\,\vert\,4 \rangle$	

not affected by ground state effects. The $3d^0 \rightarrow 2p^53d^1$ transition can be calculated from a single transition matrix $\langle A_1\,\vert\,T_1\,\vert\,T_1 \rangle$ in O_h symmetry (Table 4.27) and in D_{4h} symmetry (Table 4.28). In O_h symmetry, the ground state A_1 matrix is 1×1 and the final state T_1 matrix is 7×7, making the transition matrix 1×7. In other words, the spectrum consists of a maximum of seven peaks. The respective

TABLE 4.27
Matrix Elements in O_h Symmetry Needed for the Calculation of 2p X-Ray Absorption

General Calculation $3d^n \rightarrow 2p^53d^{n+1}$ in O_h Symmetry

Initial State	Transition	Final State
$\langle A_1\,\vert\,A_1\,\vert\,A_1 \rangle$	$\langle A_1\,\vert\,T_1\,\vert\,T_1 \rangle$	$\langle A_1\,\vert\,A_1\,\vert\,A_1 \rangle$
$\langle T_1\,\vert\,A_1\,\vert\,T_1 \rangle$	$\langle T_1\,\vert\,T_1\,\vert\,A_1 \rangle$	$\langle T_1\,\vert\,A_1\,\vert\,T_1 \rangle$
	$\langle T_1\,\vert\,T_1\,\vert\,T_1 \rangle$	
	$\langle T_1\,\vert\,T_1\,\vert\,E \rangle$	
	$\langle T_1\,\vert\,T_1\,\vert\,T_2 \rangle$	
$\langle E\,\vert\,A_1\,\vert\,E \rangle$	$\langle E\,\vert\,T_1\,\vert\,T_1 \rangle$	$\langle E\,\vert\,A_1\,\vert\,E \rangle$
	$\langle E\,\vert\,T_1\,\vert\,T_2 \rangle$	
$\langle T_2\,\vert\,A_1\,\vert\,T_2 \rangle$	$\langle T_2\,\vert\,T_1\,\vert\,T_1 \rangle$	$\langle T_2\,\vert\,A_1\,\vert\,T_2 \rangle$
	$\langle T_2\,\vert\,T_1\,\vert\,E \rangle$	
	$\langle T_2\,\vert\,T_1\,\vert\,T_2 \rangle$	
	$\langle T_2\,\vert\,T_1\,\vert\,A_2 \rangle$	
$\langle A_2\,\vert\,A_1\,\vert\,A_2 \rangle$	$\langle A_2\,\vert\,T_1\,\vert\,T_2 \rangle$	$\langle A_2\,\vert\,A_1\,\vert\,A_2 \rangle$

TABLE 4.28
Matrix Elements in D_{4h} Symmetry Needed for the Calculation of 2p X-Ray Absorption

General Calculation $3d^n \rightarrow 2p^5 3d^{n+1}$ in D_{4h} Symmetry

Initial State	Transition	Final State
$\langle A_1 \| A_1 \| A_1 \rangle$	$\langle A_1 \| E \| E \rangle$	$\langle A_1 \| A_1 \| A_1 \rangle$
	$\langle A_1 \| A_2 \| A_2 \rangle$	
$\langle B_1 \| A_1 \| B_1 \rangle$	$\langle B_1 \| E \| E \rangle$	$\langle B_1 \| A_1 \| B_1 \rangle$
	$\langle B_1 \| A_2 \| B_2 \rangle$	
$\langle E \| A_1 \| E \rangle$	$\langle E \| E \| A_1 \rangle$	$\langle E \| A_1 \| E \rangle$
	$\langle E \| E \| A_2 \rangle$	
	$\langle E \| E \| B_1 \rangle$	
	$\langle E \| E \| B_2 \rangle$	
	$\langle E \| A_2 \| A_2 \rangle$	
$\langle B_2 \| A_1 \| B_2 \rangle$	$\langle B_2 \| E \| E \rangle$	$\langle B_2 \| A_1 \| B_2 \rangle$
	$\langle B_2 \| A_2 \| B_1 \rangle$	
$\langle A_2 \| A_1 \| A_2 \rangle$	$\langle A_2 \| E \| E \rangle$	$\langle A_2 \| A_1 \| A_2 \rangle$
	$\langle A_2 \| A_2 \| A_1 \rangle$	

degeneracies of the J-values in SO_3 symmetry and the degeneracies of the representations in O_h symmetry are collected in Table 4.29.

A $2p^5 3d^1$ configuration has 12 representations in SO_3 symmetry that are branched to 25 representations in a cubic field. The overall degeneracy of the $2p^5 3d^1$ configuration is $6 \times 10 = 60$, implying a possibility of 60 transitions in a system without any symmetry. From these 25 representations, only seven are of interest for the calculation of the x-ray absorption spectral shape, because only these T_1 symmetry states obtain a finite intensity (de Groot et al., 1991a).

Table 4.30 shows the seven T_1 symmetry states calculated with a crystal field splitting of 3.04 eV. The seven peaks that are all built from the seven basis vectors

TABLE 4.29
Branching of the J Values in SO_3 Symmetry to the Representations in O_h Symmetry, Using the Degeneracies of the $2p^5 3d^1$ Final State in X-Ray Absorption

J in SO_3	Degeneracy	Branchings	Γ in O_h	Degeneracy
0	1	A_1	A_1	2
1	3	$3 \times T_1$	A_2	3
2	4	$4 \times E, 4 \times T_2$	T_1	7
3	3	$3 \times A_2, 3 \times T_1, 3 \times T_2$	T_2	8
4	1	A_1, E, T_1, T_2	E	5
Σ	12			25

TABLE 4.30

T_1 Final States of the $2p^53d^1$ Configuration with 10 Dq = 3.04 eV

	460.82	461.64	462.80	464.04	465.85	468.31	471.36
1	0.06	0.00	0.15	0.01	0.49	0.04	0.23
2	0.59	0.02	0.00	0.29	0.02	0.00	0.04
3	<u>0.00</u>	<u>0.01</u>	<u>0.11</u>	<u>0.00</u>	<u>0.18</u>	<u>0.26</u>	<u>0.41</u>
4	0.01	0.44	0.03	0.09	0.00	0.29	0.10
5	0.00	0.29	0.29	0.07	0.03	0.21	0.07
6	0.00	0.04	0.39	0.01	0.24	0.17	0.12
7	0.31	0.17	0.00	0.49	0.01	0.00	0.00
	10^{-4}	**3×10^{-4}**	**0.04**	**10^{-4}**	**0.11**	**0.24**	**0.59**

Note: The top row gives the energies and the bottom row the relative intensities of the seven final states that are built from seven basis vectors. The third row is underlined and gives the $^1P_1$ contributions to the seven states. The bottom row (in bold) gives the intensity of the seven lines.

are evident. The third row is related to $^1P_1$ symmetry and its square yields the intensity as given in the bottom row. Essentially four main peaks, peaks 3, 5, 6, and 7, are observed. Peaks 6 and 7 are essentially the L_2 edge peaks of the t_{2g} and e_g character, respectively. They are split by 3.05 eV, essentially the value of 10 Dq. Peaks 3 and 5 are the L_3 peaks of t_{2g} and e_g character, also split by 3.05 eV. Peaks 1, 2, and 4 are low-intensity peaks that originate from the "spin-forbidden transition" in the atomic multiplet calculation.

Figure 4.5 shows the LFM calculations for the $3d^0 \rightarrow 2p^53d^1$ transition in Ti^{4+}. The result of each calculation is a set of seven energies with seven intensities. These

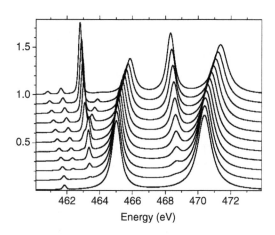

FIGURE 4.5 The LFM calculations for the $3d^0 \rightarrow 2p^53d^1$ transition in Ti^{4+}. The atomic Slater–Condon parameters and spin–orbit couplings have been used as given in Table 4.17. The bottom spectrum is the atomic multiplet spectrum. Each next spectrum has a value of 10 Dq that has been increased by 0.3 eV. The top spectrum has a crystal field of 3.0 eV. (From de Groot, F., *Coord. Chem. Rev.*, 249, 31, 2005. With permission from Elsevier Ltd.)

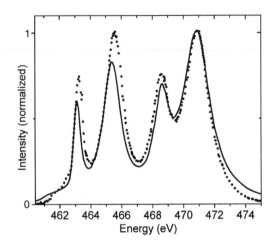

FIGURE 4.6 The 2p x-ray absorption spectrum of $FeTiO_3$ compared with a LFM calculation for Ti^{4+} with a value of 10 Dq of 1.8 eV. (Reprinted with permission from de Groot, F.M.F., Fuggle, J.C., Thole, B.T., and Sawatzky, G.A., *Phys. Rev. B*, 41, 928, 1990a. Copyright 1990 by the American Physical Society.)

seven states have been broadened by lifetime broadening and experimental resolution. From detailed comparison with experiment, it turns out that each of the four main lines has to be broadened differently. It is well known that the L_2 part of the spectrum (i.e. the last two peaks) contains an additional Auger decay that accounts for a significant broadening with respect to the L_3 part. This effect has been found to be an additional broadening of 0.5 eV half-width of half-maximum (HWHM). An additional difference in broadening is found between the t_{2g} and the e_g states. This broadening has been ascribed to differences in the vibration effects on the t_{2g} respectively the e_g states. Another cause could be a difference in hybridization effects and CTM calculations in fact (Okada and Kotani, 1993) indicate that this effect is important. Whatever the origin of the broadening, the comparison with experiment shows that if LFM calculations are performed, the e_g states must be broadened with an additional 0.4 eV HWHM for the Lorentzian parameter. The experimental resolution has been simulated with a Gaussian broadening of 0.15 eV HWHM.

Figure 4.6 compares the LFM calculation of the $3d^0 \rightarrow 2p^5 3d^1$ transition in Ti^{4+} with the experimental 2p XAS spectrum of $FeTiO_3$. The titanium ions are surrounded by six oxygen atoms in a distorted octahedron. The value of 10 Dq has been set to 1.8 eV. The calculation is able to reproduce all peaks that are experimentally visible. In particular, the two small prepeaks can be easily observed. The similar spectrum of $SrTiO_3$ has an even sharper spectral shape, related to the perfect octahedral surrounding of Ti^{4+} by oxygen.

Figure 4.7 shows the effect of the pd Slater–Condon parameters on the spectral shape of the $3d^0 \rightarrow 2p^5 3d^1$ transition in Ti^{4+}. The bottom calculation is the same as that included in Figure 4.6, and used the 80% reduction of the Hartree–Fock values in order to obtain a good estimate of the values in the free atom. In most solids, the

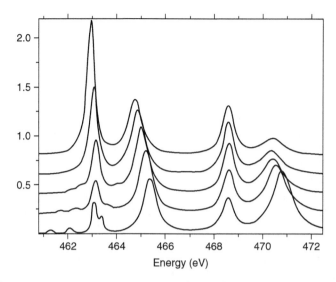

FIGURE 4.7 The LFM calculations for the $3d^0 \rightarrow 2p^53d^1$ transition in Ti^{4+}. The atomic Slater–Condon parameters and spin–orbit couplings have been used as given in Table 4.17. The bottom spectrum is the LFM spectrum with atomic parameters and corresponds to the seventh spectrum in Figure 4.5 (i.e. $10\,Dq = 1.8$ eV). The Slater–Condon parameters are rescaled from the Hartree–Fock values to 80% (bottom), 60%, 40%, 20%, and 1% (top), that is, the top spectrum is essentially the single particle (crystal field) result. (From de Groot, F., *Coord. Chem. Rev.*, 249, 31, 2005. With permission from Elsevier Ltd.)

pd Slater–Condon parameters have essentially the same values as for the free atom or, in other words, the solid-state screening of the pd Slater–Condon parameters is almost zero. The five spectra are calculated by using the same values for the 3d- and 2p-spin–orbit coupling and the same crystal field value of 1.8 eV. The Slater–Condon parameters are then rescaled to 80% (bottom), 60%, 40%, 20%, and 1% (top), respectively. The top spectrum corresponds closely to the single particle picture where four peaks are expected (L_3-t_{2g}, L_3-e_g, L_2-t_{2g}, and L_2-e_g, respectively), with respective intensities given by their degeneracies (i.e. $6:4:3:2$). This is exactly what is observed in the top spectrum where it should be noted that the intensity ratio is a little obscured by the differences in line width. It can be concluded that there is a large difference between the single particle result (top spectrum) and the multiplet result (bottom spectrum). The Slater–Condon parameters have the effect of lowering the intensity of the t_{2g} peaks and shifting intensity to the e_g peaks. At an even larger intensity, shift can be observed for the L_3 edge rather than for the L_2 edge and a very clear effect is the creation of additional peaks because of the additional transitions that become allowed.

From Figure 4.7, it can be observed that the four energy levels at the single particle limit on the left, split into seven lines if the Slater–Condon parameters are turned on. More precisely, it is only the L_3 edge that is split and its two states are split into five states. The L_2 edge is not split and because of this, the L_2 edge can be expected to stay closer to the single particle result; in particular, the energy separation between the t_{2g} and e_g level of the L_2 edge is only a little affected. In the case of 4d elements, where the multiplet effects are smaller, their L_2 edge can be expected to be more closely related to the single particle picture than the corresponding L_3 edge.

Lowering the symmetry from cubic (O_h) to trigonal (D_{3d}) or tetragonal (D_{4h}) further splits all the symmetry lines according to the rules of group theory. Since, the dipole operator also splits into two or three different operators, lower symmetry implies that ordered systems will have angular dependent XAS spectra. An overview of lower symmetries and angular dependence in XAS is given by Brouder (1990). Examples of angular dependence in lower symmetries can be found in the study of interfaces, surfaces, and adsorbates. A detailed study of the symmetry effects on calcium 2p x-ray absorption spectra at the surface and in the bulk of CaF_2 clearly showed the ability of the multiplet calculations to reproduce the spectral shapes both in the bulk as at the reduced C_{3v} symmetry of the surface (Himpsel et al., 1991). Recently, the group of Anders Nilsson performed potassium 2p x-ray absorption experiments of potassium adsorbed on Ni(100) as well as the coadsorption system CO/K/Ni (Hasselstrom et al., 2000). Further examples of reduced symmetry and its consequences for linear and circular dichroism will be given in the Chapters 6 (for XAS), 7 (for XMCD), and 8 (for RXES).

4.3.9 *AB INITIO* LFM CALCULATIONS

Recent progress on the *ab initio* use of clusters to explain core level spectra includes the work of Bagus and Ilton (2006). Within their approach, important aspects are the *ab initio* calculation of all atomic multiplet effects and the effects of covalence in the ground state and its changes in the final state. Explicit charge-transfer effects are, however, not included in this *ab initio* route. This implies that for all systems where charge transfer is important, this *ab initio* route will not yield the correct spectra. If charge-transfer effects could be implemented within this route, an *ab initio* calculation of core level spectra would become feasible. Martins et al. (2006) have reviewed the analysis of free TM ions and performed *ab initio* atomic calculations to calculate the 3p XAS and XPS spectra. In the case of free ions, there are no crystal field effects and the charge-transfer effect can be reformulated essentially as a configuration–interaction effect. An earlier combination of band structure and charge-transfer effects has been developed by Zaanen et al. (1985b).

Another route to *ab initio* multiplet calculations has been developed by Isao Tanaka and coworkers (Ikeno et al., 2004, 2005, 2006; Tanaka et al., 2005; Brik et al., 2006). In most of the calculations, they use first-principles multi-electron calculations using model clusters composed of one TM ion and coordinating six oxide ions. The atomic positions are obtained from the experimental crystal structures and the lattice is extended by point charges. The Slater integrals and crystal field strengths are calculated within this theoretical scheme. Compared to the CTM model used throughout this book, this method is "*ab initio*" in the sense that both the Slater integrals and the crystal field strengths are calculated *ab initio*, within the restrictions chosen. This approach can be seen as an extension of the method developed by Crocombette et al. (1995), who also used a real space cluster input to calculate the Slater integrals and crystal field values. Crocombette and Jollet (1994, 1996) and Crocombette et al. (1995) used a scaling factor to account for longer range interactions, where this new approach uses point charges. As yet, this *ab initio* multiplet route is limited to one $3d^n$

configuration. That is, within the context of the CTM method, it can be regarded as a LFM method. This implies that the same limitations apply as for the LFM calculations discussed previously (i.e. only ionic systems can be calculated and the effects of charge transfer, in particular satellite structures, are not included).

Looking at the calculated results, it can be concluded that the calculations on the octahedral TM oxides MnO, FeO, and CoO (Ikeno et al., 2006) approximately reproduce the experiment. It is to be noted, however, that the "empirical" LFM calculations reproduce MnO and CoO more accurately, using only one empirical parameter (i.e. the cubic crystal field value) (de Groot, 1991). The MnO spectrum is given in Section 6.3.6 and the CoO spectrum is given in Section 6.3.8. As will be discussed, in the case of CoO, the 3d spin–orbit coupling is important along with the temperature parameter. This has also been confirmed experimentally by Haverkort et al. (2005b). From the calculated spectra, it can be observed that the *ab initio* multiplet calculations include 3d spin–orbit coupling, though its implications on the spectral shape are not discussed (Ikeno et al. 2006).

One of the advantages of this *ab initio* method is its real-space input that makes it easier to combine with other calculations. Brik et al. (2006) calculated the angular dependence of 2p XAS spectrum of V_2O_3, thereby repeating the LFM calculations of Park et al. (2000). The level of agreement found in the empirical LFM calculations is significantly better than in the new *ab initio* calculations. In fact, this points to a problem of real-space calculations, that is, its approximation of anything but the nearest neighbors by point charges. It has been found by Crocombette and Jollet (1994) that the symmetry effects caused by the nearest neighbors are too small to simulate the experimental spectra, indicating that the longer-range effects cannot be simulated by charges only. Note, for example, the rather poor agreement with experiment for the 2p XAS spectrum of the strongly distorted system V_2O_5 (Brik et al., 2006). Neglecting the long-range effects imply that the crystal field parameters must be "corrected," which would turn the method into a semi-empirical method, similar to the LFM approach.

A second problem with this *ab initio* multiplet method is that, as yet, only one configuration can be included. The neglect of charge transfer implies that all systems with significant charge-transfer effects, including all higher valent ions, cannot be calculated correctly. In particular, this includes systems with negative charge transfer, for example, Cu^{3+} systems, as discussed previously. A related fundamental problem is the fact that the LDA approach that is used does not describe the strongly correlated 3d electrons correctly. Despite these problems, this *ab initio* multiplet method could become very useful after (*i*) the embedding problem is "solved," (*ii*) more than one configuration can be included, and (*iii*) the correlated ground state is implemented. The main advantage of the method is its real-space nature, which would allow the study of arbitrary surroundings.

4.4   CHARGE TRANSFER MULTIPLET THEORY

Charge transfer effects are the effects of charge fluctuations in the initial and final states. The atomic multiplet and LFM theory use a single configuration to describe the ground state and final state. This configuration can be confined with other low-lying configurations similar to the way configuration–interaction works with a combination of Hartree–Fock matrices. The interpretation of core-level XPS spectra of TM compounds has led to the development of the charge transfer model, as discussed in Chapter 3. With the aid of the charge-transfer effects, the spectral shapes can be explained and the electronic structure understood. Much of this work has centered on the late TM compounds, where the appearance of distinct satellite features in the metal 2p core-level spectra can be explained by either the single impurity Anderson model (SIAM) or a charge transfer cluster model. The physics can be described in terms of a few parameters, namely the on-site dd Coulomb repulsion energy U_{dd}, the charge-transfer energy Δ, the ligand p-TM d hybridization energy $V(\Gamma)$ and the core hole potential $-U_{dc}$. Within the impurity model, it can be found that the core-hole potential is usually 1–2 eV larger than U_{dd}; in addition, the largest core-hole potential is found for the deepest core state.

The CTM theory is an extension of the charge transfer theory in Chapter 3 by taking into account the atomic multiplet coupling effect in each configuration. For instance, the ground state of a nominally $3d^n$ TM compound is expressed as the state where the $3d^n$, $3d^{n+1}\underline{L}$, and alike configurations are coupled through the hybridization effect (i.e. the charge-transfer effect between the ligand p and TM d states), but in the CTM theory the atomic multiplet effect is taken into account in each of the $3d^n$, $3d^{n+1}\underline{L}$ configurations, and so on. In the final state of TM 2p XAS, the $2p^5 3d^{n+1}$, $2p^5 3d^{n+2}\underline{L}$, and alike configurations are coupled through the charge-transfer effect and the atomic multiplet coupling is included in each of $2p^5 3d^{n+1}$, $2p^5 3d^{n+2}\underline{L}$, and alike configurations.

The core-hole potential as used in the impurity model is assumed to act only on the 3d states. This is an approximation and the interactions of the core hole on other valence electrons are neglected. This accounts for ligand valence states as well as metal 4sp conduction states. In addition, off-diagonal interaction terms including the 3d electron and a valence electron are neglected. It is assumed that the neglect of these interactions can be taken into account by renormalizing the core-hole potential $-U_{dc}$. This implies that the value of other parameters as determined from core spectroscopies will also be influenced by the value of U_{dc}, which further implies that these parameters can be different from parameters determined from experiments that do not involve core states.

4.4.1 Initial State Effects

The CTM Hamiltonian with a cluster model is obtained from Equation 3.113 of Chapter 3 by adding the multipole component of Coulomb interaction between the 3d electrons and the spin–orbit interaction of the 3d states in the form:

$$
\begin{aligned}
H_0 = & \sum_{\Gamma,\sigma} \varepsilon_\Gamma a^+_{\Gamma\sigma} a_{\Gamma\sigma} + \sum_{\Gamma,\sigma} \varepsilon_{d\Gamma} a^+_{d\Gamma\sigma} a_{d\Gamma\sigma} + \sum_{\Gamma\sigma} V(\Gamma)(a^+_{d\Gamma\sigma} a_{\Gamma\sigma} + a^+_{\Gamma\sigma} a_{d\Gamma\sigma}) \\
& + \frac{1}{2} \sum_{\Gamma_1\sigma_1, \Gamma_2\sigma_2, \Gamma_3\sigma_3, \Gamma_4\sigma_4} g_{dd}(\Gamma_1\sigma_1, \Gamma_2\sigma_2, \Gamma_3\sigma_3, \Gamma_4\sigma_4) a^+_{d\Gamma_1\sigma_1} a_{d\Gamma_2\sigma_2} a^+_{d\Gamma_3\sigma_3} a_{d\Gamma_4\sigma_4} \\
& + \varsigma_d \sum_{\Gamma_1\sigma_1, \Gamma_2\sigma_2} (\mathbf{l}\cdot\mathbf{s})_{\Gamma_1\sigma_1, \Gamma_2\sigma_2} a^+_{\Gamma_1\sigma_1} a_{\Gamma_2\sigma_2},
\end{aligned}
\tag{4.35}
$$

where $a^+_{\Gamma\sigma}$ and $a^+_{d\Gamma\sigma}$ are electron creation operators for a ligand molecular orbital Γ with spin σ and a 3d state (Γ, σ), and the summation Γ runs over the irreducible representations of the symmetry group of the cluster, $V(\Gamma)$ is the hybridization between 3d and ligand states, $g_{dd}(\Gamma_1\sigma_1, \Gamma_2\sigma_2, \Gamma_3\sigma_3, \Gamma_4\sigma_4)$ represents the Coulomb interaction between 3d electrons, and $\xi_d(\mathbf{l} \cdot \mathbf{s})$ is the spin–orbit interaction. The g_{dd} describes all two-electron integrals, that is, it includes all terms of the multipole expansion where the lowest-order term is the spherical symmetric component given by U_{dd} in Chapter 3 and all of the other terms cause the atomic multiplet coupling (described by Slater–Condon parameters F^2 and F^4). The crystal field effect is taken into account by the Γ dependence of $\varepsilon_{d\Gamma}$.

If we extend the cluster model to the SIAM, the ligand molecular orbitals (the first term of H_0) are replaced by the valence band states. Figure 4.8 sketches the model of the valence band such as a semi-elliptical DOS with bandwidth w and a rectangular DOS with bandwidth w. Instead of such models, the actual band structure that is found from DFT calculations can be used (bottom). Actually, it has been demonstrated that the use of the real band structure instead of an approximate model band structure hardly affects the spectral shape (Jollet et al., 1997). The CTM model approximates the band usually as a rectangle of bandwidth w, where N number of points of equal intensity are used for the actual calculation. The calculation can be simplified further to $N = 1$, that is, a single state representing the band. In that case, the bandwidth is reduced to zero and the SIAM is reduced to the cluster model.

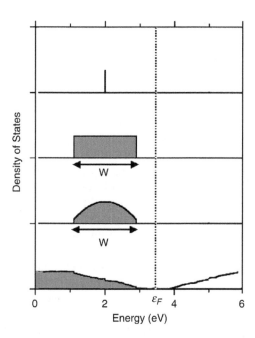

FIGURE 4.8 Representation of the delocalized bands that interact with a correlated localized state. From bottom to top: a general DOS, a semi-elliptical valence band, a square valence band, and a single valence state. (From de Groot, F., *Coord. Chem. Rev.*, 249, 31, 2005. With permission from Elsevier Ltd.)

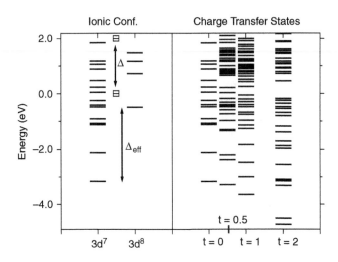

FIGURE 4.9 (Left) The LFM states of $3d^7$ and $3d^8$ configurations. The multiplet states with energies higher than +2.0 eV are not shown. Δ has been set to +2.0 eV. (Right) The CTM calculations for the combination of LFM states as indicated on the left and with the hybridization strength from 0.0–2.0 eV as indicated below the states. (From de Groot, F., *Coord. Chem. Rev.*, 249, 31, 2005. With permission from Elsevier Ltd.)

We will now study the interplay of the charge-transfer and multiplet coupling effects with the cluster model. Two configurations only are considered, namely $3d^7$ and $3d^8\underline{L}$ of Co^{2+} in the cubic crystal field. Figure 4.9 gives the LFMs for the $3d^7$ and $3d^8\underline{L}$ configurations. The $3d^7$ configurations are centered at 0.0 eV and the lowest energy state is the $^4T_1$ state, where the small splittings due to the 3d spin–orbit coupling have been neglected. The lowest state of the $3d^8\underline{L}$ configuration is the $^3A_2$ state, which is the ground state of $3d^8$. The center of gravity of the $3d^8$ configuration has been set at 2.0 eV, which identifies with a value of the charge-transfer energy Δ of 2.0 eV. The effective charge-transfer energy Δ_{eff} is defined as the energy difference between the lowest states of the $3d^7$ and the $3d^8\underline{L}$ configurations as indicated in Figure 4.9. Because the multiplet splitting is larger for $3d^7$ than for $3d^8\underline{L}$, Δ_{eff} is larger than Δ. The effect of charge transfer is to form a ground state that is a combination of $3d^7$ and $3d^8\underline{L}$. The energies of these states have been calculated on the right half of the figure. If the hybridization $V(\Gamma)$ is set equal to zero, both configurations do not mix and the eigenstates are exactly equal to $3d^7$, and at higher energy to $3d^8\underline{L}$. Turning on the hybridization, it is observed that the energy of the lowest configuration is further lowered as shown on the left side of Figure 4.9, where we assume $V(t_{2g}) = V(e_g) = t$. This state will still be the $^4T_1$ configuration, but with increasing hybridization, it will have increasing $3d^8\underline{L}$ character. It can be observed that the second lowest state is split by the hybridization and most bonding combination obtains an energy that comes close to the $^4T_1$ ground state. This excited state is essentially a doublet state and, if the energy of this state crosses the $^4T_1$ state, a charge-transfer induced spin-transition would be observed. It has been shown that the charge-transfer effects can lead to new types of ground states, for example, in the case of a $3d^6$ configuration, crystal field effects lead to a transition of $S = 2$ high-spin to $S = 0$ low-spin ground state. Charge-transfer effects are also able to lead to an $S = 1$ intermediate spin ground state.

Figure 4.9 can be expanded to Tanabe–Sugano-like diagrams for two configurations $3d^n + 3d^{n+1}\underline{L}$, instead of the usual Tanabe–Sugano diagrams as a function of only one configuration. The energies of these two-configuration Tanabe–Sugano diagrams are affected by the Slater–Condon parameters (often approximated with the B Racah parameter), the cubic crystal field $10 Dq$, the charge-transfer energy Δ and the hybridization strength t. The hybridization can be made symmetry-dependent and one can add crystal field parameters related to lower symmetries, yielding an endless series of Tanabe–Sugano diagrams. What is actually important is to determine the possible types of ground states for a particular ion, say Co^{2+}. Scanning through the parameter space of F^2, F^4, $10 Dq$, Ds, Dt, LS_{3d}, $V(\Gamma)$, and Δ, the nature of the ground state can be determined. This ground state can then be checked with 2p XAS. After the inclusion of exchange and magnetic fields, a means is then available to compare the ground state with techniques like x-ray magnetic circular dichroism (XMCD), optical magnetic circular dichroism (MCD), and EPR.

In principle, more band character can be put into this CTM picture and the first step is to make a transition from a single state to a series of $3d^8\underline{L}$ states, each with its included multiplet, but each with a different effective charge-transfer energy.

4.4.2 FINAL STATE EFFECTS

The final state of XAS includes the core hole plus an extra electron in the valence region. In the case of XPS, the core electron is excited to a free electron. One adds the energy and occupation of the 2p core hole and the Coulomb interaction between 3d and 2p states to the Hamiltonian. Then the Hamiltonian is given by explicitly including the 2p core states, as

$$
\begin{aligned}
H = &\sum_v \varepsilon_\Gamma a_v^+ a_v + \sum_v \varepsilon_{d\Gamma} a_{dv}^+ a_{dv} + \sum_\mu \varepsilon_{2p} a_{p\mu}^+ a_{p\mu} + \sum_v V(\Gamma)(a_{dv}^+ a_v + a_v^+ a_{dv}) \\
&+ \frac{1}{2} \sum_{v_1,v_2,v_3,v_4} g_{dd}(v_1,v_2,v_3,v_4) a_{dv_1}^+ a_{dv_2} a_{dv_3}^+ a_{dv_4} \\
&+ \sum_{v_1,v_2,\mu_1,\mu_2} g_{pd}(v_1,v_2,\mu_1,\mu_2) a_{dv_1}^+ a_{dv_2} a_{p\mu_1}^+ a_{p\mu_2} \\
&+ \varsigma_d \sum_{v_1,v_2} (\mathbf{l}\cdot\mathbf{s})_{v_1 v_2} a_{dv_1}^+ a_{dv_2} + \varsigma_p \sum_{\mu_1,\mu_2} (\mathbf{l}\cdot\mathbf{s})_{\mu_1\mu_2} a_{p\mu_1}^+ a_{p\mu_2},
\end{aligned}
\tag{4.36}
$$

where v are the combined indices (Γ, σ), μ are similar indices for the 2p core states, g_{pd} represents the Coulomb interaction between 2p and 3d states, and ς_p is the spin–orbit coupling parameter for the 2p state. The term g_{pd} describes all two-electron integrals and includes U_{dc} as well as the effects of the Slater–Condon parameters F^2, G^1, and G^3. This Hamiltonian can also be used for the initial state because, if we impose the condition of no core hole occupation, H (Equation 4.36) reduces to H_0 (Equation 4.35).

This final state Hamiltonian is solved in the same manner as the initial state Hamiltonian. Using the two configuration description, two final states $2p^5 3d^8$ and

$2p^5 3d^9 \underline{L}$ are found for Co^{2+}. These states mix in a manner similar to the two configurations in the ground state and, as such, give rise to a final state Tanabe–Sugano diagram. All final state energies are calculated from the mixing of the two configurations. This calculation is only possible if all final state parameters are known. The following rules are used:

1. The 2p3d Slater–Condon parameters are taken from an atomic calculation. For trivalent ions and higher valencies, these atomic values are sometimes reduced.
2. The 2p and 3d spin–orbit coupling are taken from an atomic calculation.
3. The crystal field values are assumed to be the same as in the ground state.
4. The energy difference of the two configurations (i.e. the charge-transfer energy), in the final state of XAS is given by $\Delta_f = \Delta + U_{dd} - U_{dc}$ (see Equation 3.162 of Chapter 3). In general, as U_{dc} is approximately 1–2 eV larger than U_{dd}, it is often assumed that $\Delta_f = \Delta - 1$ eV or $\Delta_F = \Delta - 2$ eV. (For XPS, $\Delta_f = \Delta - U_{dc}$.)
5. The hopping parameter $V(\Gamma)$ is often assumed to be equal in the initial and final states, but sometimes the configuration dependence of $V(\Gamma)$ is taken into account.

Detailed analysis of XAS and resonant x-ray emission spectroscopy (RXES) has shown that $V(\Gamma)$ depends on the configuration. In order to describe the configuration dependent $V(\Gamma)$, two scaling factors R_c and R_v are introduced. Since the TM 3d wavefunction is contracted when a 2p core hole is created, the hybridization strength $V(\Gamma)$ between $3d^n$ and $3d^{n+1}\underline{L}$ configurations is reduced to $R_c V(\Gamma)$. Meanwhile, the hybridization between $3d^{n+1}$ and $3d^{n+2}\underline{L}$ configurations is enhanced to $V(\Gamma)/R_v$, because the 3d wavefunction is more extended with increasing 3d electron number. This means that the hybridization strength between $2p^5 3d^{n+1}$ and $2p^5 3d^{n+2}\underline{L}$ configurations in the final state of XAS is given by $[R_c V(\Gamma)]/Rv$, where R_c and R_v are less than unity and usually taken around $R_c = 0.8$ and $R_v = 0.9$, where both are material-dependent values. In a similar way, the Coulomb interaction U_{dd} and the crystal field parameters should, in principle, depend on the configuration, but in most cases these effects are not explicitly taken into account.

4.4.3 XAS Spectrum with Charge-Transfer Effects

The essence of the charge-transfer model is the use of two or more configurations. The LFM calculations use one configuration for which it solves the LFM Hamiltonian H_{LFM} (which is the atomic Hamiltonian plus the crystal field Hamiltonian), so essentially the following matrices:

$$I_{XAS,1} \propto |\langle 3d^n | T_1(ED) | 2p^5 3d^{n+1}\rangle|^2, \tag{4.37}$$

$$E_{INIT,1} = \langle 3d^n | H_{LFM} | 3d^n \rangle, \tag{4.38}$$

$$E_{FINAL,1} = \langle 2p^5 3d^{n+1} | H_{LFM} | 2p^5 3d^{n+1}\rangle. \tag{4.39}$$

The charge-transfer model adds a configuration $3d^{n+1}\underline{L}$ to the $3d^n$ ground state. In the case of a TM oxide, in a $3d^{n+1}\underline{L}$ configuration an electron has been moved from the oxygen 2p valence band to the metal 3d band. One can continue with this procedure and add a $3d^{n+2}\underline{L}^2$ configuration, and so on. In many cases, two configurations will be enough to explain the spectral shapes of XAS, but in particular for high valence states, it can be important to include more configurations. The charge-transfer effect adds a second dipole transition, second initial, and second final states:

$$I_{\text{XAS},1} \propto |\langle 3d^{n+1}\underline{L}|T_1(ED)|2p^5 3d^{n+2}\underline{L}\rangle|^2, \tag{4.40}$$

$$E_{\text{INIT},2} = \langle 3d^{n+1}\underline{L}|H_{\text{LFM}}|3d^{n+1}\underline{L}\rangle, \tag{4.41}$$

$$E_{\text{FINAL},2} = \langle 2p^5 3d^{n+2}\underline{L}|H_{\text{LFM}}|2p^5 3d^{n+2}\underline{L}\rangle. \tag{4.42}$$

The two initial states and two final states are coupled by monopole transitions (i.e. hybridization). The mixing Hamiltonian is given by $H_{\text{MIX}} = \sum_v V(\Gamma)(a_{dv}^+ a_v + a_v^+ a_{dv})$.

$$M_{I1,I2} = \langle 3d^n|H_{\text{MIX}}|3d^{n+1}\underline{L}\rangle, \tag{4.43}$$

$$M_{F1,F2} = \langle 2p^5 3d^{n+1}|H_{\text{MIX}}|2p^5 3d^{n+2}\underline{L}\rangle. \tag{4.44}$$

The XAS spectrum is calculated by solving the equations given above. If a $3d^{n+2}\underline{L}^2$ configuration is included, its term averaged energy is $2\Delta + U_{dd}$. The formal definition of U_{dd} is the energy difference obtained when an electron is transferred from one metal site to another (i.e. a transition $3d^n + 3d^n \rightarrow 3d^{n+1} + 3d^{n-1}$). The number of interactions of two $3d^n$ configurations is one more than the number of interactions of $3d^{n+1} + 3d^{n-1}$, implying that this energy difference is equal to the correlation energy between two 3d electrons.

By analyzing the effects of charge transfer, it is found that, for systems with a positive value of Δ, the main effects on the XAS spectral shape are:

1. Formation of small satellites
2. Contraction of the multiplet structures

The formation of small satellites or even the absence of visible satellite structures is a special feature of XAS spectroscopy. Its origin is the fact that XAS is a neutral spectroscopy and the local charge of the final state is equal to the charge of the initial state. This implies that there is little screening and therefore few charge-transfer satellites. This feature is essentially the same for the cases with and without the multiplet coupling effect, and has already been explained in detail in Chapter 3.

The contraction of the multiplet structure because of the charge-transfer effect can be understood by using the 2×2 matrices (which are the same as those used in Chapter 3) of the final state Hamiltonian with the basis states of $2p^5 3d^{n+1}$ and $2p^5 3d^{n+2}\underline{L}$ configurations.

$$H = \begin{vmatrix} 0 & V(\Gamma) \\ V(\Gamma) & \Delta_f \end{vmatrix}. \tag{4.45}$$

Assume two multiplet states are split by an energy δ. They both mix with a charge-transfer state that is positioned Δ_f above the lowest energy multiplet state I. As a consequence, the charge-transfer energy of the second multiplet state is $\Delta_f - \delta$. Assuming that the hybridization $V(\Gamma)$ is the same for both these two states, the energy gain of the bonding combination is:

$$E_B(I) = \frac{\Delta_f}{2} - \frac{1}{2}\sqrt{\Delta_f^2 + 4V(\Gamma)}, \tag{4.46}$$

$$E_B(II) = \frac{\Delta_f - \delta}{2} - \frac{1}{2}\sqrt{(\Delta_f - \delta)^2 + 4V(\Gamma)}. \tag{4.47}$$

These trends are made visible in Figure 4.10. One can observe that for zero hopping, there is no energy gain. Consider, for example, a hopping of 1.5 eV. Then, the largest energy gain for the lowest value of Δ_f is observed. The higher lying multiplet states have a smaller effective Δ_f and, consequently, a larger energy gain. As such, their energy comes closer to the lowest energy state and the multiplet appears compressed.

4.4.4 SMALL CHARGE-TRANSFER SATELLITES IN 2p XAS

The treatment of charge transfer in this two-state model can explain the observation that charge-transfer yields only small satellites. Assume a $3d^5$ ground state that couples

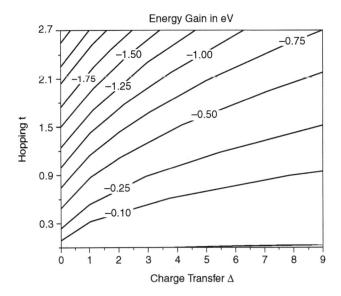

FIGURE 4.10 The energy gain of the bonding combination of a two-state problem as a function of the charge-transfer energy Δ and the hopping $V(\Gamma)$. (From de Groot, F., *Coord. Chem. Rev.*, 249, 31, 2005. With permission from Elsevier Ltd.)

to a $3d^6\underline{L}$ state via ligand–metal charge transfer (LMCT) and to a $3d^4L$ state via metal–ligand charge transfer (MLCT). The only parameters involved in this two-state model are Δ and $V(\Gamma)$. The final state value of Δ for LMCT is given as $\Delta_f = \Delta + U_{dd} - U_{dc}$. In general, it is assumed that $U_{dd} = U_{dc} - 1$ eV, yielding $\Delta_f = \Delta - 1$ eV. However, the maximum satellite intensity is found for the largest difference between Δ and Δ_f, so we modify the assumption to $U_{dd} = U_{dc} - 2$ eV, yielding $\Delta_f = \Delta - 2$ eV. In the case of MLCT, $\Delta_{f2} = \Delta - U_{dd} + U_{dc}$, because there is one 3d electron less. This yields $\Delta_{f2} = \Delta + 2$ eV. Looking in more detail, there is an additional effect on the hopping $V(\Gamma)$. In the final state, the core hole has the tendency to localize the 3d states. This implies that the hopping to the neighbors will slightly decrease (i.e. $t_f \approx 0.9t$). This small reduction in $V(\Gamma)$ counteracts the reduction in Δ, at least for the usual case of positive Δ values. In the case of LMCT, the reduced hopping slightly enhances the effect of Δ_{f2}.

The numerical results using the 2×2 determinant and the rules that $\Delta_f = \Delta - 2$ eV with the two options $V(\Gamma)_f = V(\Gamma)$ and $V(\Gamma)_f = 0.9\ V(\Gamma)$, yield the following: With the typical values for e_g-hopping in TM oxides of 2 eV, all values of Δ of 4 eV and higher yield a maximum satellite of 2%. With $\Delta \approx 0$, satellite intensity up to 6% is possible. It can be concluded that hopping values of 2 eV and higher will never yield significant satellites, essentially because the mixing of configurations is too strong to cause significant variations. Interestingly, lower hopping, for example, the typical t_{2g} hopping of 1.0 eV, is able to create a much stronger satellite structure. With $\Delta = 1$ eV, satellites are found up to 20% with $V(\Gamma)_f = V(\Gamma)$ and 22% with $V(\Gamma)_f = 0.9\ V(\Gamma)$, both for MLCT and LMCT. Reducing the hopping to $V(\Gamma) = 0.8$ with $\Delta \approx 0$ yields satellites up to 30%. In Chapter 6, we discuss many examples of TM oxides, where in almost all cases, only minor charge-transfer satellites are visible. A few cases exist for which significant satellite structure is visible, both in experiment and its simulation. Given the boundary conditions above, it can be concluded that "large" satellites need a combination of $\Delta \approx 0$ and small $V(\Gamma)$. An example of this situation is the case of the iron-cyanides, which have large MLCT satellites (Hocking et al., 2006).

4.4.5 LARGE CHARGE-TRANSFER SATELLITES IN 2p XPS

The treatment of charge transfer in this two-state model can also explain the observation that charge transfer yields large satellites in 2p XPS. Figure 4.11 shows the percentage of the well-screened peak as a function of the hopping. For this curve, we have used $\Delta = 4$ eV, but a similar curve is obtained for all values of Δ. In the final state of a 2p XPS experiment, the core hole has pulled down the 3d states with the core hole potential U_{dc}, which has been set at 8 eV. This yields a final state value of the charge-transfer energy as $\Delta_f = \Delta - U_{dc} = -4$ eV. It can be observed that for small hopping, all intensity goes to the poorly screened peak at higher energy. Increasing the hopping $V(\Gamma)$ to 2 eV yields two peaks of similar intensity. Stronger hopping moves more intensity to the well-screened peak. This phenomenon has also been discussed in Chapter 3 in the context of Ce XPS spectra.

In the case of 2p XPS, it can be concluded that larger hopping yields larger charge transfer. Hopping strengths in the order of 2 eV yield two peaks of equivalent intensity. These two peaks are the bonding and antibonding combination of the two participating states or, in other words, the well-screened and poorly-screened XPS peaks. Note that this behavior is opposite to 2p XAS. In general, 2p XPS has large

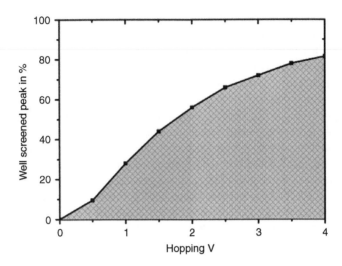

FIGURE 4.11   The percentage of the well-screened peak as a function of the hopping, using $\Delta = 4$ eV and $\Delta_f = \Delta - U_{dc} = -4$ eV.

charge-transfer satellites that are stronger for large hopping. In contrast, 2p XAS has small charge-transfer satellites that become even weaker for large hopping.

4.4.5.1   3d⁰ Compounds

We have seen that the atomic multiplet spectrum of a $3d^0$ compound has three peaks and the LFM spectrum in cubic symmetry has seven peaks. The SrTiO$_3$ spectrum is compared with a LFM calculation in Figure 4.12. The general agreement is good, but still there are significant charge-transfer effects present. The CTM effects are most noticeable in the line broadening and in satellites at higher energy. The LFM calculation as used in Figure 4.7 uses different broadenings for different peaks.

The CTM calculation uses a three configuration ground state $3d^0 + 3d^1\underline{L} + 3d^2\underline{L}^2$. The three configuration CTM calculation automatically broadens the peaks at higher energy; in particular, the e$_g$ peak of the M$_4$ edge, consistent with the experiment. The lowest energy peak consists of a single line spectrum, while the others consist of a number of lines, resulting in broader peaks. It can be concluded that the hybridization effect is important in reproducing the spectral shape, in particular the broadenings. In addition, the charge transfer calculation correctly reproduces the satellites between 10 and 20 eV. The energy separation between the main structure and the satellites is roughly the same as that in XPS and is essentially given by the energy separation between the mixed $2p^53d^1$ and $2p^53d^2\underline{L}$ states in the final state. The observed satellite structure consists of two broad peaks, which is consistent with the calculations.

The first CTM calculations for TM 2p XAS of $3d^0$ systems were made by Okada and Kotani (1993) for TiO$_2$, and the result is essentially the same as that depicted in the upper panel of Figure 8.68 in Chapter 8. The line spectra clearly shows the broadening (splitting into many lines) of the main peaks by the charge-transfer effect

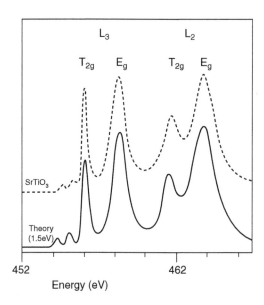

FIGURE 4.12 Titanium 2p x-ray absorption spectrum of $SrTiO_3$, compared with a ligand field multiplet calculation. The value of 10 Dq is 1.5 eV. (From de Groot, F.M.F., Figueiredo, M.O., Basto, M.J., Abbate, M., Petersen, H., and Fuggle, J.C., *Phys. Chem. Miner.*, 19, 140, 1992. With kind permission of Springer Science and Business Media.)

except for the strong low-energy peak (indicated by b). The weak satellite structures *g* and *h* are also in good agreement with those of the experimental result shown in Figure 8.70 of Chapter 8. Thus the CTM theory reproduces the satellite intensities and their energy positions, though van der Laan (1990) ascribed them to the polaronic satellites. Polaronic satellites contain a transition from the O 2p valence band to the Ti 4sp conduction band in addition to the core hole excitation.

4.4.5.2 3d⁸ Compounds

The 2×2 problem in the initial and final state explains the two main effects of charge transfer: a compression of the multiplet structure and the existence of only small satellites. These two phenomena are visible in the figures of Ni^{2+} and Co^{2+}. In the case that the charge transfer is negative, the satellite structures are slightly larger because then the final state charge transfer is increased with respect to the initial state and the balance of the initial and final state is not as good.

Figure 4.13 shows the effect of the charge-transfer energy on divalent nickel. We have used the same hopping $V(\Gamma)$ for the initial and final state and reduced the charge-transfer energy Δ by 1 eV. In the top spectrum, $\Delta = 10$, and the spectrum is essentially the LFM spectrum of a Ni^{2+} ion in its 3d⁸ configuration. The bottom spectrum uses $\Delta = -10$ and now the ground state is almost a pure 3d⁹$\underline{L}$ configuration. Looking for the trends in Figure 4.13, the increased contraction of the multiplet structure is found by going to lower values of Δ. This is exactly what is observed in the series NiF_2 to $NiCl_2$ and $NiBr_2$. Going from Ni to Cu, the atomic parameters change very little,

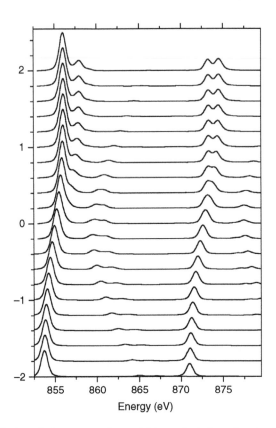

FIGURE 4.13 Series of charge transfer multiplet calculations for the Ni^{2+} ground state $3d^8 + 3d^9\underline{L}$. The top spectrum has a charge-transfer energy of +10 eV. The bottom spectrum has a charge-transfer energy of −10 eV and relates to an almost pure $3d^9$ ground state. (From de Groot, F., *Coord. Chem. Rev.*, 249, 31, 2005. With permission from Elsevier Ltd.)

except for the 2p spin–orbit coupling and the 2p binding energy. Therefore, the spectra of $3d^n$ systems of different elements are all very similar and the bottom spectrum is also similar to Cu^{2+} systems. Therefore, one can also use the spectra with negative Δ values for Cu^{3+} compounds. Examples of Cu^{3+} compounds will be given in Section 6.3.9.

5 X-Ray Photoemission Spectroscopy

5.1 INTRODUCTION

In Chapter 3, the fundamental aspects of x-ray photoemission spectroscopy (XPS) were discussed. Especially, the importance of many-body charge-transfer effects in XPS was shown. For 3d XPS of mixed valence Ce compounds and 2p XPS of transition metal (TM) compounds, the charge-transfer effect causes a characteristic splitting of the XPS spectra, which provides us with important information of the electronic state. In the analysis of these spectra, the single impurity Anderson model (SIAM) or a cluster model without the intra-atomic multiplet effect works as a model of the starting approximation, and then the multiplet effect gives some corrections.

In this chapter, we extend the analysis of XPS given in Chapter 3 to that by the charge transfer multiplet (CTM) theory, which includes both the charge transfer and multiplet effects, as has been introduced in Chapter 4. First, we give typical results of TM 2p XPS analyses for various TM compounds, and discuss a systematic variation of the charge-transfer energy Δ and the dd Coulomb interaction U_{dd} over the TM compounds. Based on the results, two types of insulators, the charge-transfer type and Mott–Hubbard type, are introduced. Then, the TM 3s XPS and 3p XPS spectra are briefly discussed.

For rare earth (RE) compounds, we first give a simplified analysis for 3d XPS of various oxide systems with the charge-transfer effect, but without the multiplet effect, and then perform the full multiplet calculation by the CTM theory. It is also shown that the multiplet effect is more important for 4d XPS of RE systems than for 3d XPS.

We also discuss resonant photoemission spectroscopy (RPES), resonant Auger electron spectroscopy (RAES), and resonant inverse photoemission spectroscopy (RIPES). These resonant spectroscopies are useful in order to take out the contribution of a specific electronic state, for instance the TM 3d or RE 4f state. RAES is also useful to reduce the lifetime broadening and to discriminate the overlapping electric dipole (ED) and electric quadrupole (EQ) transitions in XAS.

Recently, hard x-ray photoemission spectroscopy (HAXPS) has been used as a bulk sensitive probe in core level spectroscopy. HAXPS clearly detects the bulk contribution of XPS in mixed valence Ce compounds and high T_c cuprates, which is different from the surface contribution. A combination of HAXPS and RPES is one of the more interesting applications in core level spectroscopy. The core hole screening effect beyond the single cation cluster model or the SIAM is called the "nonlocal screening effect," and the first theoretical and experimental study of the nonlocal screening effect was made for the Ni 2p XPS of NiO. The importance of the nonlocal screening effect has recently been revealed in the analysis of HAXPS data of high T_c cuprates and TM compounds.

5.2 EXPERIMENTAL ASPECTS

XPS can be described with the three-step model as shown in Appendix A:

1. Absorption of the x-ray inside the solid.
2. "Transport" of the excited photoelectron to the surface.
3. Escape of the photoelectron from the surface.

More detailed accounts of the three-step model have been given by Hüfner (1995). In the second step, the electron mean free path that varies between 0.5 and 5 nm in the 5–5000 eV energy range is important, as shown in Figure 3.1 of Chapter 3. In order to get the information on the bulk electronic states, it is necessary to use the photoelectron whose mean free path is much larger than the lattice constant, as mentioned in Chapter 3. The same statement applies to Auger spectroscopy and, in general, to all experiments where an electron in this energy range escapes from the solid. In Section 5.6, hard x-ray XPS is discussed. The high energy of the x-rays is used to create higher energy electrons that have larger escape depths and are therefore less surface sensitive.

XPS experiments involve some kind of electron optics to detect the electrons that escape from the surface. The most popular device is the hemispherical analyzer, which creates a magnetic field between two spheres and, as such, separates the kinetic energy of the electrons. The electrons can be detected with a single detector or with linear or two-dimensional detectors to yield a single XPS spectrum, a line spectrum, or a 2D spectrum, respectively. Parameters that can be mapped are the electron energy and its angular variation. In addition, the position from which the electron escapes is mapped by using x-ray photoemission electron microscopy (X-PEEM) (Scholl et al., 2000).

As a major analytical technique in many research fields, XPS binding energy and intensity are often the only objects of research interest. The presence of an XPS peak signals that the related element is present and the quantification of its intensity can be used to determine the relative abundance of the element in the probed area. Combined with a variation of the x-ray energy (implying a variation in the electron kinetic energy, hence escape depth), it can be used for depth profiling (Blühm et al., 2004). The binding energy is correlated with the valence (including metal against oxide) and the position (surface against bulk) of the absorbing atom. This intensity and peak position analysis is a powerful and often used application of XPS. Its development by Kai Siegbahn resulted in the Nobel prize in 1981. We refer the reader to dedicated books and reviews on this topic, for example, the use of XPS in catalysis research by Niemantsverdriet (1993). In this book, we are not so much concerned with the energy and intensity of the XPS peaks, but with the detail in the XPS spectral shapes.

5.3 XPS OF TM COMPOUNDS

5.3.1 2p XPS

Here, we discuss the analysis of 2p XPS of TM compounds by CTM calculations with cluster models, taking into account two configurations $3d^n$ and $3d^{n+1}\underline{L}$ for Cu^{2+} compounds and three configurations, $3d^n$, $3d^{n+1}\underline{L}$ and $3d^{n+2}\underline{L}^2$ for other compounds.

As shown in Section 4.4.2, the Hamiltonian, which includes the TM 2p core states explicitly so that it is applicable to both initial and final states of XPS, is given by

$$H = \sum_{v} \varepsilon_{\Gamma} a_v^+ a_v + \sum_{v} \varepsilon_{d\Gamma} a_{dv}^+ a_{dv} + \sum_{\mu} \varepsilon_{2p} a_{p\mu}^+ a_{p\mu} + \sum_{v} V(\Gamma)(a_{dv}^+ a_v + a_v^+ a_{dv})$$

$$+ \frac{1}{2} \sum_{v_1,v_2,v_3,v_4} g_{dd}(v_1,v_2,v_3,v_4) a_{dv_1}^+ a_{dv_2} a_{dv_3}^+ a_{dv_4}$$

$$+ \sum_{v_1,v_2,\mu_1,\mu_2} g_{pd}(v_1,v_2,\mu_1,\mu_2) a_{dv_1}^+ a_{dv_2} a_{p\mu_1}^+ a_{p\mu_2}$$

$$+ \varsigma_d \sum_{v_1,v_2} (\mathbf{l} \cdot \mathbf{s})_{v_1 v_2} a_{dv_1}^+ a_{dv_2} + \varsigma_p \sum_{\mu_1,\mu_2} (\mathbf{l} \cdot \mathbf{s})_{\mu_1 \mu_2} a_{p\mu_1}^+ a_{p\mu_2}. \tag{5.1}$$

Here v denotes the combined indices (Γ, σ), μ is similar indices for the 2p core level (so that Γ is replaced by an atomic orbital angular momentum), g_{dd} and g_{pd} represent the Coulomb interaction between 3d states and that between 2p and 3d states, respectively, and ς_d and ς_p are the spin–orbit coupling parameters. It is to be noted that g_{dd} and g_{pd} include not only the spherical symmetric components, U_{dd} and U_{dc}, but also the multi-pole components, which include the Slater integrals in their explicit forms. The Slater integrals, $F^2(3d, 3d)$, $F^4(3d, 3d)$, $F^2(2p, 3d)$, $G^1(2p, 3d)$, $G^3(2p, 3d)$, and the spin–orbit coupling parameters ς_d and ς_p are calculated by an atomic Hartree–Fock program, and then the Slater integrals are scaled down to 85%.

The charge-transfer energy Δ is now redefined by

$$\Delta \equiv E[3d^{n+1}\underline{L}] - E[3d^n], \tag{5.2}$$

where $E[3d^{n+1}\underline{L}]$ and $E[3d^n]$ represent the configuration averaged energies of $3d^{n+1}\underline{L}$ and $3d^n$, respectively. Then we have the following relations:

$$\Delta + U_{dd} = E[3d^{n+2}\underline{L}^2] - E[3d^{n+1}\underline{L}], \tag{5.3}$$

$$\Delta - U_{dc} = E[2p^5 3d^{n+1}\underline{L}] - E[2p^5 3d^n], \tag{5.4}$$

$$\Delta + U_{dd} - U_{dc} = E[2p^5 3d^{n+2}\underline{L}^2] - E[2p^5 3d^{n+1}\underline{L}]. \tag{5.5}$$

For divalent Cu systems such as CuF_2, the interatomic configuration interaction (CI) is exactly treated by the two-configuration calculation with $3d^9$ and $3d^{10}\underline{L}$ states. The multiplet coupling effect is also important in the spectral shape of XPS. In Figure 5.1, we show experimental results (Nücker et al., 1987; van der Laan et al., 1981) of the Cu 2p XPS of Cu dihalides, CuF_2 and $CuCl_2$, and oxides, CuO and La_2CuO_4. As assigned in Section 3.5.5, the lower and higher binding energy peaks, respectively, of XPS in CuF_2 correspond to $2p^5 3d^{10}\underline{L}$ and $2p^5 3d^9$ final states, for each of the $2p_{3/2}$ and $2p_{1/2}$ components. This assignment also holds for $CuCl_2$, CuO, and La_2CuO_4. It is important to see that the spectral shape is different for $2p_{3/2}$ and $2p_{1/2}$ components, and this can be explained only by taking into account the intra-atomic

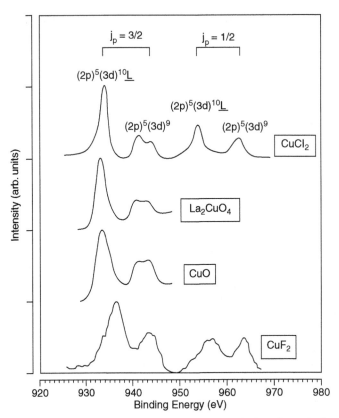

FIGURE 5.1   Experimental results of the Cu 2p XPS in $CuCl_2$, CuF_2, CuO, and La_2CuO_4. (From Okada, K., and Kotani, A. *J. Phys. Soc. Jpn.*, 58, 2578, 1989b. With permission.)

multiplet coupling effect, because the multiplet effect is different for $2p_{3/2}$ and $2p_{1/2}$ core holes. Furthermore, it is interesting to see that the spectral shape of the $2p^5 3d^9$ configuration (higher binding energy feature) of the $2p_{3/2}$ component depends strongly on the material.

The CTM calculations of XPS spectra are made with a CuX_6 cluster model (X = F, Cl, and O), as well as with the SIAM (Okada and Kotani, 1989a, b). An example is shown in Figure 5.2, where the cluster model calculation is made with fixed parameter values of $\Delta = 2.0$ eV and $U_{dc} = 8.0$ eV, but by changing the hybridization strength $V(b_{1g})$. Since the symmetry around the Cu site is D_{4h}, the irreducible representation $\Gamma = b_{1g}$, a_{1g}, b_{2g}, and e_g. For the relation between $V(\Gamma)$, it is assumed that $V(a_{1g}) = V(b_{1g})/\sqrt{3}$, $V(b_{2g}) = V(e_g)/\sqrt{2}$, and $V(b_{2g}) = V(b_{1g})/2$. It is seen that the spectral shape is different for $2p_{3/2}$ and $2p_{1/2}$ components, and especially that the spectral shape of the $2p^5 3d^9$ ($2p_{3/2}$) component depends strongly on the hybridization strength. The result for $V(b_{1g}) = 2.0$ eV is in good agreement with the experiment for La_2CuO_4. Furthermore, the spectral shape of the $2p^5 3d^9$ ($2p_{3/2}$) component is similar to that of $CuCl_2$ for $V(b_{1g}) = 1.5$ eV, CuO for $V(b_{1g}) = 2.0$ eV and CuF_2 for

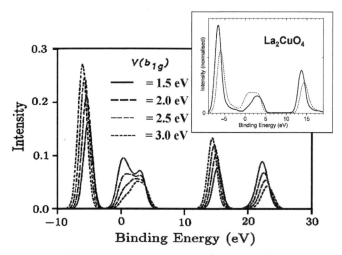

FIGURE 5.2 Cu 2p XPS spectra calculated with a cluster model for various values of $V(b_{1g})$ and a fixed value of $\Delta = 2$ eV. In the inset, the solid curve is the Cu 2p XPS spectrum of La_2CuO_4 calculated with the SIAM using the hybridization obtained by McMahan et al. (1988), and the dotted one is that obtained by reducing the hybridization strength. (From Okada, K., and Kotani, A., *J. Phys. Soc. Jpn.*, 58, 2578, 1989b. With permission.)

$V(b_{1g}) = 3.0$ eV, which suggests that the hybridization strength changes in this order for these materials although the value Δ is also different from 2.0 eV for them.

The above calculations are made with the cluster model, but the extension to the SIAM is also made for La_2CuO_4 where the dependence of the hybridization $V(\Gamma, \varepsilon)$ on the symmetry Γ and the energy ε (in the O 2p band) is taken into account (Okada and Kotani, 1989b). The solid curve in the inset of Figure 5.2 is the result calculated by using $V(\Gamma, \varepsilon)$ obtained from the *ab initio* LDA calculation by McMahan et al. (1988), but it does not agree with the experimental result. Then, a similar calculation is made by only reducing the value of $V(\Gamma, \varepsilon)$ by the factors of 0.8 and 0.6 in the initial and final states of XPS, respectively, as shown with the dashed curve in the inset, which is in good agreement with the experimental result as well as the calculated result with $V(b_{1g}) = 2.0$ eV in Figure 5.2.

Next, we consider Ni and Co compounds. As we have seen in Chapter 3, the Ni 2p XPS of NiF_2 shows two peaks for each of the $2p_{1/2}$ and $2p_{3/2}$ core levels. For those of $NiCl_2$ and $NiBr_2$, three peaks are exhibited as shown with dots (upper panel) in Figure 5.3. The data for the Ni dihalides and NiO were measured by Zaanen et al. (1986) and Bocquet et al. (1992a), respectively. In order to analyze these spectra, it is necessary to take into account the three configurations $3d^8$, $3d^9\underline{L}$, and $3d^{10}\underline{L}^2$. In Figure 5.3, the spectral shape for the $2p_{3/2}$ core level is different from that of $2p_{1/2}$, as in the case of cuprates. Furthermore, there is a difference in their intensity ratio from the statistical one. Similar experimental data for CoF_2-CoO are shown with dots (upper panel) in Figure 5.4 and were taken from Okusawa (1984), Vaal and Paulikas (1985) and Kim (1975). Experimental measurements for Ni and Co compounds were also made by Park et al. (1988) and Lee and Oh (1991).

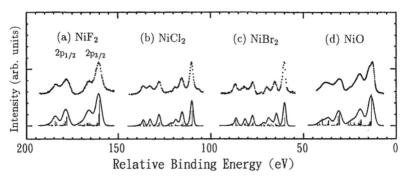

FIGURE 5.3 Theoretical (lower panel) and experimental (upper panel) results of the Ni 2p XPS for the Ni dihalides NiF_2, $NiCl_2$ and $NiBr_2$, and Ni monoxide. (From Kotani, A., and Okada, K., *Recent Advances in Magnetism of Transition Metal Compounds*, World Scientific, Singapore, 1993. Reprinted with permission from World Scientific Publishing Ltd.)

In the following, we show the analysis of the 2p XPS for Ni^{2+} and Co^{2+} compounds by Kotani and Okada (1993), taking into account three configurations by CTM calculations with O_h symmetric NiX_6 and CoX_6 clusters. The irreducible representation Γ is t_{2g} and e_g, and for hybridization strength the empirical relation $V(e_g)/V(t_{2g}) = -2.0$ is used.

In Figure 5.3a–d and Figure 5.4a–d, we show the calculated Ni and Co 2p XPS spectra, respectively, for Ni and Co dihalides and monoxides (Okada et al., 1992; Kotani and Okada, 1993). The parameter values used are listed in Table 5.1. The Ni 2p XPS of Ni dihalides show, for NiF_2, a two-peak structure for each of the $2p_{3/2}$ and $2p_{1/2}$ core levels, but a three-peak structure for $NiCl_2$ and $NiBr_2$. The peak assignment is, in the order of increasing binding energy, the $3d^9\underline{L}$ and $3d^8$ final states for NiF_2 (see Chapter 3), and $3d^9\underline{L}$, $3d^{10}\underline{L}^2$ and $3d^8$ final states for $NiCl_2$ and $NiBr_2$. In going from $NiCl_2$ to $NiBr_2$, the effect of the $3d^{10}\underline{L}^2$ configuration becomes increasingly important since the value of Δ decreases corresponding to the

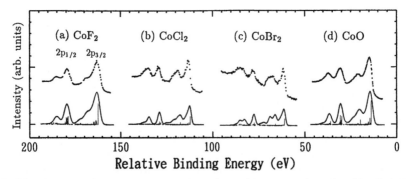

FIGURE 5.4 Theoretical (lower panel) and experimental (upper panel) results of the Co 2p XPS for Co dihalides and Co monoxide. (From Kotani, A., and Okada, K., *Recent Advances in Magnetism of Transition Metal Compounds*, World Scientific, Singapore, 1993. Reprinted with permission from World Scientific Publishing Ltd.)

TABLE 5.1
Parameter Values in the Calculation of the 2p XPS for Ni and Co Dihalides and Monoxides (in Units of eV), Taking into Account Three Configurations and the Full Multiplet Coupling. The Average 3d Electron Number n_d in the Ground State Is Also Shown

	Δ	$V(e_g)$	U_{dd}	U_{dc}	n_d
NiF_2	4.3	2.0	7.3	7.5	8.17
$NiCl_2$	1.3	1.7	7.3	7.5	8.31
$NiBr_2$	0.3	1.4	7.3	7.5	8.37
NiO	2.0	2.0	7.3	7.5	8.30
CoF_2	5.0	2.0	7.0	7.0	7.14
$CoCl_2$	2.0	1.7	7.0	7.0	7.24
$CoBr_2$	1.0	1.4	7.0	7.0	7.27
CoO	2.5	2.0	7.0	7.0	7.26

decrease of the electronegativity of the halogen ion. This peak assignment was first given by Zaanen et al. (1986) from their calculation without the multiplet effect. Even for NiF_2, the effect of the $3d^{10}\underline{L}^2$ configuration is important in renormalizing Δ and U_{dc}. Comparing Table 3.2 of Chapter 3 and the present Table 5.1, we note that the estimated values of Δ and U_{dc} decrease by 4.1 eV and 2.9 eV, respectively, by taking into account the $3d^{10}\underline{L}^2$ configuration. This renormalization is partly due to the effect of the $3d^{10}\underline{L}^2$ configuration and partly due to the effect of the multiplet coupling. The effect of multiplet coupling on the value of Δ is discussed in Section 5.3.4. Another important effect of the multiplet coupling is to cause the difference in the $2p_{3/2}$ and $2p_{1/2}$ spectral shapes, which are clearly seen in both experimental and theoretical results for $NiCl_2$ and $NiBr_2$. For NiO, the characteristic feature of the observed 2p XPS is similar to NiF_2 or $NiCl_2$, and the estimated parameter values for NiO are in between those for NiF_2 and $NiCl_2$.

In the case of CoF_2, the two peaks are the $3d^8\underline{L}$ and $3d^7$ final states in the order of increasing binding energy, though in the $2p_{3/2}$ XPS, the broadening of the spectrum due to the multiplet splitting is so marked that the satellite structure is almost a shoulder structure of the main peak. The main peaks in $CoCl_2$ to $CoBr_2$ are ascribed to the $3d^8\underline{L}$ final states and the satellite structure is a mixture of the $3d^7$ and $3d^9\underline{L}^2$ final states. In particular, the multiplet splitting of the $3d^7$ states in the $2p_{3/2}$ spectrum is so remarkable that its higher binding energy end almost reaches the $2p_{1/2}$ spectrum. If we remove the multiplet splitting, the energy of the $3d^9\underline{L}^2$ state is lower than the $3d^7$ state. Apart from this strong multiplet effect of the $3d^7$ final state of Co dihalides, the difference in the spectral features between Co and Ni comes mainly from the increase in Δ from Ni to Co dihalides. The value of Δ for CoO is estimated to be in between CoF_2 and $CoCl_2$, as in the case of NiO.

In Figure 5.5, we show the experimental (circles) and calculated (solid line) results of the Mn 2p XPS of MnO and MnF_2. The calculations were made by Taguchi et al. (1997). The parameter values used are $\Delta = 8.0$ eV, $U_{dd} = 6.8$ eV, and $V(e_g) = 1.8$ eV for MnO, and $\Delta = 9.5$ eV, $U_{dd} = 7.2$ eV, and $V(e_g) = 2.0$ eV for MnF_2.

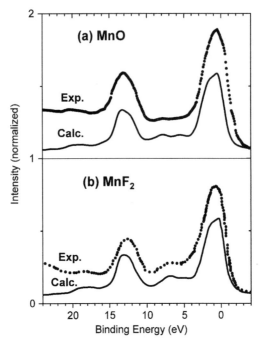

FIGURE 5.5 Theoretical (solid curve) and experimental (circles) results of the Mn 2p XPS spectra for (a) MnO and (b) MnF$_2$. (From Taguchi, M., Uozumi, T., and Kotani, A., *J. Phys. Soc. Jpn.*, 66, 247, 1997. With permission.)

5.3.2 ZAANEN–SAWATZKY–ALLEN DIAGRAM

From the systematic analysis of TM 2p XPS for many TM compounds, we can estimate the value of the key parameters Δ, U_{dd}, $V(\Gamma)$, and so on. The estimated values are somewhat different, depending on the model of the analysis (the cluster model or the SIAM, the full multiplet calculation or the calculation without the multiplet effect), but the systematic trend of their variation is recognized with respect to the change of the TM element and also the anion element. On the basis of such an analysis, Zaanen et al. (1985) pointed out that there are two categories in insulating TM compounds. When $U_{dd} < \Delta$, the insulating energy gap E_g is determined by U_{dd} (i.e. $E_g \approx U_{dd}$) corresponding to the fluctuation

$$3d_i^n 3d_j^n \longleftrightarrow 3d_i^{n-1} 3d_j^{n+1}, \tag{5.6}$$

where i and j label TM sites. A site with a configuration $3d^n$ has a correlation energy given by the total number of electron pairs $1/2n(n-1)$. This implies that the energy involved with the $3d^n 3d^n$ to $3d^{n-1} 3d^{n+1}$ transition is given as $[1/2(n-1)(n-2)] + [1/2(n+1)(n)] - [n(n-1)]$ times U_{dd}. Working out the arithmetic yields exactly one times U_{dd}. This is the well-known "Mott–Hubbard type insulator." When $U_{dd} > \Delta$, on the other hand, the charge-transfer energy Δ, which corresponds to the fluctuation

$$3d_j^n \longleftrightarrow 3d_j^{n+1}\underline{L} \tag{5.7}$$

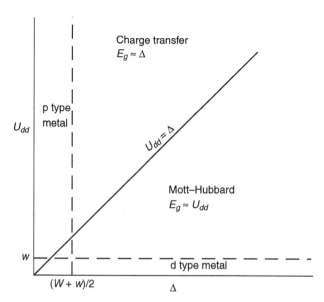

FIGURE 5.6   The Zaanen–Sawatzky–Allen diagram. (From Zaanen, J., and Sawatzky, G.A., *Can. J. Phys.*, 65, 1262, 1987. With permission from NRC Research Press.)

determines the smallest energy gap E_g (i.e. $E_g \approx \Delta$). This is a new class of insulator called "charge-transfer type insulators." Figure 5.6 shows the so-called Zaanen–Sawatzky–Allen diagram representing the two types of insulators in the two dimensional (Δ, U_{dd}).

Figure 5.7 illustrates schematically the two types of charge excitations. The closed and open circles represent the TM ions and anions, respectively. We consider, for simplicity, the limit of vanishing electron hopping between ions (ionic limit). Then, the energy of the charge transfer excitation from an anion p state to a TM d state (left of Figure 5.7) is given by Δ, while that of the Mott–Hubbard type excitation from a TM d to another TM d state (right of Figure 5.7) is given by U_{dd}.

Then we introduce a finite energy bandwidth W for the anion p band and w for the TM d band. The total energy level diagram of the two types of excitations are

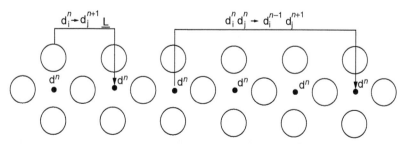

FIGURE 5.7   Schematic illustration of an ionic lattice consisting of TM ions (closed circles) and anions (open circles). The charge-transfer fluctuation (left) and the Mott–Hubbard charge fluctuation (right) are indicated. (From Zaanen, J., and Sawatzky, G.A., *Can. J. Phys.*, 65, 1262, 1987. With permission from NRC Research Press.)

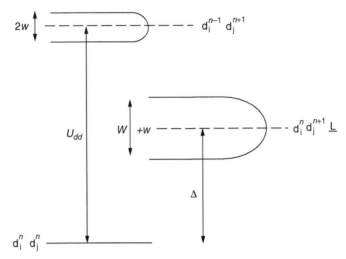

FIGURE 5.8 Total energy level diagram corresponding to a ground state and two types of excitations in TM compounds. (From Zaanen, J., and Sawatzky, G.A., *Can. J. Phys.*, 65, 1262, 1987. With permission from NRC Research Press.)

shown in Figure 5.8. Now the lowest charge transfer excitation energy is decreased from Δ to $\Delta - (W + w)/2$, and the lowest Mott–Hubbard excitation energy from U_{dd} to $U_{dd} - w$. Therefore, if the value U_{dd} decreases in the Mott–Hubbard type insulator, the system changes into a metallic phase (d type metal) for $U_{dd} < w$. This type of metal–insulator transition is well known as the Mott–Hubbard transition. If Δ decreases in the charge-transfer type insulators, on the other hand, the system undergoes the phase transition to the metallic phase (p type metal) for $\Delta < (W + w)/2$. The phase boundaries of these metal–insulator transitions are also shown in Figure 5.6.

A general consensus has so far been obtained that the late TM oxides, CuO, NiO, CoO, are the charge-transfer type insulators. High T_c superconductors are obtained by doping carriers in charge-transfer type insulators containing the CuO_2 plane. With the decrease in the atomic number of TM elements, Δ is increased and U_{dd} is decreased. Therefore, near the middle of the TM oxide series, for MnO for example, Δ is comparable with U_{dd}, and the system is considered to be in the intermediate regime between the charge-transfer and Mott–Hubbard type insulators. As an example, the values of Δ and U_{dd} for various TM compounds are shown in Figure 5.9, which is based on the data collected by Bocquet et al. (1992a,b, 1996) from the analysis of TM 2p XPS.

5.3.3 2p XPS in Early TM Systems

If we extrapolate the trend of Δ and U_{dd} in late TM oxides to those in early ones, it is expected that most of early TM oxides belong to the Mott–Hubbard regime. The early TM compounds, Ti_2O_3, V_2O_3, and Cr_2O_3 were considered to be typical materials of the Mott–Hubbard regime. However, according to the analysis of the TM 2p XPS of these compounds with the cluster model by Bocquet et al. (1996) and

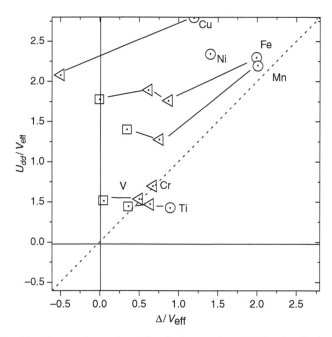

FIGURE 5.9   The Zaanen–Sawatzky–Allen diagram, where TM ions in divalent oxides are indicated with circles, trivalent oxides with triangles, and tetravalent oxides with squares. U_{dd} and Δ are normalized by the effective hybridization strength V_{eff}. (Reprinted with permission from Bocquet, A.E., Saitoh, T., Mizokawa, T., and Fujimori, A. *Solid State Commun.*, 83, 11, 1992b. Copyright 1992 by the American Physical Society.)

Uozumi et al. (1997), the estimated Δ and U_{dd} have been found to be comparable for early TM sesquioxides, from Ti_2O_3 to Mn_2O_3. In Figure 5.10, we show the experimental (dots) and calculated (solid line) results (Uozumi et al., 1997) of the TM 2p XPS of Ti_2O_3, Cr_2O_3, Mn_2O_3, and Fe_2O_3. The values of Δ, U_{dd}, and V_{eff} estimated from this analysis are shown in Figure 5.11. It can be seen that the values of Δ and U_{dd} are comparable for Ti_2O_3, V_2O_3, Cr_2O_3, and Mn_2O_3. Furthermore, for these compounds, the effect of the covalent hybridization is very strong and the value of V_{eff} is also comparable with Δ and U_{dd}. For Fe_2O_3, on the other hand, U_{dd} is considerably larger than Δ and V_{eff}, so that Fe_2O_3 is classified as being a charge-transfer type insulator. A series of vanadium oxide 2p XPS spectra has also been measured by Demeter et al. (2000).

Before closing this subsection, an analysis is given of the Ti 2p XPS of TiO_2 (rutile). The Ti ion in the ground state of TiO_2 is tetravalent, but actually the $3d^0$, $3d^1\underline{L}$, and $3d^2\underline{L}^2$ configurations are mixed by a strong covalent hybridization. The experimental result of Ti 2p XPS (Sen et al., 1976) is shown in Figure 5.12a. The main structure of XPS consists of a pair separated by the Ti 2p spin–orbit splitting (about 6 eV), each of which is accompanied by a satellite about 13 eV above the main peak.

The calculation of Ti 2p XPS was made by Okada and Kotani (1993) with the TiO_6 cluster model, and the result is shown in Figure 5.12b. The symmetry of

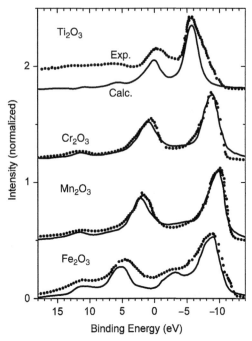

FIGURE 5.10 Theoretical (solid curve) and experimental (circles) results of 2p XPS spectra for, from top to bottom, Ti_2O_3, Cr_2O_3, Mn_2O_3, and Fe_2O_3. (From Uozumi, T., et al., *J. Electron Spectrosc. Relat. Phenom.*, 83, 9, 1997. With permission from Elsevier Ltd.)

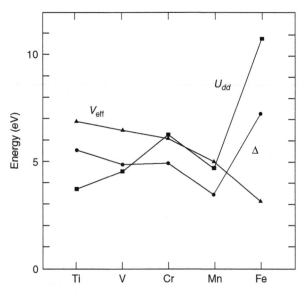

FIGURE 5.11 Systematic variation of Δ, U_{dd}, and V_{eff} for $(TM)_2O_3$ systems with TM = Ti, V, Cr, Mn, and Fe. (From Uozumi, T., et al., *J. Electron Spectrosc. Relat. Phenom.*, 83, 9, 1997. With permission from Elsevier Ltd.)

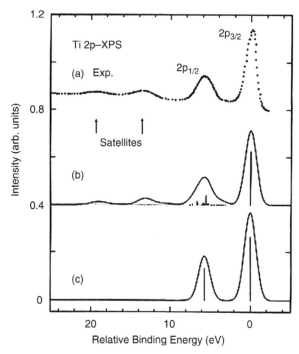

FIGURE 5.12 (a) Experimental and (b) theoretical results of the Ti 2p XPS spectra for TiO₂. The theoretical result for a Ti⁴⁺ ion is shown in (c) for comparison. (From Okada, K., and Kotani, A., *J. Electron Spectrosc. Relat. Phenom.*, 62, 131, 1993. With permission from Elsevier Ltd.)

the cluster is approximated as O_h, because the distortion from O_h is too small to be reflected clearly in the 2p XPS. The parameter values are: $\Delta = 2.98$ eV, $V(e_g) = -2V(t_{2g}) = 3.0$ eV, $U_{dd} = 4.0$ eV, $U_{dc} = 6.0$ eV, and $10\ Dq = 1.7$ eV. With these parameters, the ground state consists of $3d^0$ (39.5%), $3d^1\underline{L}$ (48.2%), and $3d^2\underline{L}^2$ (12.3%) states. The main peak and the satellite correspond mainly to the bonding and antibonding final states between the $2p^53d^0$ and $2p^53d^1\underline{L}$ configurations. The multiplet effect is small, but it is reflected in the broadening of the $2p_{1/2}$ main peak, which is much broader than the $2p_{3/2}$ main peak. Since the $2p^53d^0$ final state has no multiplet effect (except for the spin–orbit splitting of the 2p state), the spectral broadening vanishes in the 2p XPS of free Ti⁴⁺ ion, as shown in Figure 5.12c.

In Figure 5.13, we show the U_{dc} dependence of Ti 2p XPS, where Figure 5.13c is the same as Figure 5.12b. We find that the satellite intensity depends strongly on U_{dc}. For $U_{dc} = 0$, we have no satellite; for $U_{dc} = 3.0$ eV, the satellite intensity is smaller than that of the experimental result. On the other hand, the energy separation between the main peak and the satellite is not very sensitive to U_{dc}, as seen from the comparison of Figure 5.13b and c. These results are well understood from the simplified model given in Chapter 3. The effective hybridization V_{eff} in TiO₂ is given by

$$V_{eff} = \sqrt{6\, V(t_{2g})^2 + 4\, V(e_g)^2},\qquad (5.8)$$

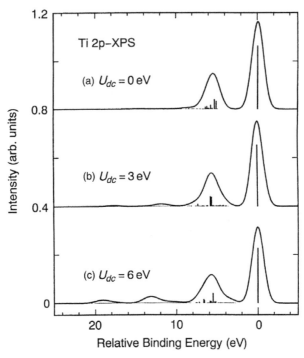

FIGURE 5.13 U_{dc} dependence of the Ti 2p XPS spectra; (c) is the same as Figure. 5.12b. (From Okada, K., and Kotani, A. *J. Electron Spectrosc. Relat. Phenom.*, 62, 131, 1993. With permission from Elsevier Ltd.)

which is about 7.0 eV. Therefore, the energy separation between the main peak and the satellite, which is given by

$$\delta E_B = E_+ - E_- = \sqrt{(\Delta - U_{dc})^2 + 4V_{eff}^2}, \tag{5.9}$$

is almost equal to $2V_{eff} \approx 14$ eV. Furthermore, the relative intensity of the satellite and the main peak, which is given by

$$\frac{I_+}{I_-} = \tan^2(\theta - \theta_0), \tag{5.10}$$

is 0 for $U_{dc} = 0$ (because $\theta = \theta_0$) and increases monotonically with increasing U_{dc}.

5.3.4 EFFECT OF MULTIPLET COUPLING ON Δ AND U_{dd}

In the CTM theory, Δ and U_{dd} are defined by configuration averaged quantities (i.e. the quantities averaged over multiplet terms within each configuration) as expressed by Equations 5.2 and 5.3. In order to discuss the low-energy characters of TM compounds, it might be more convenient to define the charge-transfer energy and dd Coulomb energy using the lowest multiplet of each configuration as shown in Figure 5.14, because

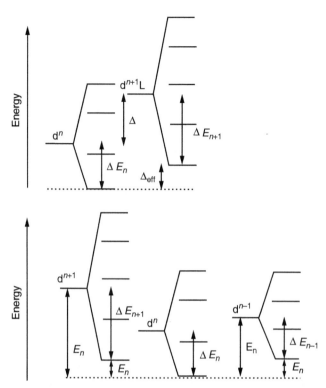

FIGURE 5.14 Multiplet correction of each configuration to define the effective charge-transfer energy and the effective Coulomb energy. (From Fujimori, A., Bocquet, A.E., Saitoh, T., and Mizokawa, T., *J. Electron Spectrosc. Relat. Phenom.*, 62, 141, 1993. With permission from Elsevier Ltd.)

the lowest multiplet states contribute more significantly to the ground state and the low-energy excitations. This was proposed by Fujimori et al. (1993), and they defined the effective charge-transfer and Coulomb energies, Δ_{eff} and $(U_{dd})_{\text{eff}}$, by

$$\Delta_{\text{eff}} = \Delta + \Delta E_n - \Delta E_{n+1}, \tag{5.11}$$

$$(U_{dd})_{\text{eff}} = U_{dd} + 2\Delta E_n - \Delta E_{n-1} - \Delta E_{n+1}, \tag{5.12}$$

where ΔE_n is the multiplet correction of the energy of the $3d^n$ configuration (Figure 5.14).

If the multiplet effect is disregarded, the boundary between the charge-transfer type insulator and the Mott–Hubbard type insulator is given by $\Delta = U_{dd}$. With the multiplet effect, it is better defined by $\Delta_{\text{eff}} = U_{\text{eff}}$ where we simply describe $(U_{dd})_{\text{eff}}$ as U_{eff}. In Figure 5.15, the multiplet corrections $\Delta_{\text{eff}} - \Delta$, $U_{\text{eff}} - U$, and $(U_{\text{eff}} - \Delta_{\text{eff}}) - (U - \Delta)$ are shown for high-spin TM compounds as a function of the 3d electron number n (Fujimori et al., 1993). From this figure, it is seen, for instance, that the multiplet effect makes the charge-transfer energy larger for $n \geq 5$ (except for $n = 9$) but smaller for $n \leq 4$. Therefore, the charge-transfer character of Ni compounds, for instance, is to some extent weakened by the multiplet effect although most of the Ni

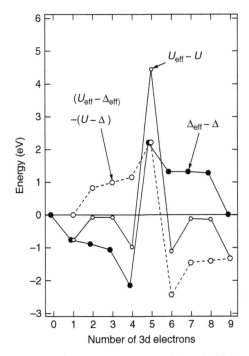

FIGURE 5.15 Multiplet corrections $\Delta_{eff} - \Delta$, $(U_{dd})_{eff} - U_{dd}$ and $((U_{dd})_{eff} - \Delta_{eff}) - (U_{dd} - \Delta)$ for high-spin TM compounds as a function of 3d electron number n. U_{dd} and $(U_{dd})_{eff}$ are indicated simply as U and U_{eff}. (From Fujimori, A., Bocquet, A.E., Saitoh, T., and Misokawa, T., *J. Electron Spectrosc. Relat. Phenom.*, 62, 141, 1993. With permission from Elsevier Ltd.)

compounds are well established to be of the charge-transfer type. This multiplet effect also explains, at least partly, the reason why the charge-transfer energies of NiF_2 and CoF_2 in Table 5.1 are smaller than those in Table 3.2 of Chapter 3. For these compounds, Δ is smaller than Δ_{eff}. If we analyze experimental TM 2p XPS spectra of these materials without taking into account the multiplet effect as done in Chapter 3, the estimated Δ would be closer to Δ_{eff} than Δ because the real charge transfer is made more effectively at the energy Δ_{eff} than at Δ.

The term averaged values Δ and U_{dd} are known to change smoothly as a function of n, but the effective values Δ_{eff} and U_{eff} are not smooth around $n = 5$ because of the multiplet corrections $\Delta_{eff} - \Delta$ and $U_{eff} - U$ as seen from Figure 5.15. Figure 5.16 shows the difference between U_{dd} and U_{eff}. Fujimori et al. (1993) also displayed Δ_{eff} and U_{eff} of various TM compounds with respect to the 3d electron number n, as shown in Figure 5.17, which also gives us interesting information on the behavior of Δ_{eff} and U_{eff} with the change of the TM valence number and the species of the anion.

5.3.5   3s XPS

The 3s XPS spectra are calculated in an equivalent manner to 2p XPS spectra, that is, by the CTM calculations with the SIAM or cluster model. All ground state

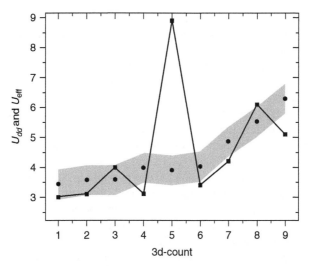

FIGURE 5.16 The trend in the value of U_{dd} (dots) and $(U_{dd})_{eff}$ (line) for divalent ions. The gray band gives an indication of the variations in U_{dd} for different compounds.

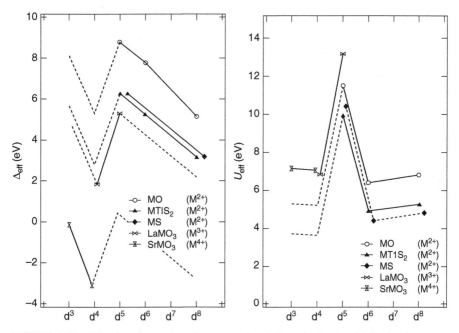

FIGURE 5.17 Effective charge-transfer energy Δ_{eff} (left) and effective dd Coulomb energy $(U_{dd})_{eff}$ indicated as U_{eff} (right) displayed with respect to the 3d electron number n. (From Fujimori, A., Bocquet, A.E., Saitoh, and Mizokawa, T., *J. Electron Spectrosc. Relat. Phenom.*, 62, 141, 1993. With permission from Elsevier Ltd.)

parameters are identical and the core hole potential of the 3s core hole is also (assumed to be) equal to the 2p core hole. The differences in the final state between the 2p core hole and the 3s core hole are (*i*) the presence of spin–orbit coupling for a 2p core hole, (*ii*) different multiplet interactions with the 3d-electrons where the 3s core hole has only an exchange interaction, (*iii*) differences in the lifetime broadenings, and (*iv*) the presence of a strong interaction channel for 3s core holes.

The analysis of 3s XPS is essentially based on the work of the paper "prediction of new multiplet structure in photoemission experiments" (Bagus et al., 1973). Bagus et al. analyzed the 3s XPS spectrum of Mn^{2+} ions in detail. The $3d^5$ ground state of an Mn^{2+} ion is ^6S and the 3s XPS transition creates $3s^1 3d^5$ final states with ^7S and ^5S symmetry, respectively. Within this model, the energy splitting between ^7S and ^5S is given as 6/5 $G^2(3s, 3d)$ and the intensity ratio 7 : 5. The 3s core hole is special due to the presence of a strong configuration interaction (CI) channel between the 3s3d and 3p3p core holes because the energy of the $3s^1 3d^5$ (^5S) configuration is rather close to that of the $3p^4 3d^6$ (^5S) configuration and they are coupled by the Coulomb interaction (intra-atomic CI). The coupling of $3s^1 3d^5$ with $3p^4 3d^6$ creates satellite structures, thereby also modifying the ^7S : ^5S intensity ratio and peak distance. Figure 5.18 shows the graphical result of such a calculation, taken from the paper of Okada and Kotani (1992a). From right to left, the ^7S peak, the main ^5S peak and two ^5S satellites are observed.

Okada and Kotani extended this analysis to the 3s XPS spectrum of MnO. In addition to the 3s3d exchange interaction and the 3s3d-3p3p CI channel, they added the usual charge-transfer channel coupling the $3s^1 3d^5$ final state to $3s^1 3d^6 \underline{L}$. The addition of the charge-transfer effect created additional structures in the satellite region, as can be seen in Figure 5.19. Okada and Kotani analyzed all 3s XPS spectra

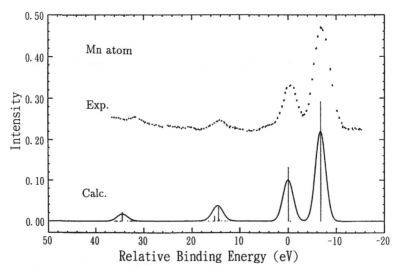

FIGURE 5.18 The 3s XPS spectrum of a Mn atom analyzed with the inclusion of the 3s3d exchange and the 3s3d-3p3p CI channel. (From Okada, K., and Kotani, A., *J. Phys. Soc. Jpn.*, 61, 4619, 1992a. With permission.)

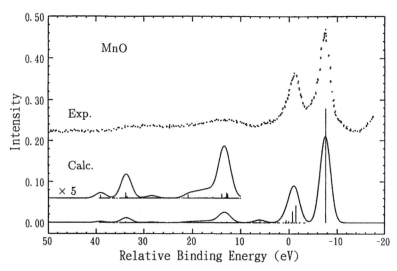

FIGURE 5.19 The 3s XPS spectrum of MnO analyzed with the inclusion of the 3s3d exchange, the 3s3d-3p3p CI channel and ligand-to-metal charge transfer. (From Okada, K., and Kotani, A., *J. Phys. Soc. Jpn.*, 61, 4619, 1992a. With permission.)

of the Mn, Fe, Co, and Ni halides. Experimental data included that of van Acker et al. (1988), Oh et al. (1992), and Gweon et al. (1993). The difference between Ni and Mn is that the 3s3d-3p3p CI channel gets weaker in going from Mn to Ni because the energy difference between the two configurations increases. The $3p^4$ CI satellites are found between 25 and 45 eV above the main peak and they are, in general, not observed in published experimental spectra. Figure 5.20 shows the comparison of the main part of the 3s XPS spectrum with the CTM calculation. The three peaks in the 3s XPS spectrum of $NiCl_2$ are caused by the final state interplay of the $3s^13d^8$, $3s^13d^9\underline{L}$, and $3s^13d^{10}\underline{L}^2$ final states, in combination with the 3s3d exchange of the $3s^13d^8$ and $3s^13d^9\underline{L}$ states. From right to left, the sequence of the ionic final states for $NiCl_2$ and NiF_2 are given as:

$$NiCl_2: 3d^9(\uparrow) > 3d^{10} \sim 3d^9(\downarrow) > 3d^8(\uparrow) > 3d^8(\downarrow), \qquad (5.13)$$

$$NiF_2: 3d^9(\uparrow) > 3d^8(\uparrow) > 3d^9(\downarrow) > 3d^{10} \sim 3d^8(\downarrow). \qquad (5.14)$$

It is observed that the ordering of the states changes from $NiCl_2$ to NiF_2 despite their rather similar spectral shape (Okada et al., 1992). The ionic state diagrams of 3s XPS are equivalent to those of 2p XPS. The absence of 2p spin–orbit coupling and the simpler 3s3d exchange versus 2p3d multiplets make the 3s XPS spectrum appear simpler in experimental spectra. However, as discussed previously, this does not imply that the approximation can be used that is observed in the 3s3d exchange splitting. The charge-transfer effects and also the 3s3d-3p3p CI channel modify the spectral shape to such a large extent that a complete CTM analysis is necessary.

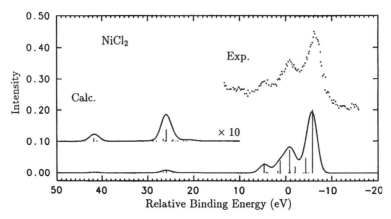

FIGURE 5.20   The 3s XPS spectrum of $NiCl_2$ analyzed with the inclusion of the 3s3d exchange, the 3s3d-3p3p CI channel and ligand-to-metal charge transfer. (From Okada, K., and Kotani, A., *J. Phys. Soc. Jpn.*, 61, 4619, 1992a. With permission.)

5.3.6   3p XPS

The 3p XPS spectra of NiO and $NiCl_2$ within the SIAM was first analyzed by Fujimori et al. (1984) and Fujimori and Minami (1984). Apart from a difference in the 2p versus 3p core hole, 3p XPS is essentially the same as 2p XPS. The theoretical calculations of 2p XPS and 3p XPS are identical with only some of the interaction parameters being different. This can still create quite different spectral shapes. The differences are:

1. The 2p core hole spin–orbit coupling is much larger than the 3p spin–orbit coupling.
2. The 2p3d multiplet interactions are smaller than the 3p3d multiplet interactions.
3. The 2p lifetime broadening is essentially constant over the spectrum, where the 3p lifetime broadening varies over the spectrum.

In first approximation, one can say that while 2p XPS shows the combination of 2p spin–orbit and charge transfer, 3p XPS shows the combination of 3p3d multiplet splitting and charge transfer.

Uozumi et al. (1997) compared the 2p XPS, 3s XPS, and 3p XPS spectra of Cr_2O_3, as given in Figure 5.21. The 2p XPS spectrum (top) shows the 2p spin–orbit splitting between the $2p_{3/2}$ and $2p_{1/2}$ peaks and, at a higher binding energy, a small charge-transfer satellite. The 3p XPS spectrum (middle) shows a main peak followed by a mixture of multiplet effects and charge-transfer effects. The 3s XPS spectrum (bottom) shows the effect of the 3s3d exchange followed by satellites due to charge transfer and CI. It has been pointed out by Uozumi et al. (1997) that these XPS spectra are not very sensitive to precise crystal field splitting because of the experimental broadening. This indicates an important difference of 2p XAS that is very sensitive to crystal field effects. In 2p XAS, the higher experimental resolution and much

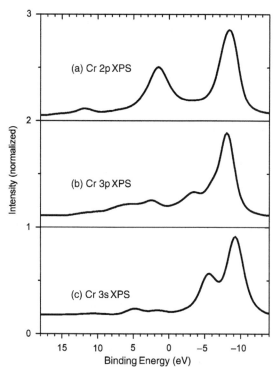

FIGURE 5.21 From top to bottom, the Cr 2p XPS, Cr 3p XPS, and Cr 3s XPS of Cr_2O_3 are given. (From Uozumi, T., et al., *J. Electron Spectrosc. Relat. Phenom.*, 83, 9, 1997. With permission from Elsevier Ltd.)

smaller charge-transfer effects create a far better situation to study crystal field effects, as will be discussed in detail in Chapter 6.

5.4 XPS OF RE COMPOUNDS

5.4.1 SIMPLIFIED ANALYSIS FOR RE OXIDES

Most RE elements form stable sesquioxides with trivalent RE ions, which are insulators. A systematic analysis of the RE 3d XPS of a RE sesquioxide series from La_2O_3 to Yb_2O_3 has been made by Ikeda et al. (1990) and others (Kotani et al., 1987a; Nakano et al., 1987), with the use of the SIAM without the multiplet coupling effect, where the three configurations $4f^n$, $4f^{n+1}\underline{L}$, and $4f^{n+2}\underline{L}^2$ are taken into account. The calculation is an almost straightforward extension from that for CeO_2 ($n = 0$) in Chapter 3 to the case of finite value of n. In the limit of $N = 1$, the Hamiltonian matrix is given by

$$
\begin{pmatrix}
0 & \sqrt{N_f - n}\, V & 0 \\
\sqrt{N_f - n}\, V & \varepsilon_f^0 - \varepsilon_v + nU_{ff}(-U_{fc}) & \sqrt{2(N_f - n - 1)}\, V \\
0 & \sqrt{2(N_f - n - 1)}\, V & 2(\varepsilon_f^0 - \varepsilon_v) + (2n+1)U_{ff}(-2U_{fc})
\end{pmatrix}
\quad (5.15)
$$

with the basis states

$$| f^n \rangle = \prod_{v=v_1}^{v_n} a_{fv}^+ | A \rangle, \tag{5.16}$$

$$| f^{n+1} \underline{L} \rangle = \frac{1}{\sqrt{N_f - n}} \sum_{v \neq v_1 \sim v_n} a_{fv}^+ a_{kv} | f^n \rangle, \tag{5.17}$$

$$| f^{n+2} \underline{L}^2 \rangle = \sqrt{\frac{2}{(N_f - n)(N_f - n - 1)}} \sum_{v > v'(v, v' \neq v_1 \sim v_n)} a_{fv}^+ a_{kv} a_{fv'}^+ a_{kv'} | f^{n+1} \underline{L} \rangle, \tag{5.18}$$

where $v_1 - v_n$ are taken arbitrarily, and the value of n is 0, 1, 2, ..., 13 for RE = La, Ce, Pr, ..., Yb. The state $| A \rangle$ is the same as that given by Equation 3.108 of Chapter 3. The results are shown in Figure 5.22 where the solid curve is the calculated result and the dashed one is the experimental data for the RE $3d_{5/2}$ XPS. The chain curve is the background contribution in the calculated spectra and obtained by the expression

$$B(E_B) = C \int_{-\infty}^{E_B} F(E_{B'}) \, dE_{B'} \tag{5.19}$$

with an adjustable constant C. In this calculation, the value N is taken to be sufficiently large so that the calculated spectra converge well. The Hamiltonian matrix and the basis states for finite N are easily obtained by extending those for $N = 1$, but these expressions are omitted here. The parameters used in these calculations are Δ, Δ_f, V_{ff}, U_{eff}, W, and Γ, where Δ, Δ_f, and V_{eff} are defined by

$$\Delta \equiv \varepsilon_f^0 - \varepsilon_v + n U_{ff}, \tag{5.20}$$

$$\Delta_f \equiv \Delta - U_{fc}, \tag{5.21}$$

$$V_{eff} = \sqrt{N_f - n} \, V, \tag{5.22}$$

and these parameter values are listed in Table 5.2.

As seen from Table 5.2, the value of Δ in $(RE)_2O_3$ decreases monotonically from La to Eu, jumps up at Gd, and decreases again from Gd to Yb. The jump at Gd is interpreted to be due to the cost of the exchange interaction energy when transferring from a $4f^7$ to a $4f^8\underline{L}$ configuration. The value of the effective hybridization V_{eff} decreases almost monotonically with increasing n. This is partly due to the decrease in V and partly due to the decrease in the prefactor $\sqrt{N_f - n}$. The decrease in V is caused by the lanthanoid contraction of the 4f wave function with n.

The splitting of the 3d XPS in La_2O_3-Nd_2O_3 and that in Eu_2O_3 and Yb_2O_3 are explained by two different mechanisms. For La_2O_3-Nd_2O_3, Δ is much larger than V_{eff} and so the ground state is in the almost pure $4f^n$ configuration. In contrast, $|\Delta_f|$ is comparable with V_{eff}, so that $3d^94f^n$ and $3d^94f^{n+1}\underline{L}$ configurations are mixed in the final state of XPS. The two peaks of 3d XPS correspond to the bonding and antibonding

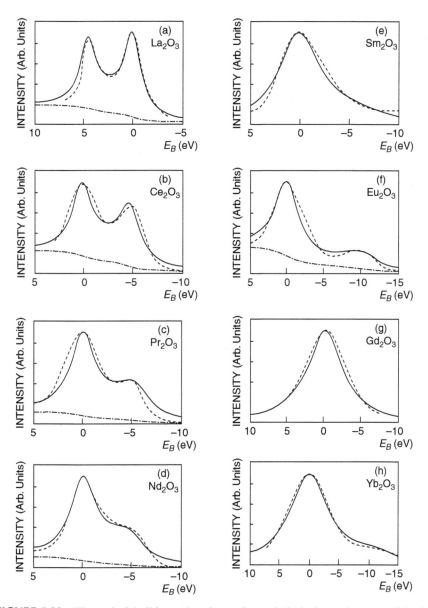

FIGURE 5.22 Theoretical (solid curve) and experimental (dashed curve) results of the 3d XPS for (a) La_2O_3, (b) Ce_2O_3, (c) Pr_2O_3, (d) Nd_2O_3, (e) Sm_2O_3, (f) Eu_2O_3, (g) Gd_2O_3, and (h) Yb_2O_3. The background contribution is shown with the chain curve. (From Kotani, A., and Ogasawara, H., *J. Electron Spectrosc. Relat. Phenom.*, 60, 257, 1992. With permission from Elsevier Ltd.)

TABLE 5.2
Parameter Values for $(RE)_2O_3$ (RE = La to Yb, Except for Pm and Tb) in the Analysis of 3d XPS (in Units of eV, Except for n_f)

	La_2O_3	Ce_2O_3	Pr_2O_3	Nd_2O_3	Sm_2O_3	Eu_2O_3
Δ	12.5	11.1	10.5	9.5	6.5	2.3
Δ_f	−0.2	−0.9	−2.4	−2.5	−4.5	−8.7
V_{eff}	2.13	2.16	1.94	1.59	0.9	0.85
U_{ff}	10.5	9.1	9.5	10.5	10.0	10.0
W	2.5	3.0	3.0	3.0	3.0	3.0
Γ	1.0	1.2	1.6	1.6	2.5	2.5
n_f	0.03	1.04	2.03	3.03	5.02	6.14
	Gd_2O_3	Dy_2O_3	Ho_2O_3	Er_2O_3	Tm_2O_3	Yb_2O_3
Δ	13.0	9.0	7.0	5.0	3.0	1.7
Δ_f	2.0	−2.0	−4.0	−6.0	−8.0	−9.8
V_{eff}	0.79	0.67	0.60	0.52	0.42	0.3
U_{ff}	10.0	10.0	10.0	10.0	10.0	10.0
W	3.0	3.0	3.0	3.0	3.0	3.0
Γ	2.8	3.8	3.8	4.0	4.0	4.0
n_f	7.0	9.0	10.0	11.01	12.02	13.05

states between $3d^94f^n$ and $3d^94f^{n+1}\underline{L}$ configurations. In the case of La_2O_3, the value of Δ_f is almost vanishes, so that the $3d^94f^0$ and $3d^94f^1\underline{L}$ configurations are mixed very strongly in the final state, resulting in the XPS peaks with comparable intensities, as seen in Figure 5.22a. In going from La_2O_3 to Nd_2O_3, the final state mixing becomes weaker because $|\Delta_f|$ increases and V_{eff} decreases. As a result, the relative intensity of the lower binding-energy peak becomes smaller because the final state of this peak has the increasingly larger weight of the $3d^94f^{n+1}\underline{L}$ configuration, which is not connected with the ground state. For Sm_2O_3, the intensity of the lower energy peak almost vanishes. In the case of Gd_2O_3, the value of Δ_f is small, similar to the case in La_2O_3-Nd_2O_3, but the 3d XPS of Gd_2O_3 has only one peak. Because the value of V_{eff} is smaller than that in La_2O_3-Nd_2O_3, the energy separation between the bonding and antibonding states of $3d^94f^7$ and $3d^94f^8\underline{L}$ configurations in the final state is small, so that they are not resolved as two peaks, as shown in Figure 5.22g.

For Eu_2O_3 and Yb_2O_3, $|\Delta_f|$ is much larger than V_{eff}, but Δ is not as large as $|\Delta_f|$ and the effect of V_{eff} is not negligible in the ground state. Therefore, the ground state is a mixed state between $4f^n$ and $4f^{n+1}\underline{L}$ configurations, although the mixing is not very large: the weight of the $4f^{n+1}\underline{L}$ configuration is 14% for Eu_2O_3 and 8% for Yb_2O_3. The two peaks in the XPS spectra are caused by the transition from the ground state to the almost pure $3d^94f^n$ and $3d^94f^{n+1}\underline{L}$ final states. It is to be mentioned that another possible origin of the XPS splitting for Eu_2O_3 and Yb_2O_3 might be due to the surface effect. For Dy_2O_3-Tm_2O_3, the 3d XPS spectrum exhibits only one peak (omitted from Figure 5.22).

As in the case of CeO_2, the ground state of PrO_2 and TbO_2 is also a strongly hybridized state between $4f^n$ and $4f^{n+1}\underline{L}$ configurations (n = 0, 1, and 7 for CeO_2,

TABLE 5.3
Parameter Values for $(RE)O_2$ (RE = Ce, Pr, and Tb) in the Analysis of 3d XPS (in Units of eV, Except for n_f)

	CeO_2	PrO_2	TbO_2
Δ	1.6	0.5	0.75
Δ_f	−10.9	−12.5	−9.75
V_{eff}	2.84	1.62	1.06
U_{ff}	10.5	10.5	10.0
W	3.0	3.0	3.0
Γ	0.7	2.0	2.5
n_f	0.52	1.56	7.61

PrO_2, and TbO_2). From the analysis of 3d XPS, the average 4f electron number is estimated to be about 1.6 for PrO_2 and 7.6 for TbO_2 (Bianconi et al., 1988; Ikeda et al., 1990). We do not show the XPS spectra for PrO_2 and TbO_2, but the parameter values of CeO_2, PrO_2, and TbO_2 are listed in Table 5.3.

5.4.2 APPLICATION OF CHARGE-TRANSFER MULTIPLET THEORY

Full multiplet calculations for the RE 3d XPS of some RE sesquioxides and dioxides were performed by Ogasawara et al. (1991b) and Kotani and Ogasawara (1992) with the SIAM including three configurations, $4f^n$, $4f^{n+1}\underline{L}$, and $4f^{n+2}\underline{L}^2$. The parameter values of the SIAM for RE sesquioxides are listed in Table 5.4, and they are almost the same as those in Table 5.2, except for Γ. Since the multiplet coupling effect does not modify 3d XPS drastically, the calculated results are in good agreement with experimental ones again, and for some spectral features, the new results are in better agreement with experiment. As examples, we show in Figures 5.23 and 5.24, the results of the 3d XPS for the sesquioxides of La, Ce, Pr, Nd, and Yb and for the Ce dioxide, respectively (Ogasawara et al., 1991b; Kotani and Ogasawara, 1992). The experimental results (Schneider et al., 1985; Allen, 1985; Ogasawara et al., 1991b) are also shown in the inset.

The parameter value Γ is fixed at 0.7 eV and is much smaller than the values of Γ in Table 5.2, which include effectively the spectral broadening due to the multiplet coupling effect. The calculated results for La ($4f^0$), Ce ($4f^1$), Pr ($4f^2$), and Nd ($4f^3$) (Figure 5.23a–d) reproduce the experimental data well, and also agree well with the results of $3d_{5/2}$ XPS in Figure 5.22a through 5.22d, which were obtained without the multiplet coupling. Therefore, we conclude that the multiplet splitting is not strong enough to give rise to additional separate peaks, but contributes to the increase in the spectral width, which can be approximately taken into account by the larger value of Γ in the analysis without the multiplet coupling. The calculated result for Yb_2O_3 (Figure 5.23e) exhibits some multiplet splitting that was not observed experimentally, but when we use a larger value of Γ (= 3.0 eV), the splitting is smeared out. For Yb_2O_3, the multiplet effect is equivalent to increasing the value of Γ from 3.0 eV (the present value) to 4.0 eV (Table 5.2).

TABLE 5.4
Parameter Values for R_2O_3 (R = La, Ce, Pr, Nd,
and Yb) in the Full-Multiplet Analysis
of 3d and 4d XPS (in eV)

Compounds	La_2O_3	Ce_2O_3	Pr_2O_3	Nd_2O_3	Yb_2O_3
Δ	12.5	11.1	10.5	9.5	1.25
Δ_f (3d)	−0.5	−2.0	−2.5	−3.5	−10.25
Δ_f (4d)	2.0	0.5	0.0	−1.0	−8.25
V_{eff}	2.13	2.16	1.94	1.59	0.3
W	2.5	2.5	2.5	—	2.5
Γ	0.7	0.7	0.7	0.7	0.7

An important effect of the multiplet coupling is to cause different spectral shapes for $3d_{5/2}$ and $3d_{3/2}$ XPS, which should have the same spectral shape only with the different intensity when we disregard the multiplet coupling. A typical example is the Pr 3d XPS of Pr_2O_3 (Figure 5.23c). The spectral splitting indicated by the main peak, m (m′) and the satellite, s (s′), is caused by the hybridization effect and is not sensitive to the multiplet coupling. However, it should be noted that the 3d XPS exhibits an intensity rise t′ only for the $3d_{3/2}$ component, resulting in the different spectral shape for $3d_{5/2}$ and $3d_{3/2}$ components. This can only be reproduced by the calculation including the multiplet effect. Also for CeO_2, shown in Figure 5.24, the agreement with experimental data is, to some extent, better than that in Figure 3.13 of Chapter 3 (without the multiplet coupling).

In 4d XPS, the effect of the multiplet coupling is much more important than in 3d XPS, because the exchange interaction between 4d and 4f states is much larger than that between 3d and 4f states. The spin–orbit interaction in the 4d state is weaker than the exchange interaction between 4d and 4f states, so that the 4d XPS spectra are not separated into $4d_{5/2}$ and $4d_{3/2}$ components. In Figure 5.25, we show the calculated 4d XPS of (a) La_2O_3, (b) Ce_2O_3, (c) Pr_2O_3, (d) Nd_2O_3, and (e) Yb_2O_3 (Ogasawara et al., 1991a; Kotani and Ogasawara, 1992) and compared them with the experimental result in the inset (Suzuki et al., 1974; Orchard and Thornton, 1978; Ogasawara et al., 1991b). The parameter values used in this analysis are listed in Table 5.4 and they are the same as those used in the analysis of 3d XPS except for Δ_f because of the reduction of U_{fc} in going from the 3d to 4d core level. The calculated results are in good agreement with the experimental ones except for Yb_2O_3. The result for Yb_2O_3 is very similar to the result for a free Yb^{3+} ion, calculated by Orchard and Thornton (1978), and we will give a detailed discussion for Yb_2O_3 later.

Detailed studies on the interplay between the interatomic hybridization and the intra-atomic multiplet coupling in 4d XPS have been made for Pr_2O_3 by Ogasawara et al. (1991a), both experimentally and theoretically. Experimentally, they performed a precise measurement of 4d XPS, which exhibits spectral shoulders a, b, d, and e and also a broad bump f, as shown in the inset of Figure 5.25c. These structures are nicely reproduced in their theoretical calculation (Figure 5.25c). In order to see the effects of the multiplet coupling and the solid-state hybridization separately,

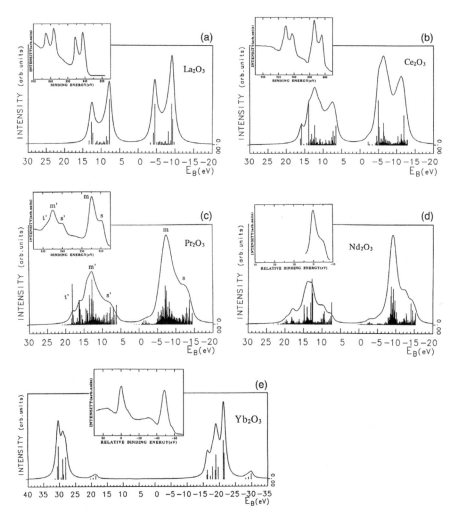

FIGURE 5.23 Theoretical results of the 3d XPS for (a) La_2O_3, (b) Ce_2O_3, (c) Pr_2O_3, (d) Nd_2O_3, and (e) Yb_2O_3. The effect of multiplet coupling is taken into account. The experimental results are shown in the inset. (From Kotani, A., and Ogasawara, H., *J. Electron Spectrosc. Relat. Phenom.*, 60, 257, 1992. With permission from Elsevier Ltd.)

Ogasawara et al. (1991a) calculated 4d XPS by changing the value of V. The results for $V = 0.0$, 0.2, 0.4, 0.56, and 0.8 eV are shown in Figure 5.26a–e, respectively. In Figure 5.26, the value N is taken to be 1 (so that $W \rightarrow 0$) for simplicity. Figure 5.26a corresponds to the 4d XPS of the free Pr^{3+} ion, which has widespread multiplet structures due to the strong Coulomb and exchange interactions. This spectrum is very different from the experimental one for Pr_2O_3. From Figure 5.26b–e, it is clearly seen that the effect of V strongly changes the atomic multiplet structure. The experimental result is reproduced well only for the value $V \approx 0.56$ eV ($V_{eff} \approx 1.94$ eV).

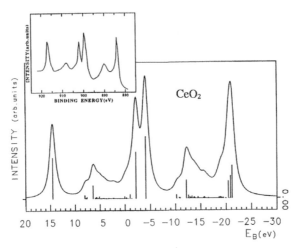

FIGURE 5.24 Theoretical result of the 3d XPS for CeO_2 with the effect of multiplet coupling. The experimental result is shown in the inset. (From Kotani, A., and Ogasawara, H., *J. Electron Spectrosc. Relat. Phenom.*, 60, 257, 1992. With permission from Elsevier Ltd.)

As seen from Table 5.4, the value of Δ_f is estimated to be 0.0 for the 4d XPS of Pr_2O_3. This means that the covalent mixing in the final state of 4d XPS is extremely strong; the mixing rate between $4d^94f^2$ and $4d^94f^3\underline{L}$ configurations should be about 0.5:0.5 for nonvanishing V. In order to see the strong covalent mixing effect on 4d XPS, the multiplet effect (except for the 4d spin–orbit interaction) is removed from Figure 5.26d. Then, the 4d XPS reduces, as shown in Figure 5.27, to four discrete lines: the m and s lines associated with the $4d_{5/2}$ core level and their $4d_{3/2}$ partner m' and s'. Figures 5.26a and 5.27 correspond to two limiting situations, in which the intra-atomic multiplet coupling and the interatomic hybridization, respectively, play a dominant role in 4d XPS. However, both of them fail in explaining the spectral features a–f of the observed 4d XPS, which are reproduced only in an intermediate situation (Figure 5.26d) between the two extremes. By comparing Figure 5.26d with $N = 1$ and Figure 5.25c with $N = 3$ and $W = 2.5$ eV, we see that the effect of the finite bandwidth W is to smear out some fine spectral structures; for instance, the kink b and some structures near f in Figure 5.26d are rounded in Figure 5.25c, in better agreement with the experimental result.

For the sesquioxides of Dy to Tm (heavy RE), the value of V_{eff} is much smaller than both Δ and Δ_f (Table 5.4), so that the effect of hybridization can be disregarded in both initial and final states of XPS. This means that the experimental results of the 3d and 4d XPS spectra of these sesquioxides should be reproduced by ionic model calculation (i.e. in the limit of $V_{eff} = 0$ in the SIAM). This is true for the 3d XPS (Kotani and Ogasawara, 1992), but the situation is more complicated for the 4d XPS. For instance, in Figure 5.28a, the experimental results of the Dy 4d XPS of Dy metal and Dy_2O_3 are shown. The spectral width of the Dy oxide is somewhat larger than that of the Dy metal, but both of these spectra are considered to reflect the multiplet structure of the $4d^94f^9$ configuration in the final state. On the other hand, the calculated result of the

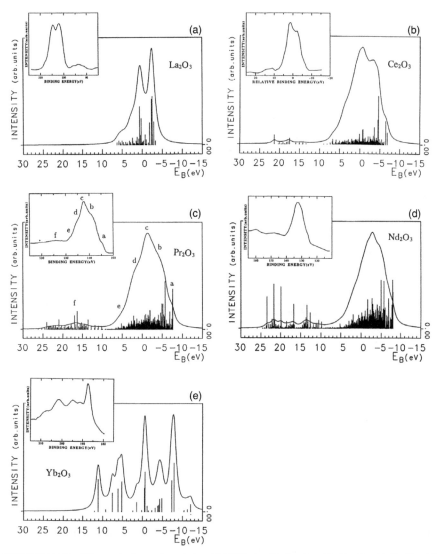

FIGURE 5.25 Theoretical results of the 4d XPS for (a) La_2O_3, (b) Ce_2O_3, (c) Pr_2O_3, (d) Nd_2O_3, and (e) Yb_2O_3. The effect of the multiplet coupling is taken into account. The experimental results are shown in the inset. (From Kotani, A., and Ogasawara, H., *J. Electron Spectrosc. Relat. Phenom.*, 60, 257, 1992. With permission from Elsevier Ltd.)

Dy 4d XPS of the Dy^{3+} free ion, which is shown in Figure 5.28d, is quite different from the experimental result. Here, the line spectra are the original result of the calculation and the solid curve is obtained by convoluting the line spectra with a Lorentzian function with a constant width (Ogasawara et al., 1994), as usual.

The reason for the disagreement between the calculated and experimental results was found to originate from the multiplet-term dependence of the lifetime of the 4d XPS final states due to the 4d4f4f super Coster–Kronig (s-CK) decay. The s-CK

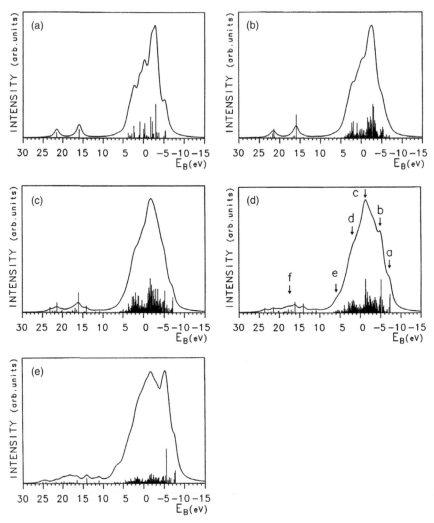

FIGURE 5.26 Theoretical results of the Pr 4d XPS spectra for Pr_2O_3 with different hybrid-ization strengths. The other parameter values are the same as those of Figure 5.25 (except for $N = 1$). In (a), (b), (c), (d), and (e), the values of V are 0.0, 0.2, 0.4, 0.56, and 0.8, respectively. (From Ogasawara, H., Kotani, A., Potze, R., Sawatzky, G.A., and Thole, B.T., *Phys. Rev. B*, 44, 5465, 1991b. With permission from Elsevier Ltd.)

decay occurs with the intra-atomic Coulomb interaction, and the lifetime broaden-ing of each XPS final state $|f\rangle$ is calculated from the expression

$$\Gamma_f = \pi \sum_F |\langle F|V_{s-CK}|f\rangle|^2 \, \delta(E_F - E_f), \qquad (5.23)$$

where V_{s-CK} denotes the Coulomb interaction describing the s-CK decay and the states $|F\rangle$ are the final states of s-CK decay with continuous energy E_F.

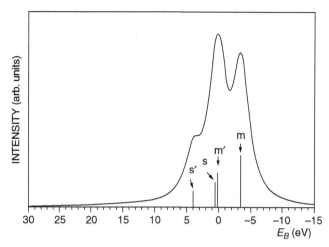

FIGURE 5.27 Theoretical result of the Pr 4d XPS spectrum for Pr_2O_3 without the multiplet coupling effect. (Reprinted with permission from Ogasawara, H., Kotani, A., Okada, K., and Thole, B.T., *Phys. Rev. B*, 43, 854, 1991a. Copyright 1991 by the American Physical Society.)

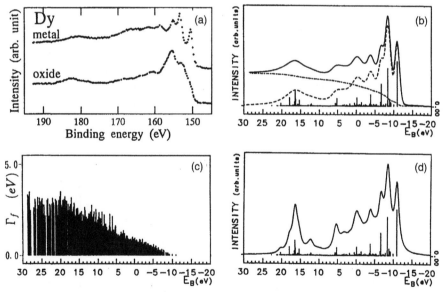

FIGURE 5.28 Theoretical and experimental results of the Dy 4d XPS spectra of Dy_2O_3. (a) Experiments for metal (upper) and oxide (lower). (b) Calculated results using the multiplet-dependent lifetime broadening Γ_f, and instrumental Gaussian broadening 0.7 eV (HWHM). The vertical bars are the original line spectra and the dashed curve is the convoluted spectra. The chain curve represents the background contribution and the solid curve is the sum of dashed and chain curves. (c) Calculated lifetime broadening Γ_f for all multiplets in the XPS final state. (d) Calculated result of XPS with a constant Lorentzian broadening 0.7 eV (HWHM). (Reprinted with permission from Ogasawara, H., Kotani, A., and Thole, B.T., *Phys. Rev. B*, 50, 12332, 1994. Copyright 1994 by the American Physical Society.)

Ogasawara et al. (1994) calculated Γ_f of each XPS final state and found that Γ_f has a strong multiplet-term dependence as shown in Figure 5.28c. The final state with a large binding energy in 4d XPS decays more rapidly than those near the threshold. If we take into account this term-dependent Γ_f, then the spectral broadening is larger for the larger binding-energy region of the 4d XPS, and agreement with the experiment is greatly improved, as shown in Figure 5.28b, where an appropriate background contribution is also taken into account. The strong term-dependence of Γ_f is a characteristic feature of the 4d XPS of heavy RE elements, whereas for light RE elements (Figure. 5.25) and also for 3d XPS (Figures 5.22 and 5.23), the term-dependence is small and can be disregarded.

In the calculated 4d XPS of Yb_2O_3 (Figure 5.25e), the effect of intra-atomic hybridization is very weak (it gives a weak satellite in the lowest binding energy region, but cannot be seen in the experimental data probably due to the large spectral broadening), so that the spectrum is almost entirely due to the free Yb^{3+} ion. The large difference between the calculated and experimental 4d XPS of Yb_2O_3 is also caused by the term-dependent Γ_f. In Figure 5.29, we show the 4d XPS of Yb_2O_3 calculated by taking into account the term-dependent Γ_f (Kotani and Ogasawara, 1992), and it is seen that the present result is in good agreement with the experimental one.

5.5 RESONANT PHOTOEMISSION SPECTROSCOPY

In valence photoemission spectroscopy (VPES), if the incident photon energy resonates with a core-electron excitation threshold, the intensity of the photoemission is resonantly enhanced. This phenomenon is called "resonant photoemission," and the spectroscopy is abbreviated as resonant photoemission spectroscopy (RPES). In a

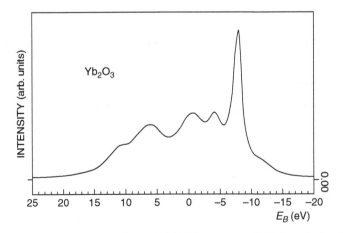

FIGURE 5.29 Theoretical result of the Yb 4d XPS spectrum of Yb_2O_3, calculated by taking into account the multiplet-term-dependent lifetime of each XPS final state. (From Kotani, A., and Ogasawara, H., *J. Electron Spectrosc. Relat. Phenom.*, 60, 257, 1992. With permission from Elsevier Ltd.)

similar way, the photoemission of core electrons (so far called XPS) is also enhanced when the incident x-ray energy resonates with the excitation threshold of a deeper core electron. We denote this spectroscopy by resonant XPS (RXPS).

5.5.1 FUNDAMENTAL ASPECTS OF RPES

The reason for the resonant enhancement is due to the Auger decay of the core-excited state that results in the same final state as VPES. For instance, let us consider the VPES of a 3d electron of transition elements with a $3d^n$ configuration and its enhancement by the incident photon energy resonating with the 3p to 3d excitation. The 3p–3d excited state has a $3p^5 3d^{n+1}$ configuration but it decays by the 3p3d3d Auger transition to the $3d^{n-1}$ final state, which is the same as the final state of VPES. The two processes, the first-order process VPES and the second-order process of the core excitation and Auger decay, with the same final states interfere with each other, resulting in asymmetry of the RPES intensity as a function of the incident energy (often called "Fano shape"). Usually, the intensity of VPES is enhanced by the resonance effect, but sometimes it can be suppressed due to the destructive interference effect and called "antiresonance."

Here we give a formula of RPES (Ogasawara et al., 1992). The total Hamiltonian H_{tot} consists of Hamiltonians describing a material system H, the radiative transition V_R and the Auger transition V_A

$$H_{tot} = H + V_R + V_A. \tag{5.24}$$

By using eigenstates of H, we can write H in the form

$$H = |g\rangle E_g \langle g| + \sum_\alpha |\alpha\rangle E_\alpha \langle\alpha| + \sum_{\varepsilon,\beta} |\varepsilon\beta\rangle E_{\varepsilon\beta} \langle\varepsilon\beta|, \tag{5.25}$$

where $|g\rangle$ with energy E_g is the ground state (for instance a $3d^n$ state), $|\alpha\rangle$ with energy E_α is the core excited state (for instance, 3p–3d excited state with $3p^5 3d^{n+1}$ configuration), and $|\varepsilon\beta\rangle$ with energy $E_{\varepsilon\beta}$ is each final state (for instance, a photoelectron and the remaining $3d^{n-1}$ state) of VPES, as well as of the Auger decay. The photoelectron state $|\varepsilon\rangle$ with energy ε and the remaining material system $|\beta\rangle$ with energy E_β can usually be treated as independent:

$$|\varepsilon\beta\rangle = |\varepsilon\rangle|\beta\rangle, \tag{5.26}$$

$$E_{\varepsilon\beta} = \varepsilon + E_\beta. \tag{5.27}$$

By the electric dipole transition, the ground state $|g\rangle$ is excited to $|\alpha\rangle$ and $|\varepsilon\beta\rangle$, respectively, with matrix elements $\langle\alpha|V_R|g\rangle$ and $\langle\varepsilon\beta|V_R|g\rangle$, and the state $|\alpha\rangle$ is coupled with the state $|\varepsilon\beta\rangle$ by the matrix element of Auger transition $\langle\varepsilon\beta|V_A|\alpha\rangle$. If we take into account the first order of V_R and up to the infinite order of V_A, the RPES spectrum for the incident photon energy is written as

$$F_{RPES}(E_B, \Omega) = \sum_\beta |\langle\varepsilon\beta|T|g\rangle|^2 \frac{\Gamma_\beta/\pi}{(E_B - E_\beta + E_g)^2 + \Gamma_\beta^2}, \tag{5.28}$$

where $\varepsilon = \hbar\Omega - E_\beta$, Γ_β is the lifetime broadening of the state $|\beta\rangle$ and T is the t-matrix defined by

$$T = V_R + V_A \left(\frac{1}{z - H - V_A} \right) T \tag{5.29}$$

with

$$z = \hbar\Omega + E_g + i\eta, \quad (\eta \to +0) \tag{5.30}$$

From Equations 5.28–5.30, after some algebra, the final expression of the RPES spectrum is obtained as:

$$F_{\text{RPES}}(E_B, \Omega) = \sum_\beta \left| \langle \varepsilon\beta|V_R|g\rangle + \sum_{\alpha, \alpha'} \langle \varepsilon\beta|V_A|\alpha\rangle \langle \alpha|G|\alpha'\rangle \right.$$

$$\left. \times \left(\langle \alpha|V_R|g\rangle + \sum_{\varepsilon', \beta'} \frac{\langle \alpha'|V_A|\varepsilon'\beta'\rangle \langle \varepsilon'\beta'|V_A|g\rangle}{z - E_{\varepsilon'\beta'}} \right) \right|^2$$

$$\times \frac{\Gamma_\beta/\pi}{(E_B - E_\beta + E_g)^2 + \Gamma_\beta^2}, \tag{5.31}$$

where G is defined by

$$G = \frac{1}{z - H - V_A}, \tag{5.32}$$

and we have used the relation

$$\langle \alpha|G|\varepsilon\beta\rangle = \sum_\alpha \langle \alpha|G|\alpha'\rangle \langle \alpha'|V_A|\varepsilon\beta\rangle \frac{1}{z - E_{\varepsilon\beta}}. \tag{5.33}$$

In order to calculate the expression of Equation 5.31 explicitly, we need to obtain the expression of $\langle \alpha|G|\alpha'\rangle$, and it is given by solving Dyson's equation

$$\langle \alpha|G|\alpha'\rangle = \frac{1}{z - E_\alpha}\delta_{\alpha\alpha'} + \frac{1}{z - E_\alpha}$$

$$\times \sum_{\varepsilon, \beta, \alpha''} \frac{\langle \alpha|V_A|\varepsilon\beta\rangle \langle \varepsilon\beta|V_A|\alpha''\rangle}{z - E_{\varepsilon\beta}} \langle \alpha''|G|\alpha'\rangle. \tag{5.34}$$

If we consider a limiting case where we only have one state $|\alpha\rangle$ in the core-electron excitation, the expression of $F_{\text{RPES}}(E_B, \Omega)$ is greatly simplified. In this case, $F_{\text{RPES}}(E_B, \Omega)$ is written as

$$F_{\text{RPES}}(E_B, \Omega) = \sum_\beta \left| \langle \varepsilon\beta|V_R|g\rangle + \langle \varepsilon\beta|V_A|\alpha\rangle \langle \alpha|G|\alpha\rangle \right.$$

$$\left. \times \left(\langle \alpha|V_R|g\rangle + \sum_{\varepsilon', \beta'} \frac{\langle \alpha|V_A|\varepsilon'\beta'\rangle \langle \varepsilon'\beta'|V_A|g\rangle}{z - E_{\varepsilon'\beta'}} \right) \right|^2 \frac{\Gamma_\beta/\pi}{(E_B - E_\beta + E_g)^2 + \Gamma_\beta^2}, \tag{5.35}$$

and we can solve Equation 5.34 to obtain

$$\langle \alpha | G | \alpha \rangle = \frac{1}{z - E_\alpha - \sum_{\varepsilon, \beta} \dfrac{\langle \alpha | V_A | \varepsilon \beta \rangle \langle \varepsilon \beta | V_A | \alpha \rangle}{z - E_{\varepsilon\beta}}} = \frac{1}{\Gamma(\hat{\Omega} + i)}, \qquad (5.36)$$

where

$$\Gamma = \pi \sum_{\varepsilon, \beta} | \langle \alpha | V_A | \varepsilon \beta \rangle |^2 \delta(\hbar\Omega + E_g - E_{\varepsilon\beta}), \qquad (5.37)$$

$$\hat{\Omega} = (\hbar\Omega + E_g - E_\alpha - \Delta)/\Gamma, \qquad (5.38)$$

with

$$\Delta = P \sum_{\varepsilon, \beta} \frac{\langle \alpha | V_A | \varepsilon \beta \rangle \langle \varepsilon \beta | V_A | \alpha \rangle}{\hbar\Omega + E_g - E_{\varepsilon\beta}}. \qquad (5.39)$$

In Equation 5.39, P represents the principal value. Furthermore, we define the so-called Fano's q-parameter by

$$q \equiv \frac{A}{B} \qquad (5.40)$$

with

$$A = \langle \alpha | V_R | g \rangle + P \sum_{\varepsilon, \beta} \frac{\langle \alpha | V_A | \varepsilon \beta \rangle \langle \varepsilon \beta | V_R | g \rangle}{\hbar\Omega + E_g - E_{\varepsilon\beta}}, \qquad (5.41)$$

$$B = \pi \sum_{\varepsilon, \beta} \langle \alpha | V_A | \varepsilon \beta \rangle \langle \varepsilon \beta | V_R | g \rangle \delta(\hbar\Omega + E_g - E_{\varepsilon\beta}), \qquad (5.42)$$

where A and B are assumed to be real. We finally obtain the expression:

$$F_{\text{RPES}}(E_B, \Omega) = \sum_\beta | \langle \varepsilon \beta | V_R | g \rangle |^2 \frac{(\hat{\Omega} + q)^2}{\hat{\Omega}^2 + 1} \cdot \frac{E_\beta/\pi}{(E_B + E_g - E_\beta)^2 + \Gamma_\beta}. \qquad (5.43)$$

If we disregard the factor $(\hat{\Omega} + q)^2/(\hat{\Omega}^2 + 1)$ in Equation 5.43, this equation reduces to the spectrum of the VPES. Therefore, $(\hat{\Omega} + q)^2/(\hat{\Omega}^2 + 1)$ represents the resonant enhancement where $\hat{\Omega}$ is the incident photon energy measured from the renormalized resonance energy $E_\alpha + \Delta - E_g$ with units of the resonance broadening width Γ. The resonance enhancement factor $(\hat{\Omega} + q)^2/(\hat{\Omega}^2 + 1)$ depends on the value of q, and in Figure 5.30, we show its behavior as a function of $\hat{\Omega}$ for various values of q. In the case of $q = 0$, we have an antiresonance, and with the increase in q, the resonant enhancement increases. In Figure 5.30, we have shown only the cases of zero and positive values of q, but the behavior of the resonant enhancement factor for the negative values of q is obtained simply by reversing the abscissa from that of the positive q.

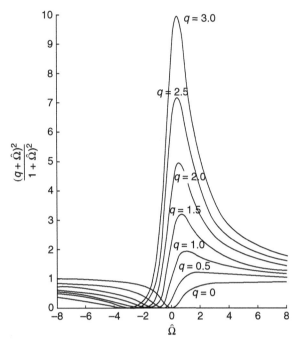

FIGURE 5.30 Spectral shapes of the Fano resonance for various values of the q parameter. (From Kotani, A., and Toyozawa, Y., *Synchrotron Radiation*, Springer-Verlag, Berlin, 1979. With kind permission of Springer Science and Business Media.)

5.5.2   RPES in Ni Metal and TM Compounds

5.5.2.1   3p RPES in Ni Metal

The first experimental observation of RPES was made by Guillot et al. (1977) for the Ni 3p threshold of Ni metal. Figure 5.31 is a sketch of the experimental result. The dashed curve is the VPES of the Ni 3d band in the off-resonance condition, which consists of the main band and a satellite at about 6 eV from the maximum of the main band. When the incident photon energy resonates with the Ni 3p–3d excitation at about 67 eV, the satellite intensity is strongly enhanced, but the main band intensity does not change very much. In the ground state of Ni metal, there are about 0.6 holes in the 3d band, but the final states of the main band and the satellite are considered, roughly speaking, to originate from the $3d^9$ and $3d^8$ configurations, respectively. The RPES result strongly supports this assignment because the intermediate state of RPES should be a $3p^5 3d^{10}$ configuration that gives rise to the $3d^8$ final state (satellite) of RPES after the 3p3d3d Auger transition. In the final state of the satellite, two 3d holes (corresponding to the $3d^8$ configuration) form a bound state called a "two-hole bound state." It should be noted that both of the $3d^9$ (one-hole state corresponding to the main band) and $3d^8$ (two-hole bound state corresponding to the satellite) configurations are not spatially localized but itinerant due to the 3d electrons in Ni metal being itinerant.

In the beginning of the 1990s, the 2p and 3p RPES spectra of Ni metal were much discussed. In particular, the issue was whether the satellite enhancement was

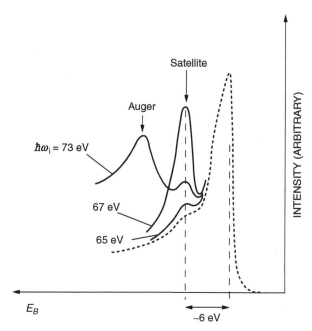

FIGURE 5.31 Schematic drawing of the resonant photoemission spectra in Ni. (From Kotani, A., *Handbook on Synchrotron Radiation Vol. 2*, 1987. With permission from Elsevier Ltd.)

due to a resonant photoemission process or was it due to an onset of a normal Auger decay process leading to overlapping spectral features. Lopez et al. (1995) measured the angular dependence of the 6 eV satellite and from the absence of the angular variation in intensity, they concluded for a normal Auger decay. Because of the complete metallic screening, the Auger-like and the photoemission-like features will appear at very similar energies for threshold excitation in metals and the two types of processes cannot, therefore, be easily separated. In both processes, the same 2p3d3d Auger channel is active and the difference is that, in RPES, this channel reaches the same final state as direct PES. As the resonant channel is angular dependent and the normal Auger is not, the absence of angular dependence indicates that the normal Auger channel dominates at the 6 eV satellite. This does not necessarily imply that there is no resonant photoemission. If in a RPES experiment, the same final state is reached via direct and resonant PES, the Fano behavior in the excitation energy dependence can be used to isolate the resonant channel. The resonant process consists of two components: the direct photoemission α and the resonant channel β. The overall intensity is given by the squared values of the direct photoemission (α^2) and the resonant photoemission (β^2), plus the interference term ($2\alpha\beta$). The interference term gives rise to the Fano line shapes if one measures the photoemission intensity as a function of the excitation energy. The relative values for α and β depends on the final state, that is, on the photoemission energy, polarization, and direction. This implies that one can measure different interference effects with the detector set at different energy intervals and/or detection angles. Another possibility to affect this ratio is to use the polarization dependence of the x-ray absorption cross section. This has been used for Ni metal by Mårtensson et al. (1997, 1999).

Figure 5.32 shows the x-ray energy dependence of the photoemission in an energy window at the 3d band (top two curves) and for an energy window at the 6 eV satellite (Weinelt et al., 1997; Mårtensson et al., 1999). The top two curves show Fano-type profiles with antiresonances before the threshold, where the top curve has the largest asymmetry, indicating the largest interference terms. Large interference occurs when the effective magnitudes of the matrix elements for the direct and resonant photoemission processes are similar. For the middle curve, the resonant process dominates and the interference term becomes less visible. At the bottom, the results for the 6 eV satellite are shown for both polarization geometries. The inset shows that the 6 eV satellite spectra also has a small antiresonance indicating interference, but this interference is small because the direct photoemission is small.

5.5.2.2 2p RPES in TM Compounds

To understand the RPES spectrum of TM compounds, the normal VPES spectrum must also be known. To explain all the features of VPES would be outside the context of this book; however, the energy positions of the main configurations is a property

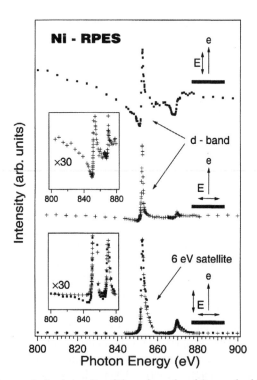

FIGURE 5.32 Photoemission intensity of the valence band (top and middle curves) and the 6 eV satellite (bottom). The measurements have been made for two different directions of the *E*-vector of the incident radiation. (From Martensson, Karis, O., and Nilsson, A., *J. Electron Spectrosc. Relat. Phenom.*, 100, 379, 1999. With permission from Elsevier Ltd.)

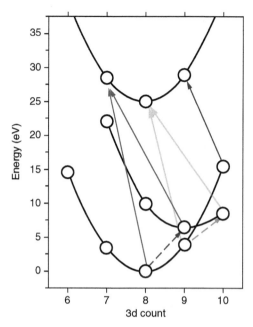

FIGURE 5.33 The energy positions of the $3d^n$ configurations of a Ni^{2+} ion using $\Delta = 4$ eV and $U_{dd} = 7.5$ eV. The bottom curve describes the $3d^8 + 3d^9\underline{L} + 3d^{10}\underline{L}^2$ ground state. The middle curve indicates the $2p^5(3d^9 + 3d^{10}\underline{L})$ XAS intermediate states and top curve the $3d^7 + 3d^8\underline{L} + 3d^9\underline{L}^2$ VPES final state. The dashed arrows indicate XAS, the solid arrows PES and AES transitions.

that is shared by valence and core excitations. Figure 5.33 shows the energy positions of the ionic configurations $3d^8$, $3d^9\underline{L}$, and $3d^{10}\underline{L}^2$, using $\Delta = 4$ eV and $U_{dd} = 7.5$ eV.

The top curve in Figure 5.33 is a copy of the bottom curve offset by 25 eV for clarity. The dark arrows indicate a VPES experiment, in which a 3d valence electron is excited. This implies the transition from $3d^8$ to $3d^7$, and alike. The ligand states are assumed not to modify the picture, which implies that exactly the same energy positions can be used for the final state. Because $\Delta = 4$ eV and $U_{dd} = 7.5$ eV (corresponding to NiO), the energy of $3d^7$ is significantly higher than $3d^8\underline{L}$. In other words, the ionic configuration at the lowest energy is dominated by the $3d^8\underline{L}$ configuration.

The energy positions of RPES at the 2p threshold excitation are also shown in Figure 5.33. The middle curve represents the intermediate state of RPES, which is the same as the final state of 2p XAS (see Chapter 3). The light arrows from intermediate to final states express the 2p3d3d Auger transitions by which the same final state as VPES is reached as in normal VPES (e.g. $3d^8$ to $3d^7$ via $2p^53d^9$).

Now we present theoretical calculations of RPES for TM oxides at 2p threshold excitation. As the resonant enhancement factor of RPES at the 2p threshold is much larger than that at the 3p threshold, we can thus extract more unambiguous resonance process. As an example, the calculated results of (a) RPES and (b) off-resonant VPES for TM oxides are shown in Figure 5.34 (Tanaka and Jo, 1994). The calculations are made so as to reproduce the corresponding experimental data (not shown here). For CuO and NiO, the resonant enhancement occurs at a high-binding energy region ($E_B \approx 10$ eV for CuO and ≈ 7 eV for NiO). This is evidence that these

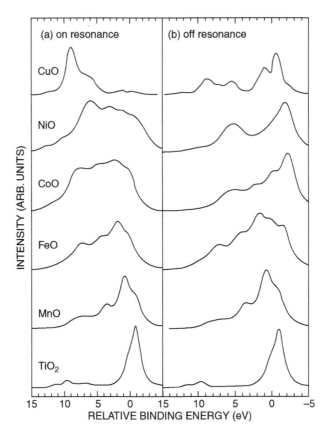

FIGURE 5.34 Calculated results of on- and off-resonance spectra of VPES in TM oxides. The RPES spectra are taken at the $2p_{3/2}$ threshold. (From Tanaka, A., and Jo, T., *J. Phys. Soc. Jpn.*, 63, 2788, 1994. With permission.)

compounds are charge-transfer type insulators because the high- and low-binding energy parts, respectively, correspond to the $3d^{n-1}$ and $3d^n\underline{L}$ configurations ($n = 9$ and 8 for CuO and NiO) whose energy difference is $U_{dd} - \Delta$ (in the limit of vanishing V), but the resonant enhancement occurs only for the $3d^{n-1}$ configuration. With a decreasing 3d electron number n, U_{dd} becomes close to Δ (corresponding to the intermediate-type of the charge transfer and Mott–Hubbard insulators), so that the resonant enhancement occurs in the whole binding energy region.

The parameter values of the SIAM estimated from this analysis are listed in Table 5.5. The trend of the variation of Δ and U_{dd} with the change of n is consistent with that discussed previously, but the values of Δ for NiO and CoO are much larger than those in Table 5.1. It will be shown in Section 5.8 that, at least for NiO, the value of Δ estimated by the analysis of XPS with the SIAM should be too small because the SIAM calculation disregards the nonlocal screening effect, which is important in the XPS of NiO. As will be mentioned in Chapter 8, the scatter of the value Δ estimated for NiO is anomalously large; from the analysis of the experimental data of XPS, valence PES, RPES, XAS, and RXES of NiO with the SIAM or

TABLE 5.5
Parameter Values of the SIAM Estimated from the Analysis of RPES

Compounds	n	Δ	U_{dd}	$V(e_g)$	10 Dq	T_{pp}	U_{dc}
CuO	9	3.0	7.8	2.7*	0.0	1.3	9.0
NiO	8	4.7	7.3	2.2	0.7	0.7	8.5
CoO	7	6.5	6.5	2.2	0.5	0.7	8.2
FeO	6	7.0	6.0	2.1	0.5	0.7	7.5
MnO	5	8.0	5.5	2.1	0.5	0.7	7.2
TiO_2	0	4.0	4.0	3.0	1.0	1.0	6.0

* The value of $V(b_{1g})$.

cluster model, the estimated value of Δ is distributed from 2.0 to 6.2 eV, depending on the spectroscopy and the model of the analysis. The value of Δ estimated from precise analysis of RXES for NiO is 3.5 eV, which is somewhat smaller that that in Table 5.5. It is interesting to see that the values of Δ by the analysis of RXES (Chapter 8) for NiO, CoO, MnO, and TiO_2 are 3.5, 4.0, 8.0, and 2.9 eV, respectively, which are smaller systematically by 0.5 to 1.0 eV than those by the analysis of RPES in Table 5.5.

5.5.2.3 3p RPES in NiO

Figure 5.35 shows the 3p RPES spectrum of NiO while scanning through the NiO 3p XAS spectrum with its peak position at 66 eV (Tjernberg et al., 1996). The normal PES spectrum is observed at 59 eV with the main peak at ~2 eV binding energy and a shoulder at ~5 eV. The main peak does not show any significant energy variations while scanning through the edge and its intensity is a little decreased before threshold and increased at threshold. This indicates a Fano behavior of the valence band PES. The 5 and 9 eV binding energy features are enhanced with the excitation of Ni 3p electrons. In addition, a number of additional peaks and shoulders become visible at around 70 eV, ~4 eV, and 7.5 eV. These additional features are related to the detailed behavior of the various states in the 3p-hole intermediate state and the final state, due to the combined effects of charge transfer, multiplets, spin–orbit coupling, and exchange. All resonances decrease after the threshold while the 9 eV peak shifts a little towards a higher binding energy. In their 1996 paper, Tjernberg et al. also showed these resonant photoemission spectra as a 2D image, similar to the RXES 2D images that are discussed in Chapter 8. An advantage of the 2D image is that one directly views the resonant channels that have constant binding energy and the nonresonant channels that have constant kinetic energy.

5.5.3 3d AND 4d RPES OF Ce COMPOUNDS

In the Ce intermetallic compounds, the localization of Ce 4f states strongly depends on the hybridization strength between the Ce 4f and conduction electron states. For instance, $CeRu_2Si_2$ is a typical material with weak hybridization and the system

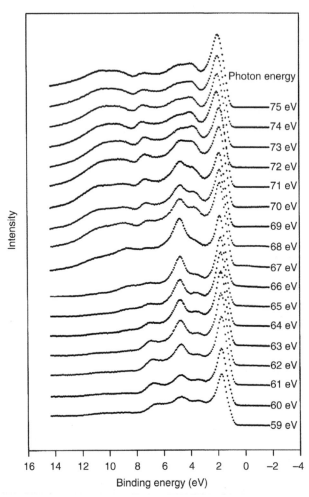

FIGURE 5.35 The resonant photoemission spectra of the NiO valence band at the Ni 3p absorption edge. (Reprinted with permission from Tjernberg, O., et al., *Phys. Rev. B*, 53, 10372, 1996. Copyright 1996 by the American Physical Society.)

behaves as a "heavy fermion" with the Kondo temperature $T_K \approx 22$ K. On the other hand, $CeRu_2$ is a material with a strong hybridization and behaves as a "valence fluctuating system" with T_K of the order of 1000 K. Therefore, the VPES spectra of the two materials are expected to behave very differently, reflecting the difference of the hybridization. In order to extract the 4f electron contribution from VPES, which includes in the off-resonance condition, the contribution of other valence electrons (p and d electrons) with considerable weight, the technique of RPES at the 4d threshold (denoted by 4d4f RPES) has usually been used because the final state of the RPES due to the 4d4f excitation followed by the 4d4f4f Auger decay is the same as the direct 4f emission in VPES. However, the 4d4f RPES measurements so far made often provided us with unexpected puzzling data probably because of the surface-sensitivity of the 4d4f RPES.

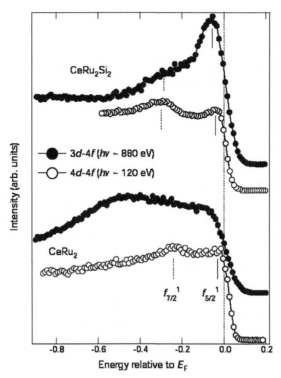

FIGURE 5.36 Experimental results of the Ce 4f spectra in the 3d4f and 4d4f RPES of $CeRu_2Si_2$ and $CeRu_2$. (From Sekiyama, A., et al., *Nature*, 403, 396, 2000. Copyright 2000. Reprinted with permission from Macmillan Publishers Ltd.)

Sekiyama et al. (2000) have succeeded in measuring the 3d4f RPES of Ce compounds having higher bulk sensitivity with very high resolution. Figure 5.36 shows their data where both 3d4f and 4d4f RPES spectra are measured for $CeRu_2Si_2$ and $CeRu_2$. It is seen that the 4d4f RPES spectra of $CeRu_2Si_2$ and $CeRu_2$ are qualitatively similar, but the 3d4f RPES of these materials are quite different and exhibit the RPES features characteristic of the heavy fermion and the valence fluctuation. The mean free path of the photoelectron of 4d4f RPES (excitation energy $\hbar\Omega \approx 120$ eV) is less than 5 Å, so that the 4d4f RPES is a surface-sensitive technique, while the mean free path in the 3d4f RPES ($\hbar\Omega \approx 880$ eV) is as long as 15 Å with much more bulk sensitivity.

5.5.4 RESONANT XPS

Instead of a valence band photoemission channel, a core XPS channel can also be studied in a resonant way. The 3p XPS spectrum can be measured while scanning through the 2p XAS spectrum. Over all excitation energies, there will be an essentially constant direct 3p XPS and at the 2p resonance the excited state can decay via the 2p3p3d Auger into a 3p XPS-like final state. Exactly like for the valence band, this will give a RXPS with the corresponding interference effects.

Experiments by Nakamura et al. (1996) have shown a similar behavior to that of the valence band. The direct 3p XPS channel is not enhanced by much; however, at resonance, new peaks arise at a higher binding energy. The direct 3p XPS channel in NiO is dominated by $3p^53d^9\underline{L}$, while the dominating resonant channel is the $3p^53d^8$ final state that is reached via the $2p^53d^9$ intermediate state. The same behavior occurs for the 3s XPS final state. More complex things are happening at resonance, but these are easier to observe in resonant Auger channels than final states that cannot be reached via direct photoemission. This is discussed subsequently.

5.5.5 RESONANT AUGER ELECTRON SPECTROSCOPY

RAES is equivalent to RPES with the difference that the final state cannot be reached by a direct photoemission channel. For example, all final states that involve two core states can only be reached at resonance via an XAS intermediate state. The equation for RAES is equivalent to that of RPES where the direct photoemission channel is omitted. If we simplify the ground state wave function to a single $3d^8$ configuration and describe the transition to a $3p^43d^9\varepsilon$ final state, the equation becomes:

$$
F_{RAES}(E_B, \Omega) \propto \left| \sum_{2p^53d^9} \frac{\langle 3p^43d^9\varepsilon | V_A | 2p^53d^9 \rangle \langle 2p^53d^9 | V_R | 3d^8 \rangle}{E_{2p^53d^9} - E_{3d^8} - \hbar\Omega - i\Gamma_{2p}} \right|^2
$$
$$
\times \delta(E_B + E_{3d^8} - E_{3p^43d^9}). \tag{5.44}
$$

This equation is similar to the RXES equations as discussed in Chapter 8. The difference is that the decay is via an Auger channel instead of a radiative x-ray emission channel, implying an electron detector instead of an x-ray emission detector. The similarity between RAES and RXES implies that one can also use the RAES channels for selective XAS experiments. In addition, the lifetime broadening can be effectively removed using RAES with a high-resolution electron detector, as discussed in Section 5.5.6.

Because they contain two core holes, the energy position changes. Figure 5.37 shows the energy position of the RAES process. We have seen that for a single core hole, the minimum of the parabola shifts from $3d^8$ to $\underline{c}3d^9$. A second core hole shifts the minimum further to $\underline{cc}'3d^{10}$. In normal Auger electron spectroscopy (NAES), the $3d^8$ state is excited to a $2p^53d^8$ state as in the XPS process, which in turn decays to a $3p^43d^8$ state. In RAES, the first step is an XAS process that brings $3d^8$ to $2p^53d^9$. Because of this, RAES spectra will be different from NAES and, in addition, RAES spectra will be energy dependent. For example, excitations into the charge-transfer satellite will produce different final states than will excitations into the main XAS peak.

Figure 5.38 shows the comparison between the RAES spectra for the 2p3s3s, 2p3s3p, and 2p3p3p transitions. Figure 5.38 shows that these transitions can be calculated using a $3d^8 + 3d^9\underline{L}$ initial state, a $2p^53d^9 + 2p^53d^{10}\underline{L}$ intermediate state and, for 2p3s3s Auger, a $3s^03d^9 + 3s^03d^{10}\underline{L}$ final state. This CTM calculation yields the correct spectral shape at resonance. Note that both the $3s^03d^9$ state and the $3s^03d^{10}\underline{L}$ configuration consist of a single state. This implies that the 2p3s3s RAES spectrum is predicted to consist of only two peaks, in agreement with experiment.

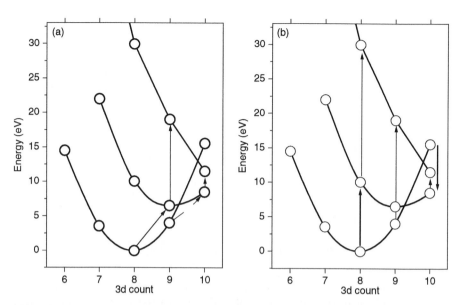

FIGURE 5.37 The energy positions of (a) RAES and (b) NAES. The bottom curve describes the $3d^8 + 3d^9\underline{L} + 3d^{10}\underline{L}^2$ ground state. The middle curve is the 2p-hole intermediate state, respectively (a) $2p^5(3d^9 + 3d^{10}\underline{L})$ XAS and (b) $2p^5(3d^8 + 3d^9\underline{L} + 3d^{10}\underline{L}^2)$ XPS. The top curve is the final state of the 2p3p3p Auger process.

Figure 5.39 shows the final-state binding energy against the photon energy dependence of the position of the 2p3s3p AES lines. Horizontal lines are observed at constant binding energy and diagonal lines at constant kinetic energy. The horizontal lines refer to RAES processes and the diagonal lines to NAES processes. Using Figure 5.37, the RAES process is given as XAS + AES and the NAES process as XPS + AES.

We can distinguish, at least, four regions characterized by a different behavior of the binding energy position of the RAES lines:

1. Before the L_3 edge.
2. Across the L_3 edge.
3. Across the L_3 charge transfer satellite.
4. Above the L_3 edge region.

(Ad. 1) Before the L_3 edge, we observe the 2p3s3p peaks at constant binding energy typical of the RAES process.

(Ad. 2) Across the L_3 edge, the main peak first continues at constant binding energy and, at ~853 eV, the peak position starts following an Auger line with constant kinetic energy. This indicates that the NAES becomes the major process at this energy.

(Ad. 3) Across the L_3 charge-transfer satellite region, the position of the peaks suddenly drops, and starts to follow an Auger line at a different kinetic energy. At ~861 eV, the peak position shifts back to the original Auger line.

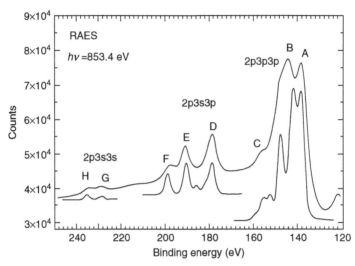

FIGURE 5.38 Comparison between experimental data and CTM calculations of the 2p3s3s, 2p3s3p, and 2p3p3p Auger lines measured at the Ni L_3 resonance (853.4 eV) of NiO. (Reprinted with permission from Finazzi, M., Brookes, N.B., and de Groot, F.M.F., *Phys. Rev. B*, 59, 9933, 1999. Copyright 1999 by the American Physical Society.)

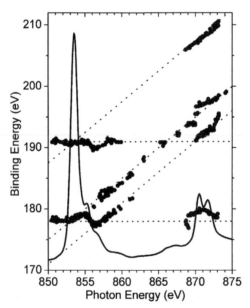

FIGURE 5.39 The XAS spectrum of NiO is given together with the binding energies of the 2p3s3p AES peaks as a function of the photon energy. The horizontal straight lines correspond to constant binding energy peaks and the diagonal straight lines correspond to constant kinetic energy peaks. (Reprinted with permission from Finazzi, M., Brookes, N.B., and de Groot, F.M.F., *Phys. Rev. B*, 59, 9933, 1999. Copyright 1999 by the American Physical Society.)

(Ad. 4) The same Auger line as observed on resonance is also the dominant peak off-resonance.

The origin of the Auger line observed at the charge-transfer satellite can be explained with Figure 5.37. Excitations to the charge-transfer satellite reach intermediate states dominated by $2p^53d^{10}\underline{L}$. The intermediate state with a $2p^53d^{10}$ character decays to the $3p^53s^13d^{10}$ final state, which according to Figure 5.37, it has the lowest energy. At resonance, ~80% of the total intensity goes into the $2p^53d^9$ intermediate state configuration and about 20% into the $3d^{10}\underline{L}$ intermediate state configuration. Above the resonance, most intensity also goes into the $2p^53d^9\underline{L}$ state via interference effects. Both at resonance and off-resonance, less than 1% goes into the $2p^53d^{10}\underline{L}$ configuration and so the decay to the $3p^53s^13d^{10}$ final state is small. This explains why the $3p^53s^13d^{10}$ Auger line has low intensity at resonance and off-resonance, but gains intensity for excitations into the charge-transfer satellite.

5.5.6 REDUCING THE LIFETIME BROADENING IN XAS

In Chapter 8, we discuss in detail how the lifetime broadening of the intermediate state is effectively removed in a RXES experiment. A similar phenomenon occurs in RPES and RAES. If we use a resonant Auger channel, the resonance is described by Equation 5.44. One can measure the x-ray absorption spectrum by detecting the Auger spectrum at a certain final state energy.

The major experimental constraint for these experiments is that the overall resolution of x-ray excitation and electron detection should be good enough to observe the removal of the lifetime broadening of the deep core state. Drube et al. (1993) have shown that, for silver and silver alloys, we can effectively assume that the lifetime broadening is reduced. More precisely, we are able to detect the Ag L_3 edge with sub-lifetime broadening resolution, as is shown in Figure 5.40. In analogy with x-ray emission, this technique is called high-energy resolution electron detection (HERED).

It is somewhat strange that the possibilities of HERED-XAS experiments are so little used. The experiments are not easy and need an intense x-ray source, but there are many advantages over normal XAS. Drube et al. (1993) demonstrated the principle of HERED-XAS on the Pd L edge and the Ag L edge. They also applied the method to the study of the Ag L edge of AgAu alloys and showed the transfer of a d-charge from gold to silver (Drube et al., 1998, 2003). Soft x-ray applications have been demonstrated by Coulthard et al. (2000) on the Co L_3 edge and the Al K edge.

5.5.7 EQ AND ED EXCITATIONS IN THE PRE-EDGE OF Ti 1s XAS OF TiO$_2$

In the pre-edge region of the Ti 1s XAS of TiO_2 (rutile), it is well known that three peaks A_1, A_2, and A_3 are observed as shown in Figure 5.41. Here we show recent experimental results by Danger et al. (2002) [see also Le Fèvre et al. (2004)], but the existence of the three peaks were known more than 20 years ago by experimental measurements (Grunes, 1983). The polarization dependence of this XAS was also measured by Poumellec et al. (1991), Uozumi et al. (1992), and Danger et al. (2002), as shown in Figure 5.41, where the linear polarization vector and the wavevector of the incident photons are both perpendicular to the c-axis, and the

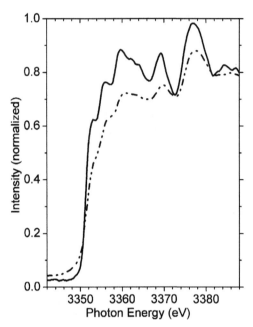

FIGURE 5.40   The silver L_3 XAS spectrum obtained via total electron yield (dash-dotted) and via the 2p3d3d RAES peak (solid). (From Drube, W., Lessmann, A., and Materlik, G., *Jpn. J. Appl. Phys.*, 32, 173, 1993. With permission.)

angle θ of the wavevector with respect to the (110) surface normal is taken as $0°$ and $45°$. As the origin of the three peaks, Uozumi et al. (1992) gave, for the first time, the interpretation that A_1 and A_2 are due to the EQ transitions to the $3d(t_{2g})$ and $3d(e_g)$ states, respectively, while A_2 and A_3 are the ED transitions to the off-site $3d(t_{2g})$ and $3d(e_g)$ states, respectively, where the off-site 3d states mean the 3d states on the Ti sites, but outside of the core excited Ti site. It should be noted that the peak A_2 is interpreted as a superposition of the EQ and ED transitions. Uozumi et al. (1992) showed that the θ-dependence of the EQ transition intensity to the $3d(t_{2g})$ state is given by

$$\frac{1}{2}+\frac{3}{4}(1+\cos 4\theta),$$

that to the $3d(e_g)$ state by

$$\frac{3}{4}(1-\cos 4\theta),$$

and the ED transition intensity does not depend on the angle θ. Therefore, the A_1 intensity is larger for $0°$, the A_2 intensity is slightly larger for $45°$, and the A_3 intensity is the same for $0°$ and $45°$, consistent with the experimental results of Figure 5.41.

In order to confirm the above interpretation, especially to confirm that the peak A_2 is the superposition of the EQ and ED transitions, the RAES is a very useful tool

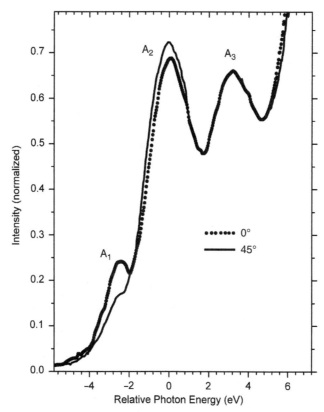

FIGURE 5.41 Three peaks observed at the pre-edge of Ti 1s XAS of TiO_2 in the 0° (points) and 45° (solid) geometries. (Reprinted with permission from Danger, J., et al., *Phys. Rev. Lett.*, 88, 243001, 2002. Copyright 2002 by the American Physical Society.)

to separate the effects of EQ and ED transitions. Experimental results of the Ti 1s2p2p RAES of the Ti 1s pre-edge excitation for TiO_2 are shown in Figure 5.42 (Danger et al., 2002). The various values of the incident photon energy are taken where the origin of the incident energy is just at the A_2 peak position, and the A_1 and A_3 peaks correspond to −2.75 eV and +3.0 eV (Figure 5.41). The RAES peaks Q_1 and Q_2 are interpreted to originate from the EQ transitions from Ti 1s to $3d(t_{2g})$ and $3d(e_g)$ states, respectively, while the RAES peak D is due to the ED transition from Ti 1s to off-site 3d states. The peak D behaves as Raman-like below the excitation A_2, but behaves as NAES-like above A_2 because the off-site $3d(t_{2g})$ and $3d(e_g)$ states are both extended in space, as will be shown later. The angle-dependence of Q_1 and Q_2 is consistent with this interpretation, and furthermore, the effects of the EQ (Q_1) and ED (D) excitations around the XAS peak A_2 are clearly separated as different RAES signals.

Theoretical calculations of Ti 1s pre-edge XAS and 1s2p2p RAES spectra were made by Uozumi et al. (2004) (see also Le Fèvre et al., 2004), by combining linear combination of atomic orbitals (LCAO) calculations for a $Ti_{445}O_{890}$ cluster and the CTM calculations for a TiO_6 cluster. The results are in good agreement with the experimental results given above and also clearly show the mechanism of the ED transition from Ti 1s to the off-site 3d states. Figure 5.43 is the calculated Ti 1s pre-edge XAS spectra for the 0° (solid curve) and 45° (dashed curve) geometries. For calculations of the EQ transitions from Ti 1s to Ti 3d(t_{2g}) and 3d(e_g) states, the CTM calculation is made with a TiO_6 cluster model, but for ED transitions, the partial density-of-states (DOS) of p-symmetric excited states obtained by LCAO calculations for a $Ti_{445}O_{890}$ cluster model is used. In the inset of Figure 5.43, we see weak p-symmetric DOS (indicated by P state) below the Ti 4p main band and superposed on the Ti 3d(t_{2g}) and 3d(e_g) bands. These p-symmetric states include weak 4p wave functions on the central Ti site (core excitation site), so that they are weakly allowed by the ED transition from the Ti 1s states in the pre-edge region, but the main weight

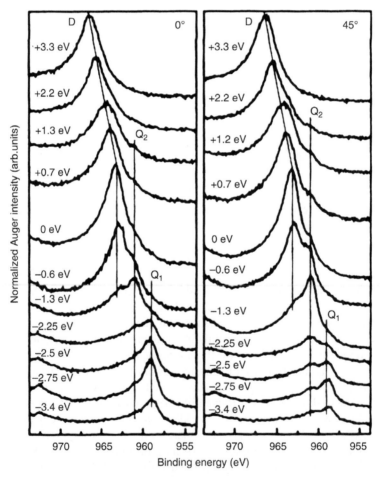

FIGURE 5.42 Experimental results of the Ti 1s2p2p RAES ($^1D_2$) spectra in the pre-edge excitation region of TiO_2 measured in the 0° (left panel) and 45° (right panel) geometries. (Reprinted with permission from Danger, J., et al., *Phys. Rev. Lett.*, 88, 243001, 2002. Copyright 2002 by the American Physical Society.)

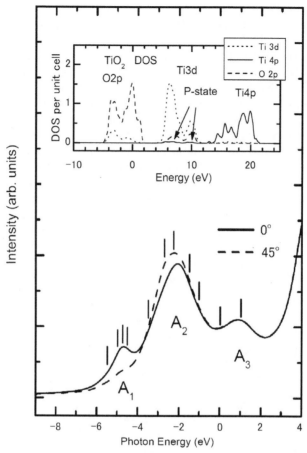

FIGURE 5.43 Calculated results of the Ti 1s pre-edge XAS spectra for 0° (solid) and 45° (dashed) geometries. The vertical bars correspond to the positions of incident photon energies used in Figures 5.42 and 5.44. The inset is the projected DOS of a $Ti_{445}O_{890}$ cluster obtained by the LCAO method with Ti 3d, 4p, and O 2p states. (From Uozumi, T., Kotani, A., and Parlebas, J.C., *J. Electron Spectrosc. Relat. Phenom.*, 137–140, 623, 2004. With permission from Elsevier Ltd.)

of the wave function consists of Ti 3d states outside of the central Ti site (therefore they are denoted by the off-site 3d states). The mixing of the off-site 3d and on-site 4p states is caused by the covalence hybridizations of Ti 4p–O 2p and O 2p–Ti 3d states. In the calculation of XAS by the ED transition, the single particle excitation of Ti 1s to off-site 3d states is made with the partial DOS of the P state in the inset of Figure 5.43, but the screening effect of the core hole potential after the single particle transition is made by the TiO_6 cluster mode (similar to the Ti 2p XPS of TiO_2). Since the photo-excited 3d electron in the EQ transition screens the core hole potential more strongly than the O 2p–Ti 3d charge transfer in the ED transition, the pre-edge XAS spectra of the EQ transition is lower in energy by about 2 eV than that of the ED transition, and this is the reason that the EQ transition to the Ti 3d(e_g) state overlaps with the ED transition to the Ti off-site $3d(t_{2g})$ state to form a single XAS peak A_2. For the calculations of the Ti 1s XAS in the pre-edge region, see Joly et al. (1999) and Shirley (2004).

The calculated results of the Ti 1s2p2p RAES are shown in Figure 5.44 (Uozumi et al., 2004). They are in good agreement with the experimental results (Figure 5.42), and it is confirmed that the RAES method is a powerful tool to discriminate the EQ and ED contributions in the pre-edge of Ti 1s XAS. It is also confirmed that since the off-site Ti 3d states are extended spatially similar to the Ti 3d band, the Auger spectra will behave as the NAES spectra (with a constant kinetic energy of the Auger electron) above the threshold (A_2), but below the threshold the Auger process is virtual and the RAES spectra will behave as the Raman-like spectra (with a constant binding energy).

Finally, it is to be remarked that the coexistence of the EQ transition to the on-site 3d state and the ED transition to the off-site 3d state in the pre-edge region of TM 1s XAS is more or less commonly observed in many TM compounds. We will discuss some examples in cuprates, and Co and Fe compounds in Chapter 8, where we will show that the RXES spectra are also a very powerful means to detect the EQ and ED excitations separately. It should be mentioned that TiO_2 is the so-called band insulator, but most other TM compounds are charge-transfer type or Mott–Hubbard type insulators, where the off-site 3d states correspond to the upper-Hubbard band above the

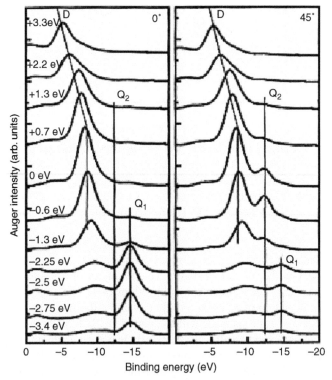

FIGURE 5.44 Calculated results of the Ti 1s2p2p RAES ($^1D_2$) spectra in the pre-edge excitation region of TiO_2 for the 0° (left panel) and 45° (right panel) geometries. (From Le Fevre, P., et al., *J. Electron Spectrosc. Relat. Phenom.*, 136, 37, 2004. With permission from Elsevier Ltd.)

correlation gap, and the electron correlation effect is essential in describing the ED transition to the off-site 3d states.

5.6 HARD X-RAY PHOTOEMISSION SPECTROSCOPY

Hard x-ray photoemission spectroscopy (HAXPS) is a recent addition to XPS spectroscopy. HAXPS is a bulk-sensitive prove, which is a very important advantage over the soft x-ray photoemission so far treated, but it has been difficult to obtain high-resolution experimental data with HAXPS. Very recently, it became possible to measure high-resolution HAXPS due to improved technology.

5.6.1 2p HAXPS OF CUPRATES

Taguchi et al. (2005a) recently measured the Cu $2p_{3/2}$ HAXPS spectra for La_2CuO_4 (LCO), hole-doped $La_{1.85}Sr_{0.15}CuO_4$ (LSCO), Nd_2CuO_4 (NCO), and electron-doped $Nd_{1.85}Ce_{0.15}CuO_4$ (NCCO) using a photon energy of $\hbar\Omega = 5.95$ keV. The results for NCCO, LCO, and LSCO are shown in Figure 5.45 with the solid curves. The results are compared with those measured with a photon energy of $\Omega = 1.5$ keV (soft x-ray) as shown by the dash-dotted curves in Figure 5.45. The HAXPS spectra are bulk-sensitive with a proving depth of about 60 Å, while the proving depth of the soft x-ray PES is about 15 Å, and so the results are surface-sensitive. The most remarkable finding in HAXPS experiments is the occurrence of an anomalously strong peak (indicated by α in Figure 5.45) in Cu 2p XPS of electron-doped $Nd_{1.85}Ce_{0.15}CuO_4$. This peak is almost absent in the soft x-ray experiments. In strong contrast to this, no such anomalous peak is observed for hole-doped $La_{1.85}Sr_{0.15}CuO_4$ where the difference between hard and soft x-ray experiments is much smaller (Taguchi et al., 2005a). For undoped La_2CuO_4, some differences are observed between hard and soft x-ray experiments. It will be interesting to investigate and learn about the origin of the

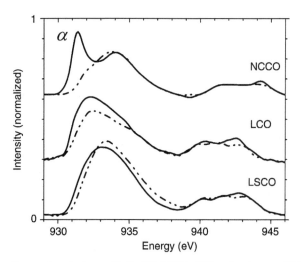

FIGURE 5.45 Comparison between Cu 2p HAXPS (solid) and soft-x-ray-PES (dash-dotted) for electron doped $Nd_{1.85}Ce_{0.15}CuO_4$, undoped La_2CuO_4, and hole doped $La_{1.85}Sr_{0.15}CuO_4$. (Reprinted with permission from Taguchi, M., et al., *Phys. Rev. Lett.*, 95, 177002, 2005a. Copyright 2005 by the American Physical Society.)

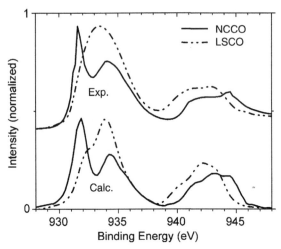

FIGURE 5.46 Comparison between calculated (lower part) and experimental (upper part) results of the Cu 2p XPS for $Nd_{1.85}Ce_{0.15}CuO_4$ (solid) and $La_{1.85}Sr_{0.15}CuO_4$ (dash-dotted). (Reprinted with permission from Taguchi, M., et al., *Phys. Rev. Lett.*, 95, 177002, 2005a. Copyright 2005 by the American Physical Society.)

anomalous peak α in $Nd_{1.85}Ce_{0.15}CuO_4$ and why such an anomalous peak is almost absent in $La_{1.85}Sr_{0.15}CuO_4$.

Taguchi et al. (2005a) undertook a simple theoretical analysis for the HAXPS data that was based on the SIAM calculations of undoped La_2CuO_4 and Nd_2CuO_4 with a charge-transfer energy $\Delta = 3.6$ and $3.0\,eV$, respectively. Also, for doped systems $La_{1.85}Sr_{0.15}CuO_4$ and $Nd_{1.85}Ce_{0.15}CuO_4$, they took into account an additional charge-transfer effect from extra metallic states at the Fermi level to the Cu 3d states with the charge-transfer energy Δ^*. The value of Δ^* (1.35 eV for $La_{1.85}Sr_{0.15}CuO_4$ and 0.25 eV for $Nd_{1.85}Ce_{0.15}CuO_4$) and the strength of the extra charge transfer were treated as adjustable parameters to fit the calculated spectra with the experimental ones. The anomalous peak in $Nd_{1.85}Ce_{0.15}CuO_4$ is thought to originate from this additional charge-transfer effect that corresponds to the metallic screening of the core hole potential by doped electrons. The calculated spectra are shown in the lower panel of Figure 5.46, compared with the experimental results in the upper panel. For hole-doped $La_{1.85}Sr_{0.15}CuO_4$ (dash-dotted), the metallic screening effect is assumed to be much weaker by taking a small charge-transfer strength; however, a weak shoulder at the highest binding energy of the calculated spectrum is due to this effect. The value of Δ^* for $La_{1.85}Sr_{0.15}CuO_4$ is slightly larger than the band gap (about 1.0 eV for both La_2CuO_4 and Nd_2CuO_4) and for $Nd_{1.85}Ce_{0.15}CuO_4$ it is much smaller than the band gap. This means that the extra metallic state in $La_{1.85}Sr_{0.15}CuO_4$ occurs near the top of the O 2p valence band while that in $Nd_{1.85}Ce_{0.15}CuO_4$ occurs near the bottom of the upper Hubbard band. See Section 5.8 for a more detailed analysis of these experimental data.

5.6.2 2p HAXPS OF V_2O_3 AND $La_{1-x}Sr_xMnO_3$

A HAXPS experiment on V_2O_3 has been performed by Taguchi et al. (2005b) and by Panaccione et al. (2006). Large variations were shown in the 2p XPS spectra of V_2O_3

through its magnetic phase transition. In addition, these HAXPS experiments have better resolved structures than the V_2O_3 spectra that were analyzed by Uozumi et al. (1997). The 2p XPS spectrum of V_2O_3 as measured with Al Kα (1486 eV) is shown in Figure 5.47c. If measured with 5934 eV in an HAXPS experiment, the $2p_{3/2}$ peak reveals significant structure, as shown in Figure 5.47a and 5.47b (Panaccione et al., 2006). Interestingly, the structure on the low-binding energy side is very sensitive to the phase transition. These sharp structures disappear for V_2O_3 samples with poorer quality, indicating that they are related to the coherent peak at the Fermi level (Panaccione et al., 2006). Taguchi et al. (2005b) made theoretical calculations of these spectra with their model being very similar to their analysis of cuprates (in the preceding subsection), and concluded that the extra structure observed in the high-temperature phase (paramagnetic metal phase) originates from the metallic screening effect.

Horiba et al. (2004) measured the Mn 2p HAXPS spectra of $La_{1-x}Sr_xMnO_3$ (LSMO) thin film and detected a clear additional feature that is absent in the Mn 2p XPS with soft x-ray (incident energy about 1 keV). The doping- and temperature-dependence of this additional feature were measured and it was found that the intensity of the feature is maximum around $x = 0.2$ and it is much sharper at 40 K than at 300 K. After simple model calculations similar to those for V_2O_3 and cuprates, they assigned this feature as originating from the metallic screening effect. Incidently, van Veenendaal (2006) gave a different interpretation based on the nonlocal screening effect calculated with a multi-Mn-site cluster model.

5.6.3 Ce COMPOUNDS: SURFACE/BULK SENSITIVITY

Before discussing the HAXPS spectra of Ce compounds, some descriptions of the surface/bulk sensitivity in XPS of Ce compounds are given. As expected from the

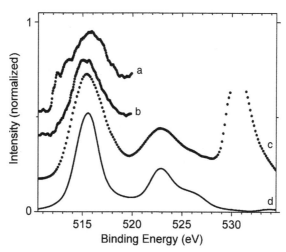

FIGURE 5.47 The 2p XPS spectrum of V_2O_3 as measured (a) with HAXPS at 190 K, (b) with HAXPS at 145 K, (c) with normal XPS, and (d) calculated with the CTM model. The normal XPS data are taken at 1486 eV and the HAXPS data at 5934 eV. (From Uozumi, T., et al., *J. Electron Spectrosc. Relat. Phenom.*, 83, 9, 1997; Panaccione, G., et al., *Phys. Rev. Lett.*, 97, 116401, 2006. With permission from Elsevier Ltd.)

experiments shown in Section 5.5.3, the 4d XPS of RE systems is surface sensitive, and the 3d XPS is more bulk sensitive. According to studies by Laubschat et al. (1990) and by Braicovich et al. (1997), the surface state of mixed valence (or valence fluctuating) Ce intermetallic compounds ($CeRh_3$, $CeFe_2$, and so on), is not in mixed valence but is almost trivalent with a nearly $4f^1$ configuration. Furthermore, the Ce 3d XPS with conventional x-ray sources still has considerable surface sensitivity. To be precise, therefore, the analysis of 3d XPS described so far in this chapter should be revised by taking into account the surface contribution. Usually, the correction is not very dramatic and, for instance, the average 4f electron number n_f in the bulk ground state of $CeNi_2$ would be changed from 0.81 to about 0.7.

In the following, an analysis is given of 3d XPS of $CePd_7$ (Iwamoto et al., 1995), which is a mixed valence intermetallic Ce compound with an extremely strong hybridization and, as a result, a severe surface effect. As it is well known that Ce L_3 XAS is bulk sensitive, so the bulk parameter values of the SIAM are first estimated by analyzing the Ce L_3 XAS of $CePd_7$. For this purpose, we add the Ce 5d conduction band states to the Hamiltonian H_0 of Equation 3.106 of Chapter 3:

$$\hat{H}_0 = H_0 + \sum_k \varepsilon_{5d}(k) a_{5d}(k)^+ a_{5d}(k), \qquad (5.45)$$

and for the final state of XAS, we use the Hamiltonian

$$\hat{H} = \hat{H}_0 - U_{fc} \sum_v a_{fv}^+ a_{fv} + \frac{U_{fd}}{N} \sum_{k,k',v} a_{fv}^+ a_{fv} a_{5d}(k)^+ a_{5d}(k') - \frac{U_{dc}}{N} \sum_{k,k'} a_{5d}(k)^+ a_{5d}(k'). \quad (5.46)$$

Here, the Coulomb interaction U_{fd} between the 5d and 4f electrons is taken into account along with the attractive core hole potential $-U_{dc}$ acting on the 5d electron, and the core hole potential $-U_{fc}$. The Ce L_3 XAS spectrum is calculated by

$$F_{XAS}(\Omega) = \sum_f \left| \frac{1}{N} \langle f | \sum_k a_{5d}(k)^+ | 0 \rangle \right|^2 \frac{\Gamma_f/\pi}{(\hbar\Omega - E_f + E_0)^2 + \Gamma_f^{\cdot 2}} \qquad (5.47)$$

where the states $|0\rangle$ (with energy E_0) and $|f\rangle$ (with energy E_f) are eigenstates of the Hamiltonian $\hat{H}_0$ and $\hat{H}$, respectively.

The calculated (solid curve) (Iwamoto et al., 1995) and experimental (squares) (Beaurepaire et al., 1993a) results are shown in Figure 5.48 where the dashed curve is the background contribution used in the calculation. With the bulk parameter values (Table 5.6), the Ce 3d XPS spectrum is calculated as shown with the dashed curve in the lower panel of Figure 5.49. Then the surface parameter values and the weight of the surface/bulk contributions are determined so as to obtain the best agreement between theory and experiment. Here the surface parameter values are chosen to keep the surface Ce ion almost trivalent instead of mixed valent, as shown in Table 5.6. Then the experimental 3d XPS spectrum (squares in the upper panel) is interpreted as a superposition of 62.5% bulk (dashed curve) and 37.5% surface (dotted curve) contributions (Iwamoto et al., 1995). For the bulk contribution of the 3d XPS spectrum, we have a three-peak structure (for each of the $3d_{5/2}$ and $3d_{3/2}$

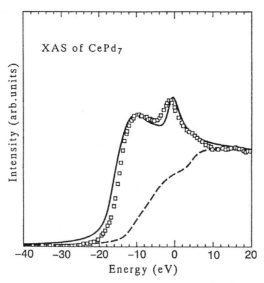

FIGURE 5.48 Calculated (solid curve) and experimental (open squares) Ce L_3 XAS spectra of CePd$_7$. The dashed curve represents the background contribution. (From Iwamoto, Y., Nakazawa, M., Kotani, A., and Parlebas, J.C., *J. Phys. Cond. Matt.*, 7, 1149, 1995. Reprinted with permission from IOP Publishing Ltd.)

components), which originates from the Ce $4f^0$, $4f^1$, and $4f^2$ configurations, while only two peaks are recognized for the Ce L_3 XAS, which are usually assigned as the Ce $4f^0$ and $4f^1$ configurations (Figure 5.48). Then, it is interesting to see where the $4f^2$ contribution is in the L_3 XAS. According to the present calculation, the $4f^2$ configuration in the L_3 XAS spectrum is merged into the $4f^1$ peak (lower energy peak in Figure 5.48), instead of being resolved as a third structure. In order to merge these two contributions easily, from a theoretical point of view, the Coulomb interaction U_{fd} and the core hole potential $-U_{dc}$ play an important role. By the effect of $-U_{dc}$, the photoexcited 5d electron is more and less localized, and then the energy separation between $4f^1$ and $4f^2$ configurations is decreased approximately by U_{fd}

TABLE 5.6
4f Level ε_f^0 (Measured from the Fermi Level ε_F) and the Hybridization Strength V Used in the Analysis of Ce L_3 XAS and 3d XPS of CePd$_7$

CePd$_7$	Surface	Bulk
$\varepsilon_f^0 - \varepsilon_F$ (eV)	−2.5	−0.7
V (eV)	0.4	0.6
n_f	0.94	0.57

Note: The average 4f electron number n_f is also listed for the bulk and surface states.

(which is chosen to be about 1 eV). The importance of U_{fd} and $-U_{dc}$ has been discussed in Chapter 3 for insulating Ce compounds. However, their effect is much less severe for intermetallic Ce compounds and a more careful check for the role of U_{fd} and $-U_{dc}$ is needed.

In the next subsection, it is shown that the hard x-ray resonant photoemission is a powerful tool for directly detecting the $4f^2$ configuration in the Ce L_3 XAS and for confirming the important role of U_{fd}.

5.6.4 Resonant HAXPS of Ce Compounds

We consider RXPS with hard x-ray (denoted by RHAXPS) in the Ce $2p_{3/2}$-5d excitation threshold and Ce 2p3d5d Auger decay of a mixed valence compound $CeRh_3$. The final state of this resonance process is the same as that of the direct process of Ce 3d XPS, thus the two processes interfere with each other. Experimental observations of this type of Ce 2p5d RHAXPS were made by Le Fèvre et al. (1998) and were analyzed theoretically by Ogasawara et al. (2000). The formula of RPES given in Section 5.5.1 can also be applied to this case simply by assigning the states $|\alpha\rangle$ and $|\varepsilon\beta\rangle$, respectively, to the intermediate states of the Ce $2p_{3/2}$-5d excitation and the final states of the Ce 3d XPS. Figure 5.50 shows (a) experimental and (b) calculated

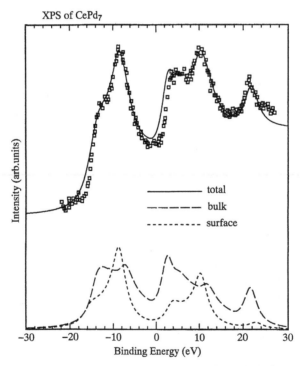

FIGURE 5.49 Calculated results of the total (solid curve), bulk (dashed curved) and surface (dotted curve) contributions to the Ce 3d XPS spectrum of $CePd_7$. The experimental result is shown with open squares. (From Iwamoto, Y., Nakazawa, M., Kotani, A., and Parlebas, J.C., *J. Phys. Cond. Matt.*, 7, 1149, 1995. Reprinted with permission from IOP Publishing Ltd.)

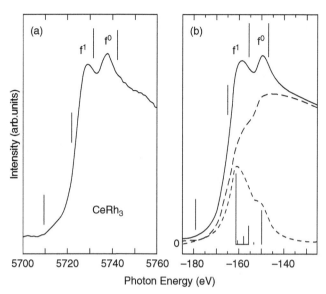

FIGURE 5.50 The Ce L_3 XAS spectra of CeRh$_3$; the experimental result is shown in (a), and the calculated one in (b). In the calculation, the line spectrum represents the calculated intensity, the dotted line their convolution with a Lorentzian of width 4.0 eV (HWHM), and the dashed curve is the background contribution. (Reprinted with permission from Ogasawara, H., Kotani, A., Le Fevre, P., Chandesris, D., and Magnan, H., *Phys. Rev. B*, 62, 7970, 2000. Copyright 2000 by the American Physical Society.)

results of the Ce L_3 XAS spectra of CeRh$_3$. The method of calculation is similar to that for CePd$_7$ in the preceding subsection, but now the effect of intra-atomic multiplet coupling is fully taken into account. To make the full multiplet calculation tractable, the Rh 4d and Ce 5d bands are approximated by a single level. In this calculation, the average 4f electron number n_f in the bulk ground state is estimated to be 0.68. The Ce 2p5d RHAXPS spectra were measured for the incident photon energies fixed at the positions shown with vertical bars in Figure 5.50a. The results are shown in Figure 5.51a and the calculated results of the Ce 2p5d RHAXPS are shown in Figure 5.51b.

From Figure 5.51, we find that the bulk sensitive Ce 2p-5d RHAXPS spectra (energy distribution curves) exhibit three peaks a, b, and c (d, e, and f) corresponding to the $4f^0$, $4f^1$, and $4f^2$ configurations for the $3d_{3/2}$ (and $3d_{5/2}$) core level, although peaks c and d are overlapping with each other. The calculated results are in good agreement with the experimental ones, and the peaks A, B, C, D, E, and F correspond to the experimental ones, a, b, c, d, e, and f, respectively. Since the characters of the Ce 2p5d RHAXPS peaks are clear, by taking the constant initial state (CIS) spectra, we can clearly assign the character of each feature in the Ce L_3 XAS, and furthermore, from the analysis of CIS we can clarify the role of the Coulomb interaction U_{fd}. In taking the CIS spectra, the binding energies of the Ce 2p5d RHAXPS are fixed at a, e, and f in experiments (at A, E, and F in the calculation), which correspond to the $4f^0$, $4f^1$, and $4f^2$ configurations, and the change of the Ce 2p5d RHAXPS intensity is measured (and calculated) as a function of the incident photon energy.

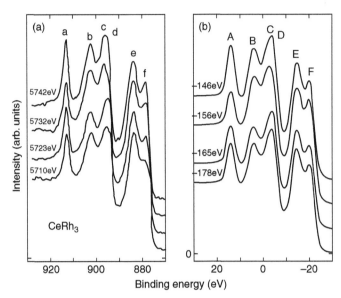

FIGURE 5.51 Energy distribution curves of Ce 3d RHAXPS in CeRh$_3$; the experimental results are shown in (a), and the calculated ones in (b). The resonance photon energies are indicated as bars in Figure 5.50. (Reprinted with permission from Ogasawara, H., Kotani, A., Le Fevre, P., Chandesris, D., and Magnan, H., *Phys. Rev. B*, 62, 7970, 2000. Copyright 2000 by the American Physical Society.)

The results are shown in Figure 5.52 where the experimental results are shown in (a) and the calculated ones in (b). The CIS spectra for a, e, and f (A, E, and F) exhibit the so-called Fano resonance at the incident photon energy corresponding to the 4f^0, 4f^1, and 4f^2 contributions in the final state of the L$_3$ XAS, as shown with the dotted lines. It is seen that the two peaks of L$_3$ XAS correspond to the 4f^0 and 4f^1 configurations, whereas the 4f^2 configuration exists just below the 4f^1 peak although it is not recognized as an additional structure in the L$_3$ XAS.

The existence of a hidden 4f^2 component on the lower energy side of the 4f^1 peak was pointed out by Bianconi et al. (1984), Schneider et al. (1985), and Kotani and Jo (1986), but this Ce 2p5d RHAXPS was the first direct observation of the 4f^2 contribution. More recently, the detection of the 4f^2 contribution in the L$_3$ XAS for Ce-Th and Ce-Sc alloys has been reported by Rueff et al. (2004), taking advantage of resonant x-ray emission spectroscopy (see Chapter 8).

It is to be remarked that the analysis of the CIS of the Ce 2p5d RHAXPS gives evidence for the existence of U_{fd}. In the calculations shown in Figures 5.50 through 5.52, the effect of U_{fd} (= 2.5 eV) was taken into account. In order to confirm the important role of U_{fd}, we show in Figure 5.53, the calculated results of XAS and CIS with U_{fd} = 0.0 eV. It is seen that the incident photon energy of the Fano resonance in the CIS of the 4f^2 component shifts, compared with Figure 5.52, to the lower energy side. It is clear that the agreement with experiment is better in Figure 5.52b than in Figure 5.53.

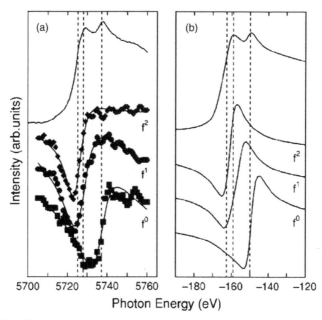

FIGURE 5.52 Constant initial state spectra of the Ce 3d RHAXPS; the experimental results are shown in (a), and the calculated ones in (b). f^0, f^1, and f^2 correspond to the peaks, a, e, and f (A, E, and F) in Figure 5.51, respectively. (Reprinted with permission from Ogasawara, H., Kotani, A., Le Fevre, P., Chandesris, D., and Magnan, H., *Phys. Rev. B*, 62, 7970, 2000. Copyright 2000 by the American Physical Society.)

Before closing this section on HAXPS, we would like to mention briefly some HAXPS measurements for mixed valence Yb compounds. Sato et al. (2004) measured the temperature-dependence of the Yb 3d HAXPS spectra of $YbInCu_4$, which exhibits an abrupt valence change at around 42 K. They have shown that the Yb 3d HAXPS is a powerful method to estimate the valence of Yb with high accuracy, and obtained the following results: the Yb valence number of $YbInCu_4$ is about 2.90 from 220 down to 50 K and about 2.74 at 30 K to 10 K. Similar measurements of the Yb valence of $YbInCu_4$ will also be shown in Chapter 8 using the RXES technique, instead of HAXPS. In addition, Suga et al. (2005b) measured the Yb 3d HAXPS spectra of $YbAl_3$ and discussed the temperature-dependence of the Yb valence number.

5.7 RESONANT INVERSE PHOTOEMISSION SPECTROSCOPY

As mentioned in Chapter 2, the inverse photoemission spectroscopy (IPES) is the inverse process of the photoemission spectroscopy (PES). In PES, a photon is incident on a material sample and an electron is emitted from the sample. In IPES, an electron is incident on the sample and a photon is emitted from that. The RIPES is the inverse process of RPES and the incident electron energy resonates with a core electron excitation energy in RIPES, while the incident photon energy resonates with a core electron excitation energy in RPES.

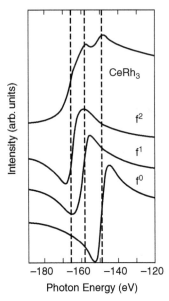

FIGURE 5.53   Calculated results of CIS spectra of the Ce 3d RHAXPS with vanishing U_{fd}. (Reprinted with permission from Ogasawara, H., Kotani, A., Le Fevre, P., Chandesris, D., and Magnan, H., *Phys. Rev. B*, 62, 7970, 2000. Copyright 2000 by the American Physical Society.)

To make the situation clearer, let us consider the IPES and PES of a RE with $4f^n$ ground state, and RIPES and RPES at the 4d to 4f excitation threshold. IPES and PES processes are represented by

$$4f^n + e \rightarrow 4f^{n+1} + \hbar\omega, \tag{5.48}$$

$$4f^n + \hbar\Omega \rightarrow 4f^{n-1} + e, \tag{5.49}$$

respectively, and RIPES and RPES processes are, respectively, represented by

$$4f^n + e \rightarrow 4d^9 4f^{n+2} \rightarrow 4f^{n+1} + \hbar\omega, \tag{5.50}$$

$$4f^n + \hbar\Omega \rightarrow 4d^9 4f^{n+1} \rightarrow 4f^{n-1} + e, \tag{5.51}$$

where e, Ω, and ω represent the incident (and also emitted) electron, the incident photon and emitted photon, respectively. It should be noted that the processes given by Equations 5.48 and 5.50 (Equations 5.49 and 5.51) interfere in the RIPES (RPES) spectrum.

Experimental measurements of RIPES of mixed valence Ce compounds ($CeRh_3$, $CeNi_2$, $CePd_3$, $CePd_7$, and so on), were made by Kanai et al. (1997, 1999, 2001) for the Ce 4d edge and by Weibel et al. (1994) and Grioni et al. (1995, 1997) for the Ce 3d edge. These experimental data were analyzed theoretically by Tanaka and

Jo (1996) and Uozumi et al. (2002) by CTM calculations with the SIAM. Here, we give the results by Uozumi et al. and compare them with the experimental data. The calculation of RIPES spectra $F_{RIPES}(\varepsilon, \omega)$ is made using the expression

$$F_{RIPES}(\varepsilon, \omega) = \sum_f \left| \langle f | \left(V_R + V_R \frac{1}{\varepsilon + E_g - H + i\Gamma} V_A^+ \right) | g \rangle \right|^2 \delta(\varepsilon + E_g - \hbar\omega - E_f), \quad (5.52)$$

where $| g \rangle$ and $| f \rangle$ are, respectively, the ground and final states of the material system with energy E_g and E_f, ε and ω are the energies of the incident electron and the emitted photon, respectively, H is the Hamiltonian of the material system, Γ is the lifetime broadening of intermediate states, and V_R and V_A are the operators of radiative and Auger transitions, respectively.

In Figure 5.54, we show the calculated RIPES spectra of CeRh$_3$ around the prethreshold region of the Ce 4d edge. For the incident electron energy ε (which is written as E_{ex} in Figure 5.54) below about 140 eV, the 4d to 4f excitation is called prethreshold resonance, which is very similar to the prethreshold excitation of Ce 4d XAS. Since the RIPES is a surface-sensitive spectroscopy, the calculation of RIPES spectra are made for the bulk (dashed curves in Figure 5.54) and surface (dotted curves) states of CeRh$_3$ separately, and the two spectra are superposed with the weight of 50% and 50% to obtain the total spectra (solid curves). As in the case of CePd$_7$ (see Section 5.6.3), $\varepsilon_f - \varepsilon_F$ and V are assumed to be different for the bulk and surface states, and taken for CeRh$_3$ as: $\varepsilon_f - \varepsilon_F = -1.2$ eV (bulk) and -1.7 eV (surface), $V = 0.54$ eV (bulk) and 0.36 eV (surface). The bulk parameter values are similar to those used in Section 5.6.4, and for both bulk and surface calculations the ground, intermediate and final states, respectively, are described by a linear combination of f^0, f^1, and f^2 configurations, by that of f^2 and f^3 configurations, and by that of f^1, f^2, and f^3 configurations.

The two RIPES structures, (1) 0–2 eV and (2) 3–9 eV above the Fermi level in Figure 5.54 originate from the 4f^1 and 4f^2 final states, respectively. In each of (1) and (2), we see that the bulk and surface contributions have somewhat different peak energy because of the different parameter values $\varepsilon_f - \varepsilon_F$ and V. Especially for the 4f^2 contribution (2), the surface contribution is in the considerably lower energy side than the bulk contribution, and this is caused mainly by the lowering of the 4f level ε_f in the surface state.

In order to see the resonance behavior of the RIPES spectra, we show in Figure 5.55, the calculated results of the integrated RIPES intensities of the 4f^1 and 4f^2 final states as a function of the excitation energy (E_{ex}) for (a) total, (b) surface, (c) bulk contributions, and (d) compare them with the experimental results. The two resonant enhancements at around $E_{ex} = 107$ eV and 112 eV of the 4f^1 final state are caused by the RIPES process of Equation 5.50 with $n = 0$. Since the weight of the 4f^0 configuration is larger for the α-like bulk state than for the γ-like surface state, the resonant enhancement of the 4f^1 final state is stronger for the bulk contribution than for the surface contribution. On the other hand, the RIPES intensity of the 4f^2 final state is due to the process of Equation 5.50 with $n = 1$, so that it is stronger for the surface contribution than the bulk contribution.

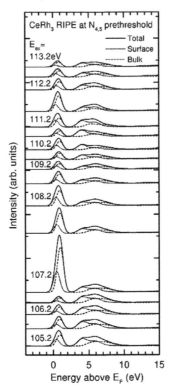

FIGURE 5.54   Calculated RIPES spectra of CeRh$_3$ around the prethreshold region of the Ce 4d edge. The total spectra (solid curves) for various excitation energies E_{ex} are obtained by a superposition of the surface (dotted curve) and bulk (dashed curves) contributions. (Reprinted with permission from Uozumi, T., et al., *Phys. Rev. B*, 65, 45105, 2002. Copyright 2002 by the American Physical Society.)

The behavior of the calculated total RIPES intensity (a) is in good agreement with the experimental one (d).

For an incident electron energy larger than about 114 eV, the strong resonant enhancement of RIPES occurs and is called the giant resonance as in the case of Ce 4d XAS. The theoretical and experimental spectra in the giant resonance region of Ce 4d RIPES are shown in (a) and (b) of Figure 5.56, and they are in good agreement with each other. Before this calculation, Kanai et al. (1999) made a theoretical analysis of the experimental data (Figure 5.56b) with the SIAM but assumed only one set of parameter values disregarding the surface contribution. In order to get the best fit with the experiment, they used the parameter values of the SIAM that is intermediate between the present bulk and surface parameters. However, there remained two problems in the calculation: (*i*) the energy position of the 4f^2 structure is higher by about 1.5 eV than the experimental result; (*ii*) the average number of 4f electrons in the ground state is about 1.0, unexpected with the mixed valence character of CeRh$_3$. The two problems are resolved by the calculations of Figure 5.56a where the bulk and surface contributions are taken into account.

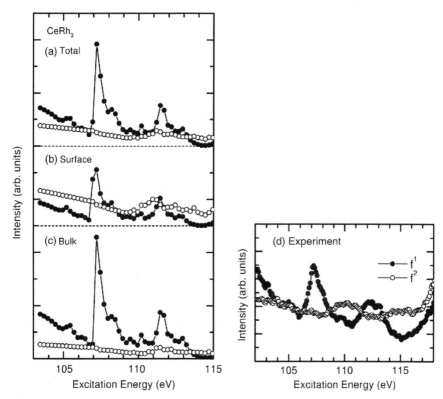

FIGURE 5.55 Excitation energy dependence of the integrated RIPES intensities of $4f^1$ (solid circles) and $4f^2$ (open circles) final states in the prethreshold region of the Ce 4d RIPES spectra. The calculated results of (a) total, (b) surface, and (c) bulk contributions are compared with the experimental results (d). (Reprinted with permission from Uozumi, T., et al., *Phys. Rev. B*, 65, 45105, 2002. Copyright 2002 by the American Physical Society.)

Uozumi et al. (2002) also analyzed, with the same SIAM model, the RIPES spectra measured by Grioni et al. (1995, 1997) around the Ce 3d edge of $CeRh_3$. The excitation energy dependence of the integrated intensities of $4f^1$ and $4f^2$ RIPES final states is shown in Figure 5.57. Since the Ce 3d RIPES is less surface-sensitive than the Ce 4d RIPES, the bulk and surface contributions are superposed with the weight of 70% and 30%, respectively, to obtain the total spectra, which are in good agreement with the experimental results shown in the inset of Figure 5.57. The most important point of this calculation is that the $4f^2$ peak position of the surface contribution is lower in energy than that of the bulk contribution. Before this analysis, Tanaka and Jo (1996) analyzed theoretically the same experimental results by Grioni et al. (1995) with the SIAM, but assuming only one set of parameter values disregarding the surface contribution. Their parameter values are almost consistent with the mixed valence character in the ground state of $CeRh_3$. However, they claimed that the hybridization strength between the $4f^2$ and $4f^3$ configurations in the intermediate state should be drastically reduced to about 50% of that between

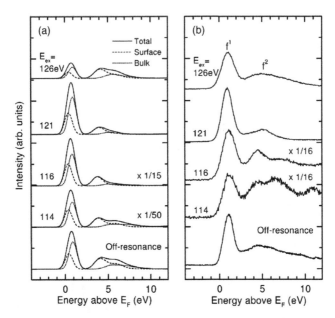

FIGURE 5.56 Calculated (a) and experimental (b) RIPES spectra of CeRh$_3$ around the Ce 4d giant resonance region. The total spectra (solid curves) in (a) are obtained as a superposition of the surface (dotted curves) and bulk (dashed curves) contributions with the weight of 50% and 50%. The off-resonance IPES spectra are also shown in the bottom of (a) and (b). (Reprinted with permission from Uozumi, T., et al., *Phys. Rev. B*, 65, 45105, 2002. Copyright 2002 by the American Physical Society.)

the 4f^0 and 4f^1 configurations (V) in the ground state. As the reason for this drastic reduction of V, they pointed out that the potential of the 3d core hole in the intermediate state makes the radial extension of the 4f wave function smaller, which causes the reduction of V. They also pointed out that if one disregards this reduction of V by the core hole effect the separation of E_{exc} between 4f^1 and 4f^2 peaks (corresponding to the 3d^{9}4f^2 and 4d^{9}4f^3 intermediate states) should be much larger than the experimental value of 3 eV (see the inset of Figure 5.57).

However, the drastic reduction of V in the intermediate state is not consistent with a first principle calculation for α–Ce by Gunnarsson and Jepsen (1998). In the SIAM calculations of XPS, XAS, and XES, the configuration-dependent V is often treated by introducing scaling factors R_c and R_v. The hybridization strength V, defined as that between 4f^0 and 4f^1 configurations, is scaled by R_c when a core hole is created and is scaled by $1/R_v^m$ when m electrons are added to 4f states. The standard values of R_c and R_v are 0.7–0.8 and 0.8–0.9, respectively, and Uozumi et al. (2002) used $(R_c, R_v) = (0.7, 0.8)$ for the analysis of RIPES of CeRh$_3$, which are the same values as those used by Ogasawara et al. (2000) in the analysis of RHAXPS of CeRh$_3$. With these values, the hybridization strength between 3d^{9}4f^2 and 3d^{9}4f^3 configurations in the intermediate state of RIPES is given by $(R_c/R_v^2)V = 1.1V$, so that V is slightly enhanced contrary to the drastic reduction.

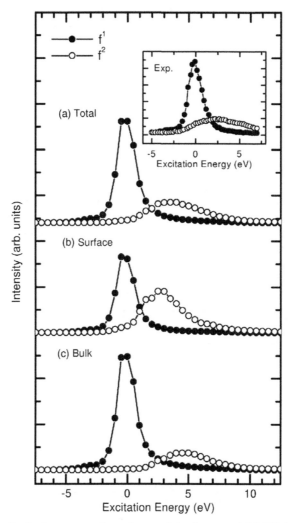

FIGURE 5.57 Excitation energy dependence of the integrated RIPES intensities of $4f^1$ (solid circles) and $4f^2$ (open circles) final states around the Ce 3d edge of CeRh$_3$. The calculated results of (a) total, (b) surface, and (c) bulk contributions are compared with the experimental results shown in the inset. (Reprinted with permission from Uozumi, T., et al., *Phys. Rev. B*, 65, 45105, 2002. Copyright 2002 by the American Physical Society.)

Actually, the difference of excitation energies ΔE_{ex} of the two peaks ($4f^1$ and $4f^2$ final states) in the RIPES is roughly given by

$$\Delta E_{ex} = \sqrt{(\varepsilon_f - \varepsilon_F - U_{fc} + 2U_{ff})^2 + 4(N_f - 2)(R_c/R_v^2)^2 V^2}. \qquad (5.53)$$

Therefore, ΔE_{ex} is larger for bulk than for surface mainly because the value of V is larger. The values of ΔE_{ex} in Figure 5.57 is 4.5 eV for bulk and 2.8 eV for surface, to reproduce the experimental value $\Delta E_{ex} = 3.0$ eV. In order to reproduce the experimental value only by the bulk parameter values, it is necessary to take a drastic reduction of V by assuming R_c/R_v^2 about 50% as concluded by Tanaka and Jo, but this is quite unphysical.

Experimental and theoretical studies of RIPES have also been made for TM compounds at the 2p edge. Here we only refer the papers by Tezuka and Shin (2004) and by Tanaka and Jo (1997).

5.8 NONLOCAL SCREENING EFFECT IN XPS

The SIAM (and the cluster model including a single metal ion) has been most successfully applied to the theoretical analysis of XPS spectra (and other core level spectra) in f and d electron systems, especially in 4f electrons of RE systems. For 3d TM compounds, the spatial extension of 3d electrons is larger than that of the RE 4f electrons, so that we need sometimes to treat a model including many TM ions, such as multiple-site cluster model.

The importance of a nonlocal screening effect with the large cluster model beyond the SIAM was first pointed out by van Veenendaal and Sawatzky (1993, 1994) and van Veenendaal et al. (1993) for NiO and cuprates. In Figure 5.58, the results of their calculations (van Veenendaal and Sawatzky, 1993) are shown. The top curve is the Ni $2p_{3/2}$ XPS of NiO calculated with the Ni_7O_{36} cluster model (see inset), compared with the experimental spectrum (solid with symbols). In order to reduce the size of the calculation, the holes of the NiO_6 clusters are frozen onto the Ni atom, except for the central Ni atom, which contains a core hole, and the multiplet coupling

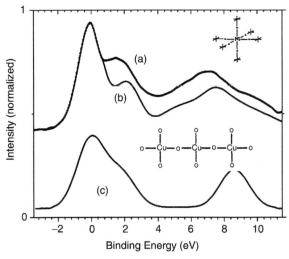

FIGURE 5.58 (a) The experimental Ni $2p_{3/2}$ XPS spectrum of NiO (solid, with circles) compared with (b) a Ni_7O_{36} cluster model calculation (solid). At the bottom (c), the calculated Cu 2p XPS spectrum for a Cu_3O_{10} cluster is given. (Reprinted with permission from van Veenendaal, M.A., and Sawatzky, G.A., *Phys. Rev. Lett.*, 70, 2459, 1993. Copyright 1993 by the American Physical Society.)

effect is disregarded. The most important point is that the main peak with the lowest binding energy splits into two peaks, while it is not split in the calculation with the NiO_6 cluster model. The experimental result (Uhlenbrock et al., 1992) in Figure 5.58 clearly shows the splitting. The main peak of Ni 2p XPS in NiO corresponds mainly to the $2p^5 3d^9 \underline{L}$ configuration, but the splitting of this peak is caused by two different screening mechanisms: one is the local screening within the central NiO_6 cluster as we have already considered and the other is the nonlocal screening where the charge transfer occurs from the neighboring NiO_6 units to the central NiO_6 unit. Therefore, in the final state of the nonlocal screening, a neighboring NiO_6 unit contains three holes because of an extra hole in addition to the two-hole ground state $e_g^2(^3A_2)$, and the three holes form a stable ^2E state. Another important point in this calculation is that by including the nonlocal screening effect, the fitting parameter values should be modified compared with those in the NiO_6 cluster model (or the SIAM). Especially, the charge-transfer energy Δ used in this calcu-lation is 5.5 eV, which is much larger than 2.0 eV in Table 5.1. Even if we consider some ambiguity in the parameter fitting, the value of Δ estimated with the present model should be larger than that with the NiO_6 cluster model. The bottom curve in Figure 5.58 is the Cu 2p XPS spectrum calculated with a Cu_3O_{10} cluster model with three holes corresponding to an undoped cuprate. If we calculate the Cu 2p XPS with a CuO_4 cluster model, the lower binding energy peak (main peak) has a single line, but in Figure 5.58, the main peak splits into two features (a peak and a shoulder) because of the local and nonlocal screening effects.

The nonlocal screening effect in the Cu 2p XPS of cuprates was studied exten-sively by Okada and Kotani (1995a, 1997a,b, 1999a,b, 2005) and Okada et al. (1996). Let us consider the chain-like (1D) and layered (2D) Cu oxide systems with nominally divalent Cu ions. Here we consider multiple-Cu-site clusters with CuO_4 plaquettes connected to each other, sharing their corners. The Hamiltonian of the system is given, with the hole picture, by

$$H = \sum_{i,\sigma} \varepsilon_d d_{i\sigma}^+ d_{i\sigma} + \sum_{j,\sigma} \varepsilon_p p_{j\sigma}^+ p_{j\sigma} + \sum_{<i,j>,\sigma} V_{pd,ij}(d_{i\sigma}^+ p_{j\sigma} + p_{j\sigma}^+ d_{i\sigma})$$

$$+ \sum_{<j,j'>,\sigma} V_{pp,jj'}(p_{j\sigma}^+ p_{j'\sigma} + p_{j'\sigma}^+ p_{j\sigma}) + U_{dd}\sum_i d_{i\uparrow}^+ d_{i\uparrow} d_{i\downarrow}^+ d_{i\downarrow} + U_{dc}\sum_\sigma d_{0\sigma}^+ d_{0\sigma} c_0^+ c_0, \quad (5.54)$$

where $d_{i\sigma}^+ (p_{j\sigma}^+)$ creates a hole with spin σ on the ith Cu $3d_{x^2-y^2}$ orbit (the jth O 2p orbit). The first and second terms on the right-hand side of Equation 5.54 represent the one body energies, the third and fourth terms the hybridizations of the Cu 3d and O 2p orbits and of the neighboring O 2p orbits, respectively, and the fifth and sixth terms are Coulomb repulsion between 3d holes and that between Cu 3d and core holes on the 0th site (i.e. the core-hole site). The multiplet coupling effect is disregarded for simplicity.

In Figure 5.59, the calculated result for a Cu_7O_{22} linear-chain cluster is shown and compared with the experimental data for Sr_2CuO_3 (Okada et al., 1996). The parameter values used in this calculation are as follows: $pd\sigma = -1.5$ eV, $pp\sigma = 1.0$ eV, $U_{dd} = 8.8$ eV, $U_{dc} = 7.7$ eV, and $\Delta(\equiv \varepsilon_p - \varepsilon_d) = 2.5$ eV. The agreement between the theoretical and

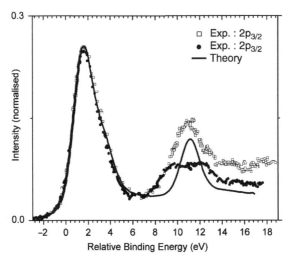

FIGURE 5.59 Calculated result of the Cu 2p XPS spectrum of Sr_2CuO_3 compared with experimental data. The open symbols indicate the $2p_{1/2}$ XPS, the closed symbols the $2p_{3/2}$ XPS and the solid line the calculation including a background. (From Okada, K., and Kotani, A., *J. Electron Spectrosc. Relat. Phenom.*, 78, 53, 1996. With permission from Elsevier Ltd.)

experimental results is satisfactory except the multiplet structure of the satellite around 11 eV. The characteristic spectral shape of the main peak, with a shoulder on the high-energy side, originates from the local and nonlocal screening effects. For the lowest final state, a hole pushed out from the core hole site moves to the neighboring CuO_4 plaquette and forms a singlet state with the hole originally existing in the plaquette. The singlet state is called a "Zhang–Rice singlet state" (Zhang and Rice, 1988), while the final state corresponding to the shoulder is due to the local screening effect.

The formation of the Zhang–Rice singlet in the final state depends on the parameter values, especially on the charge-transfer energy Δ, so that the spectral shape of the main peak changes with Δ. In Figure 5.60, we show the change in the calculated spectra with Δ for the Cu_5O_{16} two-dimensional cluster (see inset). Similar calculations were also made for corner-sharing and edge-sharing linear-chain clusters, where the nonlocal screening elect in the latter is much weaker than the former. The results of these model calculations were compared with experimental spectra for various Cu oxide systems, Bi_2CuO_4, $CuGeO_3$, La_2CuO_4, $Sr_2CuO_2Cl_2$, and so on (Okada and Kotani, 1997a).

Van Veenendaal and Sawatzky (1994), as well as Okada and Kotani (2005), studied theoretically the effect of electron and hole doping in cuprates on the Cu 2p XPS spectrum. In Figure 5.61, the results by Okada and Kotani (2005) calculated with a Cu_6O_{18} linear-chain cluster with periodic boundary condition, are shown (see inset). The parameter values are $U_{dd} = 8.0$ eV, $U_{dc} = 9.0$ eV, and $\Delta = 3.5$ eV. It is to be noted that in the electron-doped case, a remarkable peak occurs on the lower binding energy side of the nonlocal (indicated by ZRS) and local (indicated by $d^{10}\underline{L}$)

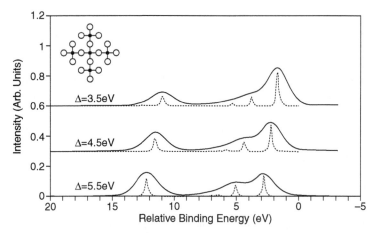

FIGURE 5.60 Calculated Cu 2p XPS spectra for a two-dimensional Cu_5O_{16} cluster by changing the charge-transfer energy Δ. The values of Δ and U_{dc} are changed with a fixed value of $\Delta - U_{dc} = 4.2$ eV. (From Okada, K., and Kotani, A., *J. Phys. Soc. Jpn.*, 66, 341, 1997b. With permission.)

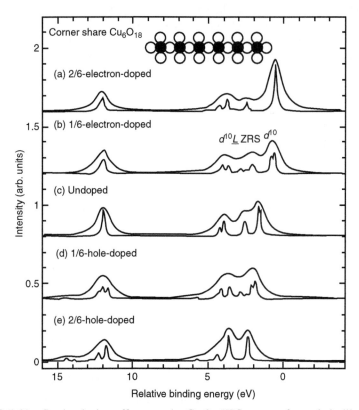

FIGURE 5.61 Carrier-doping effects on the Cu 2p XPS spectra for a chain-like Cu_6O_{18} cluster with periodic boundary condition along the chain direction. (From Okada, K., and Kotani, A., *J. Phys. Soc. Jpn.*, 74, 653, 2005. With permission.)

screening features. This strongly suggests that the prominent peak α in the experimental HAXPS spectra of $Nd_{1.85}Ce_{0.15}CuO_4$ (Figure 5.45) is caused by the charge transfer from the metallic doped-electron states to the core hole site. In the calculation by Taguchi et al. (2005a), the metallic electron states are assumed to occur near the bottom of the upper Hubbard band, but in the present large cluster calculation, the energy of the metallic doped-electron states is determined without any assumption.

More recently, Kotani and Okada (2006) have made detailed theoretical analyses of the Cu 2p HAXPS spectra with a two-dimensional Cu_6O_{17} cluster model shown in Figure 5.62b with the parameter values $pd\sigma = -1.5$ eV, $pd\sigma = 0.5$ eV, $U_{dd} = 8.0$ eV, $U_{dc} = 8.5$ eV, and Δ is taken to be 2.5 eV for Nd_2CuO_4 (NCO) and $Nd_{1.834}Ce_{0.166}CuO_4$ (NCCO) and to be 3.5 eV for La_2CuO_4 (LCO) and $La_{1.834}Sr_{0.166}CuO_4$ (LSCO). By changing the total electron number of the system, the Cu $2p_{3/2}$ XPS spectra for electron-doped, undoped, and hole-doped systems are calculated without introducing any adjustable parameters concerned with the carrier doping. The calculated XPS spectra are shown in Figure 5.62a, where two different Lorentzian convolutions are made with Γ (HWHM) = 0.2 eV and 0.7 eV, and the results are compared with the experimental HAXPS results by Taguchi et al. (2005a) shown in Figure 5.62c. The agreement between the calculated and experimental results is satisfactory, although some spectral splitting in the calculated results is spurious because of finite size cluster calculations. It is seen that the anomalous peak α occurs by electron doping, while no anomalous peak occurs by hole doping.

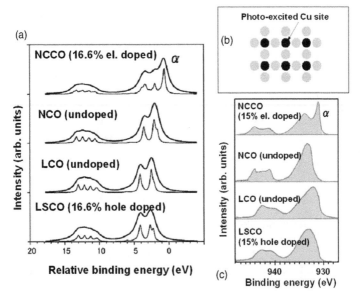

FIGURE 5.62 Calculated results (a) of the Cu $2p_{3/2}$ XPS for Nd_2CuO_4 (NCO), $Nd_{1.834}$ $Ce_{0.166}CuO_4$ (NCCO), La_2CuO_4 (LCO), and $La_{1.834}Sr_{0.166}CuO_4$ (LSCO) systems with the two dimensional Cu_6O_{17} cluster model indicated in (b). The results are compared with those of HAXPS experiments (c).

In order to clarify the assignment of the calculated XPS structures, Kotani and Okada took advantage of the "one-to-one correspondence" between the Cu 2p XPS and the Cu 3d valence PES (VPES). This one-to-one correspondence is a unique feature characteristic of the well-screened final state in XPS of Cu^{2+} systems. The well-screened Cu 3d final state on the core-hole site is in the $3d^{10}$ configuration due to the charge transfer from valence states, so that the structure of well-screened XPS spectra directly reflects the VPES structure, because the extra hole is in the valence band and the $3d^{10}$ configuration itself has no spectral structure. In Figure 5.63, the calculated results of Cu 2p XPS and VPES spectra for Nd_2CuO_4 and NCCO are shown. Since the spectral structure of VPES is well known, that of the main peak (well-screened peak) of 2p XPS can also be found from the one-to-one correspondence. In the case of undoped Nd_2CuO_4, the main features of VPES correspond mainly to the Zhang–Rice singlet band (lower binding energy) and O 2p band (higher binding energy), so that the two features of the main peak of Cu 2p XPS are found to be the final states, where the charge transfer has occurred from the Zhang–Rice singlet band and from the O 2p band to the Cu 3d state on the core hole site. By the electron doping ($Nd_{1.843}Ce_{0.166}CuO_4$), the doped electron band is formed just at the Fermi level, in addition to the Zhang–Rice singlet band and the O 2p band, so that the anomalous peak α is definitely found to originate from the metallic screening by the doped electrons. With increasing electron-doping level from 16.6% to 33.3%, the intensity of the doped electron band in VPES increases, giving rise to the increase in the intensity of the anomalous peak in 2p XPS.

Then we can understand why the anomalous peak does not occur by the hole doped case where the doped hole enters in the Zhang–Rice singlet band to decrease the weight of the Zhang–Rice singlet band in VPES. The decrease of the

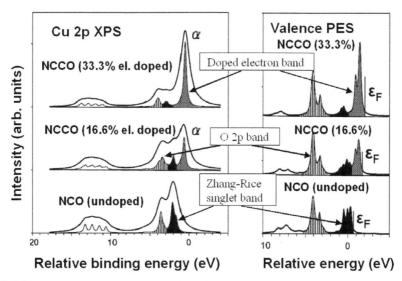

FIGURE 5.63 Comparison of the Cu 2p XPS spectra and the valence PES spectra calculated for Nd_2CuO_4 and NCCO with the Cu_6O_{17} cluster model.

Zhang–Rice singlet band intensity in VPES only weakens the intensity of the lowest binding energy feature (with the charge transfer from the Zhang–Rice singlet band to the Cu 3d state on the core hole site) of Cu 2p XPS, contrary to the occurrence of the anomalous peak. Within the calculations by Taguchi et al. (2005a), the effects of electron doping and hole doping are treated in a similar fashion to form extra metallic (coherent) states, and the difference in the effects of electron doping and hole doping on the XPS spectra results only from the difference in adjustable parameter values. The physical reason for this difference is made clear only by the multi-Cu-site cluster calculations. Taguchi et al. (2005a) also interpreted that for undoped cuprates, the Zhang–Rice singlet band should exist in La_2CuO_4 but be absent in Nd_2CuO_4. However, it is clearly shown by the Cu_6O_{17} cluster model calculations that the Zhang–Rice singlet band exists in both La_2CuO_4 and Nd_2CuO_4.

5.9 AUGER PHOTOEMISSION COINCIDENCE SPECTROSCOPY

We have seen in this chapter that there are many possible experiments using an electron analyzer. Direct photoemission of the electron in XPS can be performed or the decay of the core hole in AES can be studied. At an XAS edge, study can be made of the interference effects of the resonant XAS + AES process with the direct XPS process, and so on. Another range of experiments is possible if the simultaneous measurement of two electrons is considered. An XPS electron can be measured from a core excitation and, in coincidence, a related Auger electron. This experiment is called Auger photoemission coincidence spectroscopy (APECS). APECS experiments can be considered to be similar to RPES and RXES experiments. In RPES, an Auger electron decay energy is effectively measured in coincidence with an x-ray excitation energy and in RXES, an XES decay is measured in coincidence with an x-ray excitation. In APECS, an Auger electron decay energy is measured in coincidence with an x-ray photoelectron excitation energy. Experimentally, APECS is much more difficult than RPES and RXES. This is due to the fact that RPES and RXES do not actually measure any time coincidence because the x-ray excitation energy is set by the synchrotron. In APECS, the excited photoelectron needs to be measured and the specific photoelectron correlated with an Auger electron. This implies that subsequent photoemission processes should be distinguished from each other by the timing of the detector.

The basic notions and implementations of APECS as applied to solids have been originally developed in the late 1970s (Haak et al., 1978). Although, the adaptation of APECS to synchrotron radiation was made by Jensen et al. in 1989, only a few APECS studies have been performed since then.

Figure 5.64 shows the APECS spectra of 2p XPS and 2p3d3d AES in copper metal as measured by Haak et al. (1978). The $2p_{3/2}$ XPS peak shows coincidence with the $2p_{3/2}$3d3d Auger line at ~920 eV and the $2p_{1/2}$ XPS peak shows coincidence with the $2p_{1/2}$3d3d Auger line at ~935 eV. The $2p_{1/2}$ XPS peak shows a second coincidence peak at ~915 eV that has been assigned to decay from $2p_{1/2}$ via Coster–Kronig decay to $2p_{3/2}$ and subsequently 2p3d3d AES. The Thurgate group (Thurgate et al., 1996; Lund et al., 1997) studied the whole 3d metal series. No detailed spectra of TM

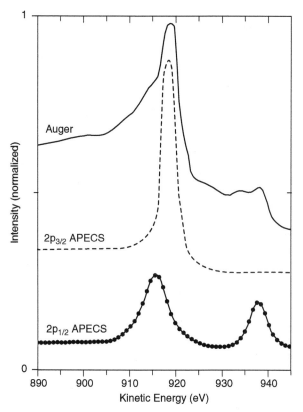

FIGURE 5.64 The spectra of normal Auger (top), APECS with the $2p_{3/2}$ XPS (middle) and APECS with the $2p_{1/2}$ XPS (bottom). (Reprinted with permission from Haak, H.W., Sawatzky, G.A., and Thomas, T.D., *Phys. Rev. Lett.*, 41, 1825, 1978. Copyright 1978 by the American Physical Society.)

oxides have been published to date. In principle, the APECS spectra of TM oxides would be very rich in structure, similar to RXES and RPES spectra.

Figure 5.65 shows the relative energies of the various configurations for MnO. The ground state is dominated by $3d^5$ character, the core excited state by a combination of $2p^53d^5$ and $2p^53d^6$. In the Auger final state, the lowest energy states are the $3p^43d^6$ and $3p^43d^7$ states. APECS studies of 2p XPS and 3p3p AES would be ideal to study the various screening mechanisms that are active during the creation of the first and the second core hole.

APECS experiments would yield equivalent 2D plots as RPES and RXES. For example, Figure 5.66 is a theoretical 2D image of the 3p XPS in coincidence with 3p3d3d AES, calculated for a pure $3d^5$ system. The experimental spectra of MnO and MnF_2 show similar behavior (Bartynski, unpublished results). As mentioned previously, this image will be strongly affected by charge-transfer effects. It would be very interesting if more APECS spectra could be measured to complement the RPES and RXES experiments.

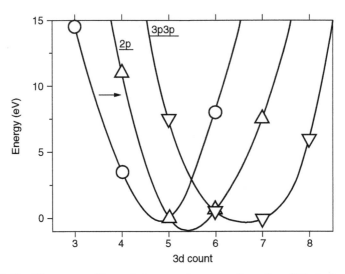

FIGURE 5.65   The energy effects of the ground state configurations 3d^5, the 2p core hole states, and the double 3p core hole states.

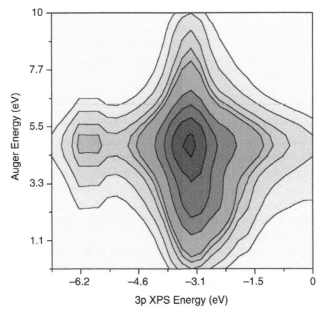

FIGURE 5.66   The 2D image of the 3p XPS (horizontal axis) versus the 2p3d3d Auger (vertical axis) for a pure 3d^5 system.

5.10   SPIN-POLARIZATION AND MAGNETIC DICHROISM IN XPS

Chapter 7 deals with x-ray magnetic circular dichroism (XMCD) effects in x-ray absorption. Also, in photoemission, circular dichroic effects are important. They will be briefly discussed in this section. An electron has its associated spin and instead of using circular polarized x-rays, magnetic effects can be also studied by measuring the spin polarized (SP) signal. Dichroism and spin polarization are strongly coupled to angular dependent effects as has been described in detail in the papers of Thole and van der Laan. This will be briefly introduced subsequently. This book does not deal with valence band photoemission where circular dichroism, spin polarization, and especially angular dependent measurements are even more important. For example, a major application of valence band photoemission is Fermi surface mapping. In particular, the Fermi surface of the copper-oxide superconductors is often studied (Damascelli et al., 2003).

5.10.1   Spin-Polarized Photoemission

Spin-polarized photoemission studies the differences in excitation of spin-up and spin-down electrons in a photoemission experiment. In first approximation, the spin polarized DOS is observed. A prerequisite is that the system under study is magnetized over the probed area, otherwise the spin-polarized signal cancels. This implies that ferromagnetic and aligned paramagnetic systems can be studied, similar to the use of circular polarized x-rays. A thorough review on spin-polarized photoemission has been written by Johnson (1997). The experimental side of spin polarized experiments is far more complex than for normal photoemission. In particular, detecting the spin of the photoelectron is not trivial. Spin detectors invoke some kind of electron scattering process that is sensitive to the spin. One can use (a) the influence of spin–orbit coupling on the scattering of electrons or, in the case of single crystals, the diffraction, and (b) the influence of the exchange interaction in reflection or transmission (Johnson, 1997).

Figure 5.67 shows a typical spin-polarized XPS spectrum for the 3s XPS of Fe. As discussed in Section 5.3.5, the 3s XPS spectral shape is determined by charge transfer, the 3s3d exchange interaction and smaller effects due to the band structure. The spin-polarized version of the 3s XPS can be analyzed within this framework of charge transfer and exchange, by performing individual calculations for an emitted spin-up and spin-down electron. The analysis in terms of the primitive and fundamental spectra as given below provides the basis for these calculations.

5.10.2   Spin-Polarized Circular Dichroic Resonant Photoemission

As already discussed, both spin-polarized photoemission and circular dichroism yield a finite result for ferromagnetically-ordered systems. In this respect, it was an interesting result when Tjeng et al. (1997) demonstrated the possibility of deriving spin-resolved valence band spectra for antiferromagnetic systems.

The method relies on the combined use of spin-polarized photoemission and XMCD in the x-ray absorption step within a resonant photoemission experiment. The principle of the experiment can be explained for the case of a Cu^{2+} ion with a $3d^9$

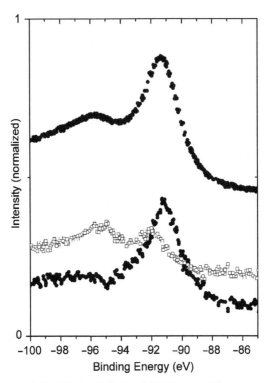

FIGURE 5.67   Spin-polarized Fe 3s XPS using 250 eV x-rays. The upper spectrum represents the spin-integrated spectrum and the lower spectra represents the majority (open symbols) and minority (closed symbols) spin-resolved spectra. (From Johnson, P.D., *Rep. Prog. Phys.*, 60, 1217, 1997. Reprinted with permission from IOP Publishing Ltd.)

ground state. In the x-ray absorption step, a 2p electron is excited to the 3d hole, which is the transition from $3d^9$ to $2p^53d^{10}$. The 2p core hole decays via various Auger channels and the 2p3p3p resonant Auger channel yields a $3p^4(3d^{10})$ final state with its characteristic multiplet structure (Tjeng et al., 1997). The basic background of the method can be explained using single particle matrix elements of a 2p to 3d transition. In Chapter 7, it is explained in detail that a 2p core electron can be excited to a spin-up 3d hole and a spin-down 3d hole with different ratios.

Table 5.7 shows first the basic transitions for the four combinations of x-ray polarization (σ^+ and σ^-) and spin detection (ε^- and ε^+). These numbers rely on the (*i*) the use of the single particle 2p to 3d transitions for a $2p_{3/2}$ core electron and (*ii*) on the fact that a spin-up 2p-hole decay emits for 100% a spin-down electron for the singlet state of the $3p^4$ RAES spectrum (Tjeng et al., 1997). The bottom half of Table 5.7 shows the consequences for XMCD and spin-polarized (SP) experiments. Magnetic circular dichroism (MCD) is the difference between σ^+ and σ^- and for a spin-up 3d hole this yields an effect of −25%. In the case of a spin-down 3d hole, the effect is opposite, which implies that for an antiferromagnetic state the combined effect is zero. This confirms the rule that antiferromagnets do not show XMCD effects. The

TABLE 5.7
Basic Transitions for the Four Combinations of X-Ray
Polarization (σ^+ and σ^-) and Spin-Detection (ε^- and ε^+)

		σ^+	σ^-		
	$3d^+\rightarrow2p^+\rightarrow\varepsilon^-$	6	14		
	$3d^+\rightarrow2p^-\rightarrow\varepsilon^+$	3	1		
	$3d^-\rightarrow2p^+\rightarrow\varepsilon^-$	1	3		
	$3d^-\rightarrow2p^-\rightarrow\varepsilon^+$	14	6		
	Σ	24	24		

		σ^+	σ^-	Δ	Δ/Σ
MCD	$3d^+(\varepsilon^- + \varepsilon^+)(\sigma^+ - \sigma^-)$	9	15	−6	−25%
SP-CD	$3d^+(\varepsilon^- - \varepsilon^+)(\sigma^+ - \sigma^-)$	3	13	−10	−42%
SP	$3d^+(\varepsilon^- - \varepsilon^+)(\sigma^+ + \sigma^-)$	20	4	16	80%
MCD	$3d^-(\varepsilon^- + \varepsilon^+)(\sigma^+ - \sigma^-)$	15	9	6	+25%
SP-CD	$3d^-(\varepsilon^- - \varepsilon^+)(\sigma^+ - \sigma^-)$	−13	−3	−10	−42%
SP	$3d^-(\varepsilon^- - \varepsilon^+)(\sigma^+ + \sigma^-)$	4	20	−16	−80%

Note: The top part shows the relative transition strengths from a $3d^9$ ground state via a $2p^5$ intermediate state to a singlet peak in the $3p^4$ RAES final state. $3d^+ \rightarrow 2p^+ \rightarrow \varepsilon^-$ indicates a spin-up 3d hole that is excited to a spin-up 2p hole and subsequently decays to a singlet $3p^4$ final state with the excitation of a spin-down free electron. The bottom part shows the resulting differences for spin-polarized and circular dichroic measurements.

spin-polarized result is the difference between a detected ε^- and ε^+ electron. This effect is bigger and yields 80% for a spin-up 3d electron. That this number is not 100% implies that the 2p spin–orbit coupling of the core hole transfers only 80% of the spin-up 3d hole to the spin-up 2p hole. In the case of a spin-down 3d hole, the spin-polarized result is again negative implying that antiferromagnets do not show a finite spin-polarized effect. A combination can be made by measuring the XMCD and spin-polarization combined. Effectively, this implies that the signals for ($\sigma^+\varepsilon^-$) and ($\sigma^-\varepsilon^+$) are added, with subtraction of the other two combinations. This yields a SP-CD effect of −42% for a spin-up 3d hole. The reason for making this SP-CD combination is revealed if we look at the effect for a spin-down 3d hole, which also yields −42%. This implies that the SP-CD effect is independent of the direction of the 3d spin and yields a finite effect for antiferromagnets. SP-CD resonant photoemission experiments have been applied to determine the nature of the Zhang–Rice singlet state in copper oxide superconductors (Brookes et al., 2001; Tjernberg et al., 2003) as well as to Ni metal (Sinkovic et al., 1997) and Ce (Tjernberg et al., 2000). For more general descriptions of combined spin polarization and magnetic dichroism effects in XPS, Thole and van der Laan (1991a,b, 1993, 1994a–c) have developed a tensor description method. An abstract of their theory is given in Appendix C.

6 X-Ray Absorption Spectroscopy

In this chapter, we will discuss x-ray absorption spectroscopy (XAS) and electron energy loss spectroscopy (EELS). XAS and EELS use a different source (respectively, x-rays and electrons), but the spectral analysis is analogous and identical under the usual approximations. For EELS, this implies that it is performed with high-energy electrons and at small scattering angles.

6.1 BASICS OF X-RAY ABSORPTION SPECTROSCOPY

A photon that enters a solid can be scattered or it can be annihilated in the photoelectric effect. All other photons will be transmitted and will leave the sample without change in the forward direction. In XAS, the absorption of x-rays by a sample is measured. The intensity of the beam before the sample I_0 and the intensity of the transmitted beam I are measured.

In Chapter 2, the interaction of x-rays with matter was analyzed. The dipole transition operator is, for the incident photon energy $\hbar\Omega$, written as (2.28)

$$T_1 = \sum_q e \sqrt{\frac{2\pi\hbar\Omega}{V_s}}\, \mathbf{e}_q \cdot \mathbf{r} \propto \sum_q \mathbf{e}_q \cdot \mathbf{r}, \tag{6.1}$$

where the operator $\mathbf{p}$ in Equation 2.29 of Chapter 2 is rewritten in terms of $\mathbf{r}$ by using the Heisenberg equation $[\mathbf{r}, H] = (i\hbar/m)\mathbf{p}$. Including this transition operator into the Fermi Golden Rule gives:

$$W_{fi} \propto \sum_q |\langle \Phi_f | \mathbf{e}_q \cdot \mathbf{r} | \Phi_i \rangle|^2\, \delta(E_f - E_i - \hbar\Omega) \tag{6.2}$$

This equation will form the basis for the rest of this chapter. So what happens in practice? If an assembly of atoms is exposed to x-rays, it will absorb some of the incoming photons. At a certain energy, depending on the atoms present, a sharp rise in the absorption will be observed. This sharp rise in absorption is called the absorption edge.

The energy of the absorption edge is determined by the binding energy of a core level. Exactly at the edge, the photon energy is equal to the binding energy, or more precisely, the edge identifies transitions from the ground state to the lowest empty state. Figure 6.1 shows the x-ray absorption spectra of manganese and nickel. The $L_{2,3}$ edges relate to a 2p core level and the K edge relates to a 1s core level binding

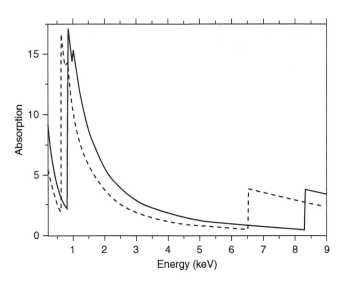

FIGURE 6.1   X-ray absorption cross sections of manganese and nickel. Visible are the $L_{2,3}$ edges at 680 and 830 eV and the K edges at 6500 and 8500 eV, respectively. (From de Groot, F., *Coord. Chem. Rev.*, 249, 31, 2005. With permission from Elsevier Ltd.)

energy. Many complicating aspects play a role here and we will come back to this issue later. In the case of solids, when other atoms surround the absorbing atom, a typical XAS spectrum is shown in Figure 6.2.

Instead of the smooth background as indicated in Figure 6.1, changes in the intensity of the absorption are observed. Oscillations as a function of the energy of the incoming x-rays are visible. The question, which arises immediately, is what causes these oscillations. An answer can be found by assuming that the electron excitation process is a one-electron process. This makes it possible to rewrite the initial state wave function as a core wave function and the final state wave function as a free electron wave function (ε). It is implicitly assumed that all other electrons do not participate in the x-ray induced transition. We will come back to this approximation below.

$$| \langle \Phi_f |e_q \cdot \mathbf{r}| \Phi_i \rangle |^2 = | \langle \Phi_i c\varepsilon |e_q \cdot \mathbf{r}| \Phi_i \rangle |^2 = | \langle \varepsilon |e_q \cdot \mathbf{r}| c \rangle |^2 = M^2 \qquad (6.3)$$

The squared matrix element M^2 is in many cases a number that has only a small variation of energy and it can often be assumed that it is a constant. The delta function of the Golden Rule implies that one effectively observes the density of empty states (ρ).

$$I_{XAS} \sim M^2 \rho \qquad (6.4)$$

Figure 6.3 shows the simplified oxygen 1s x-ray absorption transition of an oxide, where one observes the oxygen p-projected density of states (DOS). The x-ray absorption selection rules determine that the dipole matrix element M is nonzero if the orbital quantum number of the final state differs by 1 from the one of the initial state

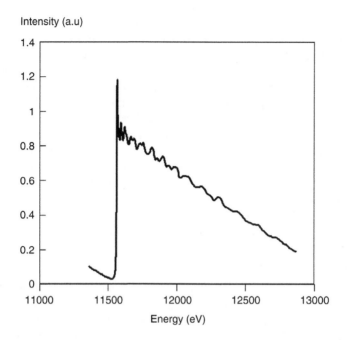

FIGURE 6.2 L_3 XAS spectrum of platinum metal. The edge jump is seen at 11564 eV and above one observes a decaying background modulated by oscillations.

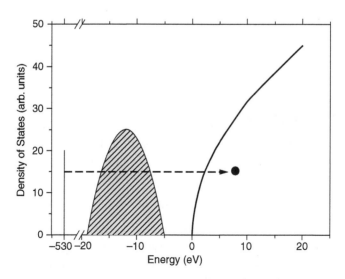

FIGURE 6.3 Schematic density of states of an oxide. The 1s core electron at 530 eV binding energy is excited to an empty state: the oxygen p-projected density of states. (From de Groot, F., *Coord. Chem. Rev.*, 249, 31, 2005. With permission from Elsevier Ltd.)

($\Delta L = \pm 1$, i.e. s $\rightarrow$ p, p $\rightarrow$ s or d, and so on) and the spin is conserved ($\Delta S = 0$). The quadrupole transitions imply final states that differ by 2 from the initial state ($\Delta L = \pm 2$, i.e. s $\rightarrow$ d, p $\rightarrow$ f or $\Delta L = 0$, i.e. s $\rightarrow$ s, p $\rightarrow$ p). They are some 100 times weaker than the dipole transitions and can be neglected in most cases. It will be shown below that they are visible, though as pre-edge structures in the K edges of 3d metals and the $L_{2,3}$ edges of the rare earths (REs). In the dipole approximation, the shape of the absorption spectrum should look like the partial density of the ($\Delta L = \pm 1$) empty states projected on the absorbing site, convoluted with a Lorentzian. This Lorentzian broadening is due to the finite lifetime of the core hole, leading to an uncertainty in its energy according to Heisenberg's principle. A more accurate approximation can be obtained if the DOS replaces the DOS in the presence of the core hole. This approximation gives a relatively adequate simulation of the XAS spectral shape, at least in those cases where the interaction between the electrons in the final state is relatively weak. This is not the case for all systems containing partly-filled d or f states.

6.1.1 METAL $L_{2,3}$ EDGES

In the case of TM $L_{2,3}$ edges, the single particle approximations as used in Equation 6.3 are not valid. The reason is that the initial state wave function contains a partly-filled 3d shell. The $L_{2,3}$ edge excites a 2p electron into the 3d shell and there are now two partly-filled shells. As discussed in Chapter 2, the overlap between these two shells is large.

$$| \langle \Phi_f | e_q \cdot \mathbf{r} | \Phi_i \rangle |^2 = | \langle \Phi_i \underline{2p} 3d | e_q \cdot \mathbf{r} | \Phi_i \rangle |^2 = | \langle \Phi_i^* 2p^5 3d^{n+1} | e_q \cdot \mathbf{r} | \Phi_i^* 3d^n \rangle |^2$$
$$= | \langle 2p^5 3d^{n+1} | e_q \cdot \mathbf{r} | 3d^n \rangle |^2 \qquad (6.5)$$

The final state wave function can be written as $\Phi_i \underline{2p} 3d$. The active orbitals are separated from the rest of the initial state wave function (Φ_i^*) and then the Φ_i^* part is omitted from the equation. What remains is the transition from $3d^n$ to $2p^5 3d^{n+1}$. This has been discussed in detail in Chapter 4.

6.2 EXPERIMENTAL ASPECTS

Consider an infinitely thin layer, of thickness dl, of the absorbing material. The intensity I_0 of the incident beam is reduced by dI on passing through dl. dI is proportional to dl and dI is also proportional to the total intensity I, that is,

$$dI = -\mu I \cdot dl \qquad (6.6)$$

μ is a proportionality constant called the linear absorption coefficient. It incorporates the combined effects of all photoelectric and scattering processes. Integrating Equation 6.6, the linear absorption coefficient μ can be written as a function of the x-ray energy, indicated with its frequency Ω:

$$I_x(\Omega) = I_0(\Omega) \cdot e^{-\mu(\Omega)x} \qquad (6.7)$$

6.2.1 TRANSMISSION DETECTION

After the beam has passed the monochromator, its intensity (I_0) is measured in an ionization chamber or by using a thin metallic foil or grid for absorbing some of the photons. The x-ray interacts with the sample of interest and the intensity after the sample is measured with a second ionization chamber or a photodiode. This is the transmission mode of the x-ray absorption. Transmission experiments are standard for hard x-rays.

An important limitation of transmission detection arises from the requirement for a homogeneous sample. Variations in the thickness or pinholes are reasons for the so-called thickness effect that can significantly affect the spectral shape by introducing nonlinear responses. This is important, in particular, for the extended x-ray absorption fine structure (EXAFS) analysis. In general, the combination of the short attenuation length with the thickness effect makes transmission experiments unsuitable for x-ray absorption below 1 keV. For the soft x-ray range, other detection modes must be used.

The advantage of transmission detection is that there can be no effects due to yield variations. Also, only with transmission detection is it possible to measure the absolute cross sections. With yield methods, signal that is proportional to the x-ray absorption cross section is always measured. To circumvent the problems associated with yield methods, thin layers of metals can be grown. For example, 5 to 7 nm thin layers of Fe and Co have been grown on parylene (C_8H_8) and the x-ray absorption in transmission mode has been measured (Chen et al., 1995).

6.2.2 ENERGY DISPERSIVE X-RAY ABSORPTION

A special technique to measure transmission experiments is to use an energy dispersive monochromator. The monochromator crystal is bent and a range of energies is sent through the sample simultaneously. Each energy has its particular angle of incidence and after the transmission through the sample the complete x-ray absorption spectrum can be measured with an array detector. Energy dispersive experiments suffer from similar limitations as normal transmission experiments with respect to attenuation length and thickness effect. A major advantage of energy dispersive x-ray absorption is that the complete XAS spectrum is obtained momentarily. It is clear that this yields interesting options for time-resolved measurements.

6.2.3 FLUORESCENCE YIELD

The decay of the core hole gives rise to an avalanche of electrons, photons, and ions escaping from the surface of the substrate. By measuring any of these decay products, it is possible to measure samples of arbitrary thickness. In this section, the different methods are discussed, specifically with regard to the conditions under which a specific yield measurement represents the x-ray absorption cross section and the related question of the probing depth of the specific yield method.

The fluorescent decay of the core hole can be used as the basis for the absorption measurement. The amount of fluorescent decay increases with energy and a comparison with the amount of Auger decay shows that the Auger decay dominates for all core levels below 1 keV. In the case of the 3d metals, the K edges show strong fluorescence and all other edges mainly Auger decay. The photon created in the fluorescent decay has a mean free path of the same order of magnitude as the incoming x-ray, which excludes any surface effect. On the other hand, it means that there will be saturation effects if the sample is not dilute.

6.2.4 Self-Absorption Effects in Fluorescence Yield Detection

For dilute materials, the background absorption μ_b dominates the absorption of the specific edge and the measured intensity is proportional to the absorption coefficient. For less dilute materials, the spectral shape is modified and the highest peaks will appear compressed with respect to the lower peaks, an effect known as self-absorption or saturation. The main effect of saturation because of self-absorption is a reduction of the peak heights and depths and, as such, a blurring of the spectral information. In principle, by using the inverse formula, the original absorption spectrum can be reconstructed from the saturated one. An uncertainty in this procedure is the exact value of μ_b; in addition, this data treatment increases the noise.

6.2.5 Nonlinear Decay Ratios and Distortions in Fluorescence Yield Spectra

An assumption for the use of decay channels is that these channels are linearly proportional to the absorption cross section. In general, this linear proportionality holds, but there are cases where the ratio between radiative and nonradiative decay varies significantly over a relatively short-energy range. The reason behind this phenomenon is a large variation in the fluorescence decay depending on the symmetry of the final state in the x-ray absorption process. It turns out that for the late 3d metal L edges and RE M edges, the fluorescence decay can vary drastically over the edge. As Auger decay dominates over fluorescence for soft x-rays, this effect will only be visible in fluorescence yield detection. This state-dependent Auger and fluorescence decay has been studied in detail for NiO. The Auger channels show small variations, while the fluorescence decay varies, and the states at higher energy of both the L_3 and L_2 edges have an approximately four times higher fluorescent decay (de Groot et al., 1995). The effects of symmetry-dependent variations in the fluorescence decay are even stronger for the $M_{4,5}$ edges of the rare earths. The Tm^{3+} $M_{4,5}$ edge has three peaks and it can be shown that the first peak hardly decays via x-ray fluorescence (Pompa et al., 1997).

6.2.6 Partial Fluorescence Yield

Recently, it became possible to use fluorescence detectors with approximately 0.3–2.0 eV resolutions to tune to a particular fluorescence channel. This could be denoted as a partial fluorescence yield. The technique is also known as selective x-ray absorption because one can select, for example, a particular valence and measure the x-ray absorption spectrum of that valence only. Other possibilities are the selectivity to the spin orientation and the chemical nature of the neighboring atoms. Partial fluorescence yield effectively removes the lifetime broadening of the intermediate state. This effect can be used to measure x-ray absorption spectra with unprecedented spectral resolution. These experiments are further discussed in Chapter 8.

6.2.7 ELECTRON YIELD

With the total electron yield method, all electrons that emerge from the sample surface are detected, independent of their energy. A number of detection devices can be
used such as the pico-ampere meter for detecting the current flowing to the sample,
and the channeltron for amplifying the emitted electrons to a detectable signal. In
addition, it can be shown that if the measurements are carried out in a gaseous atmo-
sphere with a pressure of approximately 10 mbar, the detected ionized gas molecules
yield a signal that is, under certain conditions, proportional to the absorption cross
section. The interaction of electrons with solids is much larger than the interaction of
x-rays, which implies that the electrons that escape from the surface must originate
close to the surface.

The probing depth of total electron yield lies in the range of approximately
3–10 nm, depending on the material studied. A quantitative study on the oxygen K
edge of Ta_2O_5 determined an electron escape depth of 4 nm. Studies of RE overlayers
on Ni revealed very short escape depths of the order of 1 nm. The extremely large
x-ray absorption cross sections of the RE $M_{4,5}$ edges means that, for these edges, the
x-ray penetration depth falls in the same range as the electron escape depth. This
implies that both the electron escape depth λ_e and the frequency-dependent x-ray
penetration depth $\lambda_p(\Omega)$ affect the measured x-ray absorption cross section (I_e),
according to:

$$I_{TEY}(\Omega) \propto \frac{\lambda_e}{\lambda_e + \lambda_p(\Omega)\sin\alpha} \qquad (6.8)$$

where α is the angle of incidence. This implies that if $\lambda_e \ll \lambda_p \sin\alpha$, the total electron
yield (TEY) is inversely proportional to λ_p, and hence, proportional to the absorption
coefficient. At the absorption maximum and/or at grazing incidence, saturation effects
can occur. Angular dependent measurements on Dy/Ni(110) revealed an x-ray
penetration depth of only 12 nm at the M_5 resonance of Dy (Vogel and Sacchi, 1994).
In most other cases, the x-ray penetration depth will be significantly larger.

6.2.8 PARTIAL ELECTRON YIELD

Instead of just counting the escaped electrons, their respective energies can be
detected and, as such, the partial electron yield can be measured. For example, a Ni
1s core hole creates the primary 1s2p2p Auger electrons with kinetic energies of
~6500 eV. These primary electrons can lose part of their energy in inelastic scattering phenomena after which they have energies anywhere between 6500 eV and 0 eV.
The 1s2p2p Auger channel leaves two core holes in the 2p level, each with a binding
energy of 850 eV. Both these holes can decay again via "cascade" or secondary
Auger processes, for example, 2p3d3d. The 2p3d3d Auger electron has an energy of
~850 eV and the system is left with only valence holes that decay via luminescence,
phonons, and recombination processes. The luminescence can be detected and this
technique is called x-ray excited optical luminescence (XEOL) (Soderholm et al.,
1998). Each "cascade" or secondary peak brings with it an inelastic tail. The majority
of the emitted electrons from a sample are secondary electrons with emission

energies of less than 100 eV. These electrons have lost most of their energy (due to inelastic scattering processes) on their route to the surface of the sample. If an Auger decay channel with high resolution is detected, similar experiments can be performed as with high-resolution fluorescence detection and selective x-ray absorption experiments. This has been discussed in Chapter 5.

6.2.9 ION YIELD

If the absorption process takes place in the bulk and the core hole decays via an Auger process, a positively charged ion is formed. However, due to further decay, inelastic electron scattering and screening processes, the original charge neutral situation, which will be restored after some time, typically in the femtosecond range. If, however, the absorption process takes place at the surface, the possibility exists that the atom that absorbs the x-ray is ionized by Auger decay and escapes from the surface before relaxation processes can bring it back to a bound state. If the escaping ions are analyzed as a function of x-ray energy, the signal is, in general, proportional to the absorption cross section. Because only atoms from the top-layer are able to escape, ion yield is extremely surface sensitive. The mere possibility of obtaining a measurable signal from ion yield means that the surface is irreversibly distorted. As there are of the order of 10^{15} surface atoms per cm^2, this does not necessarily mean that a statistically relevant proportion of the surface has to be affected for a measurable signal of, say, 10^7 ions. Ion yield detection has been used to probe the surface of CaF_2 single crystals and the reduced symmetry due to the surface is clearly visible (Himpsel et al., 1991).

6.2.10 DETECTION OF AN **EELS** SPECTRUM

In a transmission electron microscope (TEM), a beam of monochromatic electrons is transmitted through the sample. The EELS spectrum is measured at a small scattering angle. The low-energy losses include vibrations, electronic excitations, and plasmon excitations and in the core region, a core electron can be excited. The low-energy losses, not further discussed here, provide a tailing background on which the core excitations are visible. The tailing background is usually removed by fitting the background before and after the edge. After background subtraction, the EELS spectrum becomes visible and under the conditions as discussed in Section 2.5, the EELS spectrum is identical to the XAS spectrum. Details can be found in the review of Keast et al. (2001) and Colliex (1991, 1994).

There are a few experimental precautions that must be taken. The EELS cross section for both plasmons and core excitations implies that the sample thickness must be below ~100 nm in order to prevent too many multiple excitations taking place. For some materials and edges, the cross sections are very high and multiple excitations are visible down to a 10 nm thickness (e.g. La and Ce M edges) (Belliere et al., 2006). EELS spectra can be measured in a few modes by a TEM. Typical set-ups are the use of a magnetic electron spectrometer added to a commercial TEM. A field emission gun can be used as an electron source or an additional monochromator can be added for the electron source. A dedicated TEM-EELS set-up reaches a resolution of typically 0.3 eV for core excitations. This will be further discussed subsequently in relation to the spatial resolution.

6.2.11 Low-Energy EELS Experiments

Instead of a 100–300 keV TEM electron source, a lower energy source can be used. Typically, the same electron source is used as in inverse photoemission experiments with an energy range between 10–1500 eV. This allows measurement of the M and L edges in 3d transition metal systems. Low-energy EELS does not obey the dipole selection rule and the transition operator of the inelastic scattered electron can be rewritten into an effective Coulomb scattering operator. The effective Coulomb scattering operator contains direct scattering and exchange scattering, where there are effectively no selection rules on the orbital and spin angular momenta. This allows 3s3d transitions, which will be discussed in Section 6.4.2. An additional option is to measure low-loss features at a core resonance. This so-called resonant EELS (REELS) is equivalent to resonant inelastic x-ray scattering (RIXS), which will be discussed in Chapter 8.

6.2.12 Space: X-Ray Spectromicroscopy and TEM-EELS

X-ray absorption can be combined with a number of x-ray microscopy techniques. In addition, essentially identical XAS spectra can be measured by a TEM using EELS. In this section, we give a short introduction and overview of some of the current developments.

The following spectromicroscopy techniques are in the process of being developed:

1. Photoemission electron microscopy (PEEM), which measures the electron yield with a position-sensitive electron analyzer.
2. Transmission x-ray microscopy (TXM), which exists in scanning (STXM) and full-field modes.
3. Transmission electron microscopy (TEM), again in scanning (STEM) and full-field modes.

The main conclusions from Table 6.1 are that TEM-EELS has the highest spatial resolution of 0.5 nm (in EELS mode) and some STEM-EELS microscopes promise to go down to sub-Ångström resolution. X-PEEM is a surface technique and has typically 50 nm spatial resolution. TXM combines 20 nm resolutions with ambient

---

TABLE 6.1
Comparison of the Approximate Spatial Resolution, Energy Resolution, Pressure Range, and Thickness Constraints of TEM, TXM, and PEEM

Technique	Spatial	Energy	Pressure	Thickness
TEM-EELS	0.5 nm	0.3 eV	0	0.1 μm
In situ TEM-EELS	0.2 nm	1.3 eV	0.01 bar	0.1 μm
Scanning TXM (ALS)	20.0 nm	0.3 eV	1 bar	1 μm
Full-field TXM (ALS)	15.0 nm	1.0 eV	1 bar	1 μm
PEEM (SLS)	50.0 nm	0.3 eV	0	5 nm
PEEM-3 (ALS)	5.0 nm	0.3 eV	0	5 nm

conditions. The energy resolution of EELS, PEEM, and TXM microscopes are, in most cases, of the order of 0.3 eV or better. It remains to be confirmed whether or not these numbers can be reached on the Titan and super-STEM microscopes as well as with X-PEEM. It is obvious that these new x-ray and electron microscopes are crucial for a large range of research fields involving nanometer-sized structures. STEM-EELS is the obvious choice for the highest resolution images. In addition, a range of electron microscopy tools is available (e.g. EELS combined with electron diffraction, and so on). The obvious disadvantage is the need for vacuum conditions. There are a few TEM microscopes that run under "*in situ*" conditions, which implies a pressure of typically a few mbar, where typically the *in situ* microscopes have an energy resolution of 1.0 eV.

X-ray PEEM applications are all surface phenomena where a lot of attention is given to the magnetic effects. TXM applications involve all experiments that need *in situ* and *in vivo* conditions. This includes fields such as environmental studies, catalysis, fuel cells, and batteries. A recent STXM study showed that it is possible to measure flowing hydrogen at 1 bar and, at 500°C, the iron L edge with 0.3 eV resolution and ~25 nm spatial resolution. This allowed the study of the *in situ* development on the spatial variations in the reduction process. A major application is the study of cells and tissue where x-rays have enormous advantages against electrons with respect to damage in water. Both x-ray techniques, PEEM and TXM, can be run under XMCD conditions if circular polarized x-rays are used. The XMCD situation regarding TEM-EELS is more complex but, in principle, magnetic information is available with chiral TEM techniques (Schattschneider et al., 2005). Both TEM and TXM can be rendered into tomographic techniques, essentially by rotating the sample. TXM can be "automatically" run in resonant conditions to be element specific. TEM tomography is usually not performed with energy-filtered EELS images but with the usual TEM images, which implies the loss of chemical information.

6.2.13 TIME-RESOLVED X-RAY ABSORPTION

There are a number of different modes to perform time-resolved x-ray absorption experiments. The time scale needed depends on the process that is studied. One can distinguish two time domains: on the one hand, time-programmed reactions and catalytic reactions can use milliseconds to minutes as time-resolution and on the other hand, optical excitations need nanoseconds to femtoseconds resolution to track the electronic and magnetic system responses. This yields a wide range of time-resolved techniques.

The XAS spectrum can be measured with a monochromator that is scanning through an edge and then repeating the experiment every few minutes. Measurements in the few seconds range can use so-called quick-scanning monochromators. The use of a dispersive monochromator can further increase the speed. This allows the simultaneous measurement of a whole spectrum. The time-resolution is mainly determined by the statistics, typically between 1 ms and 1 s. An example of a time-resolved experiment with minute time resolution is the study of the valence of iron during reactions with Fe/ZSM5 using the iron L edge spectral shapes. An interesting aspect is that the valence of the sample changes in a matter of minutes at room temperature under vacuum conditions, which presents a warning for the feasibility/reliability of electron yield XAS measurements in vacuum (Heijboer et al., 2003).

Using pump-probe techniques, XAS spectra can be measured very fast; in principle, as fast as the combination of excitation and measurement/detection. The time resolution is set by the system characteristics and ultimately by the timing of the synchrotron pulses. Cavalleri et al. (2004) measured the picosecond metal-to-insulator response in VO_2. To go below the timing of synchrotron pulses, there are several methods to go into the femtosecond regime. This includes time slicing and the use of alternative x-ray sources, such as high-harmonic lasers and laser-generated plasmas. An example is given in Section 6.5.1 for the 50 ps charge-transfer excitation of $[Ru(bpy)_3]^{2+}$ (Gawelda et al., 2006). An overview is given by Bressler and Chergui (2004).

6.2.14 EXTREME CONDITIONS

All experiments, where electrons play a role (XPS, PEEM, EELS) are usually performed in a vacuum. There are a few TEM-EELS machines that go up to a few mbar and the same is true for a few XPS machines. Higher pressures are not possible due to the strong interaction of electrons with matter.

Soft x-ray experiments (below 1 keV) are usually performed under vacuum conditions. Experiments up to a few mbar are performed using electron-yield detection. Using fluorescence yield detection pressures up to 1 bar are feasible. It is also possible to use very thin windows and a transmission XAS experiment using soft x-rays, usually in combination with TXM measurements. High-pressure studies are not feasible with soft x-rays.

Hard x-ray experiments (above 3 keV) can be performed under essentially any high-temperature and high-pressure conditions. For example, diamond anvil cells can be used and the x-rays are shot through the diamonds or through Be gaskets. The larger range of extreme conditions reachable with hard x-rays is an important aspect of resonant and nonresonant XES. These experiments allow the measurement of soft x-ray absorption (like) spectra with hard x-rays [e.g. the study of the oxygen K edge of (supercritical) water]. These experiments will be discussed in Chapter 8.

6.3 $L_{2,3}$ EDGES OF 3d TM SYSTEMS

The $L_{2,3}$ edges of the 3d TM systems are described in great detail. These edges have mostly been studied over the last 20 years and, in particular, with respect to the symmetry of the ground state. The systems from $3d^0$ to $3d^9$ will now be discussed. Other issues that will be described are octahedral crystal fields and symmetry reductions at surfaces. XMCD effects are discussed in Chapter 7.

In Table 6.2, the term symbols of all $3d^n$ systems in atomic symmetry are given. Together with the dipole selection rules, this immediately sets strong limits to the number of final states that can be reached. Consider, for example, the $3d^3 \rightarrow 2p^53d^4$ transition: the $3d^3$ ground state has $J = 3/2$ and there are respectively 21, 35 and 39 states of $2p^53d^4$ with $J' = 1/2$, $J' = 3/2$, and $J' = 5/2$. This implies a total of 95 allowed peaks out of the 180 final state term symbols. From Table 6.2, some special cases can be identified: a $3d^9$ system makes a transition to a $2p^53d^{10}$ configuration that only has

TABLE 6.2
2p X-Ray Absorption Transitions from the Atomic Ground State to
All Allowed Final State Symmetries, after Applying the Dipole
Selection Rule: $\Delta J = -1, 0,$ or $+1$

Transition	Ground	Transitions	Term Symbols
$3d^0 \rightarrow 2p^5 3d^1$	$^1S_0$	3	12
$3d^1 \rightarrow 2p^5 3d^2$	$^2D_{3/2}$	29	45
$3d^2 \rightarrow 2p^5 3d^3$	$^3F_2$	68	110
$3d^3 \rightarrow 2p^5 3d^4$	$^4F_{3/2}$	95	180
$3d^4 \rightarrow 2p^5 3d^5$	$^5D_0$	32	205
$3d^5 \rightarrow 2p^5 3d^6$	$^6S_{5/2}$	110	180
$3d^6 \rightarrow 2p^5 3d^7$	$^5D_2$	68	110
$3d^7 \rightarrow 2p^5 3d^8$	$^4F_{9/2}$	16	45
$3d^8 \rightarrow 2p^5 3d^9$	$^3F_4$	4	12
$3d^9 \rightarrow 2p^5 3d^{10}$	$^2D_{5/2}$	1	2

two term symbols out of which only the term symbol with $J' = 3/2$ is allowed. In other words, the L_2 edge has zero intensity. Because of the limited amount of states for the $2p^5 3d^1$ and $2p^5 3d^9$ configurations, $3d^0$ and $3d^8$ systems only have three and four peaks, respectively.

Atomic multiplet theory is able to accurately describe the 3d and 4d x-ray absorption spectra of the rare earths. In the case of the 3d TM ions, atomic multiplet theory cannot simulate the x-ray absorption spectra accurately because the effects of the neighbors on the 3d states are too large. It turns out that it is necessary to include both the symmetry effects and the charge-transfer effects of the neighbors explicitly. Ligand field multiplet theory takes care of all symmetry effects while charge transfer multiplet theory allows the use of more than one configuration.

6.3.1   3d⁰ Systems

$3d^0$ systems are very special because they contain a band gap between the occupied ligand valence bands and the empty 3d band. This is a very stable situation, which means that a large range of ions can be found in such a situation. This applies to the whole series from K^+ to Mn^{7+}. The oxides of the covalent Cr^{6+} and Mn^{7+} systems have a large amount of metal character in the oxygen 2p valence band.

As discussed in Chapter 4, charge transfer multiplet (CTM) calculations describe the complete 2p XAS spectrum of $SrTiO_3$. The 2p XAS spectra of K^+, Ca^{2+}, and Sc^{3+} systems can be accurately described with a single $3d^0$ configuration. The higher valent systems Ti^{4+}, Cr^{6+}, and Mn^{7+} need more configurations, or large reductions in their Slater–Condon parameters. The tetrahedral $3d^0$ systems for Ti^{4+}, V^{5+}, Cr^{6+}, and Mn^{7+} are published by Brydson et al. (1993). It is apparent that the spin–orbit splitting between the L_3 and L_2 edges decreases from Mn^{7+} to Ti^{4+}. However, inspection of the spectra reveals that the general shape of the 2p XAS spectra are remarkably similar. Both the L_3 and L_2 edges show a splitting of the white line into two components, the lower energy component being of considerably lower relative intensity than the higher energy component. Furthermore, close inspection reveals the presence of a weak feature prior to the main L_3 edge. This feature is due to multiplet effects as discussed previously.

6.3.2 3d¹ Systems

6.3.2.1 VO₂ and LaTiO₃

Ti^{3+} and V^{4+} are $3d^1$ systems that have one 3d electron occupied. The atomic Hund's rule ground state is $^2D_{1/2}$. In a cubic crystal field, the ground state has one t_{2g} electron and $^2T_2$ symmetry. This ground state is affected by the 3d spin–orbit coupling and distortions to the lower symmetry split the t_{2g} state into two or three substates.

The Tjeng group recently studied two prototype $3d^1$ systems: VO₂ and LaTiO₃. VO₂ is an interesting system with regard to its metal-insulator transition and the relative role of electron-lattice interactions and corresponding structural distortions versus electron correlations. An important aspect is the possible role of the orbital structure of V^{4+} ions. Specific orbital occupation leads to the formation of one-dimensional bands, making the systems susceptible to a Peierls-like metal insulator transition. The three t_{2g} orbitals, xy, xz, and yz can be divided into linear combinations of orbitals with respect to the polarization. The 3d spin–orbit coupling is (assumed to be) quenched. The CTM calculations obtained as such are compared with experiment and the details of the polarization dependence can be fitted by a linear combination of the σ, π, and δ combinations for both the insulating and the metallic phases. Effectively, one observes a change from a three-dimensional statistical distribution of states in the metallic phase into an 80% pure one-dimensional σ-coupling in the insulating phase (Haverkort et al., 2005b).

The situation for LaTiO₃ is different. LaTiO₃ is an antiferromagnetic insulator with a pseudocubic perovskite crystal structure and a Néel temperature between T_N 130 K and 146 K, depending on the exact oxygen stoichiometry. In this system, the role of the 3d spin–orbit coupling is important. The overall situation is as follows. The dominant energy effect is the cubic crystal field. This yields a three-fold degenerate t_{2g} state. This state is affected by symmetry distortions, the 3d spin–orbit coupling, and magnetic exchange interactions. The crucial aspect for the simulations is the magnitude of the symmetry distortions. The ground state is split by this crystal field and effectively diminishes the effect of the 3d spin–orbit coupling. One obtains a single Kramers doublet that is split by the exchange interaction below the Néel temperature. The result is that there is also no temperature dependence in the 2p XAS spectrum. There will be a temperature dependence though in the polarization dependence, which will show the effect of the exchange splitting below the Néel temperature (Haverkort et al., 2005a).

6.3.3 3d² Systems

Ionic $3d^2$ systems have two 3d electrons. The atomic ground state is $^3F_2$. In a cubic crystal field, the ground state has two t_{2g} electrons and $^3T_1$ symmetry. This ground state is strongly affected by the 3d spin–orbit coupling. Distortions to lower symmetry split the t_{2g} state into two substates (e.g. a trigonal distortion creates a_{1g} or e_g substates). If the e_g state has the lowest energy, a half-filled e_g^2 situation can be created.

V_2O_3 has an interesting phase diagram that involves a monoclinic antiferromagnetic insulating state and both metallic and insulating paramagnetic rhombohedral corundum states. The metal-insulator transitions between these phases are considered to be classical examples of Mott transitions in which changes in the interplay between band formation and electron correlation causes a crossover between the metallic and insulating regimes. Due to the presence of a trigonal distortion associated with the corundum structure, the t_{2g} orbital splits into a nondegenerate a_{1g} and a doubly degenerate e_g orbital. The a_{1g} orbital has lobes directed along the c vector of the hexagonal basis, while the e_g lobes are more within the (a, b) basal plane. The XAS transition probability to an empty a_{1g} or e_g orbital depends strongly on whether the polarization vector E of the light is parallel or perpendicular to the c axis and the polarization dependence of the experimental L edge spectra provides the correct ground state, as described in detail by Park et al. (2000).

6.3.4   3d³ SYSTEMS

Ionic $3d^3$ systems have three 3d electrons. The atomic ground state is $^4F_{3/2}$. In a cubic crystal field, the ground state has three t_{2g} electrons. This half-filled t_{2g} band creates a stable magnetic ground state with $^4A_2$ symmetry. This A_2 ground state is not affected by 3d spin–orbit coupling and is also not much affected by symmetry lowering. Cr_2O_3 has a t_{2g}^3 configuration, the small distortion from cubic symmetry has no detectable effect on the spectral shape. The spectrum of Cr_2O_3 (Theil et al., 1999) can be reproduced well with a single configuration ligand field multiplet (LFM) calculation.

Figure 6.4 shows the 2p XAS spectra of Li_2MnO_3. The spectra of MnO_2 (Gilbert et al., 2003; Pecher et al., 2003) and $SrMnO_3$ (Abbate et al., 1992a) are similar but broader. The high valence of Mn^{4+} necessitates CTM calculations, using a ground

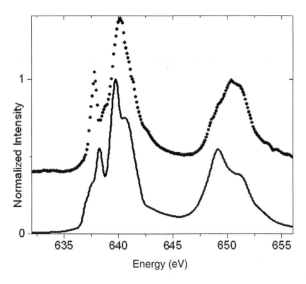

FIGURE 6.4   Manganese 2p x-ray absorption spectrum of Li_2MnO_3 compared with a CTM calculation of $3d^3$ and $3d^4\underline{L}$.

state built from $3d^3$ and $3d^4\underline{L}$. Actually, CTM calculations using parameters as expected for an Mn^{4+} system do not perfectly simulate the spectral shape. It is likely that the effects of higher lying configurations, such as $3d^5\underline{L}^2$, have an influence on the spectral shape.

6.3.5   3d⁴ Systems

Ionic $3d^4$ systems have four 3d electrons. The atomic ground state is $^5D_0$. In a small cubic crystal field, the ground state has three t_{2g} electrons and one e_g electron filled. This half-filled e_g state is sensitive to Jahn–Teller distortions. $3d^4$ systems in octahedral symmetry can be ^5E high-spin $t_{2g}^3 e_g^1$ or $^3T_2$ low spin t_{2g}^4.

The CrF_2 2p XAS spectrum has been given in Figure 6.5. The calculation is performed for $3d^4$ Cr^{2+} in D_{4h} symmetry with 10 Dq of 1.1 eV and respective Ds values of 775 meV. The atomic 3d spin–orbit coupling combined with the D_{4h} symmetry yields a zero-field splitting (ZFS) pattern for a $3d^4$ initial state with a $M_S = 2$ ground state. An important omission in the simulations is that the peaks E_1, E_2, and E_3 are missing from the simulations. They are assigned to charge transfer peaks (Theil et al., 1999).

Mn^{3+} systems have a larger crystal field than Cr^{2+} with typical values in the order of 1.8 eV for oxides. This implies that the ground state of Mn^{3+} oxides is ^5E high-spin, which makes Mn^{3+} a Jahn–Teller ion. Subsequently, we will discuss two cases, $LaMnO_3$ and $LiMnO_2$, the latter of which is close to the low-spin transition point.

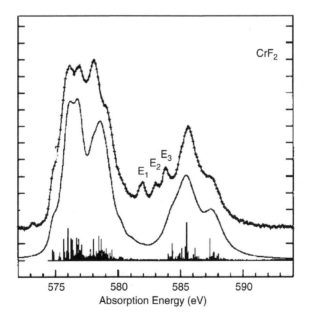

FIGURE 6.5   2p XAS spectra of $3d^4$ CrF_2 (solid lines with markers). Below the experimental spectrum, the theoretical calculation with stick diagrams is shown. (From Theil, C., van Elp, J., and Folkmann, F., *Phys. Rev. B*, 59, 7931, 1999. With permission from Elsevier Ltd.)

6.3.5.1 LaMnO₃

The most studied Mn³⁺ system is LaMnO₃. This prototype perovskite system can be doped with Sr, which yields large (giant) magnetoresistance materials. LaMnO₃ itself is an antiferromagnetic insulator with a Néel temperature of 140 K. Doping by 10–20% Sr leads to a (temperature-dependent) competition between ferromagnetism and a cooperative Jahn–Teller distorted phase in La(Sr)MnO₃. The 2p XAS spectrum of LaMnO₃ has been measured by Abbate et al. (1992a) and simulated with a ^5E ground state crystal field multiplet calculation, which was confirmed by the XMCD experiments on ferromagnetic La(Sr)MnO₃ systems by Pellegrin et al. (1997).

The doped LaMnO₃ systems are high spin, so all four electrons are parallel. The three t_{2g} spin-up electrons are spherical symmetric and the e_g electron can occupy either an $x^2 - y^2$ orbital or a z^2 orbital. The doped systems are supposed to have an orbital ordered spin-structure that orders cross-type $x^2 - y^2$ type orbitals with respect to the z-axis (i.e. a coupling of $y^2 - z^2$ and $z^2 - x^2$), or rod-type z^2 type orbitals with respect to the z-axis (i.e. a coupling of x^2 and y^2). Huang and Jo (2004) calculated the polarization dependence of both options. The calculations indicate that the ground state at 240 K has (mainly) an $x^2 - y^2$ type orbital ordering.

6.3.5.2 Mixed Spin Ground State in LiMnO₂

Like LaMnO₃, LiMnO₂ is considered to have a high-spin ground state, but the low-spin configuration is close in energy. This implies that the effects of 3d spin–orbit coupling can be large, similar to the case of PrNiO₃ as discussed subsequently. Figure 6.6 gives the spectral shapes from the LFM calculations for, from bottom to

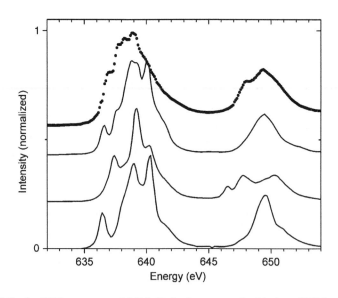

FIGURE 6.6 2p XAS spectrum of LiMnO₂ (top), compared with three LFM calculations, for respectively from bottom to top, the $^3T_1$ low-spin, ^5E high-spin and a mixed spin ground state for 3d⁴ Mn³⁺ atoms. (From de Groot, F.M.F., *J. Electron Spectrosc. Relat. Phenom.*, 67, 529, 1994. With permission from Elsevier Ltd.)

top, the low-spin $^3T_1$ configuration, the high-spin ^5E configuration, the mixed spin state and the experimental spectrum. The inclusion of 3d spin–orbit coupling yields a double group A_1 ground state for both states. The mixed spin state is a linear combination of these double group A_1 states. The situation for $LiMnO_2$ is further complicated by symmetry distortions and significant charge-transfer effects that have not been included for the calculation in Figure 6.6.

6.3.6   3d⁵ Systems

3d⁵ systems have five 3d electrons. The atomic ground state is $^6S_{5/2}$. In a cubic crystal field, the ground state has its three t_{2g} and two e_g states half-filled. This half-filled state creates a stable magnetic ground state with $^6A_2$ symmetry. This A_2 ground state is not (much) affected by symmetry lowering. In principle, it is also not affected by 3d spin–orbit coupling, but higher-order effects that mix excited configurations can split the $^6A_2$ ground state. Low-spin 3d⁵ systems have five electrons in their t_{2g} shell, yielding a spin-half $^2T_1$ ground state. This ground state occurs for Mn^{2+}-cyanides, Fe^{3+}-cyanides, and other Fe^{3+} coordination complexes.

6.3.6.1   MnO

A beautiful example of a pure 3d⁵ system is MnO. Its ground state has a large charge-transfer energy; hence, over 90% 3d⁵ character. This allows one to neglect charge-transfer effects and Figure 6.7 shows the result of a LFM calculation in O_h symmetry and with 10 Dq equal to 0.9 eV (de Groot et al., 1990b). Essentially, all features of the complex spectrum are reproduced with the correct intensity and energy. The main peak is a little too low in experiment, which could suggest a saturation effect

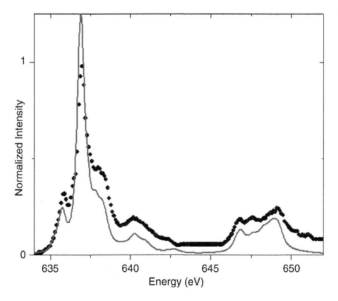

FIGURE 6.7   Manganese 2p x-ray absorption spectrum of MnO (dotted) compared with a LFM calculation (solid line). The value of 10 Dq is 0.9 eV.

in the electron yield detection due to the high x-ray absorption cross section. A large range of high-spin Mn^{2+} systems has been measured. They all have very similar spectra that can be well reproduced from LFM calculations (Cramer et al., 1991). Using thin MnO films on Ag substrates, Nagel et al. (2007) used the CTM model for the analysis of the polarization-dependent x-ray absorption measurements. They found a tetragonal distortion of the oxide films where the distortion decreased with increasing film thickness.

6.3.6.2 Fe_2O_3

Fe_2O_3, hematite, is the Fe^{3+} analog of MnO. Its ground state is also $^6A_1$ but, due to the higher valence, the crystal field splitting and the charge-transfer effects will be larger than for MnO. The LFM of Fe_2O_3 is given in Figure 6.8 with a solid line. The crystal-field strength has been set at 1.5 eV and the Slater–Condon parameters have been rescaled to 90% of their atomic value. The simulation describes the experimental spectrum well for both the L_3 edge and the L_2 edge. From a series of CTM calculations using a $3d^5 + 3d^6\underline{L}$ ground state, we found that an overall crystal-field splitting of 1.5 eV can be obtained by including an ionic crystal-field splitting of 1.2 eV, with an additional 0.3 eV splitting due to the difference of hopping energies of 1.0 eV (t_{2g}) and 2.0 eV (e_g), respectively. The charge-transfer energy has been set to 3.0 eV. This simulation is shown with the dashed line. The satellites at 713 and 717 eV are found in the CTM simulation.

A large range of high-spin Fe^{3+} systems has been measured. Compared to Mn^{2+}, there is a larger range of crystal field splittings possible and also tetrahedral Fe^{3+} oxides exist. Tetrahedral Fe^{3+} systems (e.g. $FePO_4$), have the same $^6A_1$ ground state as octahedral systems (de Groot et al., 2005). An important issue in fields as in

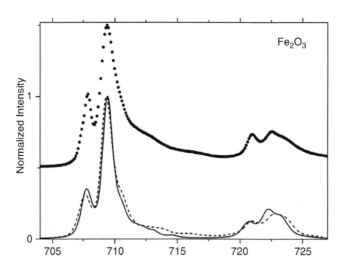

FIGURE 6.8 Fe 2p XAS of Fe_2O_3 (points) compared to a LFM calculation (solid) and a CTM simulation (dashed). (Reprinted with permission from de Groot, F. M.F., et al., *J. Phys. Chem. B*, 109, 20751, 2005. Copyright 2005 by the American Chemical Society.)

mineralogy, catalysis and materials science, is the ratio of Fe^{3+} and Fe^{2+} (van Aken and Liebscher, 2002).

6.3.6.3 $Fe^{3+}(tacn)_2$

$Fe^{3+}(tacn)_2$ is a low-spin $3d^5$ system where tacn stands for 1,4,7-triazacyclononane, which is a neutral ligand that binds three nitrogen atoms to Fe^{3+}. The two tacn molecules generate a slightly distorted, octahedral low-spin Fe^{3+} system. Low-spin $3d^5$ systems have a t_{2g}^5 ground state. The single t_{2g} hole is strongly affected by the 3d spin–orbit coupling, which is modified by distortions from octahedral symmetry. It turns out that depending on the symmetry, the single t_{2g} hole is visible at the L_2 edge. This is explained subsequently.

The 2p XAS spectrum of $Fe^{3+}(tacn)_2$ is well reproduced by the $3d^5 + 3d^6\underline{L}$ CTM calculations. In the calculation shown, the hopping parameters for t_{2g} and e_g orbitals have been optimized to the experiment and an excellent fit is obtained. The difference in t_{2g} and e_g hopping also implies that different amounts of $3d^6\underline{L}$ character are added to the ground state for both cases. The ground state of the low-spin Fe^{3+} system is t_{2g}^5 and the complete CTM ground state can then be written as $t_{2g}^5 + t_{2g}^6\underline{L} + t_{2g}^5e_g\underline{L}$ and the relative amounts of both ligand hole states can be determined. This defines the differential orbital covalence (DOC), which has been explained in detail in the discussion of the charge transfer method in Chapter 4. In the case of $Fe^{3+}(tacn)_2$, one finds that the t_{2g} mixing is small and the t_{2g} states are 99% metallic. In contrast, the e_g mixing is very strong and the e_g states are only 63% metallic. The DOC is the difference between the two orbital types and is 36% for $Fe^{3+}(tacn)_2$. These DOC values can be calculated from ground state DFT calculations as this provides an excellent comparison between density functional theory (DFT) and CTM calculations (Wasinger et al., 2003).

Figure 6.9 shows one distinct difference between experiment and calculation, namely the peak at 718 eV (indicated with an arrow). This peak is completely absent in the CTM simulations. The reason for this discrepancy is (partial) quenching of the 3d spin–orbit coupling. The t_{2g}^5 ground state has $^2T_2$ symmetry. With 3d spin–orbit coupling, this state splits into the double group symmetries Γ_7 and Γ_8, where Γ_7 is new ground state. The final state is $2p^5t_{2g}^6$ with $^2T_1$ symmetry. In double group symmetry, this yields Γ_6 and Γ_8 split by the large 2p spin–orbit coupling. The t_{2g} peak has Γ_8 symmetry at the L_3 edge, and Γ_6 symmetry at the L_2 edge. The dipole selection rules imply that a Γ_7 initial state can reach Γ_7 and Γ_8 final states, whereas Γ_6 is forbidden. From Γ_8, all final state symmetries can be reached. This implies that with the inclusion of 3d spin–orbit coupling, the t_{2g}^6 peak is not visible at the L_2 edge. Any symmetry distortion (e.g. trigonal symmetry) that mixes Γ_7 and Γ_8 in the ground state will cause the t_{2g}^6 peak to become visible at the L_2 edge as is the case in the experimental spectrum of $Fe^{3+}(tacn)_2$.

6.3.6.4 $Fe^{3+}(CN)_6$

Fe^{3+} surrounded by the six cyanide carbon atoms has the usual σ-bonding as all oxides and $Fe^{3+}(tacn)_2$, but in addition there is the metal-CN^- π-bonding. Usually the σ-bonding donates electrons and the π-bonding subtracts electrons (π-bonding is also called π-back bonding). The effects of π-bonding significantly complicate the

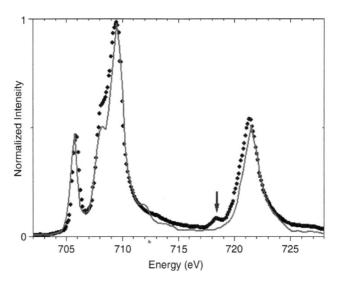

FIGURE 6.9 CTM simulation of the $^2T_2$ low-spin compound $Fe^{3+}(tacn)_2$. (Digitized from Wasinger et al., 2003. With permission.)

description of the ground state and the XAS spectra. σ-bonding can be described with ligand-to-metal charge transfer (LMCT), which is the usual charge transfer as used throughout this book. It yields a ground state described as a mixture of $3d^5$ and $3d^6\underline{L}$. Similarly, π-bonding can be described with a metal-to-ligand charge transfer (MLCT), which is the inverse process and its effect in the ground state is described as a mixture of $3d^5$ and $3d^4L$.

Using both LMCT and MLCT, the ground state of $Fe^{3+}(CN)_6$ is described as $3d^5 + 3d^4L + 3d^6\underline{L}$. The $3d^5$ state is again a t_{2g}^5 state. The LMCT has mainly the $t_{2g}^5e_g^1\underline{L}$-character due to e_g-mixing, which represents σ-bonding. The MLCT has only t_{2g}^4L-character because there are no e_g states occupied. Figure 6.10 shows the CTM simulation of the 2p XAS spectrum of $Fe^{3+}(CN)_6$. A very good description is found for all peaks observed. Comparing Figure 6.10 to the result for $Fe^{3+}(tacn)_2$, the spectrum looks essentially the same but with a large additional peak at ~712 eV. This large peak must be due to the π-bonding as this is the difference between $Fe^{3+}(CN)_6$ and $Fe^{3+}(tacn)_2$. The main mechanism behind the large satellite peak is the degeneracy between t_{2g}^5 and t_{2g}^4L, implying a close to 50% mixed ground state. This 50–50 mixture in the ground state is then excited to the bonding and antibonding combinations of the respective core hole states, and again, both are significantly mixed. In total, this yields a large satellite. The details of this process have been described by Hocking et al. (2006).

6.3.6.5 Intermediate Spin State of SrCoO$_3$

Co^{4+} is also formally a $3d^5$ system. The high valence implies larger charge-transfer effects than for Mn^{2+} and Fe^{3+} systems. Higher covalence could imply a shift to

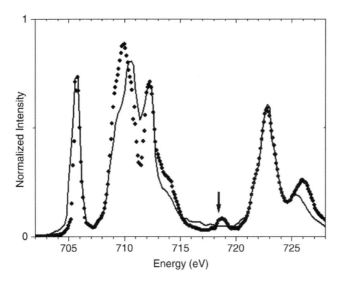

FIGURE 6.10 CTM simulations of the 2p XAS spectrum of $Fe^{3+}(CN)_6$, including both MLCT and LMCT. (Digitized from Hocking et al., 2006. With permission.)

low-spin systems, similar to the tacn and cyanide complexes for Fe^{3+}. The strong mixing of $3d^5$ and $3d^6\underline{L}$ also implies a possible new ground state that cannot be obtained with an ionic approach (i.e. an intermediate spin state). An intermediate spin state does not have a $^6A_1$ t^3e^2 or a $^2T_1$ t^5 configuration but has a $^4T_2$ t^4e^1. This state is stabilized by the $t^4e^2\underline{L}$ state. The t^4e^2 configuration is the $^5T_2$ ground state of a $3d^6$ configuration and has the lowest energy. In addition, e_g mixing is much stronger than t_{2g} mixing and together this could bring the $S = 3/2$ state below the $S = 5/2$ state. A complication, which is discussed subsequently for Ni^{3+} systems, is that if these configurations are close in energy, they can also mix due to the 3d spin–orbit coupling and, in fact, yield mixed spin ground state. A mixed spin ground state does not have a $S = 5/2$ ground state and also does not have a $S = 3/2$ ground state. Instead, it has a linear combination of both. Comparison of the CTM calculations of the three different ground states with the experimental 2p XAS spectrum of $SrCoO_3$ does not provide a clear answer with regard to the actual ground state, mainly due to the lack of structure (Potze et al., 1995).

6.3.7 3d⁶ Systems

$3d^6$ systems have six 3d electrons, one of which is always spin-down yielding a maximum spin state of $S = 2$. The atomic ground state is $^5D_2$. In a cubic crystal field, the ground state has, in addition to its three t_{2g} and two e_g states spin-up electrons, a further t_{2g} spin-down electron. In some respects, this $3d^6$ ground state is similar to the $3d^1$ ground state as a single t_{2g} electron is filled. Low-spin $3d^6$ systems will have a t_{2g}^6 configuration and $^1A_1$ symmetry, which will be immediately evident from the spectral shapes. Similar to $3d^5$ systems, one could imagine an intermediate spin state with $S = 1$ and a $t_{2g}^5e_g^1$ configuration.

6.3.7.1 Effect of 3d Spin–Orbit Coupling in Fe_2SiO_4

Fe_2SiO_4 contains Fe^{2+} in octahedral positions. The Fe^{2+} sites have a high-spin $^5T_2$ ground state. Figure 6.11 gives the iron 2p XAS spectrum of Fe_2SiO_4, in comparison with LFM calculations with a crystal-field strength of 1.2 eV. The $^5T_2$ ground state of Fe_2SiO_4 is sensitive to the effects of 3d spin–orbit coupling. The L_3, and in particular, the L_2 edge are better simulated if 3d spin–orbit coupling is not included. In particular, the feature at 719 eV is observed clearly in experiment and is absent if 3d spin–orbit coupling is included. It turns out that for Fe_2SiO_4, the 3d spin–orbit coupling is effectively quenched at room temperature, which could be caused by dispersion effects and/or by symmetry reduction. A quenched 3d spin–orbit coupling is also found for CrO_2 but not for many other systems including $LaTiO_3$, Fe^{3+}(tacn), and CoO.

The situation regarding the 2p XAS spectrum of FeO is unclear. The older spectra show a broad spectrum that is not well resolved (Colliex et al., 1991; Crombette 1995), but seems to agree with calculations. Recently, two FeO spectra were published that varied significantly. Prince et al. (2005) found a L_3 edge shoulder below the main peak, while Regan et al. (2001) found a shoulder above the L_3 main peak. This spectrum is very similar to the spectra of other octahedral Fe^{2+} systems such as Fe_2SiO_4, and to $CaFe(SiO_3)_2$ (hedenbergite) (van Aken and Liebscher, 2002) and compares well with calculations of Fe^{2+} in octahedral symmetry. Also, in the case of FeO, the calculations fit better if the 3d spin–orbit coupling is set to zero.

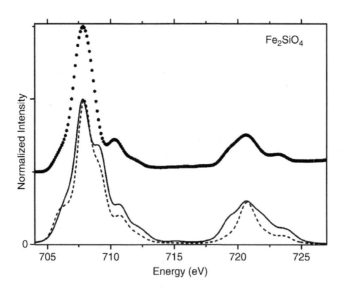

FIGURE 6.11 Fe 2p XAS of Fe_2SiO_4 (points) compared to a LFM calculation without (solid) and with 3d spin–orbit coupling (dashed). (Reprinted with permission from de Groot, F.M.F., et al., *J. Phys. Chem. B*, 109, 20751, 2005. Copyright 2005 by the American Chemical Society.)

6.3.7.2 Co^{3+} Oxides

Co^{3+} is more covalent than Fe^{2+}. This implies that it is close to a high-spin low-spin transition point for oxides. The options for the Co^{3+} $3d^6$ ground state are $^5T_2$ high-spin $t_{2g}^4e_g^2$, $^3T_1$ intermediate-spin $t_{2g}^5e_g^1$, and $^1A_1$ low-spin $t_{2g}^6e_g^0$. $LiCoO_2$ is clearly in a low-spin ground state. Its 2p XAS spectrum is essentially reproduced by a LFM calculation, as can be seen in Figure 6.12 (de Groot et al., 1993b). Other low-spin Co^{3+} oxides include $La_2Li_{0.5}Co_{0.5}O_4$ (Hu et al., 2002), $EuCoO_3$ (Hu et al., 2004), and the Co^{3+} ions in Co_3O_4 (Morales et al., 2004).

LaCoO_3 is a material with a long history regarding its spin state. In particular, its spin transition from low-spin (LS) to high-spin (HS) over a large temperature range from 50–300 K. At low temperature, $LaCoO_3$ is low spin and at higher temperatures there is a transition to a higher spin state and/or an admixture of a higher spin state. Hu et al. (2002) have shown that the branching ratio difference between the $^1A_1$ low-spin material $La_2Li_{0.5}Co_{0.5}O_4$ and $LaCoO_3$ indicates a $^3T_1$ intermediate spin state, using CTM calculations. In fact, they showed a mixed spin state between low-spin $^1A_1$ and high-spin $^5T_2$, due to the effects of 3d spin orbit coupling. For the case of $PrNiO_3$, this is discussed in Section 6.3.8.2. A mixed spin state is not the same as an intermediate spin state: an intermediate spin state has a fixed overall angular momentum of $S = 1$, while a mixed spin state between $S = 2$, $S = 1$, and $S = 0$ gradually changes its spin depending on the contributions of the three configurations. Note that a mixed spin state is a quantum-chemical single state and not a thermal population of excited configurations. We conclude that it could well be the case that $LaCoO_3$ is in a mixed spin ground state, induced by the combination of large covalence and

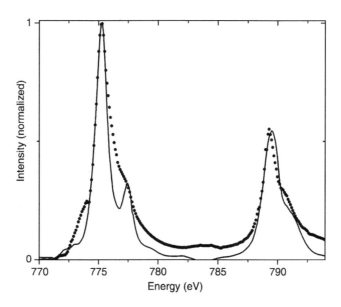

FIGURE 6.12 Co 2p XAS spectrum of $LiCoO_2$ compared with a LFM calculation for a $^1A_1$ low-spin ground state. (From de Groot, 1991, PhD thesis. With permission.)

3d spin–orbit coupling. Sr_2CoO_3Cl is a pure high-spin Co^{3+} oxide as is evident from its 2p XAS spectrum (Hu et al., 2004).

6.3.8 3d^7 Systems

3d^7 systems have seven 3d electrons with a $^4F_{3/2}$ atomic ground state. In a cubic crystal field, the ground state has, in addition to its three t_{2g} and two e_g states spin-up electrons, two additional t_{2g} spin-down electrons. Similar to 3d^1 and 3d^6 systems, 3d^7 systems are also sensitive to 3d spin–orbit coupling.

6.3.8.1 Effects of 3d Spin–Orbit Coupling on the Ground State of Co^{2+}

The ground state of a Co^{2+} system is $^4T_1$ with a partly filled t_{2g} state. This makes the ground state susceptible to 3d spin–orbit coupling. The $^4T_1$ ground state is split by the 3d spin–orbit coupling into four substates, respectively, E_2, G, another G state, and E_1. Using the atomic value for the 3d spin–orbit coupling, the relative energies of these states are: E_2 at 0 meV, the first G state at 44 meV, the second G state at 115 meV, and the E_1 state at 128 meV respectively. Figure 6.13 shows the 2p XAS spectra related to these four $^4T_1$ states. The Tanabe–Sugano diagram for 3d^7 shows that, in the whole crystal field range for high-spin ground states, the effect of 3d spin–orbit coupling is similar. At 0 K, only the ground state is occupied. However, because the energies of the excited configuration are very low, at finite temperatures the other initial state configuration will be partly occupied. Using a Boltzmann distribution,

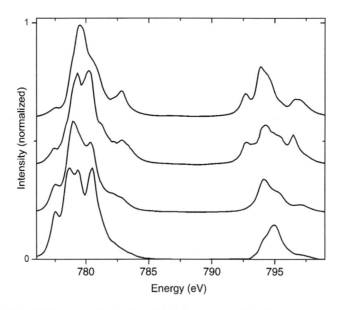

FIGURE 6.13 LFM spectra for the four low-lying states of the $^4T_1$ ground state of a high-spin Co^{2+} system. From bottom to top: E_2, the ground state; G, the first excited state; G, the second excited state; and E_1, the third excited state. (From de Groot, 1991, PhD thesis. With permission.)

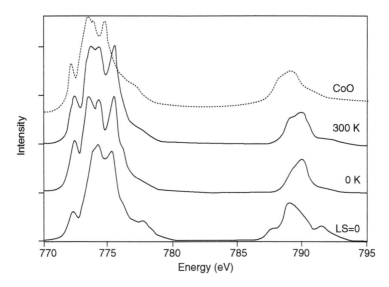

FIGURE 6.14 Comparison of the experimental 2p XAS spectrum of CoO and three simulations at 300 K, 0 K and with zero 3d spin–orbit coupling. (From de Groot, F.M.F., *J. Electron Spectrosc. Relat. Phenom.*, 67, 529, 1994. With permission from Elsevier Ltd.)

at room temperature (= 25 meV), the first excited G state has an occupation of ~20% with respect to the E_2 ground state. Since the final states that can be reached from the ground state and from the lowest excited states are different, the spectrum will change with temperature depending on how much each of the initial states is thermally populated.

Figure 6.14 shows the comparison between the experimental CoO spectrum and three simulations. Calculations are shown for the ground state only (0 K), for 300 K and for the situation that no 3d spin–orbit coupling is included. This last situation implies that all four low-lying states are added. The 300 K spectrum contains ~20% of the first G-state, which improves the agreement with experiment. Recently, this temperature dependence has been experimentally confirmed for CoO by Haverkort et al. (2005b).

6.3.8.2 Mixed Spin Ground State in PrNiO$_3$

Ni^{3+} also has a $3d^7$ ground state. Due to the increased covalence, Ni^{3+} systems are close to the high-spin low-spin transition point. The increased charge transfer also increases the effects of the 3d spin–orbit coupling. This gives the ingredients for a new type of ground state that can best be described as mixed spin. That is, the ground state is a linear combination of $^4T_1$ high-spin and ^2E low-spin. As described for Co^{2+}, the $^4T_1$ high-spin state is split by 3d spin–orbit coupling into E_2, G, G, and E_1, respectively. The ^2E low-spin state is pure and has G symmetry. At the high-spin low-spin transition point, the ground state switches from E_2 to G. Initially, this new G ground state is not a pure low-spin state, but a quantum chemical mixture of high spin and low spin. This situation has recently been confirmed experimentally in the case of $PrNiO_3$ (Piamonteze et al., 2005).

The ground state consists of a range of different symmetries that are mixed up by the 3d spin–orbit coupling and the charge-transfer effect. The contributions of these different symmetries to the ground state and to spin $\langle S_z \rangle$ and orbital $\langle L_z \rangle$ angular momenta as functions of 10 Dq are shown in Figure 6.15. The labels refer to the distribution of the 3d holes among B_1, A_1, B_2, and E orbitals. In the metallic state with 10 Dq = 2.2 eV, the ground state is composed of 38% $B_1B_1A_1$ (LS) and 8% B_1A_1E (HS), and the rest being charge transfer configurations. The ground-state transition takes place at 10 Dq = 2.1 eV where we obtain the same contributions for HS and LS states; that is, the ground state is 25% B_1A_1E and 25% $B_1B_1A_1$. The spin angular momentum assumes an intermediate value S_z = 0.7. The mixed-spin state also arises due to the strong hybridization that mixes the Ni 3d and O 2p orbitals. A detailed comparison with experimental spectra has been described by Piamonteze et al. (2005).

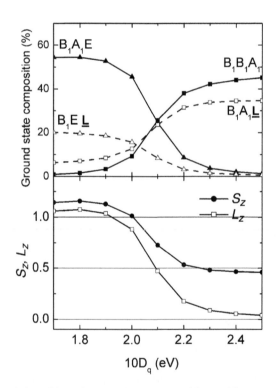

FIGURE 6.15 Variation of (top) the ground-state composition and (bottom) S_z, L_z as a function of 10 Dq. The configuration $B_1B_1A_1$ corresponds to the S = 1/2 LS state and the configuration B_1A_1E corresponds to the S = 3/2 HS state. $B_1E\underline{L}$ and other ligand hole states correspond to the charge transfer configurations. (Reprinted with permission from Piamonteze, C., de Groot, F.M.F., Tolentino, H.C.N., et al., *Phys. Rev. B*, 71, 020406, 2005. Copyright 2005 by the American Physical Society.)

6.3.9 3d⁸ SYSTEMS

3d^8 systems have eight 3d electrons with a $^3F_4$ atomic ground state. In a cubic crystal field, in addition to its three t_{2g} and two e_g states, the ground state has spin-up electrons and three additional t_{2g} spin-down electrons. This implies that the two remaining holes will be the spin-down e_g states and a stable ground state is formed that will always be high-spin in octahedral symmetry. The prototype system is NiO. In order to create a low-spin ground state, a strong symmetry distortion must be applied as will be discussed in the next section.

6.3.9.1 NiO

NiO seems to be a prototype system of a straightforward ground state, as defined by its $^3A_2$ ground state with six t_{2g} electrons plus two spin-up e_g electrons. However, in the discussion of the XMCD effect for Ni^{2+} systems, it becomes clear that, although the ground state is $^3A_2$, there is still a detectable effect of the 3d spin–orbit coupling. The 3d spin–orbit coupling is not able to split the $^3A_2$ ground state, but it is able to modify the mixing of the $^3A_2$ ground state with excited states. This is a small effect, but because the spectrum of NiO consists of only a few peaks, this effect becomes detectable.

Because the 2p XAS spectrum of NiO is well resolved, it has been shown that it is sensitive to another weak interaction, the super-exchange interaction coupling the spins on the different nickel sites.

The temperature dependence of the L_2 part of the 2p XAS spectrum of 20 monolayers of NiO on MgO(100) changes as a function of temperature [Alders et al. (1998)]. This effect can be simulated by adding an exchange field to the ground state. The ground state has $^3A_2$ (T_2) symmetry, with $S = 1$. An exchange field splits it into three M_S states (+1, 0, and –1) that each have a slightly different isotropic spectrum. A LFM calculation of the L_2 edge shows that the first peak is the highest for the ground state. Increasing the temperature gives an admixture of the first excited state and this effectively increases the second peak, reproducing the experimental trend. The inclusion of the 3d spin–orbit coupling improves the agreement with experiment. This suggests that the inclusion of the 3d spin–orbit coupling is also important for the $^3A_2$ ground state of NiO even though it has no direct effect. It still has some effect due to the admixture of excited configurations and this effect is big enough to be distinguishable. We note that the calculations by Alders et al. (1998) did include the 3d spin–orbit coupling, while the calculations for two coupled Ni sites by van Veenendaal et al. (1995) used a zero 3d spin–orbit coupling.

6.3.9.2 High-Spin and Low-Spin Ni²⁺ and Cu³⁺ Systems

Ni^{2+} and Cu^{3+} systems have a 3d^8 ground state, which implies that octahedral systems are always high-spin with the two holes in the two e_g orbitals. The only way to get a low-spin ground state is to lower the symmetry to, for example, square planar symmetry. In square planar symmetry, the energy difference between the $x^2 - y^2$ orbital and the z^2 orbital is large enough that both holes form a singlet in the $x^2 - y^2$ orbital. This is, for example, the case for K_2Ni-ditio-oxalate (van der Laan et al., 1988).

In addition to Ni^{2+} systems, Cu^{3+} systems also formally have a $3d^8$ ground state. Cu^{3+} systems have a negative charge-transfer energy, which implies that the $3d^9\underline{L}$ configuration has a lower energy than the $3d^8$ configuration. Still, these systems should be considered as Cu^{3+} because their symmetry is either $S = 1$ for the octahedral high-spin ground states or $S = 0$ for the square planar low-spin ground states.

Figure 6.16 shows the experimental spectra of high-spin Cs_2KCuF_6 (bottom) and $La_2Li_{1/2}Cu_{1/2}O_4$ (top), in comparison with charge-transfer multiplet calculations. The ground state of Cu^{3+} is similar to Ni^{2+}, both being a linear combination of $3d^8$ and $3d^9\underline{L}$. Still, the spectrum of Cu^{3+} is completely different from Ni^{2+}. The theoretical comparison between Ni^{2+} and Cu^{3+} has been discussed in Chapter 4, as an example of the charge-transfer model. The spectrum of Ni^{2+} in, for example, NiO, looks essentially like a pure $3d^8$ ground state, which is a little affected by charge-transfer satellites. The spectrum of Cu^{3+} looks completely different. In the final state, the energy of $\underline{c}3d^{10}\underline{L}$ is significantly lower than the energy of $\underline{c}3d^9$. This yields a sharp single peak at 933 eV. The $\underline{c}3d^9$ multiplet is found between 936 and 940 eV. It is clear that for negative delta systems, it would make little sense to simulate the spectrum from a single configuration LFM calculation. At least two configurations are always needed to describe the basics of the spectral shapes.

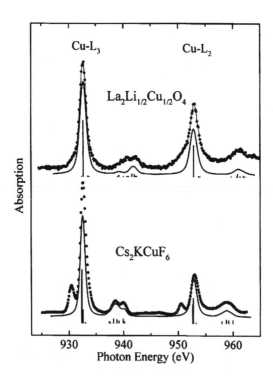

FIGURE 6.16 Results of theoretical simulations of the copper 2p x-ray absorption spectra of Cs_2KCuF_6 (bottom) and $La_2Li_{1/2}Cu_{1/2}O_4$ (top), in comparison with the experimental spectra. (From Hu, Z., Kaindl, G., Warda, S.A., Reinen, D., de Groot, F.M.F., and Muller, B.G., *Chem. Phys.*, 232, 63, 1998a. With permission from Elsevier Ltd.)

Negative delta ions are not the most common ions: they include, for example, Cu^{3+}, Ni^{3+}, and Fe^{4+}.

6.3.10 3d⁹ Systems

$3d^9$ systems have nine 3d electrons with a $^2D_{5/2}$ atomic ground state. There is only a single 3d hole. In octahedral symmetry, this hole is found in an e_g state and, in square planar symmetry, the hole has $x^2 - y^2$ symmetry.

The $3d^9$ ground state is central in the undoped cuprates that are the parent compounds for the high T_c superconductors. This initially included the systems La_2CuO_4 doped with Sr or Ca, $YBa_2Cu_3O_7$ doped with oxygen vacancies and Nd_2CuO_4 doped with Ce (Dagotto, 1994). An important experiment concerning the x-ray absorption studies on high T_c materials is to be found in a paper by Chen et al. (1992). They show the angular dependence of the oxygen K edge and copper L edge of Sr-doped $La_{2-x}Sr_xCuO_4$ ($x = 0.15$). Figure 6.17 shows the copper L edge; the oxygen K edges are discussed in Section 6.4.4. The copper 2p excitation in a $3d^9$ system has a single 2p3d transition to a $3d^{10}$ final state. Figure 6.17 shows that the sharp copper L edge shows almost complete polarization and the $3d^{10}$ final state has almost pure in-plane polarization ($E_h \perp c$, where the CuO plane is ab), indicating that the 3d hole is almost pure $x^2 - y^2$ in character.

Doping a $3d^9$ system with holes leads to the formation of singlet states that are an important aspect of high T_c superconductivity. Sr doping leads to Cu^{3+} sites that can be described with mainly a $3d^9\underline{L}$ ground state, as discussed above. The $3d^{10}\underline{L}$ peaks related to these Cu^{3+} sites are at a slightly higher excitation energy than the $3d^{10}$ peaks—931.6 eV for $3d^{10}$ and ~933 eV for $3d^{10}\underline{L}$. Chen et al. (1992) have shown

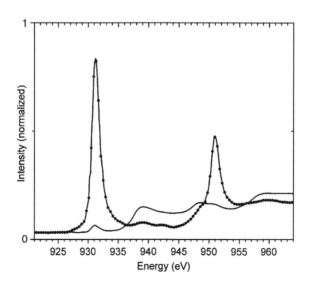

FIGURE 6.17 Cu L edge XAS of $La(Sr)CuO_4$ for the polarization in the CuO plane (solid with dots) and perpendicular to it (solid). (Reprinted with permission from Chen, C.T., et al., *Phys. Rev. Lett.*, 68, 2543, 1992. Copyright 1992 by the American Physical Society.)

that doping up to $x = 0.15$ creates essentially $x^2 - y^2$ holes, where $x = 0.35$ overdoped samples have a significant amount of z^2 holes. A more detailed discussion of the electronic structure of the copper oxides is given in Chapter 8.

6.4 OTHER X-RAY ABSORPTION SPECTRA OF THE 3d TM SYSTEMS

6.4.1 TM $M_{2,3}$ EDGES

The $M_{2,3}$ edges of the 3d TM systems are less popular than the $L_{2,3}$ edges. Their energy range is not the problem. The 3p core level binding energy increases from 28 eV in scandium to 75 eV in copper. The excitation of a 3p core electron, from the symmetry point of view, is completely analogous to the excitation of a 2p core electron. This implies that the LFM and CTM calculations are identical, with only some parameters having different values. Concerning the basic interactions, there are two major differences between 2p and 3p core states. The first is that 3p3d exchange interaction is larger than the 2p3d exchange interaction and the second is that the 3p spin–orbit coupling is smaller than the 2p spin–orbit coupling. The large 3p3d exchange interaction is in many respects a big advantage as is, for example, evident for the $K\beta$ XES spectra that have more possibilities than the $K\alpha$ XES spectra.

There are, however, two major drawbacks. The first is that the lifetime broadening of a 3p core state is larger than that of a 2p core state and the second is that the core levels are very close together and tend to overlap. Another complication of the small core level spin–orbit coupling is that most of the intensity goes to the peaks at the highest energies, the so-called delayed onset. These states tend to be broad, implying that they also decay very rapidly—in fact, so rapidly that the decay influences the spectral shape. The 3p XAS spectra have been measured with high resolution essentially since the 1960s and since then the spectral resolution has not improved. This is in sharp contrast to the 2p XAS spectra discussed previously. The 3p core states are probably more often studied with EELS in electron microscopes than at synchrotron radiation sources.

A large number of 3p XAS spectra of TM halides have been published by Nakai et al. (1974). The spectra consistently show a sharp leading edge followed by the delayed onset. It is not surprising that the initial qualitative assignment was that the sharp transitions were into the empty 3d states and the delayed onset represented the edge with initial transitions to the 4s states. Shin et al. (1981) analyzed the 3p XAS spectra of the TM halides with LFM theory. They assigned the complete $M_{2,3}$ spectra to the $3p^6 3d^n \rightarrow 3p^5 3d^{n+1}$ transitions, in agreement with the L edge analysis. Van der Laan (1991) performed a series of multiplet simulations for the $M_{2,3}$ edges. Over the last 20 years, $M_{2,3}$ edges have been measured more often with EELS than with XAS. In a TEM-EELS, it is easy to measure the whole energy-loss spectrum between the zero-loss line (0 eV) and, say, 1000 eV. This implies that the $M_{2,3}$ edges of transition metal systems can be measured in combination with the $L_{2,3}$ edges. A recent example is the work of Garvie et al. (2004) who measured the iron M_1, $M_{2,3}$, and $L_{2,3}$ edges plus the sulfur L_1, $L_{2,3}$, and K edges of FeS_2. This complete set of core spectra allows one to cross check the interpretation of the edges. In the work of van Aken et al. (1999), the $M_{2,3}$ edge was used to determine the ratio between Fe^{2+} and Fe^{3+} in a series of minerals. Compared to $L_{2,3}$ edges, the $M_{2,3}$ edges have a higher cross section but a lower resolving power. Recently, the $M_{2,3}$ edges have again been measured with x-rays. The experiments of Chiuzbaian et al. (2005) were in connection with 3p3d RXES and Berlasso et al. (2006) measured the $M_{2,3}$ edges of Fe_2O_3 and CoO. They compared the synchrotron experiments with spectra obtained from the high harmonics of a laser. The 3p core states have traditionally been used in 3p (resonant) photoemission studies that are discussed in Chapter 5. In addition, the 3p core states are often used in 1s3p (resonant) x-ray emission that are discussed in Chapter 8.

6.4.2   TM M_1 Edges

The M_1 edges of the 3d TM systems are the transitions of the 3s core state. In x-ray absorption, the 3s spectra are never used because they yield very broad spectra related to the transition from 3s to 4p states. A more interesting situation occurs for low-energy EELS where the relaxation of the dipole selection rule allows for 3s3d transitions. In addition, 3s XPS (Chapter 5) and 2p3s RXES (Chapter 8) are used. The 3s3d excitation spectra using 1000 eV electrons all look rather similar with a two-peaked spectrum for the early transition metal oxides and a single peaked spectrum for the late transition metal oxides (Steiner et al. 1996). These spectra are essentially confirmed by the calculations of Ogasawara et al. (1998). All early high-spin TM oxides show two peaks that are due to a spin-up and spin-down electron being excited, respectively. The late transition metal oxides only have spin-down holes left and, as a consequence, only one peak. It is noted that the 3s EELS spectra are different from the 3s XPS spectra. As discussed in Chapter 5, 3s XPS is a combination of 3s3d exchange and charge-transfer effects, yielding two (or more) peaks for all cases with an unpaired spin clearest for the $3d^5$ systems because of the largest number of unpaired spins.

6.4.3   TM K Edges

The energy range of the 1s core level binding energy of the 3d transition metals increases from 4492 eV in scandium to 8979 eV in copper. This hard x-ray range implies that K edges are measured solely with x-ray absorption at synchrotron radiation sources. The 1s3d exchange interactions are of the order of a few MeV and can be neglected. With the exception of the pre-edge region, multiplet effects are not important and the K edge describes the transition from the 1s core state to the empty p states. In general, the K edge spectra can be described with single particle excitation models—for example, FEFF (Ankudinov et al., 1998) or compared to the projected DOS from any band structure code. The pre-edge region of the K edges cannot be described accurately with single particle codes because of the effects of charge transfer, multiplets and, as shown recently, the combination of local quadrupole peaks plus nonlocal dipole peaks.

Actually, it is surprising that there is such good agreement between the empty p-DOS and the metal K edges of transition metal oxides because large effects from charge transfer would be expected. The K edge absorption (e.g., NiO) excites a 1s

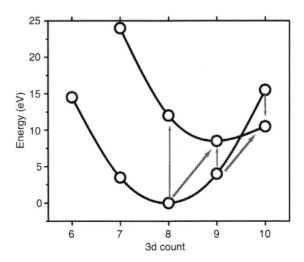

FIGURE 6.18 Energy effects on the various CTM configurations for transition metal K edges using the $3d^8$ ground state of NiO as an example. The situation for the main 1s4p edge with transitions that cause the lowest energy configuration to be shifted upwards is indicated with thin arrows. The situation for the 1s 3d quadrupole pre-edge structure with transitions that are self-screened and that maintain the ordering of the states is indicated with thick arrows. These charge-transfer effects are similar to 2p XPS for the K edge and similar to 2p XAS for the K pre-edge.

core electron to an empty 4p state and, at higher energies, to other p states. The ground state of NiO as we use throughout this book is described as $3d^8 + 3d^9\underline{L} + 3d^{10}\underline{L}^2$. The 1s4p excitation process creates a final state that can be written as $1s^13d^84p^1 + 1s^13d^9\underline{L}4p^1$, as indicated in Figure 6.18.

Bair and Goddard (1980) used this model to show that this transition creates large charge-transfer effects. Tolentino et al. (1992) described in detail the consequences of this charge-transfer effect. The 1s core hole creates an additional core charge that pulls down the 3d states by the core hole potential U_{dc}, similar to 2p XPS. The ordering of the initial states gives $3d^9\underline{L}$ at an energy Δ and, in the final state, $1s^13d^9\underline{L}4p^1$ should be at an energy $\Delta - U_{dc}$. The 4p^1 part of the configuration implies an empty 4p DOS and this implies that two (or more) superimposed DOS structures are expected. Tolentino et al. (1992) assigned the peaks in the K edge spectra of La_2CuO_4 following this interpretation. This model was also used by Collart et al. (2006) who selectively excited the $1s^13d^9\underline{L}4p^1$ peak and the $1s^13d^84p^1$ peak in La_2NiO_4 to measure the corresponding RXES spectra.

In contrast to this two-configuration model, Wu et al. (2004) found good agreement between the metal 4p DOS and the metal K edge XAS for the prototype charge transfer systems MnO, CoO, and NiO. This implies that a single configuration is able to describe the spectra and the effects of other configurations are not visible at this level of comparison. Apparently, a specific final state configuration dominates, where it is noted that the experimental resolution is, intrinsically, not very good due to lifetime broadening of the 1s core hole. The broad features can possibly hide features from other configurations. Chapter 8 discusses the possibility of effectively removing the lifetime broadening of the metal K edges and these high-resolution spectra can then be analyzed in much more detail and could show the effects of other configurations.

The pre-edge region is related to transitions to the 3d band. This includes dipole transitions from the 1s core state to 4p character that is hybridized with the 3d band, as well as quadrupole transitions from the 1s core states directly into empty 3d states. Pre-edges are difficult to analyze because they involve both band structure effects and multiplet effects. In general, one can say that there are two main reasons for their obscurity. One is the lifetime broadening and usually the applied experimental resolution will yield a total broadening on the order of 1.5 eV, thus blurring of the potential fine structure in the pre-edges. The second is the combination of both direct quadrupole transitions and dipole transitions to p-character that has hybridized with the 3d band. As a result, the pre-edge structures are in most cases analyzed only qualitatively, for example, by measuring global properties (i.e. their center of gravity and integrated intensity), to relate to properties such as their site symmetry and effective mean valence. In Chapter 8, we describe a method that effectively eliminates the 1s lifetime broadening of the pre-edge structures. This yields sharper structures and the 2D RXES images also reveal the nature of the different states in the pre-edge region.

Figure 6.19 shows the Fe 1s x-ray absorption spectra of $FeAl_2O_4$, Fe_2SiO_4, Fe_2O_3, and $FePO_4$, normalized to the edge jump with respect to the EXAFS region. One observes a pre-edge structure around 7115 eV followed by the absorption edge. The edge energy is 7118 eV for the divalent iron oxides and 7122 eV for the trivalent iron oxides. The pre-edge region of iron compounds has been investigated systematically by Westre et al. (1999). In this study, the pre-edges are analyzed in terms of pure

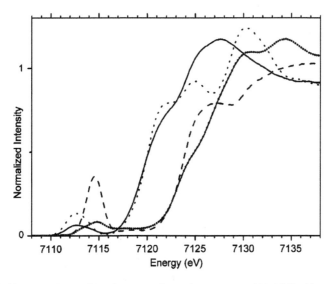

FIGURE 6.19 Experimental Fe 1s x-ray absorption spectra of $FeAl_2O_4$ (dotted), Fe_2SiO_4 (solid), Fe_2O_3 (solid with points), and $FePO_4$ (dashed). (Reprinted with permission from de Groot, F., et al., *J. Phys. Chem. B*, 109, 20751, 2005. Copyright 2005 by the American Chemical Society.)

quadrupole transitions and the spectra could be interpreted in terms of multiplet theory including the crystal field and intra-atomic electron–electron interactions. The Fe^{3+} spectra are calculated from the quadrupole transitions of $3d^5$ to $1s^13d^6$. The Fe^{2+} spectra use $3d^6$ as the ground state. The intensity of the pre-edge region is larger for compounds in which the metal site has tetrahedral symmetry than for octahedral systems. In tetrahedral systems, the local mixing of p and d nature is symmetry allowed, whereas for a system with inversion symmetry, such as octahedral symmetry, it is forbidden. This rule is relaxed in the solid and, if the DOS is calculated, then one finds small admixtures of p states in the 3d band even for perfect octahedral systems. However, this admixture is less than that for tetrahedral systems. A 4p character will be mixed into the 3d band if an octahedral metal site is distorted, resulting in an increased intensity of the pre-edge.

Figure 6.20 shows two multiplet calculations of the pre-edge of Fe_2SiO_4. The Fe site has a $3d^6$ configuration that, in spherical symmetry, has a ^5D ground state. A quadrupole excitation brings this ground state to the two possible final states of ^4F and ^4P, respectively. The addition of a cubic crystal-field (top spectrum) gives three peaks, respectively, related to $^4T_{1g}$, $^4T_{2g}$, and another $^4T_{1g}$ final state of the $(1s^1)3d^7$ configurations where the symmetry labels apply to the 3d part only. These LFM calculations do not include the 3d spin–orbit coupling, the 1s3d exchange interaction, or the effects due to charge transfer. The bottom figure includes all of these effects and instead of three peaks, a large number of peaks are visible. However, after broadening, the same spectrum is essentially obtained. Note that at the energy range between 7117 and 7123 eV, charge-transfer satellites would be visible in principle, similar to the $L_{2,3}$ edges. In this energy range, the sharply rising main K edge

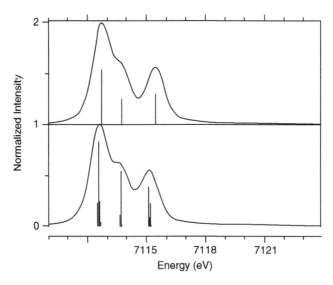

FIGURE 6.20 LFM (top) and CTM (bottom) calculations of the pre-edge of Fe_2SiO_4. (Reprinted with permission from de Groot, F., *Coord. Chem. Rev.*, 249, 31, 2005. Copyright 2005 by the American Chemical Society.)

is observed, which makes any small charge-transfer satellites impossible to distinguish. For tetrahedral symmetry, there is the well-known admixture of dipole transitions to some (but not all) multiplet states.

The above-mentioned Fe K pre-edge structure of iron oxides is explained only by the 1s3d quadrupole transition. For other TM compounds, a pre-edge structure caused by a different mechanism coexists with that due to the quadrupole transition. A typical example is the structure of Ti K pre-edge of TiO_2, as discussed in Section 5.5.7. The pre-edge of TiO_2 has three pre-peaks that are caused by two effects, namely the 1s to 3d quadrupole transition and the 1s to 3d nonlocal dipole transition. The nonlocal dipole transition occurs from Ti 1s to Ti 3d states but between different Ti sites (denoted by off-site 3d state), this transition is weakly dipole allowed because of the hybridization of Ti 4p, O 2p and off-site Ti 3d states. Both the 1s to 3d quadrupole transition and 1s to 3d nonlocal dipole transition split into T_{2g} and E_g peaks, but the E_g peak energy of the 1s to 3d quadrupole transition almost coincides with the T_{2g} peak energy of the 1s to 3d nonlocal dipole transition, so that three prepeaks are observed experimentally. Recent 1s2p RXES studies have confirmed that a similar situation occurs for iron, cobalt, and copper oxides. In all these cases, peaks due to direct quadrupole transitions into the 3d states are accompanied by nonlocal dipole transitions. These nonlocal transitions disappear for those systems that have isolated transition metal ions.

Figure 6.21 shows the K edge XAS spectrum of $LiCoO_2$, which is an oxide where Co is in a low-spin $3d^6$ ground state. The edge is visible at 7717 eV and before the edge, two separate peaks are visible at 7709 and 7711 eV. The HERFD-XAS (high-energy resolution fluorescence detection) shows two well separated peaks.

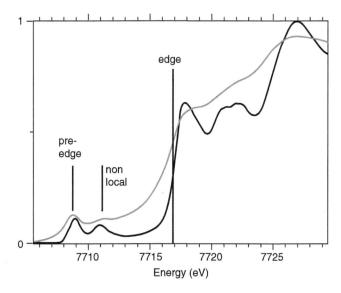

FIGURE 6.21   K edge of $LiCoO_2$ measured with HERFD-XAS (black) and normal XAS (light gray). The pre-edge indicates the quadrupole $3d^6 \rightarrow 1s^13d^7$ transition. The second peak shows the nonlocal dipole $3d^6 \rightarrow 1s^14p^1$ (3d band) transition.

Because low-spin $3d^6$ has a filled t_{2g} shell and an empty e_g shell, its quadrupole spectrum consists of only a single peak. In the XAS spectra, two peaks are visible. The 1s2p RXES images show that the first peak shows resonance behavior with decay peaks only at one excitation energy, while the second peak shows the decay that follows the excitation energy. This implies that the first peak is a $3d^6 \rightarrow 1s^1 3d^7$ quadrupole local resonance while the second peak is a nonlocal dipole $3d^6 \rightarrow 1s^1 3d^6 (3d^*)^1$ ($3d^*$ is a 3d band or an off-site 3d state) transition. For more details on HERFD-XAS and the 1s2p RXES image, see Chapter 8.

This nonlocal dipole peak is caused by a transition to the 4p states that are hybridized into the 3d band. The hybridization is via oxygen to the nearest metal neighbor, where its strength is determined by the geometry and degree of covalence. Because this final state is less localized, it is not as much affected by the core hole potential as the localized 3d states. It turns out that there is a direct relation between the relative importance of the nonlocal dipole peak and the possibilities for the coupling of the empty 4p states into neighboring 3d bands where the Co-O-Co distance and bond angle are important (Vanko et al., 2007).

6.4.4 Ligand K Edges

The single electron approximation gives an adequate simulation of the x-ray absorption spectral shape if the interactions between the electrons in the final state are relatively weak. This is the case for all ligand K edges. Assuming a TM oxide, there are no multiplet effects within the 2p final states. The ligand 2p states hybridize with the correlated 3d states. It is sometimes argued that oxygen 1s spectra are susceptible to multiplet effects as far as the 3d part of the spectrum is concerned (van Elp and Tanaka, 1999). The good agreement of the DOS with experiments shows, however, that multiplet effects are hardly (or not at all) visible in the spectrum. There are also no charge-transfer effects. The oxygen K edge excites an oxygen 1s core electron but its effect on the 2p and other oxygen valence states can be assumed to be similar (i.e. there are no localized states). The 3d states on the neighbor atoms do not feel the core hole potential on the oxygen atom. The absence of charge transfer and multiplet effects implies that one can assume the equivalence of the XAS spectrum and the DOS as a starting point. The basic aspects of the density of states as determined by band structure calculations and multiple scattering calculations are discussed in Appendix F.

An usual picture to visualize the electronic structure of a transition metal compound, such as oxides and halides, is to describe the chemical bonding mainly as a bonding between the metal 4sp states and the ligand p states (forming a bonding combination, the valence band and empty antibonding combinations). The 3d states also contribute to the chemical bonding with a valence band that causes them to be antibonding in nature. This bonding, taking place in a (distorted) cubic crystalline surrounding, causes the 3d states to be split into the so-called t_{2g} and e_g manifolds. Figure 6.22 shows a DFT calculation of the DOS of TiO_2. The total DOS is projected to oxygen (O) and titanium (Ti), respectively. Figure 6.23 shows the broadened oxygen DOS in comparison with the experimental spectrum. In line with the dipole selection rule, only the oxygen p-projected DOS is used. We assume that the DOS of complex systems is calculated using the DFT-LDA approximation.

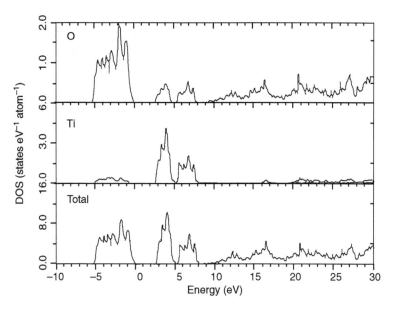

FIGURE 6.22 Density of states of rutile TiO_2, separated into the oxygen and titanium atoms, as calculated with LMTO. (Reprinted with permission from de Groot, F.M.F., Faber, J., Michiels, J.J.M., Czyzyk, M.T., Abbate, M., and Fuggle, J.C., *Phys. Rev. B*, 48, 2074, 1993. Copyright 1993 by the American Physical Society.)

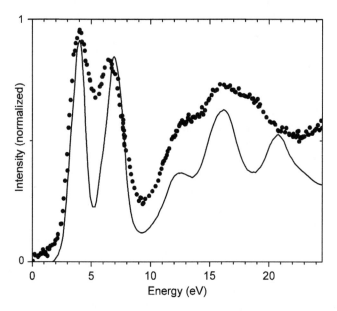

FIGURE 6.23 Comparison between the oxygen p-projected DOS and the experimental oxygen 1s XAS spectrum of rutile TiO_2. (Reprinted with permission from de Groot, F.M.F., Faber, J., Michiels, J.J.M., Czyzyk, M.T., Abbate, M., and Fuggle, J.C., *Phys. Rev. B*, 48, 2074, 1993. Copyright 1993 by the American Physical Society.)

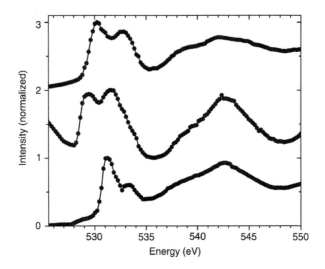

FIGURE 6.24 Oxygen 1s XAS spectra of TiO_2 (top), V_2O_3 (middle), and Cr_2O_3 (bottom). (Reprinted with permission from de Groot, F.M.F., Grioni, M., Fuggle, J.C., Ghijsen, J., Sawatzky, G.A., and Petersen, H., *Phys. Rev. B*, 40, 5715, 1989. Copyright 1989 by the American Physical Society.)

We can conclude that the oxygen K edge XAS spectrum of TiO_2 can be simulated with the oxygen p-projected DOS as calculated by DFT. Actually, in the case of the oxygen K edge of TiO_2, the core hole effect is small and can be neglected. The similarity between x-ray absorption and the projected DOS allows a qualitative analysis of the spectral shapes without the actual inclusion of calculations.

Figure 6.24 shows the oxygen K edges of three transition metal oxides. The two regions in the spectrum that can be distinguished are the oxygen 2p states related to the empty 3d band (from 530–535 eV), and the structure between 535 and 545 eV that is related to the empty 4s and 4p states of copper. The 3d band is split by the cubic crystal field and, in the first approximation, two peaks are observed that are related to the t_{2g} and e_g orbitals, respectively. Figure 6.25 shows the oxygen p-projected DOS of TiO_2, VO_2, CrO_2, and MnO_2 (from top to bottom). The calculations have been carried out with a linearized muffin-tin orbital (LMTO) code without core hole and with the oxides in their magnetic ground-state configuration. The TiO_2 DOS is the same as given in Figure 6.23 with some additional broadening. In the series from TiO_2 to MnO_2, each time one electron is added to the 3d band that is empty for TiO_2. Each added 3d electron fills a t_{2g} up-state. VO_2 has a t_{2g}^1 configuration, CrO_2 a t_{2g}^2 configuration, and MnO_2 a t_{2g}^3 configuration. The interatomic couplings of these t_{2g} electrons gives rise to a phase transition for VO_2, a half-metallic ferromagnetic system for CrO_2, and a magnetic ground state for MnO_2. Comparison with experiment shows excellent agreement for all these oxides, and it can be concluded that the oxygen 1s XAS spectra are described well with the oxygen p-projected DOS. For other ligands, such as carbides, nitrides, and sulfides, a similar interpretation as that for oxides is expected to hold. Their ligand 1s XAS spectra are expected to show close comparison to the projected spin-polarized DOS, as confirmed by calculations.

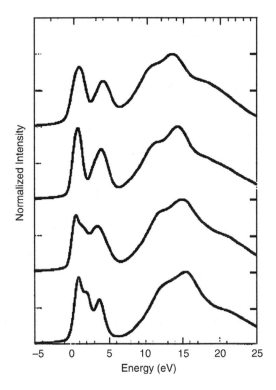

FIGURE 6.25 Series of oxygen p-projected DOS calculations of (from top to bottom) TiO_2, VO_2, CrO_2, and MnO_2 in their magnetic ground states. (Reprinted with permission from de Groot, F., *Chem. Rev.*, 101, 1779, 2001. Copyright 2001 by the American Chemical Society.)

The analysis of ligand K edges of transition metal systems shows that the intensity of the pre-edge region can be related to the ligand 2p-metal 3d overlap. This can be used to quantify the metal-ligand covalence. This is the central idea in the molecular orbital analysis of ligand K edges by Ed Solomon's group (Glaser et al., 2000; Solomon et al., 2005). The lowest unoccupied molecular orbital (LUMO) of a metal complex is described as a linear combination of metal (M_{3d}) and ligand (L_{np}) valence orbitals:

$$\Psi_{3d} = (1 - \beta^2)^{1/2} |M_{3d}\rangle + \beta^2 |L_{np}\rangle \tag{4.3}$$

The metal L edge excitation originates from a metal centered 2p orbital and only transitions to the metal-centered component of the ground state wave function have intensity. Thus, the metal L transition probes the empty metal 3d states, and the integrated intensity of the L-edge x-ray absorption spectrum can be related to the covalence $(1 - \beta^2)$ of the partially unoccupied 3d orbitals in a metal complex. Similarly, the ligand K edge originates from a ligand-centered 1s orbital and only transitions to the ligand-centered component of the ground state wave function have intensity. The integrated intensity of the K pre-edge can be related to the covalence (β^2) of the partially unoccupied 3d orbitals. Implicitly, this assumes that the effect of the core hole is neglected.

6.4.4.1 Oxygen K Edges of High T_c Copper Oxides

An important application of oxygen K edges is their use to determine the electronic structure of high T_c superconductors based on copper-oxide planes.

Figure 6.26 shows the oxygen K edge of $La_{2-x}Sr_xCuO_4$ and it can be observed that as a function of the doping, an extra state is visible below the peak at 530.3 eV. The peak at 530.3 eV is related to the oxygen holes of the Cu^{2+} state whereas the 528.8 eV peak is related to the Cu^{3+} state. It can be observed that already at $x = 0.15$, the peak at 528.8 eV is higher than the peak at 530.3 eV, caused by the larger oxygen contribution to the Cu^{3+} peak, which is reproduced by the Hubbard model calculations (Chen et al., 1991b, 1992). The polarization-dependent study of these peaks shows that they are mainly in the CuO plane for the doping range showed with approximately 15% out-of-plane character (Chen et al., 1992). Over-doping to $x = 0.35$ yields a larger out-of-plane character, up to 35%.

6.4.5 SOFT X-RAY K EDGES BY X-RAY RAMAN SPECTROSCOPY

X-ray Raman spectroscopy (XRS) is the x-ray analog of optical and UV Raman. A hard x-ray, typically with an energy of ~10 keV, impinges on the sample and the

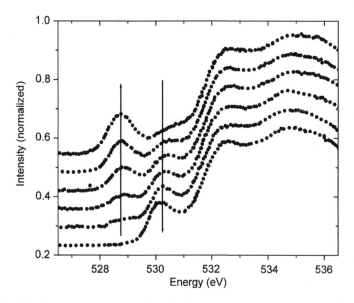

FIGURE 6.26 The oxygen K edge of $La_{2-x}Sr_xCuO_4$ as a function of the doping concentration x. The value of x is from bottom to top, respectively 0.00, 0.02, 0.04, 0.07, 0.10, and 0.15. (Reprinted with permission from Chen, C.T., Smith, N.V., and Sette, F., *Phys. Rev. B*, 43, 6785, 1991a. Copyright 1991 by the American Physical Society.)

radiation is scattered inelastically by the interaction $H_{INT(2)}$ in Equation 2.26 of Chapter 2 (denoted by Thomson scattering as described in Chapter 8), is measured. Similar to normal Raman scattering, one can study vibrations (meV range) and electronic excitations (eV range). In addition, one can study core electron excitations that relate to energy losses of several hundred eV. Core level x-ray Raman has alternatively be called x-ray energy loss spectroscopy (XELS) to stress its analogy with EELS. Exactly as is the case for EELS, the core excitation spectra from XRS can be described in analogy with x-ray absorption, at least if one remains within the dipole approximation and uses small scattering momenta. In practice, x-ray Raman is limited to the soft x-ray absorption K edge spectra of $Z = 3$ (Li) to $Z = 14$ (Si). Important applications are the ligand K edges of carbon, nitrogen, and oxygen (Bergmann et al., 2000, 2002, 2003, 2004).

It is obvious that x-ray Raman has great potential for *in situ* measurements. The complete experiment involves hard x-rays (10 keV), hence there is considerable penetrating power, and experiments under extreme conditions (e.g. high-pressure in diamond anvil cells), are possible. X-ray Raman spectroscopy represents a hard x-ray alternative to conventional XAS techniques in the study of systems with light elements and XRS is particularly useful for carbon, nitrogen, and oxygen edges. An important application is the study of H_2O under various conditions and the XRS data on liquid water have created intense discussions, especially with regard to hydrogen bonding (Wernet et al., 2004, 2005; Naslund et al., 2005).

6.4.5.1   Modifying the Selection Rules

Krisch et al. (1997) measured the Li K edge spectrum with XRS. They measured the XRS spectrum at two different scattering angles and, because the momentum transfer is different at these two angles, the selection rules are also different. Small scattering angles give small momentum transfer yielding dipole selection rules, while larger momentum transfer can bring the selection rules to a quadrupole nature. The $\Delta l = 1$ dipole selection rule (a) yields the normal Li K edge as measured with Li K edge XAS and this corresponds to the empty p-DOS. The quadrupole selection rule allows for $\Delta l = 2$ and $\Delta l = 0$ transitions and this makes the empty s-DOS and the empty d-DOS visible. In the case of Li metal, only the empty s-DOS is important at the edge (Krisch et al., 1997). In general, these large momentum transfer XRS experiments could be used to determine the s- and d-DOS versus the p-DOS for any system.

6.5   X-RAY ABSORPTION SPECTRA OF THE 4d AND 5d TM SYSTEMS

The 4d and 5d TM compounds are dominated by their partly filled 4d (5d) band. Compared with the 3d TM compounds discussed above, the main differences are that the 4d band has a much larger 4d spin–orbit coupling and that the 4d band is much less localized than the 3d band. In short, this implies that the 4d spin–orbit coupling is always important though the systems are less correlated, less affected by multiplet effects and have a larger crystal field splitting. For the spin state, this implies that 4d systems tend to be low spin.

6.5.1 $L_{2,3}$ EDGES OF 4d TM SYSTEMS

The L_2 and L_3 edges of 3d and 4d TM systems have a quite different shape. For 3d systems, the L_2 and L_3 edges are separated by an energy of only 5–20 eV. In Section 6.3, we have discussed at length the 2p XAS spectra of the 3d TM systems. The L_3 edge is completely different from the L_2 edge and their branching ratio is far from 2 : 1. The origin for these large differences has been determined as a combination of the 3d spin–orbit coupling and correlations between the 2p core hole and the 3d holes. Details on the branching ratio rules can be found in the papers of Thole and van der Laan (1988a,b,c).

For 4d TM systems, the separation between the L_3 and L_2 edges is of the order of 100 eV. Furthermore, the coupling of the 2p core wave function with the 4d valence states is much weaker, resulting in multiplet effects of the order of only 2 eV. The consequence is that the L_3 to L_2 ratio is always close to 2 : 1. Experiments have detected differences, though in the spectral shape, for a range of 4d TM systems, including compounds of Zr, Mo, Nb, and Ru. One observes that the L_3 and L_2 edges are different and that the first peak of the crystal field split doublet is more intense in the L_2 edge compared with the L_3 edge. Often it is assumed that the 4d spin–orbit coupling is the sole origin of this difference, but the main reason for the difference between the L_3 and L_2 edges of the 4d systems are in fact the multiplet effects coupling the 2p core wave function to the valence states of 4d character.

Figure 6.27 compares the experimental spectrum of MoF_6 with a LFM calculation. The calculation has been performed for Mo^{6+}, $4d^0$. The Slater integrals have been reduced to 75% of their atomic values to account for charge-transfer effects. A cubic crystal field of 4.5 eV was fitted to the experiment. For large values of the crystal field, the observed splitting directly represents the crystal field splitting. The theoretical line spectrum was broadened with a lifetime broadening of 2.0 eV. The spectrum was then convoluted with a Gaussian of 0.4 eV to simulate the experimental broadening. The different shape of the L_3 and L_2 edge can only be explained from the inclusion of multiplet effects.

Figure 6.28 shows the results of LFM calculations for the L_3 (solid) and L_2 (dashed) spectra of molybdenum for valences from Mo^{6+} ($4d^0$; bottom) to Mo^0 ($4d^6$; top). The spectra have been aligned and normalized to the peak height of the L_3 edge. The L_2 edge is shifted over the 2p spin–orbit splitting and multiplied by 2. The spectra have been calculated using a reduction of 50% for the dd Slater integrals and a reduction of 75% for the pd Slater integrals. This reduction is used to simulate the fact that the ground state is affected by charge transfer. A crystal field splitting of 4.0 eV has been used and as a result all $4d^n$ states are in a low-spin ground state. Thus, in going through the series from the bottom ($4d^0$) to the top ($4d^6$), every added electron fills one t_{2g} state. Because the number of t_{2g} holes decreases from six to zero while the number of e_g holes is four in all cases, in a single-particle interpretation the intensity of the first peak will decrease linearly with the number of holes. The multiplet effects (and the 4d spin–orbit coupling) modify this picture slightly and it can be observed that for both the L_3 and the L_2 edge, the intensity of the t_{2g} peak indeed decreases from $4d^0$ via $4d^2$ and $4d^4$ to $4d^6$. The effect is stronger for the L_2 edge and the L_2 edge t_{2g} peak intensity of the $4d^4$ system is very small. In the case of $4d^6$ systems, there is only one peak present since there are no t_{2g} holes left.

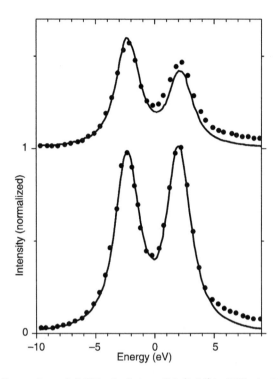

FIGURE 6.27 Comparison of a LFM calculation of Mo^{6+}, $4d^0$ (solid line) with an experimental spectrum of gas-phase MoF_6 (points). Both the L_2 edge (top) and the L_3 edge (bottom) have been aligned and centered at zero. (Reprinted with permission from de Groot, F.M.F., Hu, Z.W., Lopez, M.F., Kaindl, G., Guillot, F., and Tronc, M., *J. Chem. Phys.*, 101, 6570, 1994c. Copyright 1994 by the American Institute of Physics.)

The main reasons for this behavior are the effects of the dd multiplet and the 4d spin–orbit coupling on the initial state. Following the arguments of Sham (1983), we can illustrate this for the missing t_{2g} peak of $4d^5$. The ground state of the low-spin $4d^5$ state in octahedral symmetry is $^2T_2$. If one excites a 2p electron to the single t_{2g} hole, the final state has the configuration $2p^5t_{2g}^6$, which has the symmetry $^2T_1$. Including the 2p spin–orbit coupling, the $^2T_1$ state is split into a Γ_6 (related to the L_2 edge) and a Γ_8 (related to the L_3 edge) state. If the 4d spin–orbit coupling is also included, the $^2T_2$ ground state is split and the new ground state is the Γ_7 state, while the Γ_8 state is shifted by an energy related to the 4d spin–orbit coupling. A dipole transition is possible from the $4d^5[\Gamma_7]$ ground state to the $2p^64d^6[\Gamma_8]$ final state but not to a final state with Γ_6 symmetry. Hence the t_{2g} hole cannot be reached at the L_2 edge. Without the 4d spin–orbit coupling, both the Γ_7 and Γ_8 initial states degenerate and a dipole transition is possible for both the L_3 and the L_2 edges. Below, we discuss the effects of lower symmetries.

Figure 6.29 compares the experimental result for solid $Ru(NH_3)_6Cl_6$ with the result of a LFM calculation. The calculation was performed for $4d^5$ Ru^{3+} with Slater integrals reduced to about 25% of their atomic values and a crystal field of 3.65 eV.

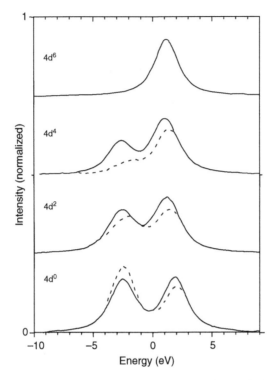

FIGURE 6.28 LFM calculations for molybdenum. The L_3 edge (solid line) and the L_2 edge (dashed) are given from bottom to top for (Mo^{6+}, 4d^0), (Mo^{4+}, 4d^2), (Mo^{2+}, 4d^4), and (Mo^0, 4d^6). The spectra have been aligned and normalized to their peak height. (Reprinted with permission from de Groot, F.M.F., Hu, Z.W., Lopez, M.F., Kaindl, G., Guillot, F., and Tronc, M., *J. Chem. Phys.*, 101, 6570, 1994c. Copyright 1994 by the American Institute of Physics.)

As discussed previously, the differences between the L_3 and L_2 spectra are special for a 4d^5 ground state because, in the spectrum of the L_2 edge, the t_{2g} peak is absent. The dashed line gives the result if the 4d spin–orbit coupling is put to zero; as discussed above, the t_{2g} peak of the L_2 edge is then present. Note the large reductions in the values of the Slater integrals, now for a system with a relatively low valence but with more covalent ligands. This strong reduction indicates that for 4d systems, the Slater integrals are reduced more than for the equivalent 3d systems.

6.5.2 PICOSECOND TIME-RESOLVED 2p XAS SPECTRA OF [Ru(bpy)$_3$]$^{2+}$

Measurements of excited states involve timescales in the order of picoseconds. An example is the excited state of [Ru(bpy)$_3$]$^{2+}$ 50 ps after the photo-excitation (Gawelda et al., 2006). The Ru^{2+} ion in the ground state of the (bpy)$_3$-complex has a 4d^6 configuration and in the presence of the large octahedral crystal field, the ground state is low spin. All six electrons fill up the t_{2g} orbitals, while the e_g orbitals are empty. The excitation of the 2p electron is only possible to the empty e_g states, which implies a single peak for both the L_3 and L_2 edges, as observed in Figure 6.30. The L_2 edge of

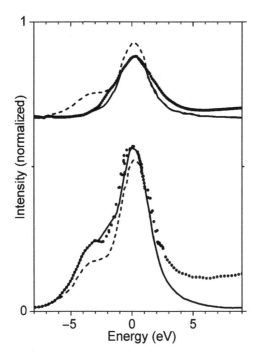

FIGURE 6.29 Comparison of LFM calculation of $4d^5$ Ru^{3+} (solid line) with an experimental spectrum of solid state $Ru(NH_3)_6Cl_6$ (points). Both the L_2 edge (top) and the L_3 edge (bottom) have been aligned at zero. The corresponding spectra with the 4d spin–orbit coupling set to zero are indicated with the dashed lines. (Reprinted with permission from de Groot, F.M.F., Hu, Z.W., Lopez, M.F., Kaindl, G., Guillot, F., and Tronc, M., *J. Chem. Phys.*, 101, 6570, 1994c. Copyright 1994 by the American Institute of Physics.)

the photogenerated species has two peaks (Figure 6.30, top), which is in contrast with the XAS spectrum of $Ru(NH_3)_6Cl_6$ that has one peak at the L_2 edge. The reason is that the trigonal symmetry of $[Ru(bpy)_3]^{2+}$ will mix the two spin–orbit split states (Γ_7 and Γ_8) of the $4d^5$ ground state. From a Γ_8 ground state, all symmetries are dipole allowed, which implies that the Γ_6 state, at the L_2 edge becomes visible.

6.5.3 HIGHER VALENT RUTHENIUM COMPOUNDS

Ruthenium oxides have recently gained considerable interest. These oxides contain higher valent $4d^4$ Ru^{4+} and $4d^3$ Ru^{5+} ions. From Figure 6.28, it can be found that these $4d^3$ and $4d^4$ ions have two peaks for both the L_2 and L_3 edges, though with different peak ratios for the quasi-t_{2g} and quasi-e_g peaks.

Figure 6.31 shows the Ru 2p XAS spectra of Sr_2RuO_4 (top) and $Sr_4Ru_2O_9$ (bottom). The lower and the higher energy component of both the L_3 and L_2 edges reflects transitions into t_{2g} and e_g states. The t_{2g} peaks are higher at the L_3 edge than at the L_2 edge for Ru^{4+} in Sr_2RuO_4 but this situation is reversed for the Ru^{5+} compound $Sr_4Ru_2O_9$. The observed spectral ratios are different from those expected for single particle models. As shown in Figure 6.31, the changes in the Ru 2p XAS spectra are

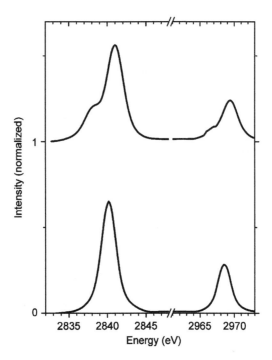

FIGURE 6.30 The experimental line shapes of the XAS L_3 and L_2 features for the Ru^{2+} ground state (bottom) and the Ru^{3+} excited state (top). (Reprinted with permission from Gawelda, W., et al., *J. Am. Chem. Soc.*, 128, 5001, 2006. Copyright 2006 by the American Chemical Society.)

well reproduced by LFM calculations. The t_{2g}/e_g intensity ratio is sensitive to the intra-atomic Coulomb interactions and to the 4d spin–orbit coupling. Considering the strong covalence between the TM 4d and the O 2p states in these systems, the Slater integrals have been reduced to 50% and 15% of their atomic values for Ru^{4+} and Ru^{5+}, respectively (Hu et al., 2002).

6.5.4 Pd L Edges and the Number of 4d Holes in Pd Metal

The situation for the late 4d transition metals is different from the early metals up to Ru. Ag ($4d^{10}$) and Pd ($4d^9$) systems have (in general) less than one hole in the 4d band. This implies that in the final state of the $2p \rightarrow 4d$ process, no 4d hole is present as the $4d^9$ contribution will be excited to $2p^5 4d^{10}$. This implies that no multiplet effects are important, in contrast to the Ru and Mo systems discussed previously. Sham (1985) analyzed Pd and Ag in detail and explained the general framework to interpret the Pd and Ag L edges. The Pd L_3 XAS spectrum corresponds to the empty DOS as calculated with band structure methods. There is a difference between the L_3 and the L_2 spectrum due to the interplay of the core hole spin–orbit coupling and the 4d spin–orbit coupling. Essentially, the empty 4d states can be divided into the $4d_{5/2}$ and $4d_{3/2}$ substates, where the $4d_{5/2}$ substates have more holes, leading to a higher white line at the L_3 edge. Another important

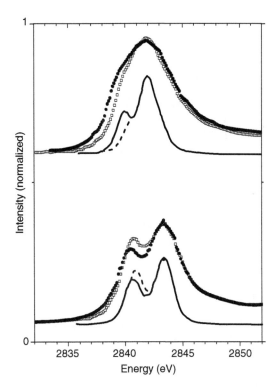

FIGURE 6.31   The 2p XAS spectra of the Ru^{4+} system Sr_2RuO_4 (top) and of the Ru^{5+} system $Sr_4Ru_2O_9$ (bottom). The Ru L_2 edge is given with open circles and the Ru L_3 edge with filled circles. The Ru L_2 data have been shifted and multiplied to overlap with Ru L_3 data. The theoretical spectra for Ru^{4+} and Ru^{5+} are shown as solid and dashed lines and for the L_3 and the L_2 edges, respectively. (From Hu, Z., et al., *Chem. Phys.*, 282, 451, 2002. With permission from Elsevier Ltd.)

aspect is that the white line intensity can be integrated and yields a number of empty 4d states.

6.5.5   X-Ray Absorption Spectra of the 5d Transition Metals

The x-ray absorption spectra of 5d TM systems can be analyzed with single particle models. Multiplet effects tend to be small except for shallow core levels such as the 4d and 5p levels.

The most studied edges for the 5d metals are the L_2 and L_3 edges. Figure 6.32 shows the L_1 (bottom), L_2 (middle), and L_3 (top) edges of Pt metal. The normal XAS spectra are rather broad due to the lifetime broadening of 5.3 eV. Making use of HERFD detection (as explained in Chapter 8), the effective lifetime broadening can be reduced to 2.0 eV. This allows a more detailed comparison to the DOS as calculated with a DFT program. The agreement between the DOS and the experimental spectra confirms that (with the 2 eV resolution) the Pt L edges correspond to the single particle DOS (de Groot et al., 2002). Pt and Au nanoparticles embedded on

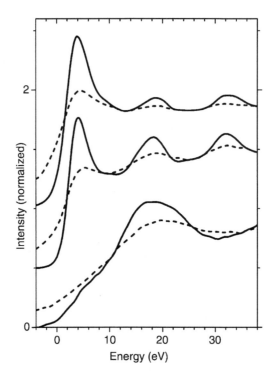

FIGURE 6.32   Bottom to top, the L_1, L_2, and L_3 x-ray absorption spectra (dashed) and corresponding HERFD L_3 spectra (solid) of Pt metal. (Reprinted with permission from de Groot, F.M.F., Krisch, M.H., and Vogel, J., *Phys. Rev. B*, 66, 2002. Copyright 2002 by the American Physical Society.)

oxide supports important present and future catalysts for a range of reactions. Important reactions involve the adsorption of H_2 and CO. The effects of these absorption processes are actively studied with XAS. The HERFD detection of these edges again improves the information. This effectively implies that the DOS is obtained with a high resolution as compared with normal Pt L edge XAS and allows for much more detailed analysis of the spectra and detailed information on the nature of the bonding of H_2, CO, and other reactant under *in situ* conditions. Safonova et al. (2006) showed the effect of CO adsorption on Pt/Al_2O_3 and van Bokhoven et al. (2006) showed the similar effect of CO on Au/Al_2O_3.

6.6   X-RAY ABSORPTION SPECTRA OF THE 4f RE AND 5f ACTINIDE SYSTEMS

In the solid state, RE atoms usually have the configuration $4f^n(5d6s)^3$. In calculations of XAS spectra, the RE trivalent ions are considered without the outermost 5d and 6s electrons that do not significantly influence the absorption spectra. The 4f electrons are localized and have little interaction with the environment. They determine the magnetic properties but do not significantly participate in the chemical bonding.

TABLE 6.3
$4f^n$ Atomic Ground States of the Rare Earths

RE	Ce	Pr	Nd	Pm	Sm	Eu	Gd
Conf.	$4f^1$	$4f^2$	$4f^3$		$4f^5$	$4f^6$	$4f^7$
Sym.	$^2F_{5/2}$	$^3H_4$	$^4I_{9/2}$		$^6H_{5/2}$	$^7F_0$	$^8S_{7/2}$

RE	Tb	Dy	Ho	Er	Tm	Yb	Lu
Conf.	$4f^8$	$4f^9$	$4f^{10}$	$4f^{11}$	$4f^{12}$	$4f^{13}$	$4f^{14}$
Sym.	$^7F_6$	$^6H_{15/2}$	$^5I_8$	$^4I_{15/2}$	$^3H_6$	$^2F_{7/2}$	^1S

The ground state of many Ce compounds is an exception. It can be $4f^1$ Ce^{3+} or $4f^0$ Ce^{4+}. A large range of Ce systems exist, in which the 4f electron is partially occupied or partially localized. This means that the explanation of most Ce systems (and some Yb and Pr systems) need to include interatomic hybridization. All other RE systems can be described with a single atomic configuration $4f^n$. The ground states for the different RE ions are given by Hund's rules and are presented in Table 6.3.

6.6.1 $M_{4,5}$ Edges of Rare Earths

The XAS spectra implying the 4f electrons, like the $M_{4,5}$ edges (3d → 4f transitions) and the $N_{4,5}$ edges (4d → 4f transitions), can be well described within the atomic multiplet theory. The absorption process can be written as $4f^n → 3d^94f^{n+1}$ (Thole et al., 1985a,c). The number of possible $3d^94f^{n+1}$ states can be very large, even though in the absorption spectrum only those reachable from the Hund's rule ground state satisfying the $\Delta J = 0, \pm1$ selection rules will be present. Still, the number of final states that can be reached increases from 3 in lanthanum to 53 in cerium, to 200 in praseodymium and to 1077 in gadolinium. In the end of the series where the number of 4f holes is reduced, it decreases again to 4 in thulium and 1 for ytterbium.

The cases of Tm and Yb involve only configurations with a maximum of one and two holes, respectively, and can be readily calculated. They will be discussed in detail. The initial state for Yb^{3+}, with 13 4f electrons has $L = 3$ and $S = 1/2$. Two J values are possible ($J = 7/2$ and $J = 5/2$), of which the first one with term symbol $^2F_{7/2}$ is the ground state according to Hund's third rule. The $^2F_{5/2}$ has an energy difference with the ground state that is given by the 4f spin–orbit coupling and from atomic calculations, one finds an energy difference of 1.3 eV. The final state, after 3d (or 4d) absorption, is given by $3d^94f^{14}$ (or $4d^94f^{14}$), with term symbols $^2D_{3/2}$ and $^2D_{5/2}$. The energy difference between these terms, corresponding to the 3d spin–orbit coupling is 49.0 eV. However, in the XAS spectrum, only the $^2D_{5/2}$ line (corresponding to the M_5-edge) is present, since the $^2D_{3/2}$ term cannot be reached from the $^2F_{7/2}$ ground state because of the ΔJ selection rules. Transitions to both final states are possible from the $^2F_{5/2}$ initial state. The relative intensities of these three transitions can be directly determined from a 2 × 2 matrix. The relative intensities of the $^2F_{5/2}$ state versus the $^2F_{7/2}$ are given by $2J + 1$ as 6 to 8. Because the relative intensities of $^2D_{3/2} : ^2D_{5/2}$ is $4 : 6$ and the transition $^2F_{7/2} → ^2D_{3/2}$ is forbidden, the other three numbers in the matrix can be immediately determined as given in Table 6.4. As was

TABLE 6.4
Relative Intensities of the Four Transitions
from a Single d to a Single f Electron

	$^2F_{5/2}$	$^2F_{7/2}$	Σ
$^2D_{3/2}$	14	0	14
$^2D_{5/2}$	1	20	21
Σ	15	20	

discussed in Chapter 2, crystal field effects will modify these numbers and effectively mix the $^2F_{5/2}$ state into the $^2F_{7/2}$ ground state.

6.6.1.1 $M_{4,5}$ Edge of Tm

For Tm, the transition is from the $4f^{12}$ ground state to the $3d^9 4f^{13}$ final state. The configurations that must be considered are the same as f^2 and $d^1 f^1$, respectively. The third of Hund's rules gives a $^3H_4$ ground state for $4f^2$. In the case of $4f^{12}$, the third of Hund's rules yields the maximum J value and the ground state is $^3H_6$. There is a complication in the calculation because there is a second term symbol with $J = 6$ (i.e. the $^1I_6$ state). In intermediate coupling, the $^3H_6$ state will mix with the $^1I_6$ state and the actual ground state will be a linear combination of both states. To discuss this point quantitatively, it is necessary to give the atomic parameters that are part of the calculation. The $4f^{12}$ configuration is set to an average energy of 0.0 and further contains three Slater–Condon parameters and the 4f spin–orbit coupling. The Slater–Condon parameters are $F^2 = 13.175$ eV, $F^4 = 8.264$ eV, and $F^6 = 5.945$ eV. The 4f spin–orbit coupling has a value of 0.333 eV. The Slater–Condon parameters determine the energies of the seven LS term symbols. The LS term symbols are split by the 4f spin–orbit coupling to a total of 13 term symbols where the lowest energy is found for the $^3H_6$ state. The $^3H_5$ state and the $^3H_4$ state are very close in energy though, being split only by 4f spin–orbit coupling. This 4f spin–orbit coupling mixes the $^1I_6$ state into the ground state according to Table 6.5. With the atomic

TABLE 6.5
Energy Matrix and Eigenvectors of the 2 × 2
Matrices of the f^2 Initial States with $J = 6$

Energy Matrix

$$\begin{vmatrix} -2.201 & -0.408 \\ -0.408 & 2.025 \end{vmatrix}$$

Eigenvectors

$$\begin{matrix} ^3H_6 \\ ^1I_6 \end{matrix} \begin{vmatrix} 0.995 & -0.095 \\ 0.095 & 0.995 \end{vmatrix}$$

Energy Levels

−2.240	$I \sim {}^3H_6$
2.064	$II \sim {}^1I_6$

parameters used, the energy of the $^3H_6$ state is -2.201 eV below the configuration average and the energy of the $^1I_6$ state lies at 2.025 eV. The small 4f spin–orbit coupling mixes in only a small part of the $^1I_6$ character and the ground state is approximately 99% pure in its $^3H_6$ character.

The $3d^9 4f^{13}$ state after absorption has symmetries that are found after multiplication of a d with an f symmetry state. $^2D \otimes {}^2F$ implies that S is 0 or 1 and L is 1, 2, 3, 4, or 5. This gives five singlet terms ($^1P_1$, $^1D_2$, $^1F_3$, $^1G_4$, and $^1H_5$) and 15 triplet terms ($^3P_{0,1,2}$, $^3D_{1,2,3}$, $^3F_{2,3,4}$, $^3G_{3,4,5}$, and $^3H_{4,5,6}$), with an overall degeneracy of $10 \times 14 = 140$. Here we use $^3H_{4,5,6}$ as a shorthand notation of $^3H_4$ plus $^3H_5$ plus $^3H_6$. The $^3H_6$ ground state has $J = 6$ and the dipole selection rules states that the final state J must be 5, 6, or 7, respectively. There are three states with $J = 5$ ($^1H_5$, $^3G_5$, and $^3H_5$) plus one state with $J = 6$ (i.e. $^3H_6$). This implies that the $M_{4,5}$ edges of Tm exist for four transitions.

The bottom spectrum in Figure 6.33 shows the three peaks at the Tm M_5 edge using atomic Slater integrals (de Groot and Vogel, 2006). The peak at the lowest

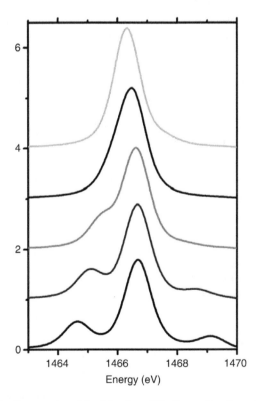

FIGURE 6.33 The three peaks of the M_5 edge of Tm^{3+} as a function of the Slater–Condon parameters. The bottom spectrum uses atomic Slater–Condon parameters, that is, 80% of their Hartree–Fock value. The magnitude of the Slater–Condon parameters is 80%, 60%, 40%, 20%, and 0% from bottom to top, respectively. (From de Groot, F., and Vogel, J., *Neutron and X-Ray Spectroscopy*, Grenoble, Springer, 2006. With permission.)

energy is the Hund's rule $^3H_6$ state. This is the only term with $J = 6$ and is therefore a pure $^3H_6$ peak also in $j \cdot j$ and thus in intermediate coupling. The other peaks all have $J = 5$ and are mixtures of the three term symbols with $J = 5$ in intermediate coupling. One can determine the exact nature of these three states by solving a 3×3 matrix. The $3d^94f^{13}$ configuration has an average energy of 1482.67 eV and further contains five Slater–Condon parameters, the 4f spin–orbit coupling and the 3d spin–orbit coupling. The Slater–Condon parameters are $F^2 = 9.09$ eV, $F^4 = 4.31$ eV, $G^1 = 6.68$ eV, $G^3 = 3.92$ eV, and $G^5 = 2.71$ eV. The 4f spin–orbit coupling has a value of 0.37 eV. The main difference with the initial state is the effect of the large 3d spin–orbit coupling of 18.05 eV that is able to strongly mix all states with equal J. This can be seen in Table 6.6. State I has a wave function $0.455 \, |^3H_5\rangle - 0.890 |^3G_5\rangle + 0.116 \, |^1H_5\rangle$. This state is thus approximately 80% pure $^3G_5$ character. The final result of the calculation is the four energies and their respective intensities.

The 3d $\rightarrow$ 4f transitions have been calculated for all the rare earths using atomic multiplets in intermediate coupling by Thole et al. (1985). The electrostatic parameters F^i_{ff} in the initial state and F^i_{ff} and F^i_{fd} in the final state, as well as the exchange parameters G^i_{fd} in the final state, were calculated as discussed in Chapter 4. The resulting line spectra were then broadened by a Lorentzian to account for the finite lifetime and an additional Gaussian to reproduce the experimental resolution. For $3d_{3/2}$ transitions, the line shape is asymmetric due to interactions between the "discrete" $3d_{3/2} \rightarrow 4f$ transitions and the transitions from $3d_{5/2}$ into the continuum (6p, 7p, 5f, and so on). This has been taken into account using a Fano line shape for the M_4-edge. The results of the calculations are in very good agreement with the experimental absorption edges.

Some general trends in the spectra are the presence of three distinct groups of lines in the calculated line spectra, giving rise to three peaks in the absorption edge. This is visible especially in the M_4-edge for the lighter rare earths and in the M_5-edge for the heavier ones. This splitting into three groups is due to the spin–orbit

---

TABLE 6.6
Energy Matrix and Eigenvectors of the 3 × 3 Matrices of the $3d^94f^{13}$ Final States with $J = 5$, after Inclusion of the Slater–Condon Parameters and the 3d and 4f Spin-Orbit Couplings. The Bottom Line Includes the Energy and Intensity of the $J = 6$ Final State

Energy Matrix			Eigenvectors			
1484.706	10.607	−19.163	$^3H_5$	0.455	−0.609	−0.649
10.607	1469.995	−10.088	$^3G_5$	−0.890	−0.302	−0.341
−19.163	−10.088	1486.965	$^1H_5$	0.116	−0.733	0.680

Energy Levels		Intensities
1464.44	I	4.11
1466.89	II	0.52
1510.33	III	0.23
1462.38	$^3H_6$	1.16

coupling, which, in intermediate coupling, tends to group lines with the same J together. Another trend is the $M_5 : M_4$ branching ratio that is almost $1 : 1$ at the beginning of the series but increases significantly for the heavier rare earths. The spin–orbit coupling in the 4f levels favor $4f_{7/2}$ holes, which are only reachable from the $3d_{5/2}$ level, with respect to $4f_{5/2}$. This is plainly evident for Yb where only $3d_{5/2}$ (M_5) absorption is possible.

6.6.1.2 $M_{4,5}$ Edge of La^{3+}

The $M_{4,5}$ edge of La^{3+} (e.g. in La_2O_3), can be described with atomic multiplets. In this closed shell system, the 3d x-ray absorption process excites a 3d core electron into the empty 4f shell and the transition can be described as $3d^{10}4f^0 \rightarrow 3d^9 4f^1$. At first, we will only look at the symmetry aspects of the problem. The $3d^{10}4f^0$ ground state contains only completely filled or completely empty shells and as such has $^1S_0$ symmetry. That is, all its quantum numbers are zero, $S = 0$, $L = 0$, $J = 0$. This automatically means that all magnetic moments are zero too.

Next, we have to determine all term symbols in the final state. Because there is again a single 4f electron (with term symbol ^2F) and a single 3d hole (with term symbol ^2D), we have to multiply $^2D \otimes {}^2F$. For the spin, this gives singlet $S = 0$ states and triplet $S = 1$ states. The maximum L value is given by $(l_{3d} = 2) + (l_{4f} = 3) = 5$. The minimum L value is $|(l_{3d} = 2) - (l_{4f} = 3)| = 1$. Thus L takes any value from 5, 4, 3, 2, or 1. This gives as term symbols $^1P_1$, $^1D_2$, $^1F_3$, $^1G_4$, $^1H_5$, and $^3P_{0,1,2}$, $^3D_{1,2,3}$, $^3F_{2,3,4}$, $^3G_{3,4,5}$, $^3H_{4,5,6}$. The overall degeneracies of the singlet term symbols are, respectively, 3, 5, 7, 9, and 11 (total 35). The overall degeneracies of the triplet states are $3 \times 35 = 105$, adding up to 140 in total, confirming the 10×14 possibilities of adding a 3d and a 4f electron. The degeneracies in J are important because the x-ray absorption calculations are always performed in intermediate coupling, that is, the degeneracies in J set the sizes of the sub-matrices to be diagonalized. The reason behind this is the large core hole spin–orbit coupling that makes any LS-nomenclature impossible for the final states.

The next important step is to realize the dipole selection rules. If one works in intermediate coupling, only the J-selection rule matters. This rule states that J is changed by -1, 0, or $+1$ with respect to the initial state J, with the addition that if J is zero in the initial state, it cannot be zero in the final state. Also, J cannot be negative; so for a $3d^{10}4f^0$ $^1S_0$ ground state, the only allowed final state values of J is 1. Looking into the table, this implies that there are only three allowed final states: $^1P_1$, $^3P_1$, and $^3D_1$. This means that the $M_{4,5}$ x-ray absorption spectrum of La_2O_3 can have a maximum of only three peaks.

Figure 6.34 shows the comparison of the $M_{4,5}$ edge of La metal between XAS and EELS, where the EELS is measured with 1450 eV primary electrons, yielding 600 eV scattered electrons. The energy of the electrons is too low to assume the dipole approximation and the full Coulomb scattering needs to be calculated. This yields the theoretical spectrum, which is in excellent agreement with experiment (Ogasawara and Kotani, 1996). In addition to the three $J = 1$ final states, a grouped range of other J values is visible at ~830 and ~848 eV. Certainly for the case of La, this implies that the low-energy EELS spectrum contains considerably more information that the XAS spectrum.

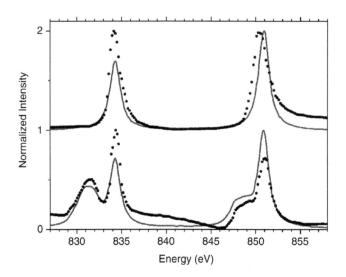

FIGURE 6.34 (Top) Experimental $M_{4,5}$ XAS spectrum of La metal (dots) compared with an atomic multiplet calculation. Three peaks are found, essentially two big peaks, the M_5 edge at 837 eV and the M_4 edge at 854 eV. In addition, a small pre-edge is visible at 833 eV. (Bottom) The experimental $M_{4,5}$ low-energy EELS spectrum of La metal (dots) compared with an atomic multiplet calculation. (Digitized from Ogasawara, H., and Kotani, A., *J. Electron Spectrosc. Relat. Phenom.*, 78, 119, 1996. With permission from Elsevier Ltd.)

6.6.1.3 $M_{4,5}$ Edge of CeO₂

Figure 6.35 shows the $M_{4,5}$ edge of CeO_2 as measured by Butorin et al. (1996b) in their study of the 3d4f RXES spectra of CeO_2. Due to strong Ce 4f–O 2p hybridization in CeO_2, the initial and final state of the 3d XAS process can be described as mixtures of $4f^0$ and $4f^1\underline{L}$ configurations and $3d^94f^1$ and $3d^94f^2\underline{L}$ configurations, where $\underline{L}$ stands for a hole in the valence band. This explains the double-line shape of the 3d XAS spectrum of CeO_2. (The CTM parameters are $V = 0.76$ eV, $\Delta = 1.6$ eV, $U_{ff} = 10.5$ eV, and $U_{fc} = 12.5$ eV.)

6.6.2 N₄,₅ EDGES OF RARE EARTHS

The $N_{4,5}$ edges of the rare earths are similar to the $M_{4,5}$ edges, but a 4d electron is excited instead of a 3d electron. The $N_{4,5}$ XAS edge of La can be described with a $4f^0$ to $4d^94f^1$ transition, which has exactly the same symmetry properties as the $M_{4,5}$ edge. So, we again expect three peaks related to the $^1P_1$, $^3D_1$, and $^3P_1$ states.

- The 4d spin–orbit coupling is only 1.1 eV, implying that the splitting between the N_5 and N_4 edge is only 2.7 eV.
- The Slater–Condon parameters are larger for the 4d4f interaction, implying that the energy separations between different (spin) states will be larger for the $N_{4,5}$ edge.

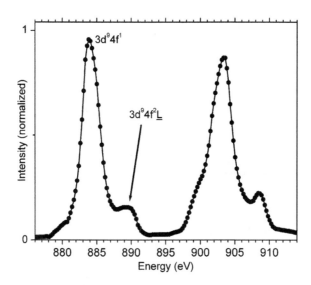

FIGURE 6.35 The 3d XAS spectrum ($M_{4,5}$ edge) of CeO_2. (Reprinted with permission from Butorin, S.M., Mancini, D.C., Guo, J.H., et al. *Phys. Rev. Lett.*, 77, 574, 1996b. Copyright 1996 by the American Physical Society.)

- Looking at the intensities, small pre-edges are found for both the $M_{4,5}$ and $N_{4,5}$ edge, but where the $M_{4,5}$ edge has an intensity ratio of 0.8 : 1.2, more than 99% of the intensity of the $N_{4,5}$ edge goes to the N_4 edge. In fact, both the pre-edge and the N_5 edge will be very small.

Figure 6.36 shows a comparison between theory and experiment (Aono et al., 1980). The $N_{4,5}$ edge has very small core spin–orbit coupling that brings it close to a *LS* coupling. In pure *LS* coupling, there is only one peak—the $^1P_1$ peak. The pre-edge region has two peaks at 99 and 103 eV that are exactly reproduced in the atomic multiplet calculation. The large $^1P_1$ peak has additional structure in experiment, mainly due to auto-ionization effects. Because the $N_{4,5}$ edges always have their large, *LS*-allowed peak at the high-energy side, these spectra are indicated with the name "delayed onset," which implies that there are a series of small pre-edges followed by the main transition. This behavior is usual for all $N_{4,5}$ edges of the rare earths. Since the 1960s, high-resolution $N_{4,5}$ edges have been available, largely collected by the group of Fomichev (Fomichev, 1967; Zimkina et al., 1984). The data were taken in their Leningrad laboratory without the use of a synchrotron. When synchrotron data became available in the 1970s, these experiments were confirmed. The $N_{4,5}$ edges of the early rare earths all show a series of small peaks. These small peaks are followed by a main peak, similar to the case of La^{3+} as already described.

Figure 6.37 shows the experimental 4d XAS spectra in the pre-edge region of CeO_2 compared with LaF_3, α-Ce, and γ-Ce. The spectra for LaF_3 and γ-Ce are in good agreement with atomic spectra calculated for the $4f^0$ and $4f^1$ ground states,

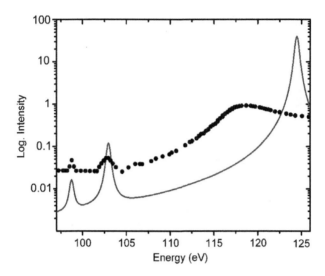

FIGURE 6.36   A logarithmic plot of the $N_{4,5}$ edge of La^{3+} calculated with atomic multiplets (solid line) compared with the experimental spectrum of La metal. (Reprinted with permission from Aono, M., Chiang, T.C., Knapp, J.A., Tanaka, T., and Eastman, D.E., *Phys. Rev. B*, 21, 2661, 1980. Copyright 1980 by the American Physical Society.)

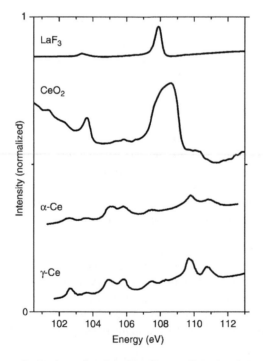

FIGURE 6.37   Experimental results of the leading small peaks of the 4d XAS for LaF_3, CeO_2, α-Ce, and γ-Ce. (Reprinted with permission from Kotani, A., Ogasawara, H., Okada, K., Thole, B.T., and Sawatzky, G.A., *Phys. Rev. B*, 40, 65, 1989. Copyright 1989 by the American Physical Society.)

respectively. The multiplet structure of α-Ce is very similar to that of γ-Ce with an almost trivalent ground state.

It is found that the multiplet structure of CeO_2 resembles LaF_3 and is quite different from that of Ce metal. The 4d XAS spectrum of CeO_2 can be calculated using the same ground state as for its 3d XAS spectrum as already given. The 4d XAS result is a superposition of the absorption spectra corresponding to transitions to $4d^94f^1$ and $4d^94f^2\underline{L}$. Their intensity ratio is the ratio of the mixing between $4f^0$ and $4f^1\underline{L}$ configurations in the ground state, which is about 6 : 4. The $4d^94f^1$ final state gives two discrete lines corresponding to $^3P_1$ and $^3D_1$ multiplets, similar to that of La^{3+}. The $4d^94f^2\underline{L}$ final state gives the continuous spectrum broadened by the finite bandwidth W of the valence hole $\underline{L}$. When the hybridization V is small, the line spectra $^3P_1$ and $^3D_1$ are broadened. When the actual hybridization $V = 0.76$ eV is used, the $^3P_1$ peak is pushed down and changes into a bound state instead of a resonance state. For more details, see Kotani et al. (1989) and Kotani and Ogasawara (1992).

6.6.3 L$_{2,3}$ EDGES OF RARE EARTHS

The $L_{2,3}$ edges of the rare earths measure the transition from the 2p core states. The 2p electrons can be excited to the empty 5d valence band and at higher energy to the other d-symmetry and s-symmetry bands. The description of the RE $L_{2,3}$ edges is in a number of ways similar to the K edges of the 3d TM systems. That is, the excitation energy of the $L_{2,3}$ edges ranges from 5483 eV in lanthanum to 8944 eV in ytterbium compared with the range from 4492 eV in scandium to 8979 eV in copper. In addition, both the RE $L_{2,3}$ edges and the TM K edges contain a pre-edge feature due to quadrupole transitions. In the case of the $L_{2,3}$ edges, a quadrupole transition is possible from the 2p core state to the partly filled 4f states.

There are also important differences with the main difference being that the $L_{2,3}$ edge is split into two by the 2p core hole spin–orbit coupling. The splitting between the L_3 and the L_2 edge is relatively large, ranging from 410 eV in La to 1030 eV in Yb. This large energy difference implies that the XMCD spin sum rule applies to the RE $L_{2,3}$ edge, as will be discussed in Chapter 7. The general shape of the $L_{2,3}$ edge spectra of the pure RE metals look rather similar, with the exception of Ce metal and Ce compounds.

In contrast to the 3d TM K edges, the 2p $\rightarrow$ 4f quadrupole pre-edge transitions are not visible as a distinguishable peak in the normal $L_{2,3}$ edge spectra. The quadrupole transitions become visible in XMCD experiments (Chapter 7) and also in HERFD-XAS experiments that will be discussed subsequently and in more detail in Chapter 8. HERFD-XAS experiments are based on the work of Hämäläinen et al. (1991), who showed that it is possible to effectively eliminate the lifetime broadening of deep core levels by using a high-resolution fluorescence detector. The $L_{2,3}$ edge of rare earths should follow the same line of reasoning as the K edges of the transition metals (as explained in Figure 6.18). The pre-edge is self-screened and can be described as a $4f^n \rightarrow 2p^54f^{n+1}$ transition, with very limited charge transfer. The main peak can be described as a $4f^n \rightarrow 2p^54f^n5d^1$ transition where the core hole pulls down the $2p^54f^{n+1}\underline{L}5d^1$ well-screened configuration. Soldatov et al. (1994) treated the $L_{2,3}$ edge of CeO_2 in detail by combining the single particle interpretation with the charge-transfer multiplet calculation. Note that the experimental resolution makes it difficult to decide on the optimal simulation.

6.6.4 $O_{4,5}$ EDGES OF ACTINIDES

The 5f electrons of actinides are less localized than the 4f electrons of RE systems, which will make more systems susceptible to hybridization effects with the neighbors. The 5f elements that are most often studied are uranium (U), plutonium (Pt), and thorium (Th).

UF$_4$ has a $5f^2$ ground state. Its L_3 edge ($2p_{3/2} \rightarrow 6d$) at ~17180 eV, M_3 edge ($3p_{3/2} \rightarrow 6d$) at ~4310 eV, M_5 edge ($3d_{5/2} \rightarrow 5f$) at ~3550 eV and N_5 edge ($4d_{5/2} \rightarrow 5f$) at ~730 eV show essentially a single peak at the edge, followed by additional structure at higher energy [see Figure 6.38 (left)]. These x-ray absorption spectra taken over a larger energy range should correspond closely to single particle calculations. The resolution of the spectra does not allow any detailed multiplet analysis of the edge structure. Only the $O_{4,5}$ spectrum ($5d \rightarrow 5f$) at ~100 eV shows significant fine structure, related to the $5d^9 5f^3$ final state multiplet. The $5d^9 5f^n$ multiplet spectrum is similar to the $4d^9 4f^n$ multiplet for the rare earths with a series of small pre-edge peaks followed by a large absorption peak (Kalkowski et al., 1987).

The interpretation of the $5d^9 5f^1$ final state is analogous to the interpretation of the $4d^9 5f^1$ final state for LaF$_3$. Figure 6.39 (top) shows the result of the atomic multiplet calculation. The main peak is dominated by the $^1P_1$ character and the two pre-edge peaks (the first peak is almost invisible but indicated with a small negative bar in the figure) are related essentially to the $^3D_1$ and $^3P_1$ states with a little admixture of $^1P_1$. The 5d XAS spectra of UF$_4$ and UO$_2$ are rather similar. It can be explained from an atomic multiplet calculation of the $5f^2 \rightarrow 5d^9 5f^3$ transition for ionic U^{4+}. The other uranium compounds show a similar 5d XAS spectrum indicating that these spectra are also dominated by atomic multiplets that separate a small pre-edge structure from the main edge. The lack of experimental details does not allow any detailed simulation.

6.6.5 $M_{4,5}$ EDGES OF ACTINIDES

Figure 6.40 shows the 4d XAS spectra of uranium and thorium materials that consist of a white line followed by an edge jump (Kalkowski et al., 1987). In case of UF$_4$ and UO$_2$ oxides, some small additional peaks are visible. These structures are related to transitions to higher-lying DOS of metal f or p character. The atomic multiplet calculations for the $5f^0 \rightarrow 4d^9 5f^1$ and $5f^2 \rightarrow 4d^9 5f^3$ transitions of $5f^0$ U^{6+} and $5f^2$ U^{4+} are shown in Figure 6.41. In the case of the $5f^0$ system, a three-peaked structure is obtained, similar to the $M_{4,5}$ edge of La^{3+}. The first peak is very small at less than 1% of the main peak and is positioned at 2.0 eV below the main peak, which renders it invisible. It is indicated with a small negative bar that is not to scale. The ratio of the M_5 to M_4 edge is 5.95 : 4.05, essentially equal to its $2J + 1$ value of 6 : 4. Figure 6.41 shows that the $M_{4,5}$ edge of $5f^2$ U^{4+} has a multiplet splitting spread of over some 5 eV for both the M_5 and M_4 edges. The lifetime broadening of 3.0 eV essentially broadens the multiplet structure to a single peak that is approximately Gaussian in shape.

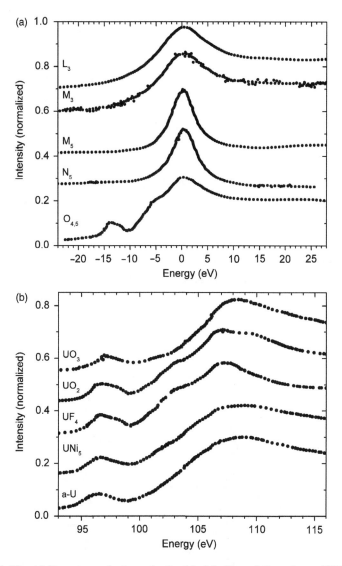

FIGURE 6.38 (a) From top to bottom, the L_3, M_3, M_5, N_5 and $O_{4,5}$ edges of UF_4. (b) The $O_{4,5}$ spectrum (5d → 5f) of a series of uranium compounds, from top to bottom UO_3, UO_2, UF_4, UNi_5, and α-U. (Reprinted with permission from Kalkowski, G., Kaindl, G., Brewer, W.D., and Krone, W., *Phys. Rev. B*, 35, 2667, 1987. Copyright 1987 by the American Physical Society.)

This again makes multiplet effects essentially invisible in the actual experimental spectrum of U^{4+} systems. Note that this calculation indicates that for multiplet effects (as well as crystal field and charge-transfer effects) to be visible, the first criterion is that they must be larger than the lifetime broadening. Even though these $M_{4,5}$ edges

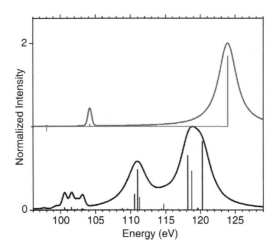

FIGURE 6.39 Atomic multiplet calculations for $5f^2$ U^{4+} (bottom) and $5f^0$ U^{6+} (top). The line spectrum is convoluted with a Lorentzian broadening (HWHM) of 0.5 (3.0) eV below (above) the threshold at 0.0 eV.

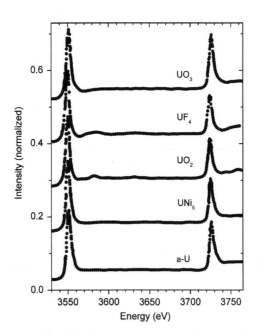

FIGURE 6.40 The $M_{4,5}$ spectrum ($4d \rightarrow 5f$) of a series of uranium compounds: UO_3, UF_4, UO_2, UNi_5, and α-U. (Reprinted with permission from Kalkowski, G., Kaindl, G., Brewer, W.D., and Krone, W., *Phys. Rev. B*, 35, 2667, 1987. Copyright 1987 by the American Physical Society.)

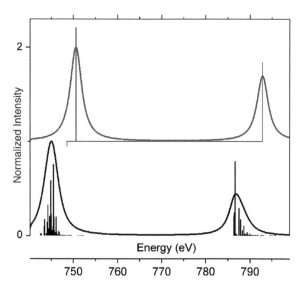

FIGURE 6.41 Atomic multiplet calculations for $5f^2$ U^{4+} (bottom) and $5f^0$ U^{6+} (top).

do not show any signs of atomic multiplets and/or charge-transfer effects, this does not mean that these effects are absent.

7 X-Ray Magnetic Circular Dichroism

7.1 INTRODUCTION

Dichroism is a property of certain objects that enables them to show different colors according to their orientation with respect to the light. It is due to the dependence of the optical response of the object on the relative orientation between the polarization direction of the light and the symmetry axes of the object. With x-rays, in some cases, a difference can be observed between the absorption of positive and negative helicity light (sometimes called left- and right-circularly-polarized light) [x-ray magnetic circular dichroism (XMCD)] or for different orientations of the polarization vector of linearly-polarized light with respect to a given quantization axis [x-ray magnetic linear dichroism (XMLD)].

For the definition of the sign of XMCD, we follow Baudelet et al. (1993). We take the axis of quantization as the $+\mathbf{z}$ direction, which is also the direction of the wave-vector of incident photons, and the magnetic field $\mathbf{B}$ is applied in the $-\mathbf{z}$ direction to align the magnetization of the material system (see Figure 7.1). Here positive helicity identifies with right circularly polarized x-rays; that is, using your right hand and tracking the magnetic field with your thumb, your fingers give the rotation direction of the positive helicity). Then, the XMCD spectrum $\Delta\mu$ is defined by:

$$\Delta\mu \equiv \mu_+(\mathbf{B}) - \mu_-(\mathbf{B}), \tag{7.1}$$

where μ_+ (μ_-) is the x-ray absorption spectroscopy (XAS) spectrum for the incident photon with $+$ ($-$) helicity. We use this convention. For the reversed magnetic field $-\mathbf{B}$, $\Delta\mu$ can also be written as:

$$\Delta\mu = \mu_-(-\mathbf{B}) - \mu_-(\mathbf{B}) = \mu_+(\mathbf{B}) - \mu_+(-\mathbf{B}) = \mu_-(-\mathbf{B}) - \mu_+(-\mathbf{B}). \tag{7.2}$$

Hereafter, we write $\mu_\pm(\mathbf{B})$ as $\mu_\pm$ unless it causes any confusion. Instead of $+$ and $-$ helicity, we can also use right (μ_R) and left (μ_L) polarizations:

$$\text{XMCD: } \Delta\mu = \mu_+ - \mu_- = \mu_R - \mu_L. \tag{7.3}$$

Similarly, the XMLD is defined as:

$$\mu_\parallel - \mu_\perp = \mu_0 - (\mu_+ + \mu_-)/2, \tag{7.4}$$

where $\mu_\parallel$ ($= \mu_0$) and $\mu_\perp$ are the XAS spectra for the incident photons with linear polarizations parallel and perpendicular to the $\mathbf{z}$ direction, respectively. Dichroism can

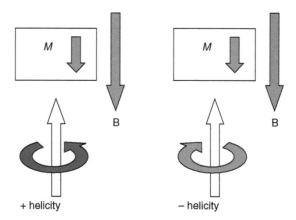

+ helicity   − helicity

FIGURE 7.1   Definition of the intensities $\mu_{\pm}$. (From Parlebas, J.C., Asakura, K., Fujiwara, A., Harada, I., and Kotani, A., *Phys. Rep.*, 431, 1, 2006. With permission from Elsevier Ltd.)

only occur when the spherical symmetry of the free atom is broken due to a magnetic field and/or an electric field such as crystal field effects. Magnetic fields can cause both circular and linear dichroism while a crystal field can only induce linear dichroism. Historically, the first predictions for an XMCD experiment were made by Erskine and Stern (1975). Later a strong XMLD was predicted (Thole et al., 1985b) and measured (van der Laan et al., 1986) in the $M_{4,5}$ edges of rare earth (RE) elements. The first XMCD experiment was performed at the K-edge of iron metal (Schütz et al., 1987, 1988, 1989) and the first L edge XMCD was reported for nickel metal (Chen et al., 1990). In this section, the question of the calculation of the XMCD signal is addressed. XMCD is the difference in absorption of the left- and right-circularly-polarized x-rays by a magnetized sample. Over the last 15 years, this field has gained much interest because of the promise of the determination of the ground state values of the spin and orbital moments using sum rules (Thole et al., 1992). The sign of the XMCD signal determines the spin orientation and, in magnetic systems that contain more than one (magnetic) element, element specific moments can be obtained. The use of XMCD is often studied for magnetic systems and devices and also for para-magnetic metal centers in proteins. In addition to MCD effects in x-ray absorption, circularly polarized x-rays can also be used to perform photoemission experiments. This yields a number of additional possibilities that are discussed in Chapter 5.

7.2   XMCD EFFECTS IN THE $L_{2,3}$ EDGES OF TM IONS AND COMPOUNDS

7.2.1   ATOMIC SINGLE ELECTRON MODEL

In a magnetic sample, the x-ray absorption is dependent on the polarization of the x-ray. This effect can be explained using an atomic, single-electron, model. This model describes the transitions of the 2p core state to the 3d valence state. Effectively, this transition occurs for systems with a $3d^9$ ground state. The initial state of Cu^{2+} contains only a single 3d-hole and has a $2p^63d^9$ configuration. This implies that the 3d3d interactions are absent and only has the crystal field effects, the 3d spin–orbit coupling, and the Zeeman exchange field or an external magnetic field. In the $2p^5 3d^{10}$ final state, there is effectively only a 2p core hole where the 2p core hole spin–orbit coupling is a crucial factor for the dichroic effects. There are no 2p3d multiplets and no effects of the crystal field, exchange field and 3d spin–orbit coupling.

The initial state has a $3d^9$ configuration and the atomic single 3d-hole ground state has $L = 2$ and $S = 1/2$. Following Hund's rules, its J-value is equal to $L + S$ and $J = 5/2$. There is an excited state with $J = 3/2$. The final state has a single 2p-hole; hence, it has a $J = 3/2$ state and a $J = 1/2$ state, at the L_3 and L_2 edges respectively. Applying an exchange field splits the J states into their respective M_J sublevels. Following the convention by Baudelet et al. (1993), we take the magnetic field **B** (and so the exchange field also) in the $-z$ direction and then the ground state takes $M_J = 5/2$ via the Zeeman splitting. The final states have M_J values between $-3/2$ and $+3/2$ and the selection rules state that $\Delta q = \Delta M_J$ is equal to -1, 0, or $+1$ dependent on the polarization of the x-ray. This implies that, for this ground state, there will be only a single transition. This is the transition from $M_J = 5/2$ to $M_{J'} = 3/2$ with the $\Delta M_J = -1$ operator. The J-value of the ground state is $5/2$ and the J-values of the final states are $J = 3/2$ and $J = 1/2$. With the dipole selection rules $\Delta J = +1$, 0, or -1, this implies that there is only a transition to the $J = 3/2$ final state. That is, the only allowed transition has $\Delta J = -1$ and $\Delta M_J = -1$. Figure 7.2 sketches the situation of

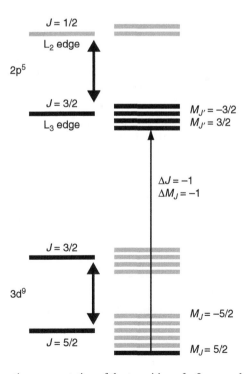

FIGURE 7.2 Schematic representation of the transition of a 2p core electron to the 3d state in the case of a $3d^9$ ground state. In a magnetic field, the ground state has $J = 5/2$ and $M_J = 5/2$. The only allowed dipole transition leads to the $J' = 3/2$ and $M_{J'} = 3/2$ state (indicated as $M_J = 3/2$ in the figure).

the various states. At finite temperatures, the other M_J levels become occupied via the Boltzmann distribution. The spectrum of an x-ray absorption transition can be calculated with:

$$W_{fi} = \frac{1}{\hbar c}\frac{4\omega^3}{3c^2} n \begin{pmatrix} J' & 1 & J \\ -M' & q & M \end{pmatrix}^2 S\delta(E_f - E_i - \hbar\Omega). \tag{7.5}$$

The squared $3j$-symbol describes the polarization dependence. In the case of a $3d^9$ ground state, $J = 5/2$. The only allowed final state has a $2p^5$ configuration, hence $J' = 3/2$. The integrated intensities for the various JJ' transitions are given by:

$$\sum_{M'} \begin{pmatrix} J' & 1 & J \\ -M' & q & M \end{pmatrix}^2. \tag{7.6}$$

This yields the numbers as given in Table 7.1. The integrated intensity for the L_3 versus the L_2 edge ($J' = 3/2$ and $1/2$) is $10 : 5$, according to the $2J' + 1$ degeneracies of the final states ($2 : 1$). The integrated intensity for $J = 5/2$ against $J = 3/2$ is $9 : 6$, according to the $2J + 1$ degeneracies of the initial states ($6 : 4$). There is a small transition strength from $J = 3/2$ to $J' = 3/2$. Note that these are atomic numbers and crystal field effects will modify these results. The x-ray absorption intensities can be found by specifically adding the values of q in the $3j$-symbol. Note that there are now several free on-line $3j$-symbol calculators available. This gives for the $|J, M_J\rangle$ ground state of $|5/2, 5/2\rangle$:

$$\sum_{M_{J'}} \begin{pmatrix} 3/2 & 1 & 5/2 \\ -M_{J'} & -1 & 5/2 \end{pmatrix}^2 = \frac{1}{6}, \quad \sum_{M_{J'}} \begin{pmatrix} 3/2 & 1 & 5/2 \\ -M_{J'} & 0 & 5/2 \end{pmatrix}^2 = 0, \quad \sum_{M_{J'}} \begin{pmatrix} 3/2 & 1 & 5/2 \\ -M_{J'} & 1 & 5/2 \end{pmatrix}^2 = 0.$$

The only nonzero $3j$-symbol is found for $q = -1$ and $M_{J'} = 3/2$ and its value is $-1/\sqrt{6}$. As the XMCD spectrum is defined by Equation 7.1, the sign of XMCD is negative.

Figure 7.3 shows the XAS (solid) and XMCD (dashed) spectra for a Cu^{2+} ion, using the atomic multiplet calculation with (bottom) and without (top) the 3d spin–orbit coupling. With 3d spin–orbit coupling, only one transition is allowed and the XMCD spectrum is identical to the XAS spectrum except for the sign. Without 3d spin–orbit coupling, more transitions are allowed and the typical two-peak $L_{2,3}$ edge spectrum is found. Omitting the 3d spin–orbit coupling effectively implies that the

TABLE 7.1
**Relative Intensities of the Various JJ
Transitions for $3d^9 \rightarrow 2p^53d^{10}$**

	$J' = 3/2$	$J' = 1/2$	Σ
$J = 5/2$	9	0	9
$J = 3/2$	1	5	6
Σ	10	5	15

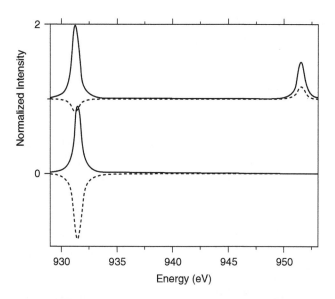

FIGURE 7.3 The XMCD spectra of a Cu^{2+} ion: (bottom) atomic multiplet calculation for $3d^9 \rightarrow 2p^5 3d^{10}$; (top) atomic multiplet calculation without the 3d spin–orbit coupling. The XAS and XMCD are shown with the solid and dashed curves, respectively.

transitions from the $J = 3/2$ state are added to those of the $J = 5/2$ state. These states will have transitions to both the L_3 and L_2 edges. This approximation identifies with the atomic, single electron, model as first used by Erskine and Stern. It describes the transitions of the 2p core state to the 3d valence state, where the 2p spin–orbit coupling is included, but the 3d spin–orbit coupling is not.

Table 7.2 gives the ratio of the XAS of the L_3 edge to the L_2 edge 2 : 1, while their XMCD ratio is $-1 : +1$. This single particle ratio is also indicated in Figure 7.4. It is found for all cases where the 3d spin–orbit coupling is zero. It is often assumed that this is the case for the 3d transition metals, which are analyzed using this model.

TABLE 7.2
XAS of the L_3 Edge to the L_2 Edge

	+Helicity	−Helicity	XAS	XMCD
L3	3/12	5/12	8/12	−2/12
L2	3/12	1/12	4/12	+2/12
L3	0.375	0.625	1.0	−0.25
L2	0.750	0.250	1.0	+0.50

Note: The top two rows give the relative intensities for the 2p3d transition assuming that the 3d spin–orbit coupling is zero. The overall intensity is normalized to 1.0. In the bottom two rows, the respective intensities of the L_3 and L_2 edges are normalized to 1.0, as also used in Figure 7.4.

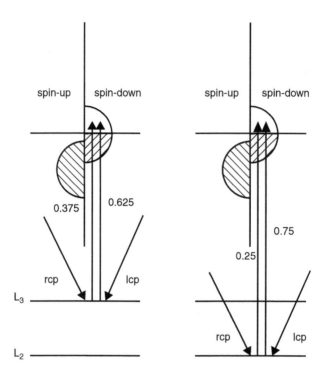

FIGURE 7.4 A sketch of the atomic, single electron, model of L-edge XMCD. The L_3 edge has a 3 : 5 ratio and the L_2 edge a 3 : 1 ratio, as given in Table 7.2. (From Funk, T., Deb, A., George, S.J., Wang, H.X., and Cramer, S.P., *Coord. Chem. Rev.*, 249, 3, 2005. With permission from Elsevier Ltd.)

Effectively, this also implies the neglect of the multiplet effects. Due to the crystal field (and other effects, for example, translation symmetry and charge-transfer effects), the $J = 5/2$ and $J = 3/2$ states in the 3d-band are mixed. For an infinitely large crystal field splitting, the 3d spin–orbit coupling is effectively quenched again. However, this quenching is different from a zero 3d spin–orbit coupling.

Figure 7.5 shows the effect of a cubic crystal field of 1.0 eV. A small L_2 edge is visible, indicating a little admixture of the $J = 3/2$ state. The effect of the crystal field is different from the calculation without 3d spin–orbit coupling. A cubic crystal field modifies the ratio of the L_3 and L_2 edges in the XAS and XMCD spectra. For a large value of the cubic crystal field, the XAS intensity ratio goes to 10 : 1 and the XMCD ratio goes to −1 : +1. The XAS ratio is different from the values when the 3d spin–orbit coupling is set to zero. The crystal field is mixing the $J = 5/2$ and $J = 3/2$ wave functions in the ground state, but it does not mix all 3d orbitals equally. These effects have been studied in detail by Arrio et al. (1995). The results on the Cu^{2+} ion indicate that the interplay between the 2p spin–orbit coupling, 3d spin–orbit coupling, crystal field effects, and the exchange splitting determine the ground state and the resulting XAS and XMCD spectra. The 3d3d interactions, 2p3d multiplet effects, and charge-transfer effects have not yet been included. They further affect the spectral shape of the XMCD spectra.

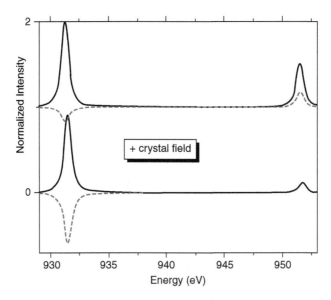

FIGURE 7.5 The XMCD spectra of a Cu^{2+} ion. (Top) Atomic multiplet calculation without the 3d spin–orbit coupling. The XAS and XMCD are shown with the solid and dashed curves, respectively. (Bottom) LFM calculation for $10\,Dq = 1.0$ eV.

7.2.2 XMCD Effects in Ni^{2+}

Next, we switch to a Ni^{2+} ground state. The difference between Ni^{2+} and Cu^{2+} is that the $3d^8$ ground state is affected by 3d3d correlation effects. In addition, the $2p^53d^9$ final state is affected by 2p3d multiplet effects. This will completely modify the 2p XAS and XMCD spectra of Ni^{2+} compared with Cu^{2+}.

Figure 7.6 shows the 2p XAS spectrum of a Ni^{2+} ion within an atomic multiplet calculation of the $3d^8 \rightarrow 2p^53d^9$ transition, including the atomic 3d spin–orbit coupling. The XMCD effect for such an atomic Ni^{2+} ion is completely negative (in Figure 7.6, the sign of XMCD is inverted to compare the amplitudes of XAS and XMCD). The reason is that there are no allowed transitions for + helicity x-rays. This can be understood as follows: The first peak relates to a transition from the $^3F_4$ ground state to a $J = 4$ final state. The transition from $J = 4$ to $J' = 4$ can have either $\Delta M_J = -1$ or $\Delta M_J = 0$ transitions, but no $\Delta M_J = +1$. The other three peaks relate to a $J = 3$ final state. The $\Delta J = -1$ transitions are purely $-$ helicity ($\Delta M_J = -1$) transitions. This implies that the whole MCD spectrum is negative.

There exists a general relationship between ΔJ and ΔM_J transitions. This is indicated in Table 7.3.

Applying Table 7.3 to the $J = 4$ ground state of a Ni^{2+} ion yields the numbers as given in Table 7.4. One can observe in Table 7.4 that there is a correlation between ΔJ and ΔM_J transitions and that most of the intensity goes to the diagonal. In the case of a Ni^{2+} ion, a $J' = 5$ final state does not exist. There is also one $J' = 4$ final state, which has mainly a $\Delta M_J = 0$ transition and there are three $J' = 3$ final states that are pure $\Delta M_J = -1$ transitions. In total, this yields the relatively simple XMCD spectrum as given in Figure 7.6.

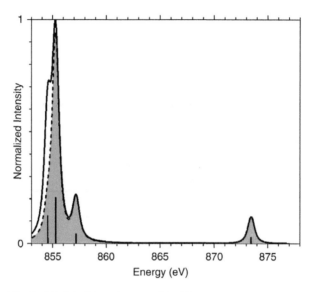

FIGURE 7.6 The 2p XAS (solid) and inverted XMCD (gray dashed) spectrum of an atomic multiplet calculation of Ni^{2+}. The bars indicate the ($\Delta m_J = -1$) transitions. For clarity the XMCD signal is shown inverted (i.e. multiplied with -1).

TABLE 7.3
Relationship between ΔJ and ΔM_J for Dipole Transitions $J_0 = J + 1$ and $J_1 = (2J + 3)(J + 1)$

	$\Delta M_J = -1$	$\Delta M_J = 0$	$\Delta M_J = +1$
$\Delta J = -1$	1	0	0
$\Delta J = 0$	$1/J_0$	J/J_0	0
$\Delta J = +1$	$1/J_1$	$(2J + 1)/J_1$	$(2J + 1)(J + 1)/J_1$

TABLE 7.4
Relationship between ΔJ and ΔM_J for a $J = 4$ Ground State

$J = 4$	$\Delta M_J = -1$	$\Delta M_J = 0$	$\Delta M_J = +1$
$\Delta J = -1$	1	0	0
$\Delta J = 0$	0.20	0.80	0
$\Delta J = +1$	0.02	0.16	0.82

The 3d spin–orbit coupling is considered to be quenched in many 3d systems. Switching off the 3d spin–orbit coupling in the atomic multiplet XMCD spectrum of Ni^{2+} implies that the $^3F_4$, $^3F_3$, and $^3F_2$ states are degenerate. This yields a number of additional transitions. There are now both $\Delta M_J = -1$ and $\Delta M_J = +1$ transitions and, in fact, the L_2 edge has a completely positive XMCD spectrum as shown in Figure 7.7. This is in contrast to the negative XMCD spectrum if the 3d spin–orbit coupling was included as indicated previously in Figure 7.6.

Applying a cubic crystal field, modifies the ground state to $^3A_2$. Figure 7.8 shows the similar 2p XAS and XMCD spectra of Ni^{2+} in an octahedral crystal field with the inclusion of 3d spin–orbit coupling and the lower panel of Figure 7.8 exhibits the two different spectra (with and without the inclusion of 3d spin–orbit coupling). It is interesting to note that though the ground state has $^3A_2$ character, there is still the visible effect of the 3d spin–orbit coupling. The main peak is a bit decreased if the 3d spin–orbit coupling is switched off and this intensity loss is compensated for at the other peaks. It is interesting that all peaks in the XMCD spectrum become more positive without 3d spin–orbit coupling. This is related to the fact that the 3d spin–orbit coupling favors high J states, which have relatively more $\Delta m_J = -1$ transitions. This effect is partly maintained after the inclusion of the cubic crystal field. Note that it is a little surprising that there is an effect of the 3d spin–orbit coupling at all, given that the ground state is $^3A_2$ (a state that is not split by the 3d spin–orbit coupling). There is still an effect because the ground state is not 100% $^3A_2$ symmetry if 3d spin–orbit coupling is included. It is 100% for a T_2 symmetry state in the double group description, but the $^3A_2$-type T_2 state suffers from the admixture of T_2 states with other T_2 states.

The next step is to include charge transfer for XMCD calculations. The procedure is analogous to the XAS calculations. Figure 7.9 gives the XAS and XMCD spectra

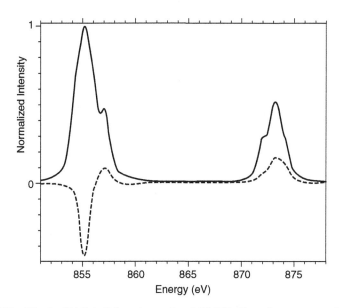

FIGURE 7.7 The 2p XAS (solid) and associated XMCD (dotted) spectrum of an atomic multiplet calculation of Ni^{2+} where the 3d spin–orbit coupling has been set to zero.

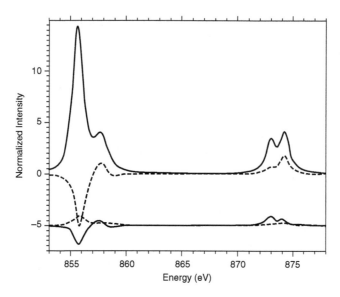

FIGURE 7.8 The 2p XAS (solid) and associated XMCD (solid) spectrum of a crystal field multiplet calculation of Ni^{2+} with 3d spin–orbit coupling on. Offset at -0.05 are the two different spectra if the 3d spin–orbit coupling is switched off, for respectively the 2p XAS (solid) and its XMCD (dotted).

for a Ni^{2+} ion (using the parameters of NiO) compared with an experimental spectrum of a magnetic Ni^{2+} ion as existing in $Cs[NiCr(CN)_6]2H_2O$ (Arrio et al., 1996a). It can be observed that the main features of the experimental spectrum are reproduced, though not exactly at the correct energies. For example, the charge-transfer

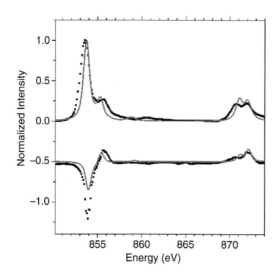

FIGURE 7.9 The $Ni^{2+}(NC)_6$ $L_{2,3}$ XAS (top) and XMCD (bottom) spectra compared with charge transfer multiplet (CTM) calculations with the parameters as given in the text. The experimental data has been digitized. (From Arrio, M.A., et al., *J. Phys. Chem.*, 100, 4679, 1996a. With permission.)

satellite at 861 eV in experiment is calculated at 859 eV. This can be easily corrected by modifying the charge transfer parameters.

7.2.3   XMCD OF CrO₂

Figure 7.10 shows the XMCD spectrum of CrO_2. A comparison with the crystal field multiplet calculation of the dichroic effects suggests that the 3d spin–orbit splitting is effectively zero. The vanishing influence of an orbital momentum can also be verified by the application of the sum rules, which gives an upper limit of L_z of less than 10^{-2} μ_B (Schütz et al., 1994).

7.2.4   MAGNETIC X-RAY LINEAR DICHROISM

Linear dichroism, or polarization dependence, is the difference in XAS spectra μ_0 and $(\mu_+ + \mu_-)/2$. The condition for the occurrence of linear dichroism is a macroscopic asymmetry in the electronic and/or magnetic structure. The symmetry criterion that determines possible polarization dependence is given by the space group of the crystal and not by the point group of the atom. Consider, for example, a crystal that has a cubic space group and the absorbing atom a tetragonal point group. In this case, the shape of the metal 2p spectrum is determined by the point group; however, no linear dichroism is found because the potential dichroic effect of the two atoms oriented horizontally and vertically, cancels. Linear dichroism can be caused by both electronic and magnetic effects, in contrast to circular dichroism that can only be caused by magnetic effects. This is a consequence of Kramers theorem, which states that the lowest state in a static electric field is always at least twofold degenerate. The only way to break the degeneracy of the Kramers doublet is by means of a time asymmetric field; in other words, a magnetic field.

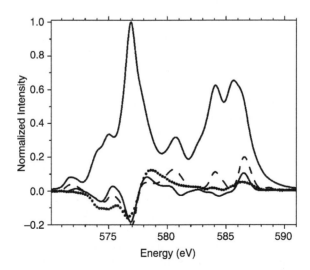

FIGURE 7.10   The experimental XMCD spectrum of CrO_2 is given with dots. It is compared to a crystal field multiplet calculation with (solid line) and without (dashed) 3d spin–orbit coupling. The theoretical XAS spectrum is also given.

Magnetic effects generate circular dichroism, but they also have a large effect on linear dichroism (van der Laan et al., 1986; Goedkoop et al., 1988a). The magnetic field determines an axial direction in the crystal and, with respect to this axis, linear dichroism occurs. Within atomic multiplet theory, this can be calculated by evaluation of the properties of the 3j-symbol as given in Table 7.3. Atomic multiplet theory can be used directly for RE systems and, because the J values for RE elements are large (between 5/2 and 8), there is a strong correlation between $\Delta J = \pm 1$ and $\Delta m_J = \pm 1$. This will be discussed subsequently for the case of the $M_{4,5}$ edges of RE systems.

7.2.5 ORIENTATION DEPENDENCE OF XMCD AND XMLD EFFECTS

Crystal field affects can cause linear dichroic effects due to the orientation difference of the 3d electrons. The linear dichroism, or angular dependence, of XAS is caused by the variations of the spatial distribution of the 3d electrons. Spherical symmetry as well as octahedral (O_h) symmetry yields no linear dichroic effect, but lower symmetries such as tetragonal (D_{4h}) or trigonal (D_{3d}), yield an angular dependence in the XAS spectral shape as discussed in Chapters 4 and 6. In this chapter, we have shown that magnetic fields cause XMCD and XMLD effects, where we have implicitly assumed that the magnetic field acts along the z-axis and for a system in octahedral symmetry. In case of octahedral symmetry, any XMLD effect is caused by the magnetic field. The general picture of an x-ray absorption experiment with a crystal field plus a magnetic field acting on a metal ion contains three different axes:

1. X-ray polarization vector (e_μ).
2. Magnetic field vector (**B**).
3. Crystal field symmetry and its related axes.

Van Elp and Searle (1997) systematically analyzed the effects on the XMCD spectrum when the magnetic field was aligned according to the [100], [111], and [110] axes in cubic and in distorted symmetries. They divide the 3d transition metal ions into three classes. The first is A-symmetry ground states in O_h symmetry that have essentially no orientation effect. The second is symmetric ground states in lower symmetries that have small orientation effects and, lastly, T-symmetry ground states in O_h symmetry (or lower symmetry) that have large orientation effects in their XMCD spectra. The main mechanism behind the orientation dependence of the XMCD is the 3d spin–orbit coupling that, in combination with symmetry distortions, will split a T-symmetry ground state.

Arenholz et al. (2006, 2007) analyzed the orientation effects on the XMLD spectra. They showed for the octahedral $^3A_2$ ground state system NiO that there is still a very large dependence on the relative orientations of the magnetic field and the x-ray polarization vector with respect to the crystal axis. In O_h symmetry, the x-, y-, and z-axis are equivalent. This leaves three axis, **B**, e_μ, and **z**, which can be varied with respect to the others. Arenholz et al. (2006, 2007) show that, in case of O_h

symmetry, there are two fundamental XMLD spectra, from which all angular dependent spectra can be derived:

1. The I_0 XMCD spectrum derived from the difference spectrum from the case where $\mathbf{B} \parallel \mathbf{z}$ and $\mathbf{e}_\mu \parallel \mathbf{z}$, and the case where the magnetic field is rotated over 90° (e.g. $\mathbf{B} \parallel \mathbf{x}$ and $\mathbf{e}_\mu \parallel \mathbf{z}$).
2. The I_{45} XMCD spectrum derived from the difference spectrum from the case where $\mathbf{B} \parallel [110]$ and $\mathbf{e}_\mu \parallel [110]$, and the case where the magnetic field is rotated over 90° (e.g. $\mathbf{B} \parallel [-110]$ and $\mathbf{e}_\mu \parallel [110]$).

All other spectra can be generated from I_0 and I_{45}. In the case of crystal field symmetries lower than O_h, more complex angular dependences can be found for the XAS spectra and the resulting XMLD spectra. Arenholz et al. (2006, 2007) note that these anisotropic XMLD effects are a property of the cubic wave functions of the 3d states with respect to the spin quantization axis and not of the anisotropic spin–orbit interaction. In contrast, the orientation dependence of the XMCD effect discussed above is mainly an effect of the anisotropic spin–orbit interaction (van Elp and Searle, 1997).

7.2.6 XMLD FOR DOPED LaMnO₃ SYSTEMS

$La_{7/8}Sr_{1/8}MnO_3$ is a system that contains orbital ordering in connection with its large magnetoresistance. X-ray MLD can be used to study the ordering of the orbitals, similar to the nonmagnetic situations for VO_2 and $LaTiO_3$ that have been described in Chapter 6. Figure 7.11 shows the XMLD spectra of $La_{7/8}Sr_{1/8}MnO_3$ at different temperatures. The two bottom spectra are CTM calculations, invoking two different orderings of the orbitals (Huang et al., 2004; Huang and Jo, 2004). The experiment clearly shows that the ground state in $La_{7/8}Sr_{1/8}MnO_3$ contains an ordering of $z^2 - y^2/ x^2 - z^2$ orbitals.

Magnetic ordering effects (including orbital ordering) have been studied intensively with soft x-ray resonant scattering (XRS) (van der Laan, 2006). XRS makes use of the charge transfer and multiplet effects of the 3d metal L edges. Using the XMCD effect of the L edges, the scattering is measured for transition metal oxides, magnetic interfaces, and devices. In the case of the $La_{7/8}Sr_{1/8}MnO_3$ system, Mirone et al. (2006) find from their XRS measurements and CTM analysis that the occupied e_g orbital at the Mn^{3+} site has $3z^2 - r^2/3x^2 - r^2$ ordering. This result is opposite to the XMLD result of Huang et al. (2004). Because the focus of this book is spectroscopy, XRS is discussed no further. Instead, we refer you to the recent review paper of van der Laan (2006) and references therein.

7.3 SUM RULES

7.3.1 SUM RULES FOR ORBITAL AND SPIN MOMENTS

In 1992, Thole et al. described a number of sum rules for x-ray absorption and later for x-ray photoemission. It was shown that the integral over the XMCD signal of a given edge allows the determination of the ground state expectation values of the orbital moment L_z and the effective spin moment $(S_{eff})_z$. The sum rules apply to

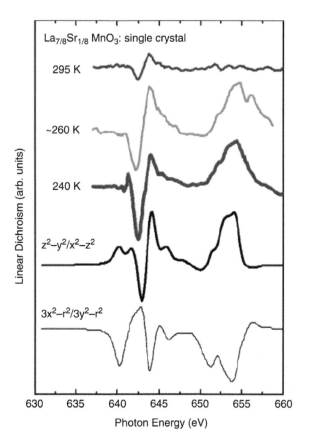

FIGURE 7.11　The XMLD spectra of the Mn L edge of $La_{7/8}Sr_{1/8}MnO_3$ at different temperatures compared with CTM calculations for Mn^{3+}. (Reprinted with permission from Kuepper, K., et al., *J. Phys. Chem. B*, 109, 15667, 2005. Copyright 2005 by the American Chemical Society.)

one specific transition, that is, the transition from a 2p core state to 3d valence states in TM systems. These 3d valence states must be separated from other final states (e.g. the 2p4s transition). It also applies that the background must be separated out. In the present subsection, we confine ourselves to the sum rules for the 2p3d (or 3p3d) transition. The general derivation of the orbital moment sum rule for 2p3d (or 3p3d) of TM systems and 3d4f (or 4d4f) of RE systems is given in Appendix D.

As also shown in Appendix D, the integrated 2p3d x-ray absorption spectrum (μ) is proportional to the number of empty 3d states $(\langle N_h \rangle)$:

$$\int \mu = \int (\mu_1 + \mu_0 + \mu_{-1}) = \frac{C}{5} \langle N_h \rangle. \qquad (7.7)$$

Here $\int \mu$ is the abbreviation of $\int_{L_3+L_2} \mu(\Omega) \, d(\hbar\Omega)$ and C is a constant factor including the radial matrix element of the dipole transition. This sum rule neglects 2p4s and other transitions. Experimentally, this implies that the 2p3d transitions must be separated out of the other transitions. In general, it is assumed that all other transitions are continuum transitions that can be described as an edge step followed by a constant cross section.

The integrated circular dichroism spectrum is defined as the absorption of + helicity x-rays (μ_+) minus the absorption of − helicity x-rays (μ_-). In the case of a 2p3d transition, this yields:

$$\int (\mu_+ - \mu_-) = -\frac{C}{10}\langle L_z \rangle. \tag{7.8}$$

This XMCD sum rule implies that one can directly determine the orbital moment from the difference of + and − helicity x-rays. Because in most soft x-ray experiments, yield detection is used, the exact absorption cross sections are not measured. A solution is to normalize the XMCD signal by the absorption edge. This defines the orbital moment sum rule as:

$$\langle L_z \rangle = -\frac{\int (\mu_+ - \mu_-)}{\int \mu} \cdot 2\langle N_h \rangle. \tag{7.9}$$

Similarly the linear dichroism spectrum can be integrated. This yields the quadrupole moment (Q_{zz}) sum rule:

$$\langle Q_{zz} \rangle = -\frac{\int (\mu_\perp - \mu_\parallel)}{\int \mu} \cdot 3\langle N_h \rangle = -\frac{\int (-2\mu_0 + \mu_+ + \mu_-)}{\int \mu} \cdot 3\langle N_h \rangle. \tag{7.10}$$

In the case that there is an asymmetry axis in the electronic states, the linear dichroism follows a $\cos^2 \alpha$ dependence with the angle α between this axis and the polarization vector of the x-rays. If α is zero, the polarization vector lies along the asymmetry axis, that is, perpendicular to the surface and one measures $\mu_\parallel$. If α is 90°, the polarization vector lies parallel to the surface and one measures $\mu_\perp$. The angular dependence can also be written in a constant term $(3\mu_\parallel - \mu_\perp)/2$ and an angular term $3/2 \cdot (\mu_\perp - \mu_\parallel) \cdot \cos^2 \alpha$.

It is important to be able to determine the spin moment, and this is indeed possible with an additional sum rule. However, this (effective) spin sum rule has some additional complications as is discussed below:

$$\langle (S_{eff})_z \rangle = \frac{\int_{L_3} (\mu_+ - \mu_-) - 2\int_{L_2} (\mu_+ - \mu_-)}{\int \mu} \cdot \frac{3}{2} \cdot \langle N_h \rangle. \tag{7.11}$$

The effective spin moment S_{eff} is given as:

$$\langle (S_{eff})_z \rangle = \langle S_z \rangle + \frac{7}{2}\langle T_z \rangle. \tag{7.12}$$

T_z is the spin-quadrupole coupling (also called the magnetic-dipole coupling). If this sum rule is used to determine the spin moment $\langle S_z \rangle$, it has to be assumed that

$\langle T_z \rangle$ is zero or $\langle T_z \rangle$ must be known from other experiments or theoretically approximated. In addition, the effective spin sum rule makes an additional approximation that the L_3 and L_2 edges are not mixed or well separated. The edges must be well separated because otherwise there is no clear method to divide the spectrum into L_3 and L_2. In addition, the two edges must be pure $2p_{3/2}$ and $2p_{1/2}$. Subsequently, we will show that this is not the case for the 3d metal L edges. A number of additional sum rules apply to photoemission. These are discussed in Chapter 6.

7.3.2 APPLICATION OF THE SUM RULES TO Fe AND Co METALS

Chen et al. (1995) measured high precision $L_{2,3}$ XAS and XMCD spectra of Fe and Co metals in transmission with *in situ* grown thin films, and verified the orbital and spin sum rules by analyzing them using these sum rules. Before their study, numerous experimental studies aimed at investigating the validity of the sum rules, had been made for transition metals but with widely different conclusions. The lack of a consensus was mainly attributed to the experimental artifacts inherent in the indirect methods of measuring XAS and XMCD, such as the total electron and fluorescence yield methods, which are known to suffer from saturation and self-absorption effects as well as the dependence of radiative and nonradiative core-hole decay probability on the symmetry and spin polarization of the XAS final states. Chen et al. made the direct observation of XAS and XMCD by a transmission method for Fe and Co thin film samples grown onto 1 μm-thick parylene substrates.

The results of $L_{2,3}$ XAS and XMCD spectra of Fe/parylene thin films are shown in Figure 7.12. Chen et al. measured the incident-photon-flux-normalized transmission of XAS spectra for + and − helicities of the incident photons as well as for the magnetization independent spectra of the parylene substrate. From these spectra, the $\mu_+ - \mu_-$ (XMCD spectra) and $\mu_+ + \mu_-$ (XAS spectra) are determined. From Equations 7.9 and 7.11, the sum rules for orbital and spin magnetic moments, respectively, are written as

$$m_{orb} = -\frac{4\int_{L_3+L_2}(\mu_+ - \mu_-)}{3\int_{L_3+L_2}(\mu_+ + \mu_-)}n_h\mu_B, \tag{7.13}$$

$$m_{spin} = -\frac{6\int_{L_3}(\mu_+ - \mu_-) - 4\int_{L_3+L_2}(\mu_+ - \mu_-)}{\int_{L_3+L_2}(\mu_+ + \mu_-)}n_h\mu_B\left(1 + \frac{7\langle T_z \rangle}{2\langle S_z \rangle}\right)^{-1}, \tag{7.14}$$

where $m_{orb} \equiv -\langle L_z \rangle\mu_B/\hbar$ and $m_{spin} \equiv -2\langle S_z \rangle\mu_B/\hbar$ represent, respectively, the orbital and spin magnetic moments, and n_h is the 3d hole number $\langle N_{3d} \rangle$. The linear polarized spectrum, μ_0, has been replaced by $(\mu_+ + \mu_-)/2$.

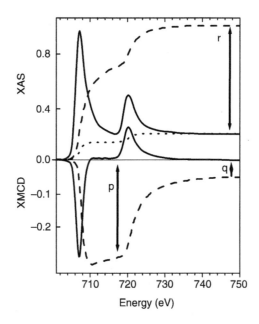

FIGURE 7.12 $L_{2,3}$ XAS (upper panel) and XMCD (lower panel) spectra of Fe metal (Fe/parylene thin films measured in transmission). The solid lines give the XAS spectrum and its XMCD. The dotted line indicates the background and the dashed lines show the integrated value of the XAS and the XMCD, respectively, where r, q indicate their values after the $L_{2,3}$ edge and p after the L_3 edge. (Reprinted with permission from Chen, C.T., et al., *Phys. Rev. Lett.*, 75, 152, 1995. Copyright 1995 by the American Physical Society.)

In order to verify the orbital and spin sum rules, we need some additional information on the value of n_h and the edge jumps of $L_{2,3}$ XAS. However, by dividing Equation 7.13 by Equation 7.14, we can obtain m_{orb}/m_{spin} only from the XMCD spectra with no additional information. Figure 7.12 shows, by the dashed curve, the energy-integration of the XMCD spectrum and the values of the integrations for the L_3 region, and the whole L edge range is indicated by p and q, respectively. Then, it is easily seen that

$$\frac{m_{orb}}{m_{spin}} = \frac{2q}{9p - 6q},$$ (7.15)

where the $\langle T_z \rangle / \langle S_z \rangle$ term in the spin sum rule is neglected because the first-principle band structure calculations (Wu et al., 1993, 1994) gave a value of -0.38% for bcc Fe (-0.26% for hcp Co). The ratio m_{orb}/m_{spin} is thus determined from XMCD data to be 0.043 for Fe in excellent agreement with that of the Einstein–de Haas gyromagnetic measurements at 0.044 (Bonnenberg et al., 1986). For Co, the ratio from XMCD is 0.095, which is in excellent agreement with that from the gyromagnetic measurements at 0.097, again.

In order to verify the applicability of the individual orbital and spin sum rules, Chen et al. (1995) adopted the edge jumps of $L_{2,3}$ XAS as shown with the dotted curve in Figure 7.12, where the height of the L_3 (L_2) step was set at 2/3 (1/3) of the average intensity of the last 15 eV of the XAS spectrum. The value r indicated in Figure 7.12 is the XAS integral after the edge-jump removal. Then, m_{orb} and m_{spin} are expressed as

$$m_{orb} = -\frac{4q}{3r} n_h \mu_B,$$ (7.16)

$$m_{spin} = -\frac{6p - 4q}{r} n_h \mu_B.$$ (7.17)

For n_h, the values of 6.61 for Fe and 7.51 for Co are used, based on the theoretical calculations of Wu et al. (1993, 1994). The values of m_{orb} and m_{spin} are thus obtained from the XMCD analysis for Fe at 0.085 μ_B and 1.98 μ_B, which are in good agreement with 0.092 μ_B and 2.08 μ_B obtained by the gyromagnetic ratio measurements of Bonnenberg et al. (1986). For Co, the values of m_{orb} and m_{spin} from XMCD are 0.154 μ_B and 1.62 μ_B, while those from the gyromagnetic ratio measurements of Bonnenberg et al. (1986) are 0.147 μ_B and 1.52 μ_B, respectively. The excellent agreement of the XMCD derived orbital to spin moment ratios and the good agreement of the XMCD derived individual moments, with those obtained from Einstein–de Haas gyromagnetic measurements, demonstrate the applicability of the orbital and spin sum rules.

7.3.3 APPLICATION OF THE SUM RULES TO Au/Co-NANOCLUSTER/Au SYSTEMS

Koide et al. (2001, 2004) studied the magnetic properties of Au/two-monolayer Co clusters/Au(111) systems. The most remarkable feature of Co clusters that are self-assembled on Au(111) is their constant height of two monolayers (ML), and independent of the nominal Co coverage $d*$ as long as $d* \leq 1.6$ ML. Taking advantage of this feature, Koide et al. succeeded in preparing Au/2ML-Co/Au samples with various $d*$, where almost all Co atoms were interfacial ones as shown in Figure 7.13. Then they measured angle-, field-, and temperature-dependent XMCD spectra at Co $L_{2,3}$ edges with the experimental arrangement shown in Figure 7.13. The circularly polarized x-ray beam is incident onto the sample with angle θ with respect to the surface-normal direction (**n**) of a quasi-two-dimensional sample, and the magnetic fields (**B**) are applied to the sample parallel and antiparallel to the θ direction.

Some typical examples of the observed Co $L_{2,3}$ XAS and XMCD spectra are shown in Figure 7.14 where the B dependence of XAS (μ_+) and XMCD ($\Delta\mu = \mu_+ - \mu_-$) is shown in Figure 7.14a for the sample with $d* \cong 1.6$ ML at 300 K and $\theta = 0°$. The temperature dependence of $\Delta\mu$ is shown in Figure 7.14b for the sample with $d* = 0.85$ ML for **B** = 0 and $\theta = 0°$. The average in-plane diameter D_{av} of clusters is proportional to $\sqrt{d*}$ due to the constant height and fixed nucleation sites [so-called "herringbone elbow" sites on the Au(111) surface], and $d* \cong 1.6$ and 0.85 ML correspond to $D_{av} \cong 8.2$ and 4.7 nm, respectively. It is seen from

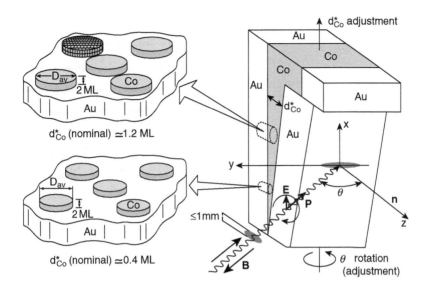

FIGURE 7.13 Nominal sample structure and arrangement for angle- and cluster-size-dependent XMCD experiments. The wedge-shaped Co with the nominal coverage d^* microscopically consists of independent clusters for the range of $d^* < 1.6$ ML. **E** stands for the electric field vector of the incident photon. (Reprinted with permission from Koide, T., et al., *J. Electron Spectrosc. Relat. Phenom.* 136, 107, 2004. Copyright 2004 by the American Physical Society.)

Figure 7.14 that the Co cluster with $D_{av} \cong 8.2$ nm exhibits a fairly strong remanent XMCD at 300 K, but that with $D_{av} \cong 4.7$ nm exhibits no remanent XMCD at 300 K. From systematic investigations of the B-dependence of XMCD intensity for Co clusters with various D_{av}, it is concluded that Co clusters with 7.2 nm $\leq D_{av} \leq 8.2$ nm retain the ferromagnetic (FM) alignment at 300 K with a single domain due to the cluster–cluster interaction, while those with $D_{av} \leq 6.7$ nm exhibit the superparamagnetism (SPM).

Koide et al. (2001, 2004) succeeded in directly determining the spin magnetic moment m_{spin}, in-plane ($m_{orb}^{\parallel}$) and out-of-plane ($m_{orb}^{\perp}$) orbital magnetic moments, and in-plane ($m_T^{\parallel}$) and out-of-plane ($m_T^{\perp}$) magnetic dipole moments of interfacial Co atoms in Au/2 ML-Co/Au(111) by applying the orbital and spin sum rules to their XMCD data. The angle-dependent XMCD sum rules for anisotropic systems (Stöhr et al., 1995; Weller et al., 1995) are given by

$$m_{orb}^{\theta} = -\frac{4\left[\Delta I_{L_3} + \Delta I_{L_2}\right]^{\theta} n_h \mu_B}{3\left[I_{L_3} + I_{L_2}\right]}, \tag{7.18}$$

$$m_{spin} + 7m_T^{\theta} = -\frac{2\left[\Delta I_{L_3} - 2\Delta I_{L_2}\right]^{\theta} n_h \mu_B}{\left[I_{L_3} + I_{L_2}\right]}, \tag{7.19}$$

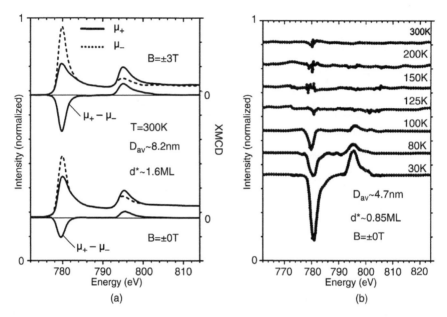

FIGURE 7.14 (a) Polarization-dependent, normal-incidence Co $L_{2,3}$-edge XAS and XMCD spectra for $D_{av} = 8.2$ nm cluster at 300 K and under $B = \pm 3$ T and ± 0 T. (b) Temperature-dependence of the Co $L_{2,3}$-edge remanent XMCD for $D_{av} = 8.2$ nm cluster. (Reprinted with permission from Koide, T., et al., *Phys. Rev. Lett.*, 87, 257201, 2001. Copyright 2001 by the American Physical Society.)

where I_{L_3} (I_{L_2}) and ΔI_{L_3} (ΔI_{L_2}), respectively, are the energy-integrated XAS and XMCD intensities at the L_3 (L_2) edge, n_h is the 3d hole number, $m_{orb}^\theta \equiv -\langle L_\theta \rangle \mu_B/\hbar$, $m_{spin} \equiv -2\langle S_z \rangle \mu_B/\hbar$ and $m_T^\theta \equiv -\langle T_\theta \rangle \mu_B/\hbar$. It was shown in the preceding subsection that the contribution of the magnetic dipole moment could be neglected in the spin sum rule for Fe and Co metals. However, but for the interfacial Co clusters, this contribution should be taken into account because the magnetic dipole moment would be enhanced by the anisotropy in the Co 3d states. The orbital magnetic moment m_{orb}^θ and magnetic dipole moment m_T^θ in the θ direction are related to those in-plane and out-of-plane components as

$$m_{orb}^\theta = m_{orb}^\perp \cos^2 \theta + m_{orb}^\| \sin^2 \theta, \tag{7.20}$$

$$m_T^\theta = m_T^\perp \cos^2 \theta + m_T^\| \sin^2 \theta, \tag{7.21}$$

Since the angular average of m_T^θ vanishes for 3d electrons (Stöhr et al., 1995), we have

$$2m_T^\perp + m_T^\| = 0, \tag{7.22}$$

and from Equations 7.21 and 7.22, we obtain

$$m_T^\theta = m_T^\|(1 - 3 \cos^2 \theta). \tag{7.23}$$

Therefore, for the magic angle $\theta = \theta_{\mathrm{mag}} = 54.7°$, m_T^θ vanishes, and Equation 7.19 reduces to

$$m_{\mathrm{spin}} = -\frac{2(\Delta I_{L_3} - 2\Delta I_{L_2})^{\theta_{\mathrm{mag}}} n_h \mu_B}{I_{L_3} + I_{L_2}},$$ (7.24)

which allows a direct determination of m_{spin}. The quantities m_{spin}, $m_{\mathrm{orb}}^\perp$, $m_{\mathrm{orb}}^\parallel$, $m_T^\perp$, $m_T^\parallel$ and $m_{\mathrm{tot}} = m_{\mathrm{spin}} + (2m_{\mathrm{orb}}^\perp + m_{\mathrm{orb}}^\parallel)/3$ were determined by applying the sum rules to the data for $\theta = 0°$ and $54.7°$, and the results are shown in Figure 7.15 as a function of D_{av} (or the number of atoms per cluster N). In this analysis, the data at $T = 300$ K and $B = \pm 3T$ were used for the ferromagnetic clusters with $D_{av} = 7.2$ and 8.2 nm, and those at $T = 30$ K and $B = \pm 5T$ were used for the superparamagnetic clusters with $D_{av} \leq 6.7$ nm. For the estimation of n_h, two independent methods [one for calculating

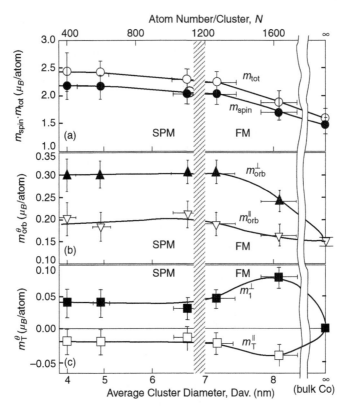

FIGURE 7.15 D_{av} and N dependence of the magnetic moments determined from the angle-resolved XMCD measurements and the angle-dependent sum rules: (a) m_{spin} and m_{tot}; (b) $m_{\mathrm{orb}}^\perp$ and $m_{\mathrm{orb}}^\parallel$; (c) $m_T^\perp$ and $m_T^\parallel$. The corresponding moments of the bulk hcp Co are shown for comparison. The hatched area denotes the FM/SPM transition region. (From Koide, T., et al., *J. Electron Spectrosc. Relat. Phenom.*, 136, 107, 2004. With permission from Elsevier Ltd.)

the values for Co_2Pt_4 multilayers and one based on determining the constant defined by $(I_{L_3}+I_{L_2})/n_h]$, are used with no appreciable resultant difference.

It is seen that m_{spin} reaches 2.0–2.15 μ_B for Co clusters with $D_{av} \leq 7.2$ nm, which is much larger than the bulk value of 1.55 μ_B. This is the first direct verification of an enhanced interfacial m_{spin} predicted theoretically (Blügel, 1992; Guevara et al., 1998). The orbital moment is anisotropic and $m_{orb}^{\perp} \sim 0.31$ μ_B and $m_{orb}^{\parallel} = 0.16$ μ_B to 0.21 μ_B for $D_{av} \leq 7.2$ nm where the value of $m_{orb}^{\perp}$ is almost twice as large as the bulk value $m_{orb} = 0.15$ μ_B. The smaller values of m_{spin} and $m_{orb}^{\parallel}$ for $D_{av} = 8.2$ nm could be due to their imperfect 2-ML height. The total magnetic moment $m_{tot} = 2.2$–2.4 μ_B for $D_{av} \leq 7.2$ nm is close to $m_{tot} = 2.4$–2.5 μ_B of free Co clusters with the smallest number of atoms, obtained by Stern–Gerlach deflection experiments (Billas et al., 1994). This indicates that the present interfacial Co 3d–Au 5d hybridization would be small, which is plausible because the 5d band of bulk Au is located well below the Fermi level.

7.3.4 LIMITATIONS OF THE SUM RULES

In Section 7.3.1, we have pointed out some approximations and experimental complications and uncertainties when using sum rules. These limitations are listed below, followed by some theoretical examples of the breakdown of the spin sum rule:

1. The polarization of the x-ray beam must be known and must be constant over time and position, including an exact inversion if the beam polarization is reversed.
2. The magnetization of the sample must be constant over time and position, including, if used, the inverse magnetization.
3. The intensity of the L_3 and L_2 edges must be normalized to the same x-ray beam intensity.
4. If electron yield is used, the detection effectiveness must be equal for spin-up and spin-down electrons. This also implies that the escape chance for spin-up and spin-down electrons must be equal and, in turn, that the electron scattering should be spin-independent, which is usually not in magnetic systems.
5. If fluorescence yield is used, there can be an angular and energy dependence of the signal distorting the XAS spectrum and also its associated XMCD signal, as has been discussed in Section 6.2.3.
6. The appropriate edge must be separated from other structures and the continuum edge jump. In general, this is a nontrivial task, with some variation in the methods used. A number of rules have been derived for the various edges.
7. The radial matrix elements must be equal for $2p_{1/2}$ and $2p_{3/2}$ and also for spin and orbital components of valence states, which is not always the case as will be discussed for the RE L edges below.
8. The radial matrix elements must be constant in energy.
9. The number of holes in the accepting band plays a role because of the normalization to the overall XAS intensity. The number of holes is not always known.

10. The XAS spectrum of μ_0 should be the average of μ_- and μ_+. This assumption is not always correct.

The effective spin sum rule suffers from a number of additional approximations:

1. The value of $\langle T_z \rangle$ must be known where the approximation that $\langle T_z \rangle$ is zero is not always valid.
2. The L_3 and L_2 edges must be pure in their respective $2p_{3/2}$ and $2p_{1/2}$ character. This is often not the case; in particular, if they are mixed by multiplet effects. This also implies that their ratio should be $2:1$, which is not the case for all 3d systems.
3. The spin–orbit coupling of the valence states can induce large deviations if the $d_{3/2}$ and $d_{5/2}$ states are populated differently.
4. The L_3 and L_2 edges must be separable in order to determine the independent integrations including the subtraction of the backgrounds.

These complications mean that it is usually a difficult, and in some cases a theoretically impossible task, to derive the quantitatively correct values of $\langle S_z \rangle$ and $\langle L_z \rangle$ from XMCD experiments.

7.3.5 THEORETICAL SIMULATIONS OF THE SPIN SUM RULE

The specific complications of the effective spin moment sum rule (shortened to "spin sum rule") can be tested theoretically for a number of elements. Just such an analysis has been performed by Teramura et al. (1996). They showed that the combination of multiplet effects, the 3d spin–orbit coupling and the value of $\langle T_z \rangle$ together mean that the spin sum rule is only valid in very special cases. In Table 7.5, new results are given with and without the 3d spin–orbit coupling where the result with the spin–orbit coupling is essentially the same as that by Teramura et al. The atomic model calculations of $\langle S_z \rangle$, $\langle S_{eff} \rangle$ are made for high-spin $3d^n$ systems ($n = 5–9$) with an octahedral crystal field $10\,Dq = 1.0$ eV, and compared with the $\langle S_{eff} \rangle$ calculated from the spin sum rule (Equation 7.11) in the cases without ($-LS$, top) and with ($+LS$, bottom) the 3d spin–orbit coupling.

Table 7.5 shows that for Cu^{2+} ($3d^9$) the spin sum rule gives exactly the correct result because there are no multiplet effects in the initial and final states. Note, however, that the large effect of $\langle T_z \rangle$ with the inclusion of 3d spin–orbit coupling raises $\langle S_{eff} \rangle$ to -1.38 from $\langle S_z \rangle$ equal to -0.50. For Ni^{2+} ($3d^8$), the spin sum rule has only small errors, but for the other systems, the deviation from the spin sum rule is large as seen from the $\langle S_{eff} \rangle$-error in Table 7.5. This implies that one should be careful when using the spin sum rule in the analysis of experimental data. For more details of the calculations shown in Table 7.5, as well as similar calculations for low-spin $4d^n$ systems, see Appendix E.

Despite this theoretical inadequacy of the spin sum rule, it is still often used. No experimental check on the validity of the spin sum rule for various TM compounds has been made systematically so far. It is, however, highly desirable. The situation might be somewhat different for the transition metals Fe and Co. Chen et al. (1995) and Koide et al. (2001, 2004) successfully applied the orbital and spin sum rules to Co and Fe metals as described in the Sections 7.3.1 and 7.3.2. Chen et al. tested the

TABLE 7.5
Results with and without the 3d Spin–Orbit Coupling

$-LS$	$\langle S_z \rangle$ Theory	$\langle S_{eff} \rangle$ Theory	$\langle S_{eff} \rangle$ Sum Rule	$\langle S_{eff} \rangle$ Error
$3d^5$	−2.50	−2.50	−1.70	−32%
$3d^6$	−2.00	−2.00	−0.63	−68%
$3d^7$	−1.50	−1.50	−1.27	−15%
$3d^8$	−1.00	−1.00	−0.91	−9%
$3d^9$	−0.50	−0.50	−0.50	−*
$+LS$	$\langle S_z \rangle$ Theory	$\langle S_{eff} \rangle$ Theory	$\langle S_{eff} \rangle$ Sum Rule	$\langle S_{eff} \rangle$ Error
$3d^5$	−2.50	−2.50	−1.70	−32%
$3d^6$	−1.97	−1.69	−2.08	+22%
$3d^7$	−1.44	−1.59	−1.18	−26%
$3d^8$	−0.99	−0.98	−0.93	−5%
$3d^9$	−0.50	−1.38	−1.38	−*

Note: The effective spin moment sum rule is checked. The theoretical values
for $\langle S_z \rangle$ and $\langle S_{eff} \rangle$ are compared with the results of the sum rule without
the inclusion of the 3d spin–orbit coupling ($-LS$, top) and with 3d spin–
orbit coupling ($+LS$, bottom). The simulations have been performed for
an octahedral system with a crystal field value of 1.0 eV.

* A $3d^9$ calculation involves no multiplet effects and is exact.

orbital and spin sum rules experimentally for Fe and Co metals. Comparison with
the gyromagnetic ratio found that the orbital (spin)-moment was 8% (10%) too small
for iron and 5% (6%) too large for cobalt (Chen et al., 1995). This implies that essentially the same error (much smaller than the aforementioned theoretical test of the
spin sum rule) is found for $\langle L_z \rangle$ and $\langle S_z \rangle$. This is a different result than obtained from
isolated Co atoms (Gambardella et al., 2003). It should be mentioned that the error of
the spin sum rule calculated for an isolated high-spin $3d^n$ ($n = 6$ and 7) configuration
with crystal field effect should be an overestimation for Fe and Co metals. The spin
moments in Fe and Co metals are not in the high-spin state because of the energy
band effect. Furthermore, the multiplet coupling effect is considered to be reduced by
the translational motion of 3d electrons. A more detailed study on the validity of the
spin sum rule for transition metals will be the subject of future investigations.

7.4 XMCD EFFECTS IN THE K EDGES OF TRANSITION METALS

In the case of TM K edges, the core hole has no orbital moment and there are no
multiplet effects that couple the 1s core hole to the valence electrons. This has important consequences for the XMCD spectrum. The K edge XMCD is usually analyzed
in terms of the Fano factor P (Fano, 1969; Schütz et al., 1987):

$$\mu_+ - \mu_- = P(R_\uparrow^2 \rho_\uparrow - R_\downarrow^2 \rho_\downarrow) \tag{7.25}$$

For K edges, the value of P is about 0.01. The physical background of the Fano factor is the spin–orbit coupling in the final state (Brouder and Hikam, 1991). If the matrix elements (R) are equal for spin-up and spin-down, Equation 7.25 implies that from the XMCD signal, the degree in spin-polarization of the density of states (ρ) is obtained directly or, in other words, that x-ray absorption is a local probe of the local magnetic moment. More precisely, there are (at least) three assumptions in the direct determination of the spin-polarization from the XMCD signal at K edges.

- The Fano factor P is constant as a function of energy.
- The radial matrix elements are constant as a function of energy.
- The radial matrix elements are equal for spin-up and spin-down.

The Fano factor itself is also defined as a function of the matrix elements for $4p_{3/2}$ states and $4p_{1/2}$ states. One can combine all matrix elements into an effective Fano parameter (P_{eff}) as:

$$\mu_+ - \mu_- = P_{eff}(\rho_\uparrow - \rho_\downarrow) \tag{7.26}$$

Concerning the energy variations of the Fano factor, it is important to note that the K edge makes a transition to the metal 4p (and higher lying p states) that have a small spin–orbit coupling. However, the 4p states that are hybridized into the 3d-band can be expected to have a much larger induced effect due to the 3d spin–orbit coupling. This makes it likely that the Fano factor is different at the 3d-band than at all other states. In Chapter 8, the possibility of measuring the spin-polarization with spin-polarized x-ray absorption is discussed. This allows the determination of the energy dependence of the Fano factor (de Groot et al., 1995)

$$P_{eff}(\Omega) = \frac{\mu_+ - \mu_-}{\mu_\uparrow - \mu_\downarrow}(\Omega) \tag{7.27}$$

Experiments on ferromagnetic MnP material have shown that the Fano factor is ~5% at the edge and decreases within a 5 eV range to a value below 1%. The 1s3d quadrupole transitions have a strong (and different) XMCD effect, which adds to the complication. In most systems, the pre-edge is situated a few eV below the edge and this effect can be separated from the effect of the effective Fano factor or, in other words, the spin- and spin–orbit-induced effect on the radial matrix elements.

7.4.1 X-RAY NATURAL CIRCULAR DICHROISM AND X-RAY OPTICAL ACTIVITY

Here, we introduce x-ray optical activity and natural dichroism effects. In principle, these effects can occur at any x-ray absorption edge. However, because the available experimental data concerns TM K edges, the effects are discussed here. X-ray optical activity is associated with transition probabilities that mix multipole moments of opposite parities. In practice, this concerns the mixing of electric dipole (ED) and electric quadrupole (EQ) transitions, whereas the XMCD effects discussed in the remainder of this chapter always concern pure ED or pure EQ transitions. Goulon et al. (2003) give a clear and detailed description of the various aspects of x-ray optical activity. The effect can be divided into the four Stokes components of the gyration tensor that results from the interference of the dipole ED_α and quadrupole $EQ_{\beta\gamma}$ transitions. The imaginary part of this tensor gives rise to x-ray magnetochiral dichroism (XMχD) and two types of other nonreciprocal dichroic effects. The real part of this tensor gives rise to the x-ray natural circular dichroism (XNCD) effect. Goulon et al. (2003) proposed that the properties related to x-ray optical activity can be either even or odd with respect to the time-reversal operator where even effects are called "natural" and odd effects are called "nonreciprocal." The XNCD effect should be zero in powders because the ED and EQ transitions are orthogonal for isotropic samples. Experimentally, the XNCD effect in the chiral Co center in Co*(en)$_3$Cl$_3$ is 0.8% of the edge jump for single crystals and 0.02% for powders. This latter effect could be due to other effects such as interference of the ED and magnetic dipole (MD) transitions (Stewart et al., 1999). More examples are given by Goulon et al. (2003).

7.5 XMCD EFFECTS IN THE M EDGES OF RARE EARTHS

7.5.1 XMCD AND XMLD EFFECTS FROM ATOMIC MULTIPLETS

The 3d4f XAS spectra of rare earth (RE) systems have been discussed in detail in Chapter 6. The $M_{4,5}$ spectral shapes can be explained from atomic multiplet theory. This implies that the XMCD and XMLD spectra can be determined directly from the J-values of the initial and final state, using Table 7.3. As an example, we treat the XMCD and XMLD spectra of Tm^{3+}. We have seen that the ground state of 4f^{12} Tm^{3+} has a $^3H_6$ character, or better, it is a mixture of 99% $^3H_6$ character and 1% $^3I_6$ character. With a J-value of six, the 3d^{9}4f^{13} final state with $J = 5$, 6, or 7 is allowed. There are three states with $J = 5$ ($^1H_5$, $^3G_5$, and $^3H_5$) and one state with $J = 6$, that is, $^3H_6$. This implies that the $M_{4,5}$ edges of Tm exist for four transitions. With large J-values, the values in Table 7.3 are almost diagonal.

With Table 7.6, we can find that the three transitions to $^1H_5$, $^3G_5$, and $^3H_5$ are pure $\Delta M_J = -1$ transitions. The transition to $^3H_6$ is 86% $\Delta M_J = 0$ and 14% $\Delta M_J = -1$. One can calculate the transitions to the final states with $\Delta J = -1$ and $\Delta J = 0$ and apply the rules that:

- XAS $= I(\Delta J = -1) + I(\Delta J = 0)$
- XMCD $= -I(\Delta J = -1) - 0.14\, I(\Delta J = 0)$

TABLE 7.6
**Relationship between ΔJ and ΔM_J for a
$J = 6$ Ground State**

	$\Delta M_J = -1$	$\Delta M_J = 0$	$\Delta M_J = +1$
$\Delta J = -1$	1	0	0
$\Delta J = 0$	0.14	0.86	0
$\Delta J = +1$	0.01	0.12	0.87

Note: In the case of Tm^{3+} only, $\Delta J = -1$ and $\Delta J = 0$ are possible.

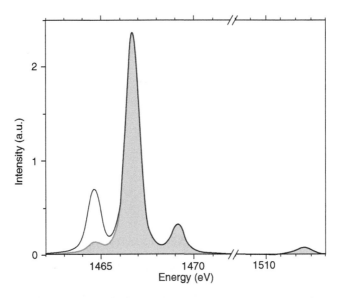

FIGURE 7.16 The XAS spectrum of Tm^{3+} (solid) compared with its XMCD spectrum (gray). The $\Delta J = -1$ spectrum identifies with the XMCD spectrum.

Figure 7.16 shows the XAS and the inverted XMCD spectra of Tm^{3+}. From the four peaks we see that the three $\Delta J = -1$ transitions have the same intensity in the inverted XMCD spectrum and one $\Delta J = 0$ transition contributes only 14% to the XMCD. Note that the M_4 edge is excited only with the – helicity photon. With the definitions of XMCD as $I(\Delta M_J = +1) - I(\Delta M_J = -1)$ and XMLD as $I(\Delta M_J = 0) - \frac{1}{2}I(\Delta M_J = -1) - \frac{1}{2}I(\Delta M_J = +1)/2$, the XMCD and XMLD spectra can be determined directly from Table 7.3. The result is given in Table 7.7.

Table 7.8 shows the XAS, XMCD, and XMLD spectra for an atomic multiplet calculation for a $J = 6$ ground state. It might appear that the integrated XMCD signal is different from zero, but this is not true. The degeneracies of the final states are equal to $2J'+1$ and adding the degeneracies, the sum over the three ΔJ transitions

TABLE 7.7
XAS, XMCD, and XMLD Spectra as a Function of the Three ΔJ Transitions

80	XAS	XMCD	XMLD
$\Delta J = -1$	1	-1	$-1/2$
$\Delta J = 0$	1	$-1/(J+1)$	$(2J-1)/2(J+1)$
$\Delta J = +1$	1	$J/(J+1)$	$-J(2J-1)/2(2J+3)(J+1)$

TABLE 7.8
XAS, XMCD, and XMLD Spectra
for $J = 6$

$J = 6$	XAS	XMCD	XMLD
$J' = 5$	1	-1	$-1/2$
$J' = 6$	1	$-1/7$	$11/14$
$J' = 7$	1	$6/7$	$-11/35$

equals zero [i.e. $(11 \times -1) + (13 \times -1/7) + (15 \times 6/7) = 0$]. The same is true for the XMLD spectrum.

7.5.2 TEMPERATURE EFFECTS ON THE XMCD AND XMLD

The XMLD spectra of all RE elements can be calculated using this scheme. The calculations as presented are only valid at 0 K. At finite temperatures, the excited M_J states become occupied and their XMCD and XMLD spectrum will be different. For a pure atomic ground state, the various M_J states are equidistantly split by the exchange interaction. This allows one to calculate the XMLD and XMCD spectra at finite temperatures. One can use a Boltzmann distribution with exponential $\exp(-M_J/\Theta)$, where Θ is the reduced temperature defined as $kT/g\mu_B H$. At 0 K, only the lowest M_J state is occupied and the formulas as defined previously apply. At temperatures $T \gg \Theta$, all ΔJ transitions reach the 1/3 limit and all XMLD and XMCD spectra become zero. At temperatures in the range of Θ, the XMLD and XMCD spectra are reduced in magnitude but their spectral shape remains equal (Goedkoop et al., 1988a).

The first XMLD spectrum published was the M_5 edge of $Tb_3Fe_5O_{12}$ (van der Laan et al., 1986). Figure 7.17 shows the M_5 edge of $Tb_3Fe_5O_{12}$ measured at 50 K. The experiments are given as a function of the angle α between the polarization vector and the [111] direction of magnetization. The theoretical curves have been calculated using Table 7.6 multiplied with a reduction factor due to the finite temperature. A more detailed analysis, including the angular variations of the magnetic moments, the so-called umbrella structure, can be found in the original publications (van der Laan et al., 1986; Goedkoop et al., 1988b).

7.6 XMCD EFFECTS IN THE L EDGES OF RARE EARTH SYSTEMS

The XAS at RE L edges are described in Sections 3.6.5 and 6.6.3. They correspond to transitions from the $2p_{3/2}$ and $2p_{1/2}$ core states, which are split from 410 eV in La up to 1030 eV in Yb. The 2p core electrons are excited to the empty 5d valence states. At the edge, quadrupole transitions are possible to the empty 4f states. This can be made visible with HERFD XAS, but XMCD can also be used to show the presence of the quadrupole transitions. The quadrupole transitions are excited directly into the 4f states and, as such, show large XMCD effects. In addition, the quadrupole transitions have an angular dependence in cubic symmetry, which makes them different in different directions (Giorgetti et al., 1993, 2001).

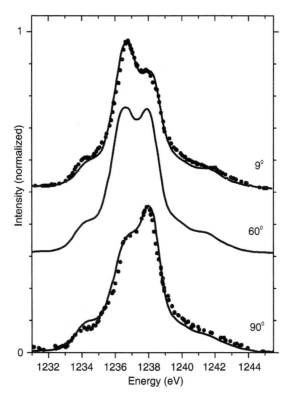

FIGURE 7.17 The M_5 edge of $Tb_3Fe_5O_{12}$ is given as a function of the angle α between the polarization vector and the [111] direction of magnetization. The experimental spectra are compared to atomic multiplet calculations for 9° (top) and 90° (bottom). The middle spectrum is the theoretical curve for 60° that identifies with experiment. The theoretical curves have been calculated using Table 7.6 multiplied with a reduction factor due to the finite temperature. (Reprinted with permission from Goedkoop, J.B., Fuggle, J.C., Thole, B.T., van der Laan, G., and Sawatzky, G.A., *J. Appl. Phys.*, 64, 5595, 1988b. Copyright 1988 by the American Institute of Physics.)

7.6.1 EFFECTS OF 4f5d EXCHANGE INTERACTION

Jo and Imada (1993) gave an interpretation for a systematic trend of XMCD in L_2 and L_3 edges of trivalent RE atoms from Ce to Tm, assuming a 2p5d dipole transition and the 4f states in the Hund's rule ground state. They took into account the spin and orbital moments of the 5d state induced through the intra-atomic 4f5d exchange interaction within a molecular field approximation. The exchange energy of the 5d state, specified by the z components of the orbital and spin quantum numbers, m_d and s_d, respectively, is given by

$$E_{d\mu} \equiv E(m_d, s_d) = -\sum_{k=1,3,5} \sum_{m_f,s_f} |\,c^k(2m_d, 3m_f)\,|^2 G^k n(m_f, s_f)\,\delta(s_d, s_f), \quad (7.28)$$

where μ denotes the combined indices of m_d and s_d, c^k is proportional to the Clebsch–Gordan coefficient, G^k represents the 4f5d Slater integrals, $n(m_f, s_f)$ is the number of

4f electrons corresponding to m_f and s_f, and δ is the Kronecker delta function. Note that the energy $E_{d\mu}$ depends on the RE element where $n(m_f, s_f)$ is obtained for the Hund's rule ground state. In order to determine the spin and orbital moments of the 5d states, Jo and Imada assumed that each of the $d\mu$ states forms an energy band with semi-elliptical density of states (DOS), and with its central energy at $E_{d\mu}$ and that the Fermi level is determined so that the number of the 5d electrons per atom is unity. Then they calculated the integrated XMCD intensity of L_2 and L_3 edges for each RE element. The calculated result was mostly consistent with experiments in the systematic variation of the integrated XMCD amplitude over the RE elements, but the result failed to reproduce the exact sign of the experimentally observed integrated XMCD intensity (Baudelet et al., 1993).

The opposite XMCD sign in the calculation by Jo and Imada is caused by their assumption that the exchange interaction (giving rise to the exchange energy using Equation 7.28), does not modify the dipole transition amplitude to the state $d\mu$. Carra et al. (1991) pointed out through energy band calculations that in the case of Gd metal, in which the orbital magnetic moment is completely quenched, the spin dependent enhancement of the dipole matrix element is crucial for reproducing the correct XMCD sign. This is a consequence of the contraction of the radial part of the 5d orbital due to the 4f5d exchange interaction. Matsuyama et al. (1997) and van Veenendaal et al. (1997) improved the theory by Jo and Imada by taking into account that the enhancement of the dipole matrix element by the exchange interaction depends not only on the 5d spin moment (s_d) but also on the 5d orbital moment (m_d) for the entire series of RE elements. Matsuyama et al. considered a finite occupation in the 5d electron band, as will be described subsequently in some detail, but van Veenendaal et al. (1997) assumed that the 5d band is empty in the ground state.

Matsuyama et al. (1997) introduced a factor α to describe the enhancement of the dipole matrix element and wrote the XAS spectrum in the form:

$$\mu_j^{\pm}(\Omega) = \sum_{\mu, j_z} |M^{\pm}(pjj_z; d\mu)|^2 (1 - \alpha E_{d\mu}) \int_{\varepsilon_F}^{W + E_{d\mu}} d\varepsilon\, \rho_{d\mu}(\varepsilon) \frac{\Gamma/\pi}{(\hbar\Omega + E_{pj} - \varepsilon)^2 + \Gamma^2}, \quad (7.29)$$

where $\mu_j^{\pm}$ represents the XAS spectrum with $\pm$ helicity for the angular momentum $j = (3/2, 1/2)$ of the 2p core state, $M^{\pm}(pjj_z; d\mu)$ is the dipole transition matrix element for $\pm$ helicity before taking into account the enhancement effect, $\rho_{d\mu}(\varepsilon)$ is the semi-elliptical DOS of the 5d ($d\mu$) band with the width W expressed as

$$\rho_{d\mu}(\varepsilon) = \frac{2}{\pi W^2} \sqrt{W^2 - (\varepsilon - E_{d\mu})^2}, \quad (7.30)$$

ε_F is the Fermi level and Γ is the lifetime broadening of the 2p core hole. The XAS and XMCD spectra are given, respectively, by $\mu_j^+ + \mu_j^-$ and $\Delta\mu \equiv \mu_j^+ - \mu_j^-$. If we put $\alpha = 0$, these results reduce essentially to those by Jo and Imada (1993). It should be noted that $M^{\pm}(pjj_z; d\mu)$ is expressed, apart from a constant factor, as

$$M^{\pm}(pjj_z; d\mu) = \sum_{m_p, s_p} \langle 2m_d\, 1 \pm 1 | 1\, m_p \rangle \langle 1\, m_p\, \tfrac{1}{2}\, s_p | jj_z \rangle\, \delta(s_d, s_p), \quad (7.31)$$

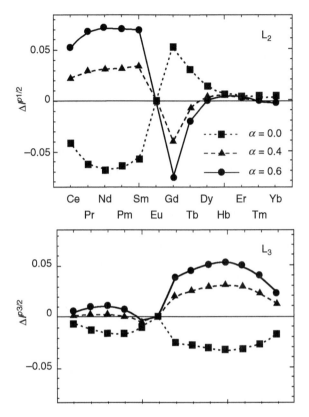

FIGURE 7.18 Calculated integrated intensities of XMCD spectra at the $L_{2,3}$ edges for an entire series of RE elements. The parameter values of α are set to 0.0, 0.4, and 0.6. (From Matsuyama, H., Harada, I., and Kotani, A., *J. Phys. Soc. Jpn.*, 66, 337, 1997. With permission.)

where $\langle j_1 m_1 j_2 m_2 | JM \rangle$ is the Wigner coefficient.* Matsuyama et al. (1998) calculated the integrated intensity of the XMCD spectrum (normalized by the integrated XAS intensity) ΔI^{pj}, for the entire series of trivalent RE elements. The calculated results for $\alpha = 0.0$, 0.4, and 0.6 (1/eV) are shown in Figure 7.18. The result for $\alpha = 0.0$ reproduces the result found by Jo and Imada (1993), but by introducing $\alpha \neq 0$ (e.g., $\alpha = 0.4$ or 0.6 1/eV), the sign of the integrated XMCD changes in accord with experimental observations.

Let us first consider the case of Gd. By taking into account the enhancement effect, the sign of ΔI^{pj} is positive and negative for $j = 3/2$ and $1/2$, respectively. This is easily understood from the following. With our convention, the magnetic field **B** is in the $-\mathbf{z}$ direction, so that all of the seven 4f electrons occupy spin-up states in the

---

*The relation between the Wigner 3j symbol and the Wigner coefficient, $\langle j_1 m_1 j_2 m_2 | JM \rangle$ is given by

$$\begin{pmatrix} j_1 & j_2 & J \\ m_1 & m_2 & M \end{pmatrix} = (-1)^{j_1 - j_2 - M}(2J + 1)^{-1/2} \langle j_1 m_1 j_2 m_2 | J - M \rangle$$

Hund's rule ground state, where the spin quantization axis is in the +z direction. Therefore, by the 4f5d exchange interaction, the energy of 5d spin-up states is lower than that of 5d spin-down states, and the electric dipole transition amplitude is enhanced for the 2p to 5d (up) transition instead of the 2p to 5d (down) transition. As can be seen from Figure 7.19, where we show the relative intensity of the dipole transition, $|M^{\pm}(pjj_z; d\mu)|^2$, for each pair of pjj_z and $d\mu$, the sign of XMCD is determined for the L_3 edge ($j = 3/2$) mainly by the competition between the 2p to 5d (up) transition [the transition to $(m_s, s_d) = (2, \uparrow)$ is most dominant] by the + helicity and the 2p to 5d↓ transition [most dominant is to $(-2, \downarrow)$] by the − helicity. For the L_2 edge ($j = 1/2$), the sign of XMCD is determined mainly by the competition between the 2p to 5d↓ transition [most dominant is to $(2, \downarrow)$] by the + helicity and the 2p to 5d↑ transition [most dominant is to $(-2, \downarrow)$] by the − helicity. Therefore, by the enhancement of the 2p to 5d↑ transition matrix element, the transition by the + and − helicity become dominant for L_3 and L_2 edges, respectively, giving rise to the positive and negative signs of XMCD. This is exactly the same situation as that shown by Carra et al. (1991) through the energy band calculation. From comparison with these results, the value of α is estimated to be 0.4 to 0.6 (1/eV).

For other RE elements, it is seen from the results of $\alpha = 0.4$ and 0.6 (1/eV) that ΔI^{pj} has large positive values for the L_3 ($j = 3/2$) of heavy RE elements (Tb-Yb) and for the L_2 ($j = 1/2$) of light RE elements (Ce-Sm), but it has much smaller values for the L_3 of light RE elements (Ce-Sm) and for the L_2 of heavy RE elements (Tb-Yb). The mechanism of this behavior is more complicated than that of Gd, but the trend is understood in the following way. Looking at Figure 7.19, let us confine ourselves, for simplicity, to the following cases of largest dipole transition intensity: for the L_3 edge, the transition to $(m_s, s_d) = (2, \uparrow)$ by the + helicity light and that to $(-2, \downarrow)$ by the − helicity one, and for the L_2 edge, the transition to $(2, \downarrow)$ by the + helicity light and that to $(-2, \uparrow)$ by the − helicity one. For light rare RE elements, the dipole

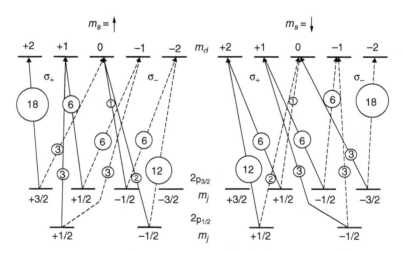

FIGURE 7.19 The relative intensity of the dipole transition $|M^{\pm}(pjj_z; d\mu)|^2$. (Reprinted with permission from Koide, T., *Oyo Butsuri*, 63, 1210, 1994. Copyright 1994 by the Japanese Society of Applied Physics.)

transition intensity to $(m_s, s_d) = (2, \downarrow)$ is strongly enhanced by the exchange interaction with 4f states in the Hund's rule ground state, but the dipole transition intensities to $(-2, \downarrow)$, $(-2, \uparrow)$, and $(2, \uparrow)$ are not much enhanced. Therefore, in the L_2 edge, the XAS intensity by the + helicity light is much larger than that by the – helicity light, and the XMCD has a large positive value. However, in the L_3 edge, the XAS intensities by the + and the – helicities are small and comparable, so that the sign of XMCD is determined by higher-order effects. From a similar consideration, for heavy RE elements, the XAS intensity by the + helicity light is much stronger than the – helicity light for L_3, but those by the + and the – helicity lights are small and comparable for L_2.

7.6.2 CONTRIBUTION OF ELECTRIC QUADRUPOLE TRANSITION

The quadrupole contribution is also inevitable for a quantitative comparison of the calculated spectra with experiment. Fukui et al. (2001c) calculated it with the atomic multiplet model. The calculated integrated intensities of XAS and XMCD are shown in Figure 7.20. The atomic calculation is a reasonable method, since the 4f electrons directly concerned with the quadrupole process (i.e. the initial state $2p^6 4f^n$ and the final state $2p^5 4f^{n+1}$), are well localized. Intra-atomic multiplet coupling effects are crucial in this process, especially the Coulomb interaction between the photo-excited 4f electron and the core hole left behind as well as between the other 4f electrons. However, the lifetime effect of the 2p core hole smears out the detailed structure of the spectra.

The contribution of the quadrupole transition to XMCD is generally weaker than that of the dipole transition. However, the quadrupole contribution is somewhat lower in energy than the dipole contribution, and especially for the L_3 XMCD of light RE elements and for the L_2 XMCD of heavy RE elements, the quadrupole contribution can be important because the dipole contribution is rather weak.

7.6.3 EFFECT OF HYBRIDIZATION BETWEEN RE 5d AND TM 3d STATES

In order to discuss the XMCD spectra of RE elements in intermetallic compounds containing TM elements [e.g. $R_2Fe_{14}B$ or RFe_2 (R is RE element)], it is important to take into account another polarization effect of the 5d states due to the hybridization with spin-polarized 3d states of surrounding transition metal (TM) ions, which occurs through the 2p5d dipole transition. Actually, it is well known that the magnetic coupling between Fe 3d spin and R 4f spin always presents an antiferromagnetic alignment (via the R 5d conduction electrons). This makes them either ferromagnetic compounds for light RE elements (Ce-Sm) or ferrimagnetic compounds for half-filled (Gd) and heavy RE elements (Tb-Yb), because the R 4f spin moment is antiparallel and parallel, respectively, to the R 4f orbital moment (which is larger than the R 4f spin moment) for light and heavy RE elements.

On the contribution of R 5d-Fe 3d hybridization to XMCD, this effect must dominate the spectra for La or Lu compound, since there is no effect from the 4f electrons. For other RE elements, this effect is generally weaker than that of the dipole contribution due to the R 4f–5d interaction. However, for the L_3 XMCD of light RE elements and for the L_2 XMCD of heavy RE elements, this contribution can be important because the dipole contribution by 4f–5d interaction is rather weak. In the case of $R_2Fe_{14}B$, where Fe constitutes the majority of the magnetic moment,

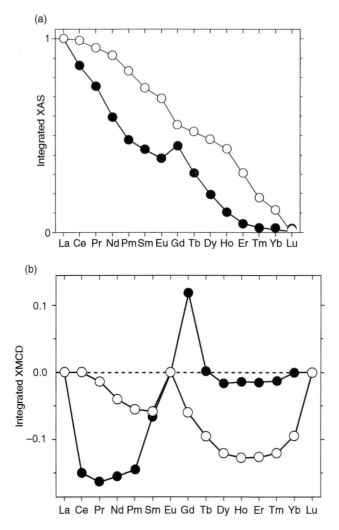

FIGURE 7.20 Integrated intensities of RE $L_{2,3}$ XAS and its XMCD by the 2p4f quadrupole transition. The incident angle is taken to be 45°, and the intensities are normalized at the values of La XAS for L_2 (closed circles) and L_3 (open circles), respectively. From Figure 3 of Fukui et al. (2001c), an error for the value of XMCD in L_2 of Sm has been corrected. (Reprinted with permission from Fukui, K., et al., *Phys. Rev. B*, 64, 104405, 2001c. Copyright 2001 by the American Physical Society.)

the contribution of R 5d-Fe 3d hybridization is large and determines almost the total L_2 XMCD of heavy RE elements as will be shown later.

7.6.4   XMCD at L Edges of $R_2Fe_{14}B$ (R = La−Lu)

In order to obtain R 5d and Fe 3d electronic states, Asakura et al. (2002) performed cluster model calculations. It is known that $R_2Fe_{14}B$ has tetragonal symmetry and its unit cell contains 68 atoms. Then, a cluster consisting of 10 R atoms including the central R and 16 Fe atoms was adopted. Actually, the considered cluster contains all R atoms up to the fifth nearest neighbors from site 0 and all Fe atoms up to the sixth nearest neighbors from site 0.

The Hamiltonian of the cluster model is expressed as

$$H = \sum_i \sum_\mu \varepsilon_{5d}(\mu) d^+_{\mu,i} d_{\mu,i} + \sum_j \sum_v \varepsilon_{3d}(v) D^+_{v,j} D_{v,j} + \sum_{i \neq i'} \sum_{\mu,\mu'} [t_{i,i'}(\mu,\mu') d^+_{\mu,i} d_{\mu',i'} + \text{h.c.}]$$

$$+ \sum_{j \neq j'} \sum_{v,v'} [t_{j,j'}(v,v') D^+_{v,j} D_{v',j'} + \text{h.c.}] + \sum_{i \neq j} \sum_{\mu,v} [t(\mu,v) d^+_{\mu,i} D_{v,j} + \text{h.c.}], \qquad (7.32)$$

where the operators $d^+_{\mu,i}$ ($i = 0$–9) and $D^+_{v,j}$ ($j = 1$–16) represent, respectively, the creation of an electron in the 5d state μ of the R ith site and in the 3d state v of the Fe jth site. The indices μ and v denote (m_d, s_d) of R 5d and Fe 3d states, respectively. The 5d energy level ε_{5d} is essentially the same as the exchange energy $E_{d\mu}$ but now we take into account the reduction factor R_E of the Slater integral G^k:

$$\varepsilon_{5d}(\mu) = R_E E_{d\mu}. \qquad (7.33)$$

The 3d energy level is expressed as

$$\varepsilon_{3d}(v) = \Delta_{3d5d} - E_{\text{exc}} s_{3d}, \qquad (7.34)$$

where Δ_{3d5d} is the Fe 3d-R 5d energy separation in nonmagnetic state and E_{exc} is the exchange splitting of the Fe 3d state. The value of Δ_{3d5d} is taken to be -2.6 eV, and the Fermi energy is fixed at -2.0 eV with respect to the center of the 5d levels. The electron transfer integrals are obtained from the Slater–Koster integrals given by the empirical formula by Pettifor (1977) [for more details, see Asakura et al. (2002)].

XMCD spectra are calculated by Equation 7.29, putting $\alpha = 0.4$ (1/eV) and using the partial DOS of R 5d states $\rho_{d\mu}(\varepsilon)$ calculated with the cluster model. In order to compare quantitatively with the experimental results measured at room temperature, it was necessary to take the reduction factor R_E as shown in Table 7.9, while the value of E_{exc} was fixed at 0.2 eV.

The calculated XAS and XMCD for L_3 and L_2 edges are shown in Figure 7.21a and b, respectively, together with the experimental data. The overall agreement

TABLE 7.9
Reduction Factor (R_E) of the 4f-5d Exchange Energy

	La	Pr	Nd	Sm	Gd	Tb	Dy	Ho	Er	Tm	Yb
	$4f^0$	$4f^2$	$4f^3$	$4f^5$	$4f^7$	$4f^8$	$4f^9$	$4f^{10}$	$4f^{11}$	$4f^{12}$	$4f^{13}$
R_E	—	0.12	0.12	0.16	0.40	0.12	0.12	0.20	0.12	0.08	0.04

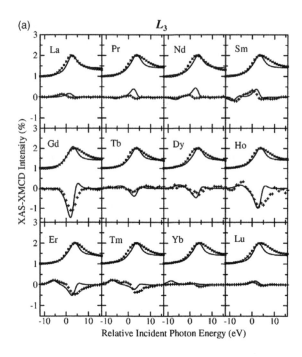

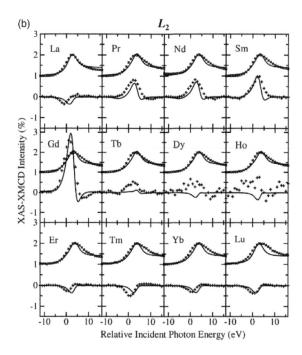

FIGURE 7.21 Calculated (solid curves) and experimental (crosses) results of XAS and XMCD at (a) L_3 and (b) L_2 edges of $R_2Fe_{14}B$. (From Asakura, K., et al., *J. Phys. Soc. Jpn.*, 71, 2771, 2002. With permission.)

between calculated and experimental results is quite satisfactory in both XMCD spectral shape and amplitude. The calculated XMCD involves the contributions of dipole transitions with the effects of interatomic hybridization and intra-atomic exchange interaction, as well as quadrupole contributions. Note that, since each contribution to XMCD spectra is sometimes different in shape and sign, the total XMCD spectra show a variety of shapes.

For the L_2 edges of the light RE elements Pr, Nd, and Sm, the L_2 and L_3 edges of Gd, and L_3 edges of heavy RE elements from Tb to Tm, the contribution from the 4f-5d exchange interaction is large, as seen from Figure 7.18, and it predominantly determines the XMCD spectra. For La and Lu, on the other hand, we only have the mechanism of the R 5d-Fe 3d hybridization. For L_2 edges of heavy RE elements from Dy to Yb, the calculated XMCD spectra are almost the same as that of Lu, which means that the mechanism of the R 5d-Fe 3d hybridization is dominant.

For L_3 edges of light RE elements, it can be shown that XMCD spectra consist of the following three contributions with comparable intensity: (*i*) the dipole contribution from RE 4f-5d exchange interaction, (*ii*) that from the RE 5d-Fe 3d hybridization, and (*iii*) the quadrupole contribution. As an example, the calculated XMCD spectra at the L_2 and L_3 edges of $Sm_2Fe_{14}B$ are decomposed into three contributions, as shown in Figure 7.22, where the contributions (*i*), (*ii*), and (*iii*) are shown with a dashed, dotted, and dash-dot curve, respectively. The solid curve is the sum of the three contributions. It is seen that, for the L_3 edge, the three contributions have comparable intensities while, for the L_2 edge, the contribution (*i*) is much larger than (*ii*) and (*iii*).

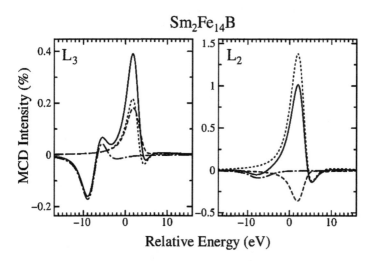

FIGURE 7.22 XMCD spectra of the Sm $L_{2,3}$ XAS in $Sm_2Fe_{14}B$. Solid curves are the sum of the three contributions: the dipole contribution due to the 4f-5d exchange interaction (dotted curves), the dipole contribution due to the Sm 5d-Fe 3d hybridization (dashed) and the quadrupole contribution (chain curves). (From Asakura, K., et al., *J. Phys. Soc. Jpn.*, 71, 2771, 2002. With permission.)

Let us comment in a small way on the reduction factor R_E of the exchange energy in Table 7.1 with respect to the calculated results of Figure 7.21a and b. There are two different origins for R_E. One is the correction from the Hartree–Fock calculation of the exchange interaction between 4f and 5d states, namely the correction by the intra-atomic configuration interaction that results in the value of $R_E \approx 0.8$ at zero temperature. The other one is the reduction of the exchange energy by thermal fluctuations of 4f magnetic moment. In the comparison of calculated XMCD spectra with experiment at room temperature, this effect is important because the magnetic polarization of 5d states due to 4f-5d exchange interaction is reduced in proportion to the reduction of 4f magnetization. Actually, the magnetization of $R_2Fe_{14}B$ at room temperature is mainly carried out by the Fe 3d electrons. Also, the magnetization of R 4f electrons is strongly reduced by thermal fluctuations (Herbst, 1991), except for Gd. In the case of Yb and Tm, the spectral shape of XMCD is similar to that of Lu, which has no 4f magnetic moment. Therefore, Asakura et al. (2002) suggested a dramatic reduction of the 4f magnetization and extremely small values of R_E (0.04 and 0.08). For Gd, on the other hand, a considerable contribution from 4f magnetization to XMCD was seen, reflected in the reduction factor 0.4. It should be noted that the rough trend of reduction factor R_E (which takes a maximum at Gd and decreases in going away from Gd) is in qualitative agreement with the behavior of T_c in $R_2Fe_{14}B$ (Herbst, 1991).

Finally, it should be mentioned that Asakura et al. (2004b) have also performed theoretical calculations of XMCD in $R_2Fe_{14}B$ with the use of the partial DOS of R 5d and Fe 3d states obtained with the tight-binding energy-band model instead of the cluster model. They have shown that the results of XMCD calculated with the tight-binding model coincide almost perfectly with those calculated with the cluster model if we take into account the spectral broadening due to the 2p core hole lifetime. At the same time, they have shown that more details of the R 5d band structure obtained with the tight-binding model should be reflected in the XMCD spectra if the spectral broadening could be suppressed.

7.6.5 Mixed Valence Compound $CeFe_2$

Let us consider a Laves-phase compound $CeFe_2$, which behaves as a ferromagnetic mixed valence system. $CeFe_2$ has anomalously small magnetic moment and a low Curie temperature ($T_c = 230$ K) as compared to other RFe_2. We ascribe the considered anomalous behavior to the mixed valence character of Ce, which can be well described by the single impurity Anderson model (SIAM) (Kotani et al., 1988). Including the various Coulomb interactions, Asakura et al. (2004a) combined SIAM with a LCAO ($Ce_{17}Fe_{12}$) cluster calculation for the Ce 5d and Fe 3d conduction states and calculated XMCD.

Experimental results of XAS and XMCD for the Ce L_3 and L_2 edges of $CeFe_2$ are shown in Figures 7.23 and 7.24, respectively (Giorgetti et al., 1993). The double-peak structure of XAS is a characteristic feature of the mixed valence Ce compounds and usually explained by the $4f^0$ (higher energy peak) and $4f^1$ (lower energy peak) components in the final state. Corresponding to them, the XMCD also exhibits the double-peak structure. The sign of XMCD is positive for L_3 and negative for L_2, similar to

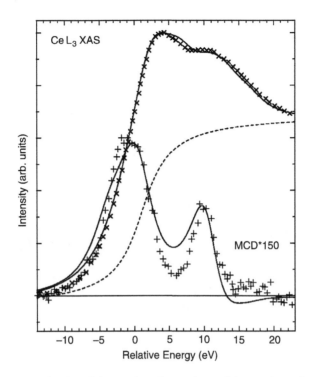

FIGURE 7.23 Calculated (solid curves) and experimental (crosses) results of XAS and XMCD at the Ce L_3 edge of CeFe$_2$. The dashed curve represents the background contribution assumed in the calculation. (From Asakura, K., et al., *J. Phys. Soc. Jpn.*, 73, 2008, 2004a. With permission.)

that in LuFe$_2$, so that we expect that the XMCD in CeFe$_2$ is caused mainly by the spin polarization of the Ce 5d band due to the hybridization with the spin-polarized Fe 3d band, as in the case of LuFe$_2$. This is reasonable because the spin and orbital polarization of the Ce 4f state would be very small due to the mixed valence character. However, there remain some features to be solved theoretically. The first is that the energy positions of the two peaks in XAS are different from those of XMCD. The difference is larger for the lower energy peak (4f^1 peak) than the higher energy one (4f^0). Second, the widths of the two XMCD peaks are different in that the lower energy peak is broader than the high energy one. Lastly, it is not clear where the contribution from the 4f^2 configuration can be seen. To solve these problems, theoretical calculations have been made by Asakura et al. (2004a).

Asakura et al. considered the Ce$_{17}$Fe$_{12}$ cluster model to describe the Ce 5d and Fe 3d states because the cluster with finite size is more convenient to be combined with the SIAM. The Hamiltonian of the cluster (similar to Equation 7.3.2) is diagonalized in the form

$$H = \sum_{\ell} \varepsilon_{v\ell} a_{v\ell}^{+} a_{v\ell} \qquad (7.35)$$

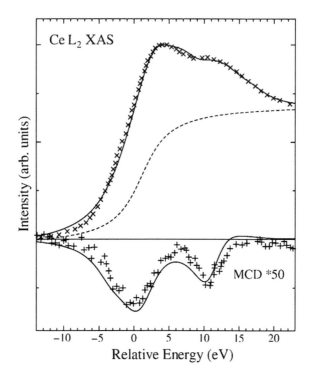

FIGURE 7.24 Similar to Figure 7.23, but for the Ce L_2 edge of CeFe$_2$. (From Asakura, K., et al., *J. Phys. Soc. Jpn.*, 73, 2008, 2004a. With permission.)

by the transformation

$$a_{v\ell}^+ = \sum_{\mu,i} \langle \mu,i | \ell \rangle a_{\mu i}^+ + \sum_{v,j} \langle v,j | \ell \rangle a_{vj}^+. \tag{7.36}$$

The mixed Ce 5d-Fe 3d states are combined with Ce 4f states in the frame of the extended SIAM. The Hamiltonian of the extended SIAM in the initial state of XAS is written as

$$H_g = \varepsilon_{4f} \sum_\lambda a_{f\lambda}^+ a_{f\lambda} + U_{ff} \sum_{\lambda > \lambda'} a_{f\lambda}^+ a_{f\lambda} a_{f\lambda'}^+ a_{f\lambda}$$

$$+ \sum_\ell \varepsilon_{v\ell} a_{v\ell}^+ a_{v\ell} + \sum_{\ell,\lambda} (V_{\ell\lambda} a_{v\ell}^+ a_{f\lambda} + \text{h.c.}), \tag{7.37}$$

where λ represents the combined spin and orbital quantum numbers of the Ce 4f state, U_{ff} is the Coulomb interaction of Ce 4f states and $V_{l\lambda}$ is the hybridization between 4f(λ) and v(l) states. In the final state of XAS, the 2p core electron is excited to the 5d band and the 4f level is pulled down due to the core hole potential $-U_{fc}$. Furthermore, it is taken into account that the 5d electrons on the core hole site ($j = 0$) interact with the 4f electron through the Coulomb interaction U_{fd} and the core hole through $-U_{dc}$ (see Chapter 3). Therefore, the Hamiltonian in the final state is given by

$$H_f = H_g - U_{fc} \sum_\lambda a_{f\lambda}^+ a_{f\lambda} - U_{dc} \sum_{k,k'} C_{k,k'} a_{\mu k}^+ a_{\mu k'} + U_{fd} \sum_{k,k',\lambda} C_{k,k'} a_{f\lambda}^+ a_{f\lambda} a_{\mu k}^+ a_{\mu k'}, \quad (7.38)$$

where

$$C_{k,k'} = \sum_v \langle \mu k | v, 0 \rangle \langle v, 0 | k' \mu \rangle, \quad (7.39)$$

and $a_{\mu k}^+$ is the creation operator of a Ce 5d electron in the state (μ, k). The Hamiltonians H_g and H_f are diagonalized by taking into account the three configurations $4f^0$, $4f^1$, and $4f^2$. In diagonalizing H_f, it is assumed that only a photo-excited electron is affected by the Coulomb interactions $-U_{dc}$ and U_{fd}. By the calculation, $CeFe_2$ is shown to be in the mixed valence state with the average 4f electron number in the ground state $n_f = 0.64$. The calculated results of XAS and XMCD are shown in Figures 7.23 and 7.24 by solid curves where the dashed curve is the background contribution in XAS. The values of U_{dc} and U_{fd} are taken to be 2.0 and 1.7 eV, respectively. The agreement between the calculated and experimental results is satisfactory both for XAS and XMCD at both L_2 and L_3 edges.

In order to see the mechanism determining the relative peak positions of XAS and XMCD, the effect of the core-hole potential $-U_{dc}$ at the L_3 edge of XAS and XMCD is calculated for the $4f^0$ peak, disregarding the $4f^1$ and $4f^2$ contributions. The results are shown in Figure 7.25. In the case of $U_{dc} = 0$, the peak of XMCD is shifted by about 4 eV towards lower photon energy with respect to that of XAS, and its width is smaller. The reason for this is that, since the XAS peak occurs due to optical transition of the $2p_{3/2}$ core electron to the Ce 5d conduction band above the Fermi level, the position and the width of the XAS peak correspond to the center and the width of the unoccupied part of the Ce 5d band. On the other hand, XMCD is only caused by the spin-polarized part of the unoccupied Ce 5d band. The spin polarization of the Ce 5d band is induced by the Fe 3d spin polarization due to the hybridization between Ce 5d and Fe 3d states. However, the Fe 3d band is mainly located in the lower energy side of the Ce 5d band, so that the spin polarization of the empty states of the Ce 5d band is limited to the states near the Fermi level. With increasing U_{dc}, on the other hand, the oscillator strength of XAS is transferred to the lower energy side (toward the Fermi level, which is the peak position of XMCD), and therefore the energy difference of the XAS and XMCD peaks decrease. This explains why the positions of the higher energy peak ($4f^0$ peak) of XAS and XMCD are close to each other in Figures 7.23 and 7.24.

For the lower energy peak (mainly $4f^1$), on the other hand, the effect of the Coulomb interaction between 4f and 5d states U_{fd} has to be taken into account. Then, the effects of the attractive $-U_{dc}$ and the repulsive U_{fd} almost cancelled each other so that the situation becomes similar to the case of $U_{dc} = 0$ in Figure 7.25. Therefore, the effects of $-U_{dc}$ and U_{fd} are essential to the understanding of the relative positions of the two peaks in XAS and XMCD in $CeFe_2$.

The difference in the spectral widths of the two XMCD peaks, as well as the difference in the spectral width of XAS and XMCD peaks, is also mainly caused by

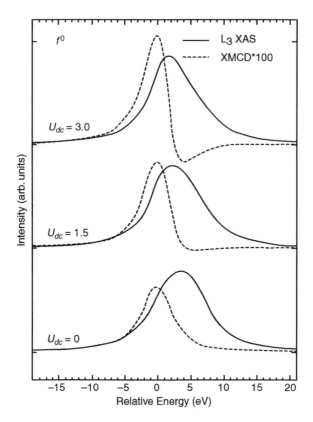

FIGURE 7.25 Calculated results of XAS and XMCD at L_3 edge for various values of U_{dc} in the limit of $4f^0$ configuration. The peak position of XMCD is taken as the origin of the relative energy. (From Asakura, K., et al., *J. Phys. Soc. Jpn.*, 73, 2008, 2004a. With permission.)

the effects of $-U_{dc}$ and U_{fd}. For more details on the contents of Section 7.6.1 to Section 7.6.5, see a review article by Parlebas et al. (2006).

7.6.6 MULTIELECTRON EXCITATIONS

Each core state can have multielectron-excitations (MEE) associated with it, but in solids they are usually not visible as separate structures. Because of their effects, they are well known in extended x-ray absorption fine structure (EXAFS) analysis (Di Cicco and Filipponi, 1994; van Dorsen et al., 2002). In the case of the 2p excitations in RE systems, the following processes are possible. The ground state of a RE can be described as $4f^n(5d6s6p)^3$. By neglecting the 6s and 6p electrons, $4f^n5d^\delta$ is obtained where δ is of the order of one. The normal L edge describes an excitation from $4f^n5d^\delta$ to $2p^5 4f^n 5d^{\delta+1}$ and the quadrupole pre-edge to $2p^5 4f^{n+1} 5d^\delta$. In addition, one can imagine that a second core electron is coexcited. This is only significant for shallow core levels and only if they couple strongly to the 4f valence states. This is the case for the 4d core states that are located between 100 and 190 eV binding energy. This yields the 4d5d shake-up MEE to $2p^5 4d^9 4f^n 5d^{\delta+2}$ final states. Note that there are two core holes now, so one can imagine significant screening effects due to $4f^{n+1}$ contributions. These 2p4d MEE are approximately located at an energy given by the addition of their binding energies, implying that they can be found some 100 to 190 eV above the L edges. They can be made visible with XMCD because they have a large spin-polarization due to the exchange interaction with the 4f electrons. This was discussed with respect to the $N_{4,5}$ edges in Chapter 6. It turns out that the 2p4d MEE are clearly visible in the XMCD signal (Dartyge et al., 1992). In addition, these 2p4d MEE peaks can be made visible with resonant x-ray emission measurements.

7.7 APPLICATIONS OF XMCD

The main applications of XMCD experiments are studies of magnetic materials, including magnetic oxides, thin magnetic (multi)layers, interface and surface effects, single atom magnetism and adsorbates magnetic nanoparticles and catalyst materials, molecular magnets, and metal centers in proteins. Important tools that can be developed using the XMCD effects include the determination of magnetic moments under extreme conditions, element specific moments in multielement systems, the study of magnetic coupling, element specific magnetization curves, its use in x-ray microscopy, and the refinement of XAS for site symmetry and electronic structure.

7.7.1 MAGNETIC OXIDES

Recently there has been an increased interest in the magnetic properties of TM oxides. Binary bulk TM oxides are usually antiferromagnetic. This includes the systems such as NiO, CoO, MnO, and α-Fe_2O_3. Historically, the most famous (ferri)-magnetic oxide is magnetite Fe_3O_4, which has an inverse spinel structure. Recently, a number of new magnetic oxide families have been (re)discovered that include the giant magnetoresistance materials such as the doped $LaMnO_3$ perovskites, and the vanadium oxide-based spin ladders.

One of the first applications of XMCD was devoted to iron garnets. At room temperature, the primary Fe XMCD signal is negative while the Gd signal is positive. This indicated that the bulk magnetic moment was dominated by the contribution from the Fe spins and that the Gd was antiferromagnetically coupled to the Fe. At low temperature, the Gd moment becomes the dominant factor, while positive Fe L_3-edge XMCD again indicates antiferromagnetic coupling (Rudolf, 1992).

Because XMCD usually adds fine structure not visible in XAS, it can be a useful additional technique to determine the nature of para- or ferrimagnetic phases and impurities in intrinsically diverse systems such as natural minerals and heterogeneous catalysts. By comparison with CTM calculations, the site occupancies of the cations can be determined. This principle has been applied to a series of iron spinel systems. The analysis yields similar information to that of the Mössbauer spectroscopy, though some differences were found that lead to a better understanding of these systems (Pattrick et al., 2002).

An important development has been the combination of XMCD and XMLD with microscopy using photoemission electron microscopes (PEEMs). The Stöhr group published a range of elegant studies (e.g. the study of the $LaFeO_3/SrTiO_3$ system). The $LaFeO_3$ layer is an epitaxial thin film and its antiferromagnetic (AF) domains are studied using XMLD at the iron L edge (Scholl et al., 2000). Nolting et al. (2000) extended this study by adding a 1.2 nm-thick Co layer on top of the $LaFeO_3/$ $SrTiO_3$ system. The PEEM images of thin ferromagnetic (F) Co films grown on anti-ferromagnetic $LaFeO_3$ show a direct link between the arrangements of spins in each material. This implies that the alignment of the ferromagnetic spins is determined, domain-by-domain, by the spin directions in the underlying antiferromagnetic layer (Nolting et al., 2000). The AF magnetization angle does not have to be exactly perpendicular to the surface. Actually, the XMLD signal allows for the determination of this angle by measuring the angular dependencies of the XAS intensities, as discussed previously for Equation 7.10. The detailed measurement for various in-plane and out-of-plane intensities has been applied to a $LaFeO_3$ thin film and it was found that the AF axis is tilted 20° out of the surface plane (Czekaj et al., 2006).

7.7.2 THIN MAGNETIC (MULTI)LAYERS, INTERFACE, AND SURFACE EFFECTS

Over the last 10 years, magnetic multilayer systems have been extensively studied because of the discovery of giant magnetoresistance (GMR) and of oscillatory inter-layer exchange coupling. Oscillatory coupling between two ferromagnetic layers occurs in essentially all TM multilayer systems in another transition or noble metal, which forms the nonferromagnetic layer. GMR occurs if the two layers are coupled antiferromagnetically, which occurs in systems such as Fe/Cr and Co/Cu. In addi-tion, Co/Pd and Co/Pt magnetic multilayers are potentially important systems to increase the bit size in magneto-optical recording (Stöhr and Nakajima, 1998).

XMCD experiments have been performed on a range of magnetic multilayers, where the sum rules have often been applied. Because of the limitations of the spin sum rule as discussed previously, the spin moments obtained for 3d systems might have some ambiguity. In the case of 4d-systems, the spin sum rule works much better and detailed results have been obtained for Pd/Fe multilayers (Cros et al., 1997).

Table 7.10 indicates the values for $\langle L_z \rangle$ and $\langle S_z \rangle$ for an increasing thickness of the Pd layers. For thicker layers, $\langle L_z \rangle$, $\langle S_z \rangle$, and their ratio all decrease. Analysis has shown that the induced moment on the Pd layer includes the first four Pd layers at the interface. A large number of XMCD studies have been performed on magnetic multilayers (Wende, 2004; Poulopoulos, 2005).

XMCD can be used to measure the element-selective hysteresis curves. During a hysteresis loop, it can be assumed that the peak intensity of the XMCD is proportional to the magnetic moment. As such, the normalized XMCD peak intensity can be tracked during the changes in the applied magnetic field and a hysteresis loop can be created. In a multielement system, this can be done for the various magnetic elements present. Figure 7.26 shows the case of a Fe/Cu/Co trilayers system, for which different Fe and Co hysteresis curves were obtained. From these curves, the individual mag-netic moments for the Fe and Co layers can be determined (Chen et al., 1995).

The group of Alain Fontaine (CNRS, Grenoble) have developed a nanosecond resolved XMCD measurement that allows for the study of the magnetization reversal

TABLE 7.10
$\langle L_z \rangle$ and $\langle S_z \rangle$ for Increasing Thickness of the Pd Layers

Pd AL	$\langle L_z \rangle$	$\langle S_z \rangle$	Ratio
2	0.04	0.17	0.25
4	0.02	0.15	0.13
8	0.02	0.12	0.16
14	0.01	0.07	0.09

Note: XMCD sum rule determined values for orbital and spin moments per Pd atom for the different multilayers. The first column gives the number of Pd atomic layers (AL) where there are 8 Fe AL (Cros et al., 1997).

dynamics of thin films with element selectivity. The response of the XMCD in pump-probe mode of a $Co_{5\,nm}Cu_{X\,nm}(Ni_{80}Fe_{20})_{5\,nm}$ trilayer system excited with 30 ns pulses has been studied. They showed that for the 10 nm Cu interlayer, the cobalt magnetization does not switch for pulses up to 28 mT. The $Ni_{80}Fe_{20}$ switches completely for fields around 10 mT and its speed of reversal increases with the pulse amplitude and

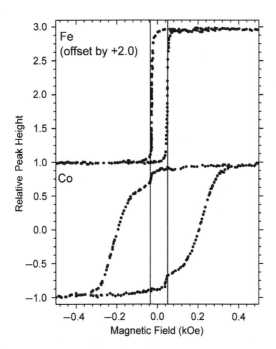

FIGURE 7.26 The Fe and Co XMCD intensity at the L_3 edge maximum as a function of the applied magnetic field. (Reprinted with permission from Chen, C.T., et al., *Phys. Rev. B*, 48, 642, 1993. Copyright 1993 by the American Physical Society.)

the two magnetic layers are clearly decoupled. In contrast, for the 6 nm Cu interlayer, the Co and $Ni_{80}Fe_{20}$ layers are strongly coupled and relax with the same switching times. The 8 nm Cu interlayer shows more complex intermediate behavior (Bonfim et al., 2001).

7.7.3 IMPURITIES, ADSORBATES, AND METAL CHAINS

Novel growth techniques allow for the preparation of well-characterized systems that contain localized impurities, small clusters, islands, chains, and suchlike on single crystal surfaces. The magnetic properties of these systems can be studied easily with XMCD thanks to its element selectivity and high cross section.

Fe, Co, and Ni impurities on potassium and sodium films have been studied with XAS and XMCD. The multiplet structure indicates that Fe, Co, and Ni have localized atomic ground states with a predominantly $3d^7$, $3d^8$, and $3d^9$ character. This implies that the effective valence can be written as Fe^{1+}, Co^{1+}, and Ni^{1+}, in line with Fe^{2+} having a $3d^6$ configuration. The XMCD spectra in Figure 7.27 show that the impurity states possess large, atomic-like, magnetic orbital moments that are progressively quenched as clusters are formed (Gambardella et al., 2002). Co impurities can also be deposited on a Pt(111) surface. This creates a large magnetic anisotropy energy of 9 meV per atom arising from the combination of its orbital moment and the spin–orbit coupling induced by the platinum substrate. Sum rule analysis of the

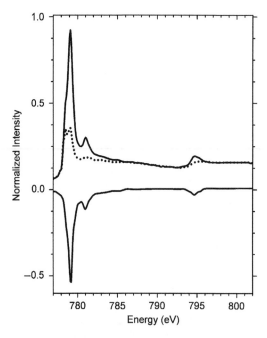

FIGURE 7.27 $L_{2,3}$ XAS spectra with parallel (solid line), antiparallel (dashed line) alignment and the resulting XMCD spectrum for the 0.015 monolayers Fe and Co and the 0.004 monolayer Ni. (Reprinted with permission from Gambardella, P., et al., *Phys. Rev. Lett.*, 88, 2002. Copyright 2002 by the American Physical Society.)

XMCD spectra show that the orbital moment $\langle L_z \rangle$ is -1.1 and $\langle S_z \rangle$ is -0.9 where $\langle S_z \rangle$ is calculated to be -1.05 from the density functional theory (DFT) calculations. In the case of cobalt clusters with eight atoms, the value of $\langle L_z \rangle$ has already decreased to below -0.4, close to its bulk value. The large orbital moment directly implies a large magnetic anisotropy energy (Gambardella et al., 2003).

One-dimensional linear magnetic chains are popular in theoretical modeling, and, recently, it has been possible to actually make and characterize them with XMCD. One-dimensional chains of Co were constructed on a Pt substrate. Depending on the temperature, these chains change from fluctuating segments of ferromagnetically coupled atoms into a ferromagnetic long-range ordered state. The Co chains are characterized by large localized orbital moments and correspondingly large magnetic anisotropy energies compared to two-dimensional films and bulk Co (Gambardella et al., 2002).

7.7.4 MAGNETIC NANOPARTICLES AND CATALYST MATERIALS

Magnetic nanoparticles (e.g. iron oxide nanoparticles), are components of biomedical materials, catalysts, and magnetic recording media. Below a critical diameter of typically ~20 nm, nanoparticles are single magnetic domains, and can exhibit a superparamagnetic behavior. Because these systems are small, their surface to bulk ratio is large and the effect of surface properties is large. This includes the presence of defects and adsorbates that also affect the spin-canting effects.

XMCD at the Fe $L_{2,3}$ edges of γ-Fe_2O_3 (maghemite) nanoparticles allows to separate the contributions of the magnetic moments of Fe^{3+} ions in tetrahedral and octahedral sites. It turns out that, under high magnetic fields, a reduction of the magnetic contribution of the octahedral Fe^{3+} ions occurs for 8 nm phosphate-coated particles by comparison with the uncoated ones. A similar reduction appears at lower magnetic fields for 2.7 nm particles. The results jointly show the existence of a preferential spin canting of octahedral Fe^{3+} spins at the surface. A possible explanation is that surface spins experience weakened exchange interactions with their neighbors, yielding an increased disorder of spins at low (<500 mT) magnetic fields for small particles (Brice-Profeta et al., 2005).

Heterogeneous catalysts contain low-loaded metal sites that form isolated sites, binuclear clusters and larger clusters, and nanoparticle oxide phases. Assuming a paramagnetic system of isolated moments, full magnetization can be obtained. In contrast, a binuclear (Fe^{3+}) center will couple antiferromagnetically, thereby cancelling its XMCD effect. Also, iron-oxide nanoparticles will be mainly antiferromagnetic. This implies that the magnitude of the XMCD effect can be related to the amount of single iron sites. In addition to the determination of the valence and site-symmetry, this approach has been used to show the clustering of iron during calcination and steaming treatments of Fe/ZSM5 materials (Heijboer et al., 2005).

7.7.5 MOLECULAR MAGNETS

Molecular magnets are systems where a permanent magnetization can be achieved (usually at low temperatures) within a single molecule. In most cases, this is achieved by a combination of different TM and RE ions. Knowledge of the internal magnetic structure of molecular magnets is of great importance to understanding the magnetic properties. Crucial aspects are the high zero-field splitting and magnetic anisotropy, that is, the combination of spin–orbit coupling, (low) crystal field symmetries and exchange interactions. XMCD, in combination with multiplet analysis, is an ideal tool to determine the nature of these systems.

Molecular magnet systems that have been studied with XMCD include cyanide-bridged Prussian Blue analogs that contain infinite three-dimensional $M_1(CN)M_2$ chains (e.g. $Fe^{3+}(CN)Fe^{2+}$ or $Cr^{3+}(CN)Ni^{2+}$). A single $M_1[(CN)M_2]_6$ magnetic core can also be created (e.g. $Cr^{3+}Ni_6^{2+}$ or $Cr^{3+}Mn_6^{2+}$). Another important class of molecular magnets are the oxo-bridged complexes. This includes the Fe_4O_4 ferric star, Fe_6O_{12} ferric wheel and $Mn_{12}O_{12}$. These oxo-bridged systems bear a close resemblance to the similar metal cores in proteins such as Mn_4O_4 in photosystem II. The Mn ions in $Mn_{12}O_{12}$ are mixed-valent between Mn^{3+} and Mn^{4+}. They are compared with Mn^{3+} and Mn^{4+} references. The analysis of the XMCD measurements (at different fields and temperatures) revealed that ^5E Mn^{3+} ions had anisotropic magnetic properties, whereas the $^4A_2$ Mn^{4+} were essentially isotropic (Moroni et al., 2003).

7.7.6 METAL CENTERS IN PROTEINS

Metal centers in proteins are paramagnetic and can be aligned in a strong magnetic field at low temperature. The nature of the ground state of the metal centers in the proteins can be studied along with the coupling of the moments in multi-metal sites. Steve Cramer and his group have measured a large range of Ni, Fe, and Mn metal centers in proteins, which have recently been reviewed (Cramer et al., 1996; Funk et al., 2005).

8 Resonant X-Ray Emission Spectroscopy

8.1 INTRODUCTION

In the x-ray emission process, a core hole is first created (say, in core level c) by incident x-ray or electron beam, and then a valence electron or a core electron in the core level higher than c makes a radiative transition to the core level c by emitting an x-ray photon. When the incident x-ray energy resonates with the excitation threshold of the core electron, this x-ray emission is called "resonant x-ray emission" (or resonant x-ray Raman scattering), while if the incident energy is large enough to excite the core electron to a high energy continuum well above the excitation threshold, it is called "normal x-ray emission" (or ordinary x-ray emission, or x-ray fluorescence).

Both resonant x-ray emission spectroscopy (RXES) and normal x-ray emission spectroscopy (NXES) are coherent second-order optical processes where the excitation and de-excitation processes are coherently correlated by the Kramers–Heisenberg formula. The information obtained from RXES and NXES is much greater than the first-order optical processes of x-ray absorption spectroscopy (XAS) and x-ray photoemission spectroscopy (XPS), but the intensity of the signal of RXES and NXES is much weaker than XAS and XPS. The recent remarkable progress in the study of RXES and NXES owes much to the implementation of undulator radiation from third generation synchrotron radiation sources, as well as highly efficient detectors. Especially with RXES, selected information can be obtained connected directly with a specific intermediate state to which the incident x-ray energy is tuned (Gel'mukhanov and Ågren, 1994; Nordgren and Kurmaev, 2000; Kotani and Shin, 2001; Kotani, 2005).

In this chapter, we mainly concentrate on the theoretical and experimental study of RXES. It is one of the most important core-level spectroscopies, providing us with such information as both x-ray absorption and emission processes and their correlation. Furthermore, RXES provides bulk-sensitive, element-specific, and site-selective information. The RXES technique can be applied equally to metals and insulators and can be performed in applied electric or magnetic fields as well as under high pressure, since it is a photon-in and photon-out process.

RXES is divided into two categories depending on the electronic levels participating in the transition of x-ray emission. In the first category, the transition occurs from the valence state to the core state, and no core hole is left in the final state of RXES. Typical examples are the 3d to 2p radiative decay of transition metal (TM) elements following the 2p to 3d excitation by the incident x-ray (denoted as 2p3d(3d) RXES or just 2p3d RXES), because it is obvious that the 2p electron is excited to the 3d states at resonance. In this case, the difference between the incident and the emitted x-ray energies (denoted as Raman shift) corresponds to the energy of the electronic elementary excitations, such as the crystal field level excitation, charge-transfer excitation, correlation gap excitation, and so on. In this case, RXES is also denoted as resonant inelastic x-ray scattering (RIXS), and the energy transfer corresponds to the energy of elementary excitations of valence electrons. If the initial and final electronic states are the same in the RXES process, the incident and emitted x-ray energies are the same and this x-ray scattering process is no more resonant inelastic, but resonant "elastic" x-ray scattering.

Compared with the (nonresonant) inelastic x-ray scattering (IXS), RXES has a larger intensity and depends on each intermediate state, which is convenient to identify the character of electronic excitations. IXS is alternatively known as x-ray Raman scattering (XRS) or x-ray energy loss spectroscopy (XELS). Compared with optical absorption spectroscopy, the selection rule of detecting electronic excitations in RXES is different in that RXES and optical absorption give complementary information on elementary excitations. In the hard x-ray region, it can be observed that the momentum transfer in RXES, which corresponds to the wave number of the excitation mode, provides us with important information on the spatial dispersion of elementary excitations.

The second category of RXES is the case where the radiative decay occurs from a core state to another core state, so that a core hole is left in the final state of RXES. A typical example is the 3p to 1s radiative decay following the 1s to 4p excitation in TM elements. In general, the lifetime of a shallow core hole is longer than that of a deeper one, and furthermore, the lifetime broadening of RXES is determined by the core hole in the final state, instead of the intermediate state. Taking advantage of these facts, a weak signal of core electron excitations can be detected using RXES measurements, which cannot be detected by the conventional XAS measurements because of the large lifetime broadening of a deep core hole. Information on the spin-dependence of core electron excitation can also be obtained by RXES measurements.

The polarization-dependence in RXES gives important information on the symmetry of electronic states. For linearly polarized incident x-rays, two different polarization geometries (polarized and depolarized geometry) are often used. In both geometries, as shown in Figure 8.1, the angle of the incident and emitted x-ray directions is fixed at 90°, and the polarization of the emitted x-ray is not detected

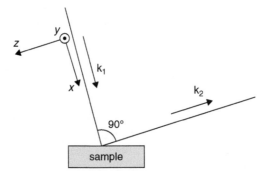

FIGURE 8.1   Polarized and depolarized geometries in RXES. (From Kotani, A., *Eur. Phys. J. B*, 47, 3, 2005. With permission.)

(because detection of the emitted x-ray polarization is very difficult in the x-ray region). In the polarized geometry, the incident x-ray polarization is perpendicular (y direction) to the scattering plane, while in the depolarized geometry it is parallel (z direction) to the scattering plane. Since the incident x-ray polarization is parallel to the x-ray emission direction in the depolarized geometry, the polarization of the x-ray should necessarily be different before and after the scattering. In the polarized geometry, on the other hand, the polarization of the incident x-ray can be the same as that of the emitted x-ray. Sometimes a geometry that is different from the polarized and depolarized geometries is also used. For instance, the angle between the incident and emitted x-rays is to some extent changed from 90°, and as such, the geometries where the incident polarization is perpendicular and parallel to the scattering plane are called the V (vertical) and T (transverse) geometries, respectively. Another important polarization-dependence in RXES is the magnetic circular dichroism (MCD) in ferromagnetic samples. The difference in RXES for circular polarized incident x-rays with + and − helicities gives important information on the magnetic polarization of electronic states in ferromagnetic materials (Kotani and Shin, 2001; Kotani, 2005).

Now, a few words about terminology. The terms of both RXES [or resonant soft x-ray emission spectroscopy (RSXES)] and RIXS are widely used in describing the resonant photon-in and photon-out processes. In this chapter, we mainly use the term RXES rather than RIXS. Strictly speaking, RXES includes more processes than RIXS; the RIXS spectroscopy is a part of RXES, which disperses linearly with the incident photon energy (i.e. the part with a constant Raman shift). The term RXES also includes resonant elastic x-ray scattering with zero energy loss and "fluorescence" spectroscopy with a constant (or almost constant) x-ray emission energy. If the incident x-ray energy is increased far above the excitation threshold of a core electron, "fluorescence" spectrum is changed into NXES (or "ordinary fluorescence" spectrum) with a strictly constant x-ray emission energy. The spectral shape of NXES is quite independent of the incident photon energy. It is to be mentioned that the "fluorescence" spectra near the threshold excitation is sometimes called "normal x-ray emission-like spectra."

As mentioned previously, the term of RIXS is only appropriate when used for the first category of RXES, where the Raman shift corresponds to the energy of an elementary excitation in the valence electron states. For a second category of RXES, we can denote the RXES due to a 3p to 1s radiative decay following a 1s to 4p excitation in TM elements by "$K\beta$ RXES" or "$K-M_{23}$ RXES" because the NXES of the 3p to 1s radiative decay is traditionally called "$K\beta$ XES" or "$K-M_{23}$ XES." The term "$K\beta$ RIXS" is also sometimes used but we consider that $K\beta$ RXES is more appropriate. The terminology for RXES has not yet been fully established because it is a new field that has only very recently been developed. It is, therefore, desirable to establish an appropriate and unambiguous terminology.

8.1.1 Experimental Aspects of XES (RXES and NXES)

NXES experiments need an x-ray analyzer to detect the emitted x-rays from the sample. A RXES experiment needs to be able to use a variable x-ray source for the excitation process. In practice, most RXES and NXES experiments are performed with synchrotrons where, in principle, any beamline used for XAS can also be used for RXES. A special situation arises for RXES experiments with hard x-rays, where the lifetime broadening of the deep core hole can effectively be removed from the spectral shape. This implies that the traditional hard x-ray XAS beamlines that use Si(111) monochromator crystals have to be replaced with crystals with higher $h^2 + k^2 + l^2$ surfaces such as Si(311) or Si(440).

8.1.1.1   Detectors for Soft X-Ray XES

In the case of soft x-rays, the monochromator is based on artificial gratings. Also, the soft x-ray detectors are (mainly) based on gratings. A typical soft x-ray XES detector is based on a number of fixed gratings coupled to a two-dimensional detector. Only ~0.1% of the core holes decay via fluorescence and this fluorescence is emitted in all directions. In addition, the solid angle of soft x-ray gratings is relatively small and taken together, this implies a relatively low signal, even at a brilliant third generation soft x-ray beamline (Nordgren et al., 1989; Ghiringhelli et al., 1998; Hague et al., 2005).

8.1.1.2   Detectors for Hard X-Ray XES

Hard x-ray XES uses crystal monochromators for excitation and also crystal monochromators as x-ray analyzers. The most popular detector system presently in use is based on a spherically bent crystal in Rowland geometry that selects the fluorescence energy from the sample and directs it onto a photon detector (Bergmann et al., 2001, 2003). In order to enlarge the solid angle, two or more analyzer crystals can be used in parallel, each with its own photon detector or coupled into a combined photon detector. This detection scheme is standard for all XES spectra, including synchrotron excitation, x-ray tube excitation, electron excitation, and K capture where, in principle, all NXES, RXES, and XRS can use the same detector (Glatzel and Bergmann, 2005).

8.1.1.3   X-Ray Raman Allows Soft X-Ray XAS under Extreme Conditions

An important experimental aspect is that XRS experiments measure a soft x-ray core excitation (e.g. an oxygen K edge), while using the inelastic scattering of hard x-rays (Wernet et al., 2005). This implies that it is possible to measure soft XAS edges of C, N, and O under any conditions essentially. This includes high-pressure measurements using diamond anvil cells (Badro et al., 2004) and measurements of, for example, liquids, supercritical materials, and encapsulated systems.

8.1.2   BASIC DESCRIPTION AND SOME THEORETICAL ASPECTS

Let us consider the XES process (either of RXES and NXES processes) where an x-ray photon with energy $\hbar\Omega$ (wavevector $\mathbf{k}_1$) and polarization λ_1 is incident on a material and then an x-ray photon with energy $\hbar\omega$ (wavevector $\mathbf{k}_2$) and polarization λ_2 is emitted as a result of the electron–photon interaction in the material. We take into account the electron–photon interaction of the form $[e^2/(2mc^2)] \sum_n \mathbf{A}(\mathbf{r}_n)^2$ by the lowest-order perturbation, and that of $[e/(mc)] \sum_n \mathbf{p}_n \mathbf{A}(\mathbf{r}_n)$ by the second-order perturbation where $\mathbf{A}(\mathbf{r})$ is the vector potential of the photon. Then the differential cross section of the photon scattering (with respect to the solid angle $\Omega_{\mathbf{k}_2}$ and energy $\hbar\omega$ of the scattered photon) is expressed as (Kramers and Heisenberg, 1925; Heitler, 1944)

$$\frac{d^2\sigma}{d\Omega_{\mathbf{k}_2} d(\hbar\omega)} = \frac{\omega^2 V_s}{\hbar c^4} \left(\frac{1}{2\pi}\right)^3 W_{12},$$ (8.1)

where the transition rate W_{12} is given by

$$W_{12} = \frac{2\pi}{\hbar} \sum_j \frac{(2\pi)^2}{\Omega\omega} \left(\frac{e^2\hbar}{mV_s}\right)^2 \delta(E_g + \hbar\Omega - E_j - \hbar\omega)$$

$$\times \left\| \left[\langle j|\rho_{\mathbf{k}_1 - \mathbf{k}_2}|g\rangle (\mathbf{e}_1 \cdot \mathbf{e}_2) + \frac{1}{m} \sum_i \left(\frac{\langle j|\mathbf{p}(\mathbf{k}_2)\cdot\mathbf{e}_2|i\rangle\langle i|\mathbf{p}(-\mathbf{k}_1)\cdot\mathbf{e}_1|g\rangle}{E_i - E_g - \hbar\Omega} \right. \right. \right.$$

$$\left. \left. \left. + \frac{\langle j|\mathbf{p}(\mathbf{k}_1)\cdot\mathbf{e}_1|i\rangle\langle i|\mathbf{p}(-\mathbf{k}_2)\cdot\mathbf{e}_2|g\rangle}{E_i - E_g + \hbar\omega} \right) \right] \right\|^2.$$ (8.2)

Here $|g\rangle, |i\rangle$, and $|j\rangle$ are initial, intermediate, and final states of the material system, respectively, E_g, E_i, and E_j are their energies, and $\mathbf{e}_1$ and $\mathbf{e}_2$ are polarization directions (unit vectors) of incident and emitted photons (for instance, $\mathbf{e}_1$ is an abbreviation of $\mathbf{e}_{\mathbf{k}_1\lambda_1}$). We define $\mathbf{p}(\mathbf{k})$ and $\rho_{\mathbf{k}}$ by

$$\mathbf{p}(\mathbf{k}) = \sum_n \mathbf{p}_n \exp(-i\mathbf{k}\cdot\mathbf{r}_n),$$ (8.3)

$$\rho_{\mathbf{k}} = \sum_n \exp(-i\mathbf{k}\cdot\mathbf{r}_n).$$ (8.4)

The three terms in the square bracket of W_{12} are shown in Figure 8.2. The first term comes from the first-order perturbation of the A^2-type interaction, and this x-ray scattering is called Thomson scattering. If we only take into account the

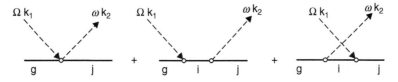

FIGURE 8.2 Schematic representations of three scattering terms. (Reprinted with permission from Kotani, A., and Shin, S., *Rev. Mod. Phys.*, 73, 203, 2001. Copyright 2001 by the American Physical Society.)

Thomson scattering, the scattering cross section is rewritten as the standard expression:

$$\frac{d^2\sigma}{d\Omega_{k_2}\,d(\hbar\omega)} = \frac{\omega}{\Omega}\left(\frac{e^2}{mc^2}\right)^2 (\mathbf{e}_1 \cdot \mathbf{e}_2)^2 S(\mathbf{k}_1 - \mathbf{k}_2, \Omega - \omega), \tag{8.5}$$

where $S(\mathbf{k}, \nu)$ is the dynamical structure factor defined by

$$S(\mathbf{k},\nu) = \frac{1}{2\pi\hbar}\int_{-\infty}^{\infty} dt\,\exp(i\nu t)\langle g|\rho_{\mathbf{k}}(t)\rho_{-\mathbf{k}}|g\rangle. \tag{8.6}$$

Therefore, the Thomson scattering gives us direct information on the elementary excitation caused by the charge fluctuation in materials. When the incident photon energy is close to or above the core electron excitation threshold as in the case of RXES and NXES, the contribution of the second term becomes dominant. Above the threshold, the denominator $E_i - E_g - \hbar\Omega$ vanishes so that the second-order perturbation calculation breaks down. However, if we take into account that the intermediate state has a finite lifetime τ_i $(= \hbar/\Gamma_i)$ because of the lifetime of a core hole, then the energy E_i is replaced by a complex number $E_i - i\Gamma_i$ and the divergence is removed (Sakurai, 1967; Tulkki and Åberg, 1980). Taking into account these facts and removing unimportant factors in Equation 8.2, we can describe the most essential part of the spectrum of RXES (and NXES) in the form

$$F(\Omega,\omega) = \sum_j \left|\sum_i \frac{\langle j|T_2|i\rangle\langle i|T_1|g\rangle}{E_g + \hbar\Omega - E_i + i\Gamma_i}\right|^2 \delta(E_g + \hbar\Omega - E_j - \hbar\omega), \tag{8.7}$$

where the operators T_1 and T_2 represent the radiative transitions by incident and emitted photons, respectively, and Γ_i represents the spectral broadening due to the core-hole lifetime in the intermediate state. The quantity Γ_i is a result of the Auger and radiative decays of the core hole, and in most cases, it can be taken approximately to be constant, independent of the index i. If we consider the optical dipole transition (long wavelength limit of the photon), T_i $(i = 1, 2)$ is given by

$$T_i = \mathbf{p}(0) \cdot \mathbf{e}_i, \tag{8.8}$$

taking the limit of $\mathbf{k}_i \rightarrow 0$ in Equation 8.2. In the optical quadrupole transition, which is the next lowest-order term of T_i with respect to $\mathbf{k}_i$, T_i is given by

$$T_i = -i(\mathbf{p}(0)\cdot\mathbf{e}_i)(\mathbf{k}_i \cdot \mathbf{r}), \tag{8.9}$$

where we took the lowest-order term of the exponential factor in Equation 8.3. In the soft x-ray range, the optical quadrupole transition can be disregarded but, in the hard x-ray range, it sometimes plays an important role.

Equation 8.7 can be directly applied to RXES, but for NXES, some modification of the expression is possible. In the case of NXES, $\hbar\Omega$ is well above the x-ray absorption threshold and the electron, which is excited from the core level to a high-energy continuum, can be treated as independent of other electrons in the intermediate and final states. Then we put

$$|i\rangle = |\phi_\varepsilon\rangle|i'\rangle, \quad E_i = E_{i'} + \varepsilon, \tag{8.10}$$

$$|j\rangle = |\phi_\varepsilon\rangle|j'\rangle, \quad E_j = E_{j'} + \varepsilon, \tag{8.11}$$

into Equation 8.7 and obtain

$$F(\Omega,\omega) = \sum_{j'}\int d\varepsilon \rho(\varepsilon)t^2 \left|\sum_{i'}\frac{\langle j'|T_2|i'\rangle\langle i'|a_c|g\rangle}{E_g + \hbar\Omega - E_{i'} - \varepsilon + i\Gamma_{i'}}\right|^2 \delta(E_g + \hbar\Omega - E_{j'} - \varepsilon - \hbar\omega), \tag{8.12}$$

where t ($\cong$ constant) is the dipole transition amplitude from a core state to the photo-electron state, a_c is the annihilation operator of the core electron, and $\rho(\varepsilon)$ is the density of states (DOS) of the photoelectron. Performing the integration over ε and putting $\rho(\varepsilon) \cong$ constant, we obtain

$$F(\Omega, \omega) = \rho t^2 \sum_{j'}\left|\sum_{i'}\frac{\langle j'|T_2|i'\rangle\langle i'|a_c|g\rangle}{E_{j'} - E_{i'} - \hbar\omega + i\Gamma_{i'}}\right|^2. \tag{8.13}$$

It is found that $F(\Omega, \omega)$ of NXES does not depend on Ω. Actually ρ^2 might depend on Ω but the spectral shape of NXES is independent of Ω. Also, it is to be stressed that NXES is still a coherent second-order optical process, where the process of core hole creation (i.e. the photoelectron excitation process) is correlated, in general, with the x-ray emission process (Tanaka et al., 1989a, 1989b). If we assume that the system is well described by the one-electron approximation (for instance, with the band model), then the effect of the coherence plays no important role. In that case, Equation 8.13 reduces to

$$F(\omega) = \rho t^2 \sum_{i}^{occ}\frac{|\langle\phi_c|t|\phi_i\rangle|^2}{(\hbar\omega - \varepsilon_i + \varepsilon_c)^2 + \Gamma_i^2}, \tag{8.14}$$

where $\langle\phi_c|t|\phi_i\rangle$ is the one-electron transition matrix element of the x-ray emission. Therefore, if we assume that $|\langle\phi_c|t|\phi_i\rangle|^2 \cong$ constant and Γ_i is infinitesimally small, the NXES spectrum is proportional to the DOS of the occupied states i. Actually, the NXES spectrum gives the partial DOS, which is symmetry-selected by the dipole transition $|\langle\phi_c|t|\phi_i\rangle|^2$ and broadened by the lifetime broadening Γ_i of the intermediate state.

If the incident photon energy $\hbar\Omega$ is decreased to the x-ray absorption threshold, t is no longer a constant and the excited electron couples with other electrons. Then $F(\Omega, \omega)$ depends strongly on Ω and this is nothing but RXES (or RIXS). Thus, RXES and NXES are two different aspects of the XES spectrum $F(\Omega, \omega)$, which are caused by the different character of intermediate states due to the different choice of Ω. In between RXES and NXES, it is sometimes possible to observe a "NXES-like" spectrum where the spectral shape of $F(\Omega, \omega)$ is almost independent of Ω but slightly depends on it.

We consider the first category of RXES (i.e. RIXS). If the final state $|j\rangle$ is the same as the initial state $|g\rangle$, then the spectrum of Equation 8.7 describes the resonant elastic x-ray scattering. If $|j\rangle$ is not the same as $|i\rangle$, it gives the RIXS spectrum. As in the case of the Thomson scattering, RXES provides us with important information on the charge excitations in material systems. Furthermore, RXES is much more useful than the Thomson scattering. Usually the Thomson scattering is too weak to obtain precise information about electronic excitations, but the intensity of RXES is stronger because of the resonance effect. Thomson scattering depends on Ω and ω only through $\Omega - \omega$, but RXES depends on both of them independently. Therefore, we can obtain more detailed information about the electronic excitations by tuning Ω to different intermediate states. Since the intermediate states are different for different atomic species, the information given by RXES depends on the atomic species. Since the Raman shift, $\hbar\bar\Omega - \hbar\bar\omega$, in RXES corresponds to the energy of electronic elementary excitations, so RXES gives the information on the elementary excitations projected on atomic species, atomic sites, and the intermediate electronic states.

Before closing this subsection, let us briefly mention two pioneering works. Sparks (1974) observed RXES experimentally as a resonant scattering of Cu $K\alpha$ XES on various target metals (Ni, Cu, Zn, and so on). In the intermediate state of the experiment, a 1s electron was excited virtually above the Fermi level ε_F, and in the final state a 2p electron made a radiative transition to the 1s state. Therefore, the emitted photon was observed at $\hbar\omega \cong \hbar\Omega - (\varepsilon_F - \varepsilon_L)$, where $\varepsilon_F - \varepsilon_L$ corresponds to the binding energy of the 2p electron. It was found that the intensity of this RXES was consistent with the resonant enhancement factor produced for each material.

Eisenberger et al. (1976a,b) were first to experiment with RXES utilizing synchrotron radiation. Their experiment was performed for Cu $K\alpha$ RXES of Cu metal and the schematic representation of the energy level scheme is given in Figure 8.3. They changed the incident photon energy $\hbar\Omega$ continuously around $\varepsilon_F - \varepsilon_K$ and observed the emitted photon in the neighborhood of $\hbar\omega \cong \varepsilon_L - \varepsilon_K$, where ε_K and ε_L are the Cu 1s and 2p core levels (more exactly, $2p_{3/2}$ in their experiments). The experimental result was consistent with what is expected from the second-order optical formula $F(\Omega, \omega)$ given by Equation 8.7. If we disregard electron–electron interactions, $F(\Omega, \omega)$ is written, apart from unimportant factors, as

$$F(\Omega,\omega) \approx \sum_{\mathbf{k}(k>k_F)} \left| \frac{1}{\hbar\Omega - \varepsilon_{\mathbf{k}} + \varepsilon_K - i\Gamma_K} \right|^2 \delta(\hbar\omega + \varepsilon_{\mathbf{k}} - \varepsilon_L - \hbar\Omega)$$

$$= \begin{cases} \rho/[(\Delta E_0)^2 + \Gamma_K^2] & (\Delta E_0 \le \Delta E_R), \\ 0 & (\Delta E_0 > \Delta E_R), \end{cases} \tag{8.15}$$

where

$$\Delta E_R = \hbar\Omega - (\varepsilon_F - \varepsilon_K), \tag{8.16}$$

$$\Delta E_0 = \hbar\omega - (\varepsilon_L - \varepsilon_K). \tag{8.17}$$

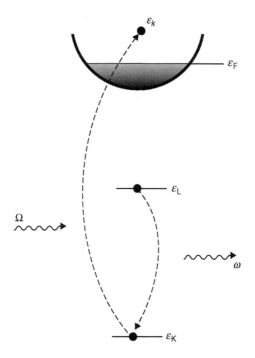

FIGURE 8.3 Schematic representation of RXES in Cu metal. (Reprinted with permission from Kotani, A., and Shin, S., *Rev. Mod. Phys.*, 73, 203, 2001. Copyright 2001 by the American Physical Society.)

Here Γ_K and ρ are, respectively, the lifetime broadening of the 1s core level and the DOS of the conduction band. In the above expression of $F(\Omega, \omega)$, the effect of the lifetime broadening of the 2p core level (Γ_L) is disregarded for simplicity, but if it is taken into account, the discontinuity at $\Delta E_0 = \Delta E_R$ is blurred out by the width Γ_L. In any case, the above calculation shows that for $\Delta E_R < 0$, the peak of $F(\Omega, \omega)$ occurs at $\Delta E_0 = \Delta E_R$, but for $\Delta E_R > 0$, it occurs at $\Delta E_0 = 0$. Furthermore, the half width at the half maximum of the peak should be minimized for $\Delta E_R = 0$. The experimental data by Eizenberger et al. were well explained by these facts.

8.2 RARE EARTH COMPOUNDS

8.2.1 Effect of Intra-Atomic Multiplet Coupling

In rare earth (RE) systems, the crystal field energy (typically of the order of 10 meV) for 4f states is smaller than the spin–orbit interaction energy (order of 100 meV) of 4f states. In contrast, the crystal field energy (order of 1 eV) is larger than the spin–orbit interaction energy (order of 10 meV) for 3d states of TM atoms. Therefore, for most theoretical analyses of experimental RXES spectra in RE systems, the effect of the crystal field can be disregarded (for the effect of the crystal field, see the Dy 4d4f RXES shown in Figure 8.9). Furthermore, in many RE compounds, except for some Ce, Pr, Tb, and Yb compounds, the 4f states are well localized so that the effect of hybridization between 4f and ligand states can be disregarded. Then, the single impurity Anderson model (SIAM) of RE compounds reduces to the free atom model, and RXES spectra reflect the atomic multiplet (AM) structures (f-f excitations) caused by the multipole components of Coulomb interaction and the spin–orbit interaction. However, charge-transfer excitations do not occur.

Here, we consider the case where the system is well described by the atomic Hamiltonian and the solid-state effect can be disregarded. One of the simplest examples is the Ce 3d XAS and 3d4f RXES of CeF_3. Experimental and calculated results (Butorin et al., 1996b; Nakazawa et al., 1996) are shown in Figure 8.4 where the Ce 3d XAS is shown on the top and the RXES spectra with the incident energies A, B, and C are shown in the lower panel. The Ce ion in CeF_3 is well described by an isolated Ce^{3+} free ion and the RXES process is described by the transitions $3d^{10}4f^1 \rightarrow 3d^9 4f^2 \rightarrow 3d^{10}4f^1$. The 3d XAS shows the multiplet structure (for instance, structures A, B, and C), which is characteristic of the atomic $3d^9 4f^2$ configuration for each of the $3d_{3/2}$ and $3d_{5/2}$ core levels. On the other hand, the RXES spectra always shows a single peak. This peak corresponds to an elastic x-ray scattering with missing IXS. This is because the final state ^2F in the $4f^1$ configuration is the same as the initial state (note that there is no multiplet splitting for the single electron $4f^1$ configuration). More exactly, the initial state is the $^2F_{5/2}$ ground state, while the final state includes a $^2F_{7/2}$ excited state in addition to the ground state. However, the energy difference between $^2F_{5/2}$ and $^2F_{7/2}$ states is about 0.3 eV (4f spin–orbit splitting), which is too

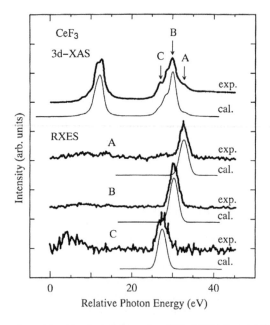

FIGURE 8.4 Experimental and calculated results of Ce 3d XAS and 3d4f RXES for CeF_3. The results A, B, and C of RXES spectra are obtained by tuning the incident photon energy to A, B, and C of the XAS. (From Nakazawa, M., Tanaka, S., Uozumi, T., and Kotani, A., *J. Phys. Soc. Jpn.*, 65, 2303, 1996. With permission.)

small (much smaller than the experimental resolution) to be resolved as an IXS spectrum in this experimental observation.

As another example of the AM effect in RXES, we again consider the Tm 3d XAS and 3d4f RXES of Tm metal where Tm is well approximated by a free Tm^{3+} ion. In Figure 8.5, the calculated results of Tm 3d XAS and 3d 4f RXES are shown and in Figure 8.6, the energy level diagram of the system is shown (Nakazawa et al., 1998). The ground state of Tm^{3+} (with $4f^{12}$ configuration) is the $^3H_6$ state. The Tm 3d XAS shows four peaks (A, B, C, and D) that correspond to the multiplet terms of the $3d^94f^{13}$ configuration. Peak A corresponds to a pure $^3H_6$ term. Peak B corresponds mainly to a $^3G_5$ term. However, peaks C and D correspond to strongly mixed states between $^3H_5$ and $^1H_5$ terms due to the spin–orbit interaction of the 3d core state. The RXES spectra with the incident photon energy tuned to A, B, C, and D are shown in Figure 8.5. It is found that for A and B, the spectra are mostly given by an elastic x-ray scattering peak where the final state is the same multiplet term $^3H_6$ as the ground state, while for C and D, we have a strong IXS peak corresponding to a spin flip excited state $^1I_6$, in addition to the elastic scattering peak. This is because the XAS final states C and D are mixed states between $^1H_5$ and $^3H_5$ and then the $^1H_5$ component decays to the $^1I_6$ final state while the $^3H_5$ component decays to the $^3H_6$ final state.

There has been no direct experimental observation of these RXES spectra for Tm^{3+}, but the fluorescence yield (FY) spectrum of the Tm 3d4f RXES has been measured as shown in Figure 8.7 (Pompa et al., 1997). The FY spectrum is the intensity

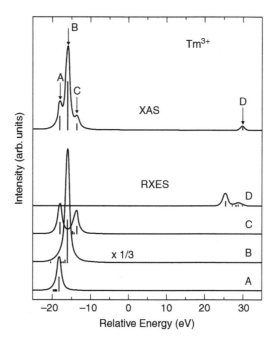

FIGURE 8.5 Calculated results of 3d XAS and 3d4f RXES for Tm^{3+} ion. The results A, B, C, and D of RXES spectra are obtained by tuning the incident photon energy to A, B, C, and D of the XAS. (From Nakazawa, M., Ogasawara, H., Kotani, A., and Lagarde, P., *J. Phys. Soc. Jpn.*, 67, 323, 1998. With permission.)

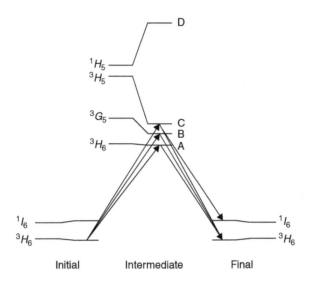

FIGURE 8.6 Schematic energy-level diagram of 3d4f RXES for Tm³⁺ ion. (From Nakazawa, M., Ogasawara, H., Kotani, A., and Lagarde, P., *J. Phys. Soc. Jpn.*, 67, 323, 1998. With permission.)

of RXES (both elastic and inelastic components) integrated over the emitted x-ray energy and then measured as a function of the incident x-ray energy. The observed FY spectrum is similar to the XAS spectrum but there is a clear difference between them: the Tm $3d_{5/2}$ XAS spectrum exhibits a three-peak structure and the lowest energy peak is almost missing in the FY spectrum. This result strongly supports the RXES process shown in Figure 8.7. The calculated FY, which is obtained by

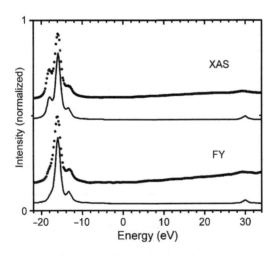

FIGURE 8.7 Experimental and calculated results of Tm $3d_{5/2}$ XAS and the corresponding total fluorescence yield (FY) of Tm metal. (From Nakazawa, M., Ogasawara, H., Kotani, A., and Lagarde, P., *J. Phys. Soc. Jpn.*, 67, 323, 1998. With permission.)

integrating the calculated RXES spectra over the emitted x-ray energy, is in good agreement with the experimental result as shown in Figure 8.7 (Pompa et al., 1997; Nakazawa et al., 1998). From this calculation, it becomes clear why the lowest peak A of XAS is missing in FY. If we only take into account the elastic x-ray scattering, we then have almost a single peak B in FY. This is because the intensity ratio of the three FY peaks for the elastic x-ray scattering is roughly given by the square of that of the three XAS peaks. Then the contribution from the IXS to FY gives a peak at the position C, but no peak at A because of almost vanishing IXS. In other words, the difference between FY and XAS originates from a radiative decay rate that is not constant but which is dependent on each intermediate state, as discussed by de Groot et al. (1994b) for different systems.

Before closing this subsection, an example of where the AM structure was observed in RXES with a considerably high resolution, is shown. Figure 8.8a shows the experimental results of Dy 4d XAS and Dy 4d4f RXES of DyF$_3$ (Butorin et al., 2000), and the calculated RXES spectra for the Dy^{3+} free atom (Nakazawa, 1998) are shown in Figure 8.8b. In order to precisely analyze the 4d XAS and 4d4f RXES of RE systems, it is necessary to take into account the Fano effect (the details of the calculation are omitted). In each RXES spectrum from a to m, the sharp peak at the

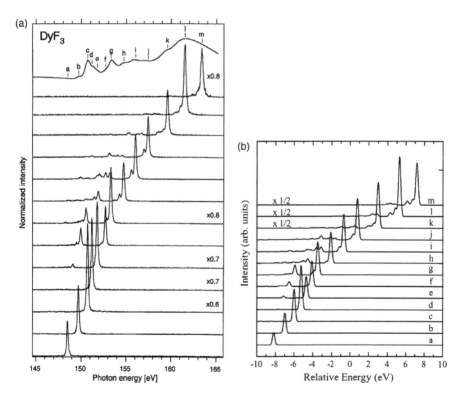

FIGURE 8.8 (a) Experimental results of Dy 4d XAS and 4d4f RXES for DyF$_3$; (b) Calculated results of Dy 4d4f RXES for Dy^{3+} ion. (Reprinted with permission from Kotani, A., and Shin, S., *Rev. Mod. Phys.*, 73, 203, 2001. Copyright 2001 by the American Physical Society.)

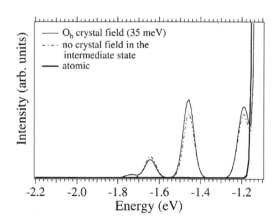

FIGURE 8.9 The energy-loss region between 1.0–2.2 eV of the Dy 4d4f RXES spectrum, calculated using a pure atomic approach (solid) and crystal field multiplet theory in O_h symmetry (dashed and thin solid). (From Butorin, S.M., *J. Electron Spectrosc. Relat. Phenom.*, 110, 213, 2000. With permission from Elsevier Ltd.)

highest emitted x-ray energy corresponds to the elastic x-ray scattering, while the structures on the lower energy side are IXS spectra, which are caused by the AM excitations. The calculated result is in good agreement with the experimental data.

Figure 8.9 shows a small region from the Dy 4d4f RXES spectrum, calculated with an atomic model (thick solid line) and a ligand field multiplet (LFM) model (thin solid and dash-dotted lines) for the incident energy 1 of Figure 8.8. It is important to note that without crystal field splitting, the small peak between 1.2 and 1.8 eV would not be present. Adding a cubic crystal field of 35 meV simulates the mixing of the $J = 13/2$ and $J = 11/2$ ground states, which induces the presence of these small peaks between 1.2 and 1.8 eV. In the experimental data of Figure 8.8a, very weak RXES features are seen in this energy region (if the intensity scale is much extended). This simulation shows that the 4f4f excitations in 4d4f RXES can be used to determine the amount of J-mixing in the ground state. Assuming J-mixing is induced by the crystal field effects only, the crystal field strengths on the 4f electrons can be determined (Butorin, 2000).

8.2.2 Effect of Interatomic Hybridization in CeO_2 and PrO_2

CeO_2 is a nominally f^0 system, but actually the $4f^0$ and $4f^1\underline{L}$ configurations are strongly mixed by the covalency hybridization in the ground state, forming the bonding, nonbonding, and antibonding states (see Chapters 3 and 5). In the intermediate state of the Ce 3d4f RXES of CeO_2, the $3d^54f^1$ and $3d^54f^2\underline{L}$ configurations are also strongly mixed. The schematic transition scheme of the Ce 3d4f RXES of CeO_2 is shown in Figure 8.10.

The results of Ce 3d XAS and 3d4f RXES are shown in Figure 8.11 (Butorin et al., 1996b; Nakazawa et al., 1996). The polarization geometry was taken as the polarized one, and the calculations were done with the SIAM. The main peak B and the satellite A of 3d XAS correspond to the transitions to bonding and antibonding of the intermediate state (final states of XAS), respectively, while the transition to the

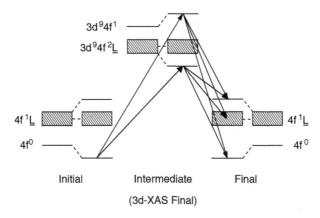

FIGURE 8.10 Schematic energy-level diagram of the Ce 3d4f RXES of CeO$_2$. (Reprinted with permission from Kotani, A., and Shin, S., *Rev. Mod. Phys.*, 73, 203, 2001. Copyright 2001 by the American Chemical Society.)

nonbonding state is almost forbidden. When Ω is fixed at the main peak position, which corresponds to selecting the bonding intermediate state, we have a strong transition to the bonding state, a somewhat weaker transition to the nonbonding state and a weak transition to the antibonding state in the final state. On the other hand,

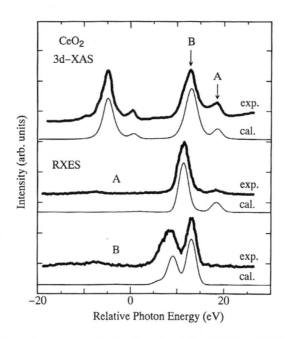

FIGURE 8.11 Experimental and calculated results of Ce 3d XAS and 3d4f RXES for CeO$_2$. The results A and B of RXES spectra are obtained by tuning the incident photon energy to A and B of the XAS. (Reprinted with permission from Nakazawa, M., Tanaka, S., Uozumi, T., and Kotani, A., *J. Phys. Soc. Jpn.*, 65, 2303, 1996. Copyright 1996 by the American Chemical Society.)

when Ω is fixed at the satellite peak position, which corresponds to the antibonding intermediate state, we have a strong transition to the antibonding state and a weak transition to the bonding state. This is the reason why the spectrum in CeO_2 has the two-peak structure. The difference in the relative intensity of the two peaks for different Ω originates from the different character of the intermediate states.

It can be clearly seen from the XAS and the RXES (curve A) that the energy difference of the two peaks in XAS is smaller than that of RXES. Therefore, the energy difference between the antibonding and bonding states with the core hole is smaller than that without the core hole. It is to be noted that these energy differences are mainly determined by the covalency hybridization. Therefore, the covalency hybridization strength V is smaller with the core hole than without the core hole. This is because the 4f wavefunction is contracted by the attractive potential of the core hole. The reduction factor R_c is defined by

$$V \text{ (with core hole)} = R_c \times V \text{ (without core hole)}.$$

The value of R_c, as well as the other parameters of the SIAM, can be estimated by the analysis of the RXES and XAS (Nakazawa et al., 1996). The estimated values are: $R_c = 0.6$, V (without core hole) $= 1.0$ eV, $\Delta = 2.0$ eV, $U_{ff} = 9.0$ eV, and $U_{fc} = 12.6$ eV.

There have been much experimental data for XAS and XPS associated with various core levels in CeO_2, and most of them were successfully reproduced by charge transfer multiplet (CTM) calculations using the SIAM with $R_c = 1.0$ (the other parameters are $V = 0.76$ eV, $\Delta = 1.6$ eV, $U_{ff} = 10.5$ eV, and $U_{fc} = 12.5$ eV). Only the exception is the analysis of valence XPS (VXPS) and Bremsstrahlung isochromat spectra (BIS). According to the experimental results of valence band XPS and BIS for CeO_2, the energy difference between the lowest affinity state and the first ionization state is about 4.5 eV, which corresponds to the insulating energy gap of CeO_2. A previous calculation of VXPS and BIS with $R_c = 1.0$ gave an insulating gap (about 2.0 eV) much smaller than the experimental value. Then, Nakazawa et al. (1996) calculated VXPS and BIS with their new parameters with $R_c = 0.6$ and found that the experimental value of the insulating energy gap was well reproduced by their calculation. It was also found that the other experimental data, Ce 3d XPS, 4d XAS, and 4d XPS for CeO_2, were well reproduced by the new parameter values. In this way, the RXES gives us important information on the hybridization strength in both the ground state configuration and the core electron excited state.

The polarization dependence in RXES of CeO_2 is similar to that of TiO_2 (described in detail in Section 8.6.2) and was confirmed both theoretically and experimentally. Nakazawa et al. (2000) showed theoretically the dramatic polarization dependence in CeO_2, where only the nonbonding final state is allowed for the depolarized geometry. Conversely, the bonding, nonbonding, and antibonding final states are allowed for the polarized geometry. This was confirmed experimentally by Watanabe et al. (2002).

For PrO_2, Butorin et al. (1997a) measured a similar RXES spectra and analyzed the result with the SIAM. The experimental and theoretical results are shown in Figure 8.12. The ground states of PrO_2 is a bonding state between the $4f^1$ and $4f^2\underline{L}$ configurations, and the final states of 3d XAS consist of bonding and antibonding states between the $3d^9 4f^2$ and $3d^9 4f^3\underline{L}$ configurations. The main peak in regions A, B, and C (and also

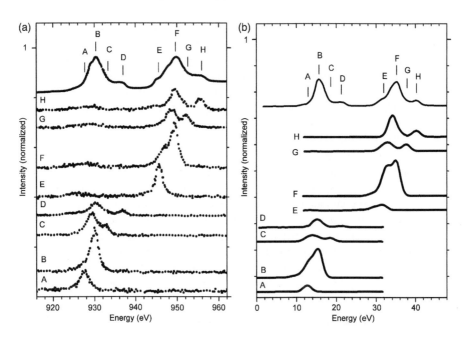

FIGURE 8.12 (a) Experimental data of the Pr 3d XAS and 3d4f RXES spectra of PrO_2. In the RXES results A–H, the incident photon energy is tuned at the XAS energy positions A to H, respectively; (b) Calculated results of the Pr 3d XAS and 3d4f RXES spectra of PrO_2. (From Butorin, S.M., et al., *J. Phys. Cond. Matt.*, 9, 8155, 1997a. Reprinted with permission from IOP Publishing Ltd.)

E, F, and G) is the bonding state with some multiplet structure and the satellite D (and H) corresponds to the antibonding state. The RXES A–H are obtained by tuning the incident photon energy to the XAS positions A–H and the calculated results are in fair agreement with the experimental ones. The used parameter values are $V = 0.8$, $\Delta = 0.5$ and $U_{fc} - U_{ff} = 5.0$ (in units of eV), and the reduction factor $R_c = 0.7$ is essential in reproducing the experimental results. It should be stressed that the interplay between the interatomic hybridization and the intra-AM coupling plays an essential role in the RXES of PrO_2, while the latter is less important in CeO_2.

8.2.3 METALLIC Ce COMPOUNDS WITH MIXED-VALENCE CHARACTER

The mixed-valence and heavy fermion properties of metallic Ce compounds have attracted much attention for their unusual and interesting many-body effects. The hybridization between Ce 4f and conduction band states causes, at a finite temperature, a crossover between the localized Ce 4f magnetic state and the singlet bound state of coupled 4f and conduction electrons. For this phenomenon, the effect of electron-hole pair excitations across the Fermi level due to the higher-order terms of the hybridization plays an important role. When the hybridization strength is large and, therefore, the crossover temperature is high, the singlet bound state is called a mixed-valence state (or fluctuating valence state). When the hybridization strength is small (low crossover temperature), it is called the Kondo resonance state. The Kondo resonance state on each Ce site becomes coherent at a temperature lower than the crossover temperature and the system is denoted by a heavy fermion system.

It is interesting to measure the RXES spectra of these Ce compounds. As an example, the Ce 3d XAS and Ce 3d4f RXES spectra measured for the heavy fermion compound CeB_6 are shown in Figure 8.13 (Magnuson et al., 2001). The experiment was made at room temperature with the depolarized geometry. The XAS spectrum is essentially the same as that of the typical Ce^{3+} system like CeF_3, but the RXES spectra are not a single peak as in CeF_3.

Theoretical calculation of the RXES with the SIAM in metallic mixed-valence (or Kondo resonance) Ce compounds is essentially the same as that in insulating mixed-valence Ce compounds like CeO_2 where the completely-filled valence band (O 2p band for CeO_2) is replaced by the metallic conduction band below the Fermi energy (see Chapters 3 and 5 where this is explained for the calculation of XPS but where the situation is the same for the calculation of RXES). The effect of electron-hole pair excitations across the Fermi level, which is characteristic of metallic systems, gives only higher-order corrections with respect to the $1/N_f$ expansion,

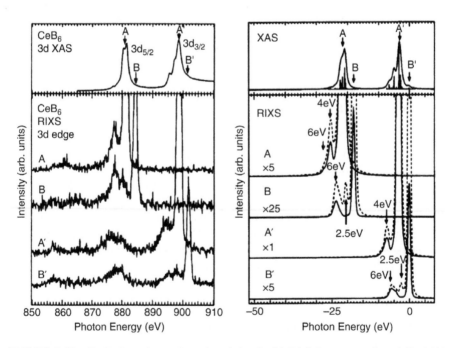

FIGURE 8.13 (Left) Experimental results of the Ce 3d XAS (upper panel) and Ce 3d4f RXES (lower panel) spectra of CeB_6. The measurements were made at room temperature with the depolarized geometry; (right) calculated results of the Ce 3d XAS (upper panel) and Ce 3d4f RXES (lower panel) spectra of CeB_6. The solid and dashed curves of the RXES spectra were obtained with the depolarized and polarized geometries, respectively. (Reprinted with permission from Magnuson, M., et al., *Phys. Rev. B*, 63, 75101, 2001. Copyright 2001 by the American Physical Society.)

where N_f (= 14) is the spin and orbital degeneracy of the 4f state. Quantitatively, however, the parameter values are different for insulating and metallic Ce compounds. Most important is that the 4f level is above the O 2p band in CeO_2, but it is below the Fermi level in metallic mixed-valence Ce compounds. Therefore, the charge transfer excitations originating from the $4f^2$ configuration play an important role in the RXES of metallic Ce compounds as relatively low-energy excitations, whereas the energy of the $4f^2\underline{L}^2$ configuration in CeO_2 is too high to be important in the RXES spectra.

The calculated spectra of the Ce 3d XAS and 3d4f RXES of CeB_6 are shown in Figure 8.13 where the solid and dashed curves are obtained in the depolarized and polarized geometries (Magnuson et al., 2001). The calculation was made with the SIAM and disregarded the effect of coherent heavy fermion state (the energy scale of which is too small to be detected in the present RXES experiment). However, some features of the Kondo resonance are taken into account. From this analysis, the observed 4 eV and 6 eV RXES structures are interpreted as originating from the charge transfer excitation to the $4f^2$ final state. Theoretical calculation predicts an additional 2.5 eV RXES structure ($4f^0$ final state) that cannot be observed in the depolarized geometry but is expected to be observed in the polarized geometry (as an indication of the Kondo singlet ground state).

Another interesting experiment for detecting a $4f^2$ signal was done by Rueff et al. (2004) for the Ce $L\alpha_1$ RXES ($2p_{3/2}3d$ RXES) of Ce-Th and Ce-Sc alloys. The Ce-Th and Ce-Sc alloys are known to show the so-called $\alpha-\gamma$ transition, across which the hybridization strength between Ce 4f and conduction electron states changes due to the change of the lattice spacing with the crystal structure unchanged. On the lower temperature side of the $\alpha-\gamma$ transition, the Ce is in the mixed valence state (similar to the α Ce), while on the higher temperature side, it is nearly in the Ce^{3+} state (similar to the γ Ce). In the Ce $L\alpha_1$ RXES process, the Ce $2p_{3/2}$ core electron is excited to the 5d band, and then a Ce 3d core electron makes a radiative transition to the $2p_{3/2}$ state.

In Figure 8.14, the experimental results for Ce L_3 XAS (at 60 and 300 K, which are below and above the transition temperature) and the $L\alpha_1$ RXES spectra (at 60 K) of $Ce_{0.9}Th_{0.1}$ are shown. In XAS spectra, the main peak and a higher energy feature correspond to the $4f^1$ and $4f^0$ configurations, but no signal from the $4f^2$ configuration is seen although it is expected to be located at the pre-edge region. As seen from Figure 8.14, however, when the incident energy is taken in the pre-edge region, the RXES spectrum exhibits two structures (indicated as f^2 and f^1) corresponding to the core electron excitation to the $4f^2$ and $4f^1$ configurations. Namely, the excitation to the $4f^2$ configuration is invisible in XAS because of the large spectral broadening, but it can be detected by measuring the RXES. Then, Rueff et al. (2004) fixed the incident energy at a position to make the $4f^2$ intensity of RXES spectrum maximum, and measured the intensity ratio f^1/f^2 of the RXES spectra by changing the temperature across the $\alpha-\gamma$ transition. The results of the measured intensity ratio f^1/f^2 for three different alloy systems are shown as a function of temperature on the left-hand side of Figure 8.15. It is seen that the f^1/f^2 ratio exhibits a sharp change with a hysteresis at the $\alpha-\gamma$ transition temperature. Furthermore, the behavior of the f^1/f^2 ratio is found to closely resemble the magnetization loop, where the temperature-dependence of the magnetization at the magnetic field $H = 2$ T is also shown on the right-hand side of Figure 8.15, for comparison. Rueff et al. (2004) claimed that these

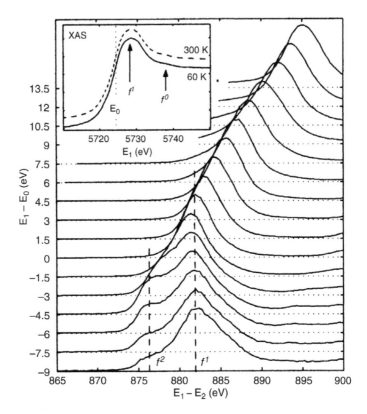

FIGURE 8.14 Experimental results of Ce $L\alpha_1$ RXES of $Ce_{0.9}Th_{0.1}$ measured at 60 K where the incident and emitted photon energies are written as E_1 and E_2, respectively. In the inset, the Ce L_3 XAS spectra measured at 60 and 300 K are shown. (Reprinted with permission from Rueff, J.P., et al., *Phys. Rev. Lett.*, 93, 067402/1, 2004. Copyright 2004 by the American Physical Society.)

measurements confirmed recent dynamical mean-field calculations that predicted significant $4f^2$ occupancy in the ground state. This finding is very interesting but we should be careful because the f^1/f^2 intensity ratio observed experimentally is not the quantity in the ground state; rather, it is that in the intermediate state of RXES (the final state of XAS). It is not very clear at present what the relationship of the f^1/f^2 intensity ratios is in the ground and core excited states.

8.2.4 KONDO RESONANCE IN Yb COMPOUNDS

In addition to metallic Ce compounds, metallic Yb compounds often behave as mixed valence materials or heavy fermion systems. Here, the Yb $4f^{13}$ (Yb^{3+}) and $4f^{14}$ (Yb^{2+}) configurations are mixed in the ground states, instead of the Ce $4f^0$ (Ce^{4+}) and $4f^1$ (Ce^{3+}) configurations in Ce compounds. Dallera et al. (2002) observed the valence change of $YbInCu_4$ and $YbAgCu_4$ as a function of temperature by measuring the excitation spectra of the Yb $L\alpha$ RXES [they called this "XAS in the partial fluorescence yield (PFY) mode"]. The valence change in Yb compounds has been measured by

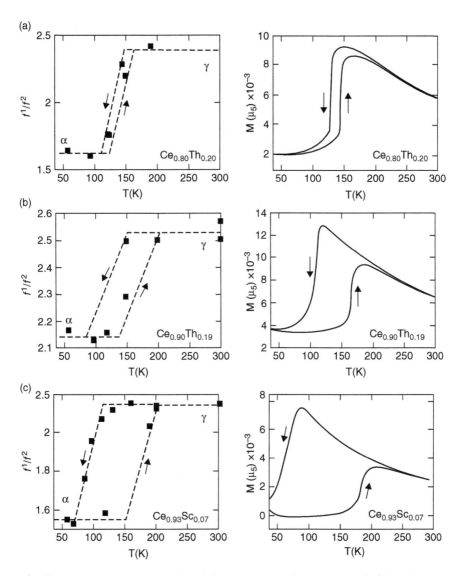

FIGURE 8.15 Experimental results of the temperature-dependence of f^1/f^2 intensity ratio measured by the Ce Lα_1 RXES for (a) Ce$_{0.80}$Th$_{0.20}$, (b) Ce$_{0.90}$Th$_{0.19}$, and (c) Ce$_{0.93}$Sc$_{0.07}$ (left panels), compared with the temperature-dependence of the magnetization at $H = 2$ T (right panels). (Reprinted with permission from Rueff, J.P., et al., *Phys. Rev. Lett.*, 93, 067402/1, 2004. Copyright 2004 by the American Physical Society.)

photoemission spectroscopy (PES) but, since PES spectra are surface-sensitive, the interpretation of the results have proved to be controversial. The experiments by Dallera et al. (2002) detected the unambiguous bulk behavior of the Yb valence change and solved the problem. It is shown that the valence change in YbAgCu$_4$ occurs continuously in a consistent manner with the prediction of the SIAM (Kondo temperature of 70 K), while that in YbInCu$_4$ occurs suddenly at a phase transition temperature.

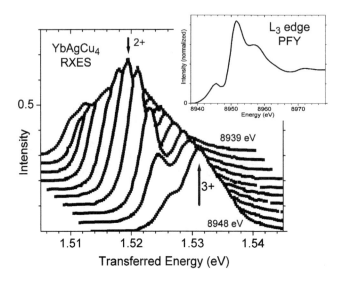

FIGURE 8.16 Experimental results of Yb Lα RXES spectra of YbAgCu$_4$ at 15 K with the change of the incident photon energy at 1 eV intervals along the pre-edge of Yb L$_3$ XAS. The inset is the intensity of the Yb^{2+} peak in the RXES spectrum as a function of the incident energy. (Reprinted with permission from Dallera, C., et al., *Phys. Rev. Lett.*, 88, 196403, 2002. Copyright 2002 by the American Physical Society.)

Figure 8.16 is an example of the Yb Lα RXES of YbAgCu$_4$ taken at 15 K with the incident energy changed from 8939–8948 eV at 1 eV intervals in the pre-edge region. Although the signal of the excitation to the 4f^{14} configuration (Yb^{2+} component) is almost invisible in the conventional XAS spectrum (not shown here), the signal of both Yb^{2+} and Yb^{3+} components is clearly seen in the Yb Lα RXES. Therefore, the excitation spectrum (ES) of the Yb Lα RXES taken at the emitted photon energy 7415 eV exhibits a peak corresponding to the Yb^{2+} component in the pre-edge region (shown with the solid line in the inset of Figure 8.16). Then Dallera et al. (2002) measured the change of the relative intensity of this Yb^{2+} peak by changing the temperature, and confirmed that the change of the intensity is continuous around the crossover temperature (Kondo temperature) at 70 K.

For YbInCu$_4$, on the other hand, the relative intensity of the Yb^{2+} component in the Yb Lα RXES is very much different above and below the valence transition temperature 42 K, as shown in Figure 8.17. The fractional intensity of the Yb^{2+} component is a direct measure of the average hole number in the 4f shell $\langle n_h \rangle$; if the Yb^{2+} fractional intensity changes from 0 to 1, the quantity $1 - \langle n_h \rangle$ also changes from 0 to 1. From the experimental results of the Yb^{2+} intensity, the temperature dependence of $1 - \langle n_h \rangle$ is shown in the inset of Figure 8.16, where the Yb^{2+} intensity is obtained by the bandwidth of the analyzer shown with the dash-dotted lines and the value of $\langle n_h \rangle$ is assumed to be 0.83 at 15 K as estimated by XAS.

Dallera et al. (2003) also measured the valence change of the mixed valence compound YbAl$_2$ under external pressure. They found that the Yb valence number increased from 2.25 at ambient pressure to 2.9 at 385 kbar. These measurements

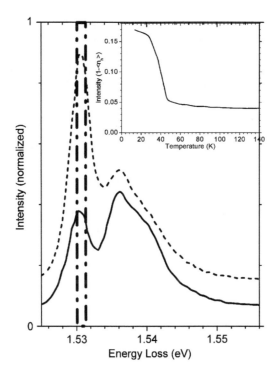

FIGURE 8.17 Experimental results of Yb $L\alpha$ RXES spectra of $YbInCu_4$ at 50 K (solid) and 15 K (dashed) excited at the maximum of the Yb^{2+} resonance. The temperature-dependence of the Yb^{2+} intensity is shown in the inset (solid), by monitoring the divalent RXES signal (the dash-dotted lines indicate the bandwidth of the analyzer). (Reprinted with permission from Dallera, C., et al., *Phys. Rev. Lett.*, 88, 196403, 2002. Copyright 2002 by the American Physical Society.)

demonstrated clearly that RXES is the new powerful tool to probe the bulk electronic configuration in strongly correlated systems. In those experiments by Dallera et al. for Yb compounds, as well as those by Rueff et al. for Ce compounds mentioned in the preceding subsection, the signals ($4f^{14}$ for Yb compounds and $4f^2$ for Ce compounds), which are invisible in the conventional L_3 XAS spectra, can be detected by measuring $L\alpha$ RXES spectra. This novel and powerful technique was first introduced by Hämäläinen et al. (1991).

8.2.5 Dy 2p3d RXES Detection of the 2p4f EQ Excitation

In general, in the conventional L_3 XAS of RE systems, the spectrum is broadened by the short lifetime of the 2p core hole. This means that fine structures, such as the 2p4f EQ transition structure, are often smeared out. A pioneering experiment to detect the electric quadrupole (EQ) excitation was done by Hämäläinen et al. (1991), who measured the so-called excitation spectrum (ES) of 2p3d ($L\alpha$) RXES for Dy compounds with very high-spectral resolution. They first measured the NXES spectrum where a Dy $2p_{3/2}$ core electron was excited to the high-energy continuum and a Dy $3d_{5/2}$ electron made a radiative transition to the $2p_{3/2}$ level, and fixed the emitted photon energy at the maximum position of the NXES spectrum with a high analyzer resolution (as high as 0.3 eV). Then they measured the change of this emitted photon intensity as a function of the incident photon energy near the Dy L_3 threshold. Therefore, what they measured was the ES of the Dy 2p3d RXES with high-energy resolution. The observed ES is shown in Figure 8.18a together with the conventional Dy L_3 XAS measured by the transmission method. It is seen that the spectral width of ES is smaller than XAS, and some weak structures are observed in the pre-edge region, which are invisible in the conventional XAS. The observed pre-edge structure of ES is shown in Figure 8.18b with the extended scale, and this pre-edge structure is interpreted to originate from the EQ transition.

Theoretical analysis of these experimental data was made by Tanaka et al. (1994) with the Dy^{3+} atomic model combined with the Dy 5d band approximated by a semi-elliptical DOS (see also Kotani, 1993). The ES is calculated, as a direct application of Equation 8.7, by

$$F(\Omega,\omega) = \sum_j \left| \sum_i \frac{\langle j|T_2|i\rangle\langle i|T_1|g\rangle}{E_g + \hbar\Omega - E_i + i\Gamma_L} \right|^2 \frac{\Gamma_M/\pi}{(E_j + \hbar\omega - E_g - \hbar\Omega)^2 + \Gamma_M^2} \qquad (8.18)$$

as a function of $\hbar\Omega$ with a fixed value of $\hbar\omega$ at the NXES peak position (which corresponds to the energy separation of the core levels $\Delta\varepsilon = \varepsilon_{3d(5/2)} - \varepsilon_{2p(3/2)}$).

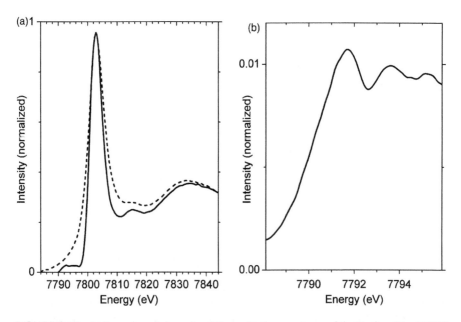

FIGURE 8.18 (a) Experimental results of the excitation spectrum of the Dy $2p_{3/2}3d_{5/2}$ RXES (solid) and the normal Dy $2p_{3/2}$ XAS (dashed) of $Dy(NO_3)_3$; (b) the enlarged pre-edge structure of the excitation spectrum. (Reprinted with permission from Hämäläinen, K., Siddons, D.P., Hastings, J.B., and Berman, L.E., *Phys. Rev. Lett.*, 67, 2850, 1991. Copyright 1991 by the American Physical Society.)

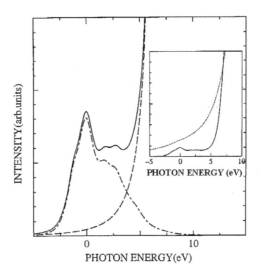

INTENSITY(arb.units)

PHOTON ENERGY (eV)

PHOTON ENERGY(eV)

FIGURE 8.19 Calculated result of the excitation spectrum in the pre-edge region of L_3 XAS for a Dy^{3+} system. The total spectrum (solid curve) consists of the EQ (chain curve) and ED (dashed curve) contributions. In the inset, the calculated excitation spectrum (solid curve) is compared with the theoretical result of the conventional XAS (dotted curve). (From Tanaka, S., Okada, K., and Kotani, A., *J. Phys. Soc. Jpn.*, 63, 2780, 1994. With permission.)

In Equation 8.18, T_1 represents both $2p_{3/2}4f$ EQ transition and $2p_{3/2}5d$ electric dipole (ED) transition, T_2 is the $3d_{5/2}2p_{3/2}$ ED transition, and Γ_L and Γ_M represent the lifetime broadening of the $2p_{3/2}$ and $3d_{5/2}$ core holes, respectively. The values of Γ_L and Γ_M are taken to be 2.1 and 0.7 eV, respectively. The intra-atomic multiplet coupling between the 4f electrons [that between the 4f electron and the 2p hole (in the intermediate state) and that between the 4f electron and the 3d hole (in the final state)], are fully taken into account.

The calculated ES in the L_3 pre-edge region is shown in Figure 8.19 by the solid curve, where the energy $\hbar\omega$ is fixed at the NXES peak position $\Delta\varepsilon = \varepsilon_{3d(5/2)} - \varepsilon_{2p(3/2)}$. The spectrum consists of two contributions: the EQ excitation (chain curve) and the low-energy tail of the $2p_{3/2}$-5d ED excitation (dashed curve). The conventional XAS is also calculated with the same model, and the result is shown in the inset of Figure 8.19 with the dotted curve and compared with the ES (solid curve). The pre-edge structure, which is clearly seen in the ES is invisible in conventional XAS. These results are in good agreement with the experimental results by Hämäläinen et al. (1991) (Figure 8.18b and c). The calculated ES corresponds almost to a fictitious XAS with a spectral width smaller than that of conventional XAS. A similar calculation was also made by Carra et al. (1995).

Here, we show why the ES of RXES corresponds to a less broadened version of XAS. In the following discussion, we assume for simplicity that $|j\rangle$ and E_j can be written as

$$|j\rangle = a_{3d(5/2)} a^+_{2p(3/2)} |i\rangle, \tag{8.19}$$

$$E_j = \Delta\varepsilon + E_i, \tag{8.20}$$

where $\Delta\varepsilon = \varepsilon_{3d(5/2)} - \varepsilon_{2p(3/2)}$ as defined above. This assumption does not necessarily mean that we have confined ourselves entirely to the one electron approximation. What it means is that the x-ray emission process occurs simply by the core electron transition between $3d_{5/2}$ and $2p_{3/2}$ states, leaving the other electronic states unchanged. We can take into account the many body 4f4f and 4f3d interactions, but the 4f3d interaction (3d core hole effect) is assumed to be the same as the 4f2p interaction (2p core hole effect). Substituting Equations 8.19 and 8.20 into Equation 8.18 and using

$$\hbar\omega = \Delta\varepsilon \tag{8.21}$$

we obtain, apart from an unimportant factor, the expression of the ES

$$F(\Omega, \Delta\varepsilon/\hbar) = \sum_i \frac{|\langle i | T_1 | g \rangle|^2}{[(\hbar\Omega - E_i + E_g)^2 + \Gamma_L^2][(\hbar\Omega - E_i + E_g)^2 + \Gamma_M^2]} . \tag{8.22}$$

Since Γ_L (~2.1 eV) is much larger than Γ_M (~0.7 eV), we obtain

$$F(\Omega, \Delta\varepsilon/\hbar) \sim I_{\Gamma_M}(\Omega), \tag{8.23}$$

where $I_{\Gamma_M}(\Omega)$ is the L_3 XAS spectrum with the broadening Γ_L replaced by Γ_M:

$$I_{\Gamma_M}(\Omega) = \sum_i |\langle i | T_1 | g \rangle|^2 \frac{\Gamma_M/\pi}{(\hbar\Omega - E_i + E_g)^2 + \Gamma_M^2} . \tag{8.24}$$

Equations 8.19 and 8.20 do not hold exactly. Consequently, Equation 8.23 also does not hold exactly but holds as a good approximation.

8.2.6 EQ EXCITATIONS IN LIGHT RARE EARTH ELEMENTS

After the work by Hämäläinen et al. (1991), the structure of the EQ excitation has been studied for various RE systems. As a typical example, Bartolomé et al. (1997) measured $L\alpha$ RXES for various RE systems, and found that the EQ excitation of light RE systems is split into two peaks. According to their results, the energy separation of the two peaks is almost proportional to the atomic number. Furthermore, they also observed the energy splitting of the EQ excitation in ferromagnetic light RE systems by x-ray magnetic circular dichroism (XMCD) measurements, and showed that the energy separation measured by RXES coincides with that by XMCD within the experimental accuracy.

More recently, new theoretical and experimental developments have been made on the result by Bartolomé et al. (1997). Figure 8.20a,b displays the energy separation of the EQ excitation peaks in light RE elements (from La to Eu) determined by XAS (and XMCD) and $L\alpha$ RXES, respectively. The open circles are the results by Bartolomé et al. (1997) and the dotted lines are guides for the eye, which indicate that the energy separation of the two peaks is roughly proportional to the atomic number. The open squares are new experimental results by Journel et al. (2002) for LaF_3 and CeF_3. It is clear that the energy separations given by open squares are not proportional to the atomic number and, furthermore, they are different in XAS and RXES.

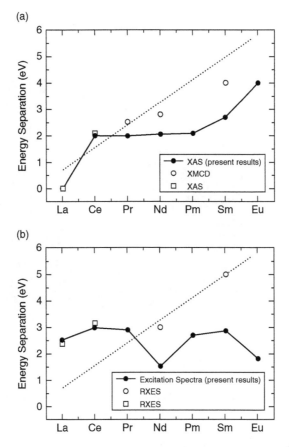

FIGURE 8.20 Calculated energy separation (solid circles) for (a) $2p_{3/2}4f$ XAS and (b) the excitation spectrum of $2p_{3/2}3d_{5/2}$ RXES under the $2p_{3/2}4f$ excitation. The experimental results of the energy separation are shown with open circles (Bartolomé et al., 1997) and open squares (Journel et al., 2002). The dashed lines are guides for the eye. (Reprinted with permission from Nakazawa, M., et al., *Phys. Rev. B*, 66, 113104, 2002. Copyright 2002 by the American Physical Society.)

Theoretical calculations for these energy separations have been performed by Nakazawa et al. (2002, 2003) with an atomic model including full multiplet coupling effects. An example of calculated $L\alpha$ RXES spectra is shown in Figure 8.21a for the Nd $2p_{3/2}4f$ EQ excitation and $3d_{5/2}2p_{3/2}$ ED transition in a Nd^{3+} system, displayed as a contour map in a two-dimensional plane spanned by incident and emitted photon energies. This type of figure is called a "2D RXES image." The RXES spectrum is the vertical cross section of this 2D RXES image for a fixed incident photon energy: an example of the RXES spectrum is depicted in Figure 8.21b for an incident energy indicated by the vertical line in Figure 8.21a. On the other hand, the ES at the fluorescence peak position is the horizontal cross section of the 2D RXES image for the emitted photon energy at −169 eV (which corresponds to $\Delta\varepsilon = \varepsilon_{3d(5/2)} - \varepsilon_{2p(3/2)}$, but the origin of the photon energy is shifted somewhat arbitrarily).

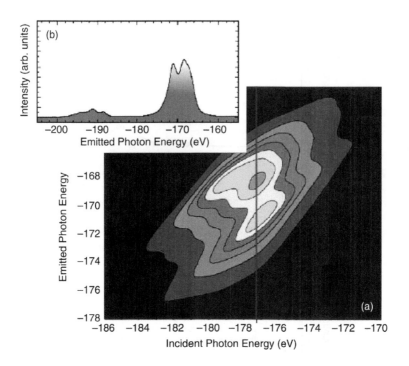

FIGURE 8.21 (a) The calculated result of the Nd $2p_{3/2}3d_{5/2}$ RXES intensity due to the Nd $2p_{3/2}4f$ EQ excitation in the Nd^{3+} system, displayed as a contour map in a two-dimensional plane spanned by incident and emitted photon energies. In the calculation, the full-multiplet coupling effect is taken into account. Panel (b) is the RXES spectrum where the incident photon energy is fixed at the vertical line in (a), so that it corresponds to the cross section of the contour map. (From Kotani, A., *Eur. Phys. J. B*, 47, 3, 2005. With permission.)

The results of the energy separation calculated by Nakazawa et al. (2002, 2003) are shown by closed circles in Figure 8.20, and are found to reproduce well the experimental results by Journel et al. (2002), but not those by Bartolomé et al. (1997). The main points clarified here are as follows: (*i*) The energy separation in XAS is determined by the 4f4f interaction. The 4f4f exchange interaction causes the energy separation (due to the up- and down-spin excitation) almost proportional to the atomic number, as mentioned by Bartolomé et al., but the 4f4f multipole Coulomb interaction also influences the energy separation, giving rise to considerable deviation from the proportionality relation. (*ii*) The energy separation in RXES is determined not only by the 4f4f interaction, but also by the 4f3d interaction in the final state of RXES. The situation is very clear in the case of La; since we have no 4f electron in the ground state of La, the energy separation vanishes in XAS because of no 4f4f interaction, but we have a finite energy separation in RXES due to the 4f3d interaction in the final state.

The technique of ES with high-energy resolution, as well as 2D RXES images, is a powerful method to extract hidden structures in conventional XAS spectra, and has been applied widely to the detection of various excitations both for RE and TM systems. Some more examples will be given later for systems including Cu, Co, and Fe. The ES with high-energy resolution is also denoted by "XAS in the PFY mode" or "high-energy resolution fluorescence detected XAS (HERFD-XAS)."

8.3   HIGH T_c CUPRATES AND RELATED MATERIALS

Study of RXES for high T_c cuprates and related materials has been extensively made both from a theoretical and experimental side. As a typical example of the application of RXES, we describe here in some detail, the Cu 2p3d, 1s4p and 1s2p RXES, and the O 1s2p RXES of these materials.

8.3.1   Cu 2p3d RXES

Based on the SIAM, we first consider what kind of information on the electronic states can be obtained from the Cu 2p3d RXES spectra of La_2CuO_4. The Cu ion in cuprates is nominally in the Cu^{2+} state, but actually the Cu $3d^9$ configuration is strongly mixed with the $3d^{10}\underline{L}$ configuration where an electron is transferred from the O 2p band to the Cu 3d state through the p-d hybridization. The situation is explained in Figure 8.22. By the hybridization effect, we have the bonding, nonbonding, and antibonding states, as shown in the figure, and further, the bonding states split into crystal field levels.

The local symmetry around the Cu ion in La_2CuO_4 is D_{4h}, and the crystal field states are represented by the irreducible representation of the D_{4h} group: $\Gamma = b_{1g}$, a_{1g}, e_g, and b_{2g}. The ground state is a strongly mixed state (bonding state) between $3d^9$ (Γ) and $3d^{10}\underline{L}$ (Γ) configurations with $\Gamma = b_{1g}$. We take the coordinate axes so that the z axis is perpendicular to the CuO_2 plane, and then the b_{1g} orbit is represented by $d(x^2 - y^2)$, and the weight of the $3d^9(b_{1g})$ configuration is about 60% (Kotani and Okada, 1990). Above the ground state, there are bonding states with other irreducible

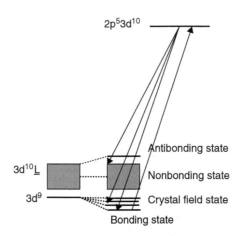

FIGURE 8.22   Schematic transition diagram of the Cu 2p3d RXES in cuprates.

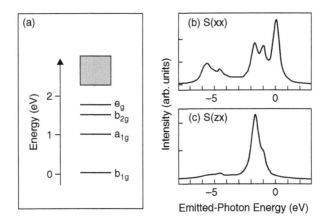

FIGURE 8.23   (a) Energy level diagram of crystal field levels and charge transfer continuum in La$_2$CuO$_4$. (b) Calculated result of the Cu 2p$_{3/2}$3d RXES in La$_2$CuO$_4$ with the incident and emitted photon polarizations in the x direction. (c) Same as (b), but with the incident and emitted photon polarizations in the x and z directions, respectively. (From Tanaka, S., and Kotani, A., *J. Phys. Soc. Jpn.*, 62, 464, 1993. With permission.)

representations (crystal field levels) and nonbonding and antibonding states (denoted by charge transfer states). The energy level scheme of crystal field states and the nonbonding states are also shown in Figure 8.23a.

The intermediate state of the present RXES (which is the same as the final state of Cu 2p XAS) is in the single configuration of 2p^{5}3d^{10} (actually it splits into two levels due to the spin–orbit interaction of the 2p states), and the incident photon energy is tuned to the energy of this intermediate state excitation. By the Cu 3d to 2p radiative transition, this intermediate state changes to final states of RXES. If the final state is the same as the ground state, we have an elastic x-ray scattering peak, whereas if it is some excited state above the ground state, we have IXS structures.

Theoretical calculation of the Cu 2p$_{3/2}$3d RXES for La$_2$CuO$_4$ was made with the SIAM by Tanaka and Kotani (1993). Their results are shown in Figure 8.23b and 8.23c. Here, the polarization dependence of RXES is taken into account, and the RXES spectra S(xx) and S(zx) are shown to exhibit different spectral features where the incident photon polarization is taken in the x direction and that of the emitted photon is in the x [for S(xx)] and z [for S(zx)] directions. In these figures, the origin of the emitted photon energy is taken at the position of the elastic scattering (so that the abscissa corresponds to $\omega - \Omega$), and it is seen that the inelastic scattering spectrum exhibits strong crystal field excitations for -2.0 eV $< \omega - \Omega < 0$, while for $\omega - \Omega < -2.0$ eV it shows weak and broad RXES features due to the charge-transfer excitation. From this calculation, they showed that the polarization dependence of RXES provides us with important information on the symmetry of electronic states. However, a direct experimental measurement of the polarization dependence of RXES is very difficult because of the problem of detecting the polarization of the emitted photon.

A few years after this theoretical calculation, the RXES of La$_2$CuO$_4$, including its polarization dependence, was observed experimentally by using a sophisticated geometrical arrangement where a spectrometer can be rotated around the incident

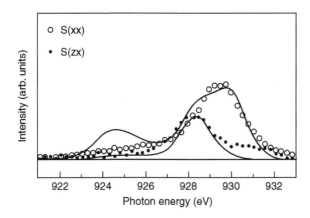

FIGURE 8.24 Experimental data of polarization dependence in Cu 2p3d RXES of La_2CuO_4. The open and closed circles are $S(xx)$ and $S(zx)$, respectively. Calculated results are also shown with the solid curves, which are a more broadened version of Figure 8.23 (b) and (c). (Reprinted with permission from Duda, PhD thesis, 1996. Copyright 1996 by the American Physical Society.)

beam axis and a sample can also be rotated around its surface normal (Duda, 1996, Duda et al., 1998). The result (Duda, 1996) is shown in Figure 8.24, where the open circles and dots are the experimental data corresponding to $S(xx)$ and $S(zx)$, respectively. Unfortunately, the resolution is not very good. In order to compare them with the theoretical results, $S(xx)$ and $S(zx)$ in Figure 8.23b and 8.23c are more broadened and plotted in Figure 8.24 with solid curves. The experimental and theoretical results are in fair agreement, although the intensity of the charge-transfer excitation is weaker in the experimental data. These studies suggested that the polarization-dependent RXES would become a powerful tool in the study of the electronic state symmetry in various materials with improved resolution in future.

The polarization dependence of RXES has also been measured by Kuiper et al. (1998) for the Cu 3p3d RXES in $Sr_2CuO_2Cl_2$. They observed RXES by changing the angle between the x-ray emission direction and the sample normal (z axis), where the scattering angle is kept at 90° and the polarization direction of the incident photon is always in the xy plane. The relative intensity of various crystal field level excitations in RXES changes with the angle, which made it possible to estimate the crystal field excitation energies. In addition, spin-flip transitions are observed. These spin-flip transitions are described in detail for NiO in Section 8.4.3.

More recently, Ghiringhelli et al. (2004) measured the Cu 2p3d RXES for various cuprates: insulating compounds CuO, La_2CuO_4, $Sr_2CuO_2Cl_2$, and optimally doped superconductors $La_{1.85}Sr_{0.15}CuO_4$, $Bi_2Sr_2CaCu_2O_{8+\delta}$, and $Nd_{1.85}Ce_{0.15}CuO_4$. The resolution is 0.8 eV. Figure 8.25 is an example of the results where the experimental data for CuO, La_2CuO_4 (polycrystalline), and $La_{1.85}Sr_{0.15}CuO_4$ (single crystal) are shown in the upper panel and the calculated results with a simple crystal field model (point charge model) are shown in the lower panel. The incident x-ray is of 10° grazing incidence with the polarization perpendicular to the scattering plane while the scattering angle is 70°. Here the crystal field excitation in RXES is the central

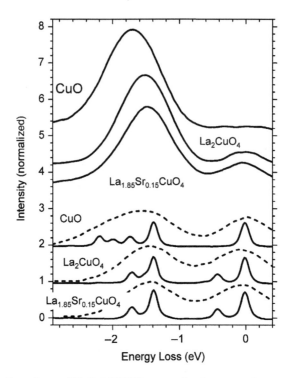

FIGURE 8.25 Experimental Cu 2p3d RXES spectra for, from top to bottom, CuO, La$_2$CuO$_4$, and La$_{1.85}$Sr$_{0.15}$CuO$_4$ (upper panel) and calculated ones (lower panel) with a 0.05 eV Lorentzian broadening (solid curves) and after a 0.8 eV Gaussian broadening (dashed curves). (Reprinted with permission from Ghiringhelli, G., et al., *Phys. Rev. Lett.*, 92, 117406, 2004. Copyright 2004 by the American Physical Society.)

issue, and the calculated results (with the LFM model disregarding the hybridization effect) after a 0.05 eV Lorentzian broadening are shown with the solid curves and those after a 0.8 eV Gaussian broadening are with the dashed curves.

It is seen that the RXES energy of the crystal field excitation of La$_2$CuO$_4$ is, after 0.8 eV broadening, almost unchanged by doping and somewhat smaller than that of CuO in agreement with experiments. For instance, the crystal field excitation energies referred to the b$_{1g}$ state for La$_{2-x}$Sr$_x$CuO$_4$ are 0.41 eV (for a$_{1g}$), 1.38 eV (b$_{2g}$) and 1.51 eV (e$_g$), and the solid curve in Figure 8.25 represents these states and their spin-flip states with the flipping energy 0.2 eV. They also measured the polarization dependence (incident polarizations perpendicular and parallel to the scattering plane for both grazing and normal incidence) of the RXES spectra due to the crystal field excitation and showed that the polarization dependence is well reproduced by their LFM calculations. Here, it is to be remarked that there are two origins, that determine the crystal field excitation energy: the static crystal field effect and the anisotropic hybridization between Cu 3d and O 2p states. Therefore, more careful studies will be necessary to fully understand the crystal field excitation mechanism. In any case, the polarization dependence (selection rule) of the crystal field excitations in RXES requires only the symmetry argument, which is the same for both static crystal field and hybridization mechanisms.

8.3.2   Cu 1s4p RXES

Hill et al. (1998) measured the Cu $1s4p_z$ RXES of Nd_2CuO_4 where a Cu 1s electron is excited to Cu $4p_z$ conduction band by the incident photon with nearly z polarization, and then the Cu $4p_z$ electron makes a radiative transition to the 1s level. The local electronic structure around the Cu ion in Nd_2CuO_4 is almost the same as that of La_2CuO_4. The intermediate state of the present RXES consists of two configurations: $1s^13d^{10}\underline{L}4p_z$ (main peak of Cu 1s XAS) and $1s^13d^94p_z$ (satellite), where the former is about 7 eV lower in energy than the latter. The experimental result (open circles) of the Cu 1s4p XAS is shown in Figure 8.26a, where the spectral features A and B correspond to the bonding and antibonding states of $1s^13d^{10}\underline{L}4p_z$ and $1s^13d^94p_z$ configurations. According to their RXES experiments, the spectra show a broad peak at about 6 eV from the elastic scattering energy as shown in Figure 8.27. The intensity of the 6 eV peak is strongly enhanced at the XAS feature B but almost no resonance at A. The observed intensity of the 6 eV peak is shown as a function of the incident photon energy with the open circles in Figure 8.26b.

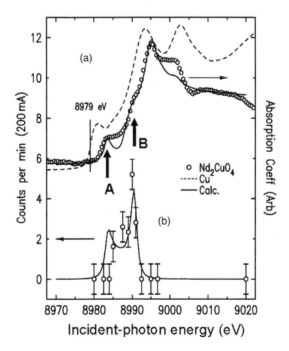

FIGURE 8.26 Experimental data (open circles) and calculated results (solid curves) of (a) the Cu 1s XAS and (b) the intensity of 6 eV feature of the Cu $1s4p_z$ RXES in Nd_2CuO_4. Experimental result of the Cu 1s XAS for Cu metal is also shown with the dashed curve. (Reprinted with permission from Hill, J.P., et al., *Phys. Rev. Lett.*, 80, 4967, 1998. Copyright 1998 by the American Physical Society.)

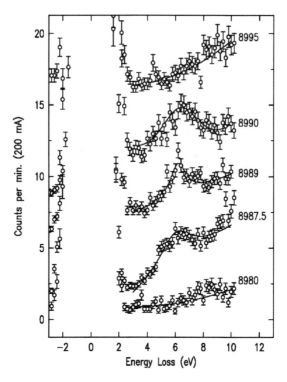

FIGURE 8.27 Experimental data of the Cu 1s4p$_z$ RXES of Nd$_2$CuO$_4$ shown as a function of energy loss. Data are offset vertically for clarity and the solid curves are guides to the eye. (Reprinted with permission from Hill, J.P., et al., *Phys. Rev. Lett.*, 80, 4967, 1998. Copyright 1998 by the American Physical Society.)

Theoretical analysis of this RXES data has been made with the SIAM (Hill et al., 1998) and the 6 eV excitation is assigned to the charge-transfer excitation [more exactly, the transition from the ground state to the antibonding state between 3d^9(b$_{1g}$) and 3d^{10}L(b$_{1g}$) configurations]. The calculated result is shown with the solid curve in Figure 8.26b. Within the SIAM, however, the resonance enhancement occurs for both the main peak and the satellite. In order to remove this disagreement, it is necessary to extend the model from the SIAM to a model including multiple Cu sites, such as a multi-Cu-site cluster model. The calculation of RXES with a Cu$_5$O$_{16}$ cluster model was done by Idé and Kotani (1999). In Figure 8.28, we show the Cu$_5$O$_{16}$ cluster. In the ground state, each CuO$_4$ plaquette includes one hole and orders antiferromagnetically with up-spin on the central plaquette and down-spin on the neighboring plaquettes. In the intermediate state, where a Cu 1s electron on the central Cu site is excited to the lower energy feature A, an electron is transferred from the neighboring plaquette to the central one to screen the core hole potential. As a result, we have two holes with singlet coupling [denoted by Zhang–Rice singlet state (Zhang and Rice, 1988)] in the neighboring plaquette, as shown in Figure 8.28. Then in the final state of RXES, we have the state with Cu d^{10} state (with no hole) on the central plaquette and a Zhang–Rice singlet state on the neighboring

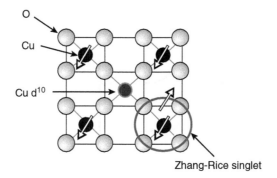

Zhang-Rice singlet

FIGURE 8.28 Schematic illustration of intermediate and final states of Cu $1s4p_z$ RXES of cuprates with the Cu_5O_{16} cluster model. (From Kotani, A., *Eur. Phys. J. B*, 47, 3, 2005. With permission.)

plaquette. This final state is almost orthogonal with the 6 eV antibonding final state, so that the intensity of the 6 eV peak is strongly suppressed, in agreement with the experiment.

The calculated results of XAS and RXES by Idé and Kotani (1999) are shown in Figure 8.29a, where the incident photon energies 1–7 of RXES are indicated in the XAS spectrum. The used parameter values are $\Delta = 2.5$ eV, $V(b_{1g}) = 2.4$ eV, $U_{dd} = 8.8$ eV, and $U_{dc} = 7.5$ eV. In these results, the background contribution in XAS and the elastic scattering contribution in RXES are omitted. It is seen that for the incident photon energy 2 (feature A), the RXES intensity at about 6 eV is strongly suppressed and instead the resonant enhancement occurs at about 2 eV, which corresponds to the pair excitation of the $3d^{10}$ and Zhang–Rice singlet state. The intensity of the RXES spectra at about 6 eV excitation (along the dotted line in Figure 8.29a is shown in Figure 8.29b) as a function of the incident photon energy with the solid curve, which is in good agreement with the experimental results (open circles). The final state of the 2 eV excitation is the low-lying charge-transfer excitation state across the correlation gap, but it cannot be recognized in the experimental data by Hill et al. (Figure 8.27) because it is superposed on the strong high-energy tail of the elastic line.

Due to the improved experimental resolution of RXES, the measurements of the 2 eV excitation have become possible, as well as its momentum dependence. Within the Cu_5O_{16} cluster model, the 2 eV excitation corresponds to the pair excitation of the $3d^{10}$ and the Zhang–Rice singlet states on the two neighboring CuO_4 plaquettes. In larger clusters or an infinite system, the $3d^{10}$ state and the Zhang–Rice singlet state have a spatial dispersion due to the translational motion of these elementary excitations to form the upper Hubbard band (UHB) and the Zhang–Rice singlet band (ZRB), respectively. Therefore, the RXES measurements can detect the electronic excitation from the occupied ZRB to the empty UHB across the correlation gap (denoted by UHB-ZRB pair excitation). Furthermore, by measuring the dependence of RXES spectra on the momentum transfer $\mathbf{q}$, the dispersion of this UHB-ZRB pair excitation can be obtained. In Figure 8.30, we show a typical example of experimental results for the $\mathbf{q}$ dependent RXES spectra for $Ca_2CuO_2Cl_2$ observed by Hasan et al. (2000). In this figure, the RXES peak around 5.8 eV is due to the antibonding state

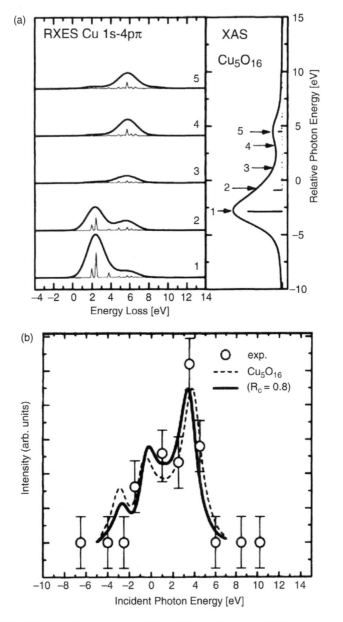

FIGURE 8.29 Calculated results of (a) Cu 1s XAS and Cu 1s4p$_z$ RXES spectra and (b) the intensity of the 6-eV feature with the Cu$_5$O$_{16}$ cluster model. The experimental result is also shown in (b). (From Idé, T., and Kotani, A., *J. Phys. Soc. Jpn.*, 68, 3100, 1999. With permission.)

excitation, which is well localized with almost no energy dispersion, and those on the lower energy side (indicated by the vertical bars) are UHB-ZRB pair excitations. The dispersion of the UHB-ZRB pair excitation is anisotropic: in the ⟨110⟩ direction the excitation energy increases from 2.5 eV at $(\mathbf{k}_x, \mathbf{k}_y) = (0, 0)$ to 3.8 eV at (π, π), while in the ⟨100⟩ direction it changes from 2.5 eV at $(0, 0)$ to 3.0 eV at $(\pi, 0)$.

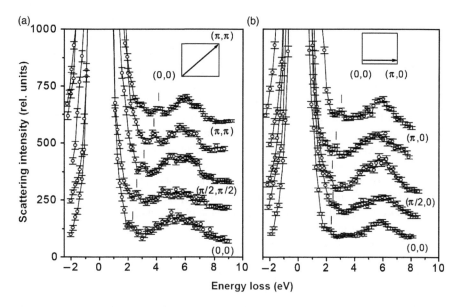

FIGURE 8.30 Experimental results of the **q** dependent Cu 1s4p RXES spectra for Ca$_2$CuO$_2$Cl$_2$, where **q** is in the ⟨100⟩ direction (a) and ⟨100⟩ direction (b). The incident photon energy is taken as 8996 eV. (From Hasan, M.Z., et al., *Science*, 288, 1811, 2000. Reprinted with permission from AAAS.)

Theoretical calculations of these energy dispersions were made by an exact diagonalization method with a 4 × 4 site cluster model described by a single band Hubbard Hamiltonian (instead of the periodic Anderson Hamiltonian) taking into account the first, second, and third nearest neighbor hoppings (Hasan et al., 2000). The calculated results are in good agreement with the experimental ones. It is to be mentioned that the single band Hubbard model does not include the O 2p states, so that the calculated RXES spectra are due to a pair excitation of an UHB electron and a lower Hubbard band (LHB) hole. Note that the calculated LHB states effectively represent the ZRB states in the case of the periodic Anderson model. It should also be mentioned that the energy dispersion of the ZRB of various cuprates has been extensively measured by angle-resolved photoemission spectroscopy (ARPES) experiments and the results have been analyzed theoretically. However, what is measured in RXES is not the single particle (a hole in ZRB) excitation, but the UHB-ZRB pair excitation. On this point, RXES provides more information than ARPES and, in addition, is a more bulk sensitive probe than ARPES.

Hasan et al. (2002) also measured the momentum-resolved Cu 1s4p RXES spectra in the one-dimensional cuprate systems SrCuO$_2$ and Sr$_2$CuO$_3$, where the CuO$_4$ plaquettes are connected as a one-dimensional chain with shared corners. The observed momentum dependence of the UHB-ZRB pair was found to be more dispersive than that of the two-dimensional cuprate Ca$_2$CuO$_2$Cl$_2$. This result is very interesting because the trend is quite the opposite of that in a simple energy band picture of uncorrelated electron systems; within a one-body energy band model, the dispersion of one-dimensional energy band is smaller than that of two-dimensional energy band due to the smaller number of neighboring atoms if the hopping integrals are the same (actually the Cu 3d-O 2p hopping strength should be very similar for $SrCuO_2$, Sr_2CuO_3, and $Ca_2CuO_2Cl_2$). In strongly correlated cuprates, on the other hand, the hopping of charge carriers (an electron in UHB and a hole in ZRB) is more difficult in a two-dimensional lattice with an antiferromagnetic 1/2 spin arrangement than in a one-dimensional one, because the hopping of the charge carrier induces a change of the spin arrangement around the carrier in the two-dimensional lattice, but this is not the case in the one-dimensional lattice due to the separation of spin and charge degrees of freedom. Theoretical calculations of RXES spectra in one-dimensional and two-dimensional insulating cuprates have been made by Tsutsui et al. (1999, 2000) using the exact diagonalization technique for the extended Hubbard model, and the results are consistent with the experimental ones by Hasan et al. (2002). It is considered that a photo-created hole in the one-dimensional ZRB cannot exist as a quasiparticle, but changes into two collective excitations, a spinon and a holon, as discussed by Kim et al. (1996).

Further measurements of the momentum-resolved Cu 1s4p RXES spectra and their interpretations were made by Kim et al. (2002, 2004a, b) for two-dimensional cuprate La_2CuO_4, one-dimensional corner-sharing cuprate $SrCuO_2$, and one-dimensional edge-sharing cuprate Li_2CuO_2. The result for La_2CuO_4 exhibits some features, which have not been observed in $Ca_2CuO_2Cl_2$ by Hasan et al. (2000). Kim et al. (2002) observed for La_2CuO_4, two highly dispersive charge-transfer excitations, in addition to the antibonding excitation at about 7.3 eV. The low-energy mode has a gap of 2.2 eV and bandwidth of 1.0 eV, and shows a strong **q**-dependent intensity variation, while the second peak shows a smaller dispersion of about 0.5 eV with a zone-center energy of about 3.9 eV. The former corresponds to the UHB-ZRB pair excitation (forming an exciton-like state), while the latter might be another UHB-ZRB pair with different symmetry or another exciton-like mode. Kim et al. (2004c) also made similar RXES measurements for hole-doped $La_{2-x}Sr_xCuO_4$ with $x = 0.05$ and 0.17. The result for the $x = 0.05$ sample is similar to the undoped La_2CuO_4 ($x = 0$), but for the sample with $x = 0.17$, they observed the appearance of a continuum of intensity below 2 eV and the spectral weight transfer from the lowest-lying charge transfer excitation of the $x = 0$ sample to the continuum intensity below the gap. The gap-filling continuum excitation is considered to arise from the incoherent particle-hole pairs creation near the Fermi surface. In contrast to this, the second peak and the highest antibonding excitation are not very much affected by doping except for some change of the excitation energy. More recently, some more experimental data have been reported by Collart et al. (2006) for $La_{2-x}Sr_xCuO_4$ in comparison with those for $La_{2-x}Sr_xNiO_4$. Experimental data for a one-dimensional edge-sharing system $CuGeO_3$ and corner-sharing systems Sr_2CuO_3 have also been reported by Suga et al. (2005a).

For a twin-free optimally doped $YBa_2Cu_3O_{7-\delta}$ system, Ishii et al. (2005a) measured the contributions to RXES spectra both from the two-dimensional CuO_2 plane and the one-dimensional CuO chain. The former gives broad excitations at 1.5 to 4 eV that are almost independent of the momentum transfer, and the latter gives the enhancement of the RXES intensity at about 2 eV near the zone boundary of the one-dimensional Brillouin zone. Ishii et al. (2005b) also measured the RXES spectra for electron-doped $Nd_{1.85}Ce_{0.15}CuO_4$, and detected the interband excitations (UHB-ZRB pair) at about 2 eV near the Brillouin zone center and intraband excitations (within UHB), which have strong $\mathbf{q}$-dependence. They made theoretical calculations for these RXES spectra with the single-band Hubbard model and obtained results consistent with the experimental results.

Kim et al. (2004d) made systematic measurements of the antibonding excitation energy for a wide variety of cuprate compounds, including doped and undoped La_2CuO_4. They observed for about ten different cuprate compounds, a systematic trend that the excitation energy increases with the decrease in the Cu-O bond length d. For instance, the excitation energy is about 5.7 eV for Nd_2CuO_4 with $d = 1.97$ Å, about 7.3 eV for La_2CuO_4 with $d = 1.90$ Å, and about 7.7 eV for $La_{1.83}Sr_{0.17}CuO_4$ with $d = 1.88$ Å. This trend is reasonable from a simple picture shown in Figure 8.22, because the decrease in d causes the increase of the Cu 3d-O 2p hybridization strength, resulting in an increase in the energy separation of the antibonding and bonding states. At the same time, they observed some features showing a deviation from the simple picture; the observed excitation energy is proportional to d^{-8}, while the simple picture and a simple relation (see Harrison, 1989) between the hybridization strength and d predicts that it is proportional to $d^{-3.5}$. Furthermore, the observed excitation energy shows some momentum dependence, which increases significantly with the decrease of d, indicating that the antibonding–bonding excitation is not necessarily well localized, but has some energy dispersion due to the effect of the translational symmetry.

8.3.3 Cu 1s2p RXES

We consider the Cu 1s2p RXES ($K\alpha$ RXES) of cuprates, where a Cu 2p electron decays radiatively to the Cu 1s level after the resonant excitation of a Cu 1s electron. The Cu 1s2p RXES is a powerful tool to detect fine structures in the pre-edge region of the Cu 1s XAS, as in the cases of Ce $L\alpha$ RXES (see Section 8.2.3) and Dy $L\alpha$ RXES (see Section 8.2.5). Hayashi et al. (2002, 2004) made experimental observations of the Cu 1s2p RXES for CuO and $Nd_{2-x}Ce_xCuO_4$. In Figure 8.31a, we show the Cu 1s XAS of CuO and in Figure 8.31b, its related Cu 1s2p RXES. The Cu 1s XAS exhibits a weak structure indicated by f due to the EQ transition from Cu 1s to 3d states. The Cu 1s2p RXES was measured by changing the incident energy from f to d and the observed spectra are shown as a function of the emitted x-ray energy. The RXES peaks C and C′ are caused by the Cu 1s3d EQ excitation, and C and C′ are the spin–orbit partner of the Cu 2p core level; that is, they correspond to the RXES final states with a hole in the Cu $2p_{3/2}$ and $2p_{1/2}$ levels. Here we confine ourselves to the Cu $2p_{3/2}$ component. It is interesting to see that a peak B occurs very close to peak C for the excitations e and d, which are only 2 or 3 eV higher than the EQ excitation f. Similar measurements of peak B have also been made by Döring et al. (2004). The occurrence of B strongly suggests that the Cu 1s XAS has some specific structure just above f, but in the XAS spectrum we cannot see any structure for e and d probably because its intensity is too small and its spectral broadening due to the 1s core hole is too large. The microscopic origin of peak B was not clarified, but in any case, this experiment indicates that the Cu 1s2p RXES is a very sensitive probe of the fine structure (or hidden structure) in the pre-edge of the Cu 1s XAS.

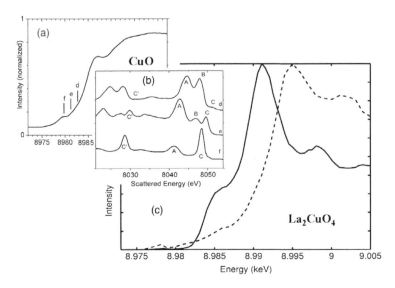

FIGURE 8.31 (a) Cu 1s XAS of CuO; (b) Cu Kα RXES spectra of CuO at the excitation energies d, e, and f; (c) Cu 1s XAS spectra (partial fluorescence yield) of La₂CuO₄ measured with two different polarization geometries. (From Kotani, A., *J. Molecular Structure: Theochem*, 777, 17, 2006. With permission.)

More recently, Shukla et al. (2006) measured the polarization-dependent Cu 1s2p RXES for La₂CuO₄ by changing the incident photon energy in the vicinity of the EQ excitation, and observed an RXES peak corresponding to peak B in CuO. Furthermore, they performed a theoretical analysis of this RXES peak using the SIAM combined with the Cu 4p DOS obtained by *ab initio* energy band calculations. The dashed and solid curves in Figure 8.31c are their Cu 1s XAS experimental results for the incident photon polarization parallel to the CuO₄ plane and almost perpendicular to it (15° from the plane normal), respectively. The spectra were measured by the PFY method and the two XAS curves were normalized, so that their maximum intensities were the same. A small structure on the lowest edge (at about 8.978 eV) of the dashed curve is due to the Cu 1s3d EQ transition. Then they measured the Cu 1s2p RXES by changing the incident photon energy in the pre-edge region from 8.976 to 8.981 eV and the result is displayed in the center panel of Figure 8.32. The baseline of each RXES spectrum corresponds to the incident energy at which it was measured as seen on the XAS spectrum on the left panel. The RXES spectra have been arbitrarily shifted such that the lowest energy transfer feature (due to the EQ excitation) corresponds to the zero of the energy scale and have been normalized to the most intense feature in each spectrum. The sharp peaks at 0 eV and −20 eV correspond to peaks C and C′ of CuO (note that the peak energy of C and C′ in Figure 8.31b shifts with the shift of the incident energy because they are plotted as a function of the emitted x-ray energy, in contrast to Figure 8.32). It should be noted that for the highest three RXES curves, the second peak (indicated by vertical bars) occurs corresponding to peak B in CuO. These two peaks (corresponding to B and C) are allowed only for the incident polarization parallel to the CuO₄ plane.

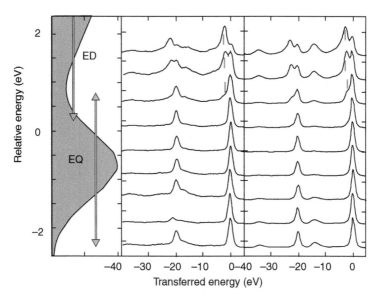

FIGURE 8.32 Left panel: Cu 1s XAS spectrum in the pre-edge region of La_2CuO_4 measured with the incident polarization parallel to the CuO_2 plane. Center panel: measured Cu 1s2p RXES spectra for the incident energy in this pre-edge region. Right panel: Calculated Cu 1s2p RXES spectra. (Reprinted with permission from Shukla, A., et al., *Phys. Rev. Lett.*, 96, 77006, 2006. Copyright 2006 by the American Physical Society.)

The result of theoretical calculations is shown in the right panel of Figure 8.32. In the theoretical calculation, the projected DOS of Cu $4p_\sigma$ (namely $4p_x$ and $4p_y$) and $4p_\pi$ ($4p_z$) states, which are allowed, respectively, from the Cu 1s state by the ED transition with polarization parallel and perpendicular to the CuO_4 plane, is first obtained by the *ab initio* local density approximation + U (LDA + U) band calculation. As seen from the result shown in Figure 8.33, the main DOS peak is higher for the $4p_\sigma$ state than for the $4p_\pi$ state, but the bottom of the 4p DOS is also formed by the $4p_\sigma$ state. The $4p_\sigma$ state in the bottom of the 4p DOS is caused by the hybridization of the Cu $4p_\sigma$ state with off-site Cu 3d ($x^2 - y^2$) states via O 2p states, so that we denote it as "off-site 3d" state, while the EQ transition is allowed to the "on-site 3d" state. The single electron wave function of the on-site and off-site 3d states is schematically shown in Figure 8.34 with a Cu_5O_{16} cluster model. The left-hand side is the on-site 3d state, which mainly consists of the Cu 3d ($x^2 - y^2$) state on the central Cu site (the core hole site) but hybridized with the neighboring O 2p states. On the other hand, the off-site 3d state with the $4p_x$ symmetry, which is shown on the right-hand side, consists of the Cu $4p_x$ state on the central Cu site, O $2p_x$ states on the nearest neighboring O sites and Cu 3d ($x^2 - y^2$) states on the next neighboring Cu sites (off-site Cu). The main weight of the wave function is on the off-site Cu 3d ($x^2 - y^2$) states, but this state is weakly allowed by the ED transition due to the small weight of the central Cu $4p_x$ state.

The calculation of the Cu 1s2p RXES is made by the SIAM, which is combined with the Cu $4p_\sigma$ projected DOS. For instance, the screening effect of the core hole

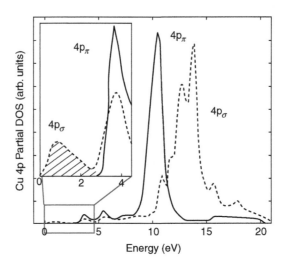

FIGURE 8.33 Projected Cu $4p_\pi$ and $4p_\sigma$ DOS obtained by *ab initio* energy band calculation. The inset shows the low-energy region in detail, and the hatched peak is due to the off-site 3d states. (Reprinted with permission from Shukla, A., et al., *Phys. Rev. Lett.*, 96, 77006, 2006. Copyright 2006 by the American Physical Society.)

potential by the hybridization between Cu 3d and O 2p states is taken into account by the SIAM in the intermediate and final states of the RXES. The lowest edge of the Cu 1s XAS consists of the EQ transition to the on-site 3d state and the ED transition to the off-site 3d state, the energy of the former is about 1.5 eV lower than the latter, and both of them are allowed only for the incident polarization parallel to the

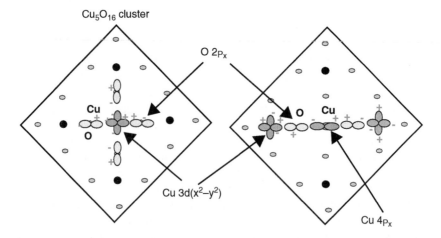

FIGURE 8.34 Schematic representation of the on-site Cu 3d state (left-hand side) and the off-site Cu 3d state (right-hand site), which are allowed by the EQ and ED transitions, respectively, with the Cu_5O_{16} cluster model.

CuO_4 plane. The first and second peaks in the Cu 1s2p RXES (corresponding to C and B) are caused by the former and latter transitions by the incident x-ray, and the calculated result is in good agreement with the experimental one.

It is to be mentioned that the projected 4p DOS of the off-site 3d state is about two orders of magnitude smaller than the main $4p_\sigma$ DOS as shown in Figure 8.33, and the XAS structure due to the ED transition to the off-site 3d state cannot be recognized in the Cu 1s XAS spectrum due to its weak intensity and the superposition on the low-energy tail of the main XAS peak, as well as on the EQ transition structure. However, the Cu 1s2p RXES has ultra-high sensitivity to this weak ED excitation; by resonating with this excitation, a clear peak occurs in the RXES spectrum. After this theoretical analysis, more detailed calculations of the Cu 1s2p RXES was made by Kotani et al. (2007) with Cu_5O_{16} cluster model, and the distribution of spin-dependent valence holes (Cu 3d and O 2p holes) are calculated on each of 5 Cu and 16 O sites and for both ground state and excited states by the on-site 3d and off-site 3d transitions.

In Figure 8.35a, the hole distribution in the ground state is shown for up-spin (left side) and down-spin (right side) states. It is seen that the up-spin hole is mainly located around the central Cu site and four down-spin holes are located mainly around the Cu sites in the neighboring CuO_4 plaquettes, corresponding to the antiferro-magnetic spin arrangement. In Figure 8.35b, we show the hole distribution in the excited state by the 1s3d EQ transition of up-spin electron. By this transition, the valence holes with up-spin are filled, so that almost no up-spin hole is left, while the distribution of down-spin holes is almost the same as that in the ground state. It is interesting to see the hole distribution in the excited state due to the ED transition of a down-spin electron from Cu 1s to the off-site 3d states, as shown in Figure 8.35c. By the off-site ED transition due to the x polarized incident photon, the Cu 3d hole population in the neighboring CuO_4 plaquettes along the x direction is remarkably decreased. At the same time, the distribution of up-spin holes is drastically changed; the holes on the central Cu site are mainly transferred to the neighboring plaquettes to screen the core hole potential, which is denoted by the nonlocal screening effect. The up-spin holes transferred to the neighboring plaquette couple with the down-spin holes on the same plaquette to form the Zhang–Rice singlet state. It should be noted that in the EQ excitation, the core hole potential is well screened by the Cu 3d elec-tron excited from the Cu 1s state, while in the ED excitation, it is screened by the charge-transfer effect (nonlocal screening).

Usually, the x-ray transition occurs within a single atom as the on-site transi-tion, but the ED transition to the off-site 3d states treated here is the novel nonlocal transition, which is very weak but it is possible to observe it by the 1s2p RXES technique. A similar nonlocal ED transition to the off-site 3d states will be discussed later for Co^{3+} and Fe^{3+} systems by using the 2D RXES images, which is the best way to visualize such a weak XAS transition that is invisible to the conventional XAS technique.

8.3.4 O 1s2p RXES

Recently, theoretical studies of RXES for cuprates were extended from that on the Cu site to that on the O site (Duda et al., 2000; Okada and Kotani, 2002; Harada

(a)   **Ground state**

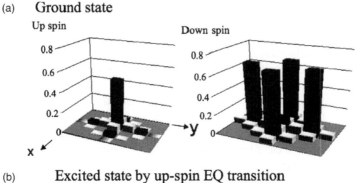

(b)   **Excited state by up-spin EQ transition**

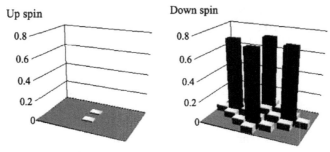

(c)   **Excited state by down-spin ED transition**

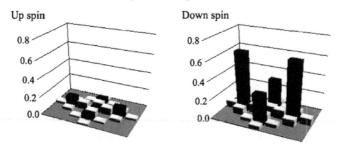

FIGURE 8.35   Spin-dependent valence hole distribution in the ground state and the EQ and ED excited states. [From Kotani, A., Okada, K., Calandra, M., and Shukla, A., *AIP Conference Series (XAFS13)*, 2007. With permission.]

et al., 2002; Okada and Kotani, 2003). The experimental results of the O site derived RXES have so far been interpreted based on the partial O 2p DOS obtained by energy band calculations. It was Duda et al. who first pointed out for their O 1s2p RXES spectra, which were measured for $CuGeO_3$, that the crystal field excitation signal was similarly recognized to that in the RXES at the Cu 2p and 3p edges. Okada and Kotani showed that even in RXES by the O site excitation (O 1s2p RXES), the many-body effects such as the Cu crystal field excitation, the charge-transfer excitation and the UHB-ZRB pair excitation, should be reflected in RXES because of the strong hybridization between Cu 3d and O 2p states.

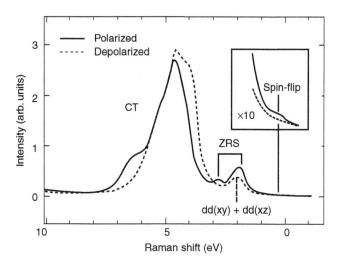

FIGURE 8.36 Calculated result for the O 1s2p RXES with a Cu_4O_{13} cluster model. In the inset, the intensity is enlarged 10 times. (Reprinted with permission from Harada, Y., et al., *Phys. Rev. B*, 66, 165104, 2002. Copyright 2002 by the American Physical Society.)

In Figure 8.36, the calculated result for the O 1s2p RXES with Cu_4O_{13} cluster model (corresponding to an undoped cuprate) is shown. The incident x-ray energy is tuned at the peak position of O 1s XAS. In this calculation, two different polarization geometries (polarized and depolarized), with the incident x-ray normal to the CuO_2 plane, is taken into account. It is shown that the 6 eV antibonding charge transfer excitation and the UHB-ZRB pair excitation, as well as the crystal field excitation [indicated as "dd(xy) and dd(xz)"] and a spin-flip excitation, can be clearly seen, in addition to the most prominent RXES peak, which corresponds to the nonbonding charge-transfer excitation. The charge-transfer excitation and the UHB-ZRB pair excitation depend on the polarization direction of the incident x-ray because of the symmetry selection rule. Experimental observation corresponding to this calculation has also been made by Harada et al. for $Sr_2CuO_2Cl_2$ with almost the same geometry (see Figure 8.37). The experimental result is in satisfactory agreement with the calculated one.

Here, we would like to mention the difference in the UHB-ZRB pair excitation mechanism in RXES with the Cu site excitation (Cu 1s4p RXES) and that with the O site excitation (O 1s2p RXES). In the former RXES, the UHB-ZRB pair excitation is formed in the intermediate and final states as seen in Figure 8.28. In the latter RXES, on the other hand, the UHB-ZRB pair excitation is not formed in the intermediate state and occurs only in the final state of RXES. In Figure 8.38, we show schematically, the process of the UHB-ZRB pair excitation. In the ground state (top panel), we have a hole (hybridized Cu 3d and O 2p hole with b_{1g} symmetry) in each CuO_4 plaquette (this hybridized state is written symbolically as "b_{1g}" state), and their spin is ordered antiferromagnetically ("b_{1g}" ↑ and "b_{1g}" ↓ states in the right- and left-hand plaquettes in Figure 8.38). We assume, as shown in the figure, that an O 1s electron

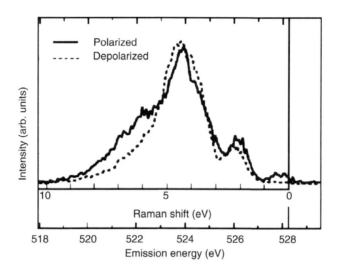

FIGURE 8.37 Experimental results of the O 1s2p RXES for $Sr_2CuO_2Cl_2$. (Reprinted with permission from Harada, Y., et al., *Phys. Rev. B*, 66, 165104, 2002. Copyright 2002 by the American Physical Society.)

with ↑ spin on the central O site is excited to the O $2p_x$↑ state by the x-polarized incident x-ray. Then in the intermediate state (middle panel), we have an O 1s ↑ hole and a "b_{1g}" ↓ hole. In the x-ray emission process, an O $2p_x$ or $2p_y$ or $2p_z$ electron with ↑ spin on the central O site can recombine with the O 1s ↑ hole, and we have various final states depending on the emitted x-ray polarization. If the emitted polarization is in the x direction, for example, one of the typical final states is that shown on the left-hand side of the bottom panel, where we have two holes with "b_{1g}" ↑ and "b_{1g}" ↓ states bound on the right-hand plaquette and no hole on the left-hand plaquette. This is nothing but the UHB-ZRB pair excitation. If the emitted polarization is in the y direction, however, the UHB-ZRB pair final state is forbidden, but two holes with "b_{2g}" ↑ and "b_{1g}" ↓ states bound on the right-hand plaquette can be formed. In this way, the final states of RXES strongly depend on the incident and emitted x-ray polarizations. It can also be seen, as another example of the selection rule, that the elastic scattering where the final state is the same as the ground state is allowed for x-polarized incident and x-polarized emitted x-rays, but is forbidden for the x-polarized incident and y-polarized emitted x-rays. Theoretical calculations of Cu 1s4p RXES and O 1s2p RXES spectra of cuprates with multi-Cu-site cluster model taking into account the effect of orbital degeneracy have been made by Okada and Kotani (2006), and more detail of the comparison of Cu 1s4p RXES and O 1s2p RXES spectra are given.

8.4   NICKEL AND COBALT COMPOUNDS

8.4.1   Ni 2p3d RXES in NiO: Charge Transfer Excitations

NiO is a prototype material of the charge transfer type insulator. The Ni 2p3d RXES was measured by Ishii et al. (2001), Magnuson et al. (2002a), and Ghiringhelli et al.

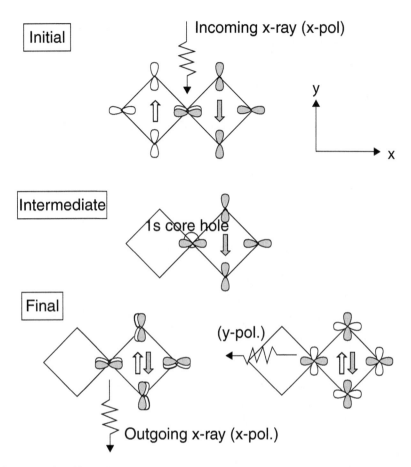

FIGURE 8.38 Illustration of the O 1s2p RXES process. With x-polarized incident and emitted x-rays, the Z-R singlet state is formed in the right-hand side CuO_4 plaquette (see the left-hand side of the lowest panel.) (From Kotani, A., *J. Electron Spectrosc. Relat. Phenom.*, 137–140, 669, 2004. With permission from Elsevier Ltd.)

(2005). The last study has the highest resolution and will be discussed in the next section where we focus on the dd excitations. The result by Ishii et al. is shown in Figure 8.39 where (a) and (b) are the Ni 2p XAS (measured by total electron yield method) and 2p3d RXES measured in the depolarized geometry, respectively. The RXES spectra at about 1.1, 1.6, and 3.0 eV from the elastic line position (indicated with the vertical bar) are due to the dd excitations, although not very well resolved, while the peaks with almost constant emission energy (shown with the dotted line) look like NXES. The dd excitations will be discussed in detail in the next subsection.

Magnuson et al. made the calculations of these RXES spectra with the SIAM, and succeeded in reproducing the NXES-like spectra for incident energies 6 and 7, but failed for 8–13. It is generally known that the SIAM calculation cannot reproduce the NXES spectra. Therefore, these spectra for 8–13 are interpreted as the NXES spectra where the core electron is excited to the continuum band mainly

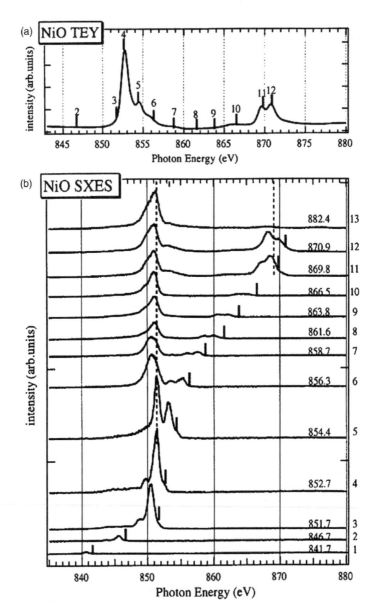

FIGURE 8.39 Experimental results of (a) Ni 2p XAS obtained by the total electron yield (TEY) method and (b) Ni 2p3d RXES for NiO. (From Ishii, H., et al., *J. Phys. Soc. Jpn.*, 70, 1813, 2001. With permission.)

consisting of the s and p electrons. The mechanism of the NXES-like spectra for 6 and 7 was first proposed by Butorin (2000), who considered that they might be the charge-transfer excitation, but behave like NXES spectra, because the incident x-ray energy resonates with the charge-transfer satellite of XAS with continuous excitation energy. Butorin referred to a theoretical calculation by Tanaka et al. (1990), who predicted the NXES-like behavior in CeO_2 when the intermediate state is the charge-transfer satellite of XAS with continuous excitation.

Recently, Matsubara et al. (2005) examined this problem carefully with the SIAM. They not only confirmed Butorin's mechanism for the NXES-like spectra, but also showed what the most appropriate value of the charge-transfer energy Δ is in order to reproduce the experimental charge-transfer excitations in RXES. The parameter values of the SIAM or cluster model have so far been estimated for NiO by analyzing experimental XPS, valence PES, XAS and so on. However, the estimated value of Δ is distributed from 2.0–6.2 eV, depending on the spectroscopy and the method of the analysis. In the analysis by Matsubara et al., the values were $V(e_g) = 2.2$ eV, $U_{dd} = 7.2$ eV and $U_{dc} = 8.0$ eV, which are mostly consistent with the estimations so far made, but the value of Δ is changed from 2.0 to 6.5 eV by steps of 0.5 eV.

The calculated results of Ni L_3 XAS and 2p3d RXES by Matsubara et al. are shown in Figure 8.40a and 8.40b, respectively, together with the experimental RXES spectra by Ishii et al. (2001) in Figure 8.40c, where the incident energies are limited to 4–6, and the RXES spectra are plotted as a function of the Raman shift. The value of Δ is taken to be 3.5 eV where the calculated results are in best agreement with the experimental ones. The incident x-ray energies 6 and 7 correspond to the lower and upper edges of the charge-transfer satellite of XAS (although the intensity is very

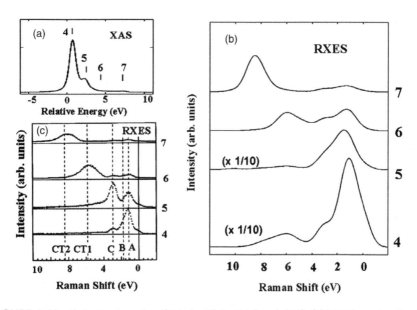

FIGURE 8.40 Calculated results of (a) the Ni L_3 XAS and (b) 2p3d RXES spectra (in the depolarized geometry) for NiO with the SIAM, compared with (c) experimental results by Ishii et al. (2001). The incident energies from 4 to 7 are indicated in (a). The spectral features indicated in (c) by CT1, CT2 and A, B, C are due to the charge transfer excitations and dd excitations. (From Matsubara, M., Uozumi, T., Kotani, A., and Parlebas, J.C., *J. Phys. Soc. Jpn.*, 74, 2052, 2005. With permission.)

TABLE 8.1

Energy Difference of the Charge Transfer Satellites 6 and 7 from the Main Peak 4 in the Ni L_3 XAS and Raman Shift Energies of the Structures CT1 and CT2 in RXES Calculated with Charge-Transfer Energies (Δ) from 2.0 to 6.5 eV

Δ	2.0	2.5	3.0	3.5	4.0	4.5	5.0	5.5	6.0	6.5	Exp.
6-4	2.3	2.7	3.1	3.5	3.9	4.3	4.7	5.2	5.7	6.1	3.6
7-4	5.4	5.6	5.9	6.1	6.4	6.8	7.4	7.6	8.1	8.6	6.0
CT1	4.7	5.2	5.5	5.9	6.3	6.7	7.1	7.5	7.9	8.3	5.8
CT2	7.7	7.9	8.2	8.5	8.8	9.1	9.5	9.9	10.3	10.7	8.4

Note: Values obtained experimentally by Ishii et al. (2001) are also shown. All values are in units of eV.

small), and the RXES peaks at 5.8 eV (for 6) and at 8.4 eV (for 7) are caused by the charge-transfer excitations, and are in good agreement with CT1 and CT2 in the experimental result. In order to show that the calculated results with Δ = 3.5 eV give the best agreement with the experimental results, both for the charge-transfer excitation energies in RXES, CT1, and CT2, and the energy differences between 6 and 4 and between 7 and 4 in XAS, the comparison of the experimental values and the calculated ones for Δ changed from 2.0 to 6.5 eV is given in Table 8.1.

Matsubara et al. (2005) have also calculated the behavior of the charge-transfer excitation by changing the incident x-ray energy from below CT1 to above CT2 with a step of 0.1 eV. The result is shown in Figure 8.41. It is seen that the charge-transfer excitation energy changes continuously and gradually with the change of the incident x-ray energy. Furthermore, this behavior of RXES is similar to that of NXES, in agreement with the old theoretical prediction by Tanaka et al. (1990), as well as the interpretation of the NXES-like spectra by Butorin (2000).

8.4.2 Ni 2p3d RXES IN NiO: dd Excitations

High-resolution 2p3d RXES spectra of NiO have been measured by Ghiringhelli et al. (2005). The results are shown in Figure 8.42. The incident x-ray energy is taken from C to H as indicated in the Ni L_3 XAS (lower right panel of Figure 8.42), where F is the same as 4 of Figures 8.39 and 8.40 and G is close to 5. The total experimental resolution is 0.75 eV (FWHM), which is much better than that of Figure 8.40c, as seen from the comparison of the RXES spectra for F and for 4 of Figure 8.40c. The polarization geometry of these experiments is close to the polarized geometry but the scattering angle is 70° and denoted by the vertical (V) polarization (see Figure 8.1, as well as the geometry V in the inset of Figure 8.44).

From Figure 8.42, it is seen that the RXES spectra with the incident energy below F have almost the same spectral shape with only the change in the intensity, while the spectral shape changes for those above F. In all of these dd excitation spectra, we find the dd excitation energies at about 1.1, 1.8, and 3.2 eV, which correspond to the dd excitations with $^3T_{2g}$, ^1E_g (superposed on $^3T_{1g}$), and $^3T_{1g}$ multiplet terms, respectively.

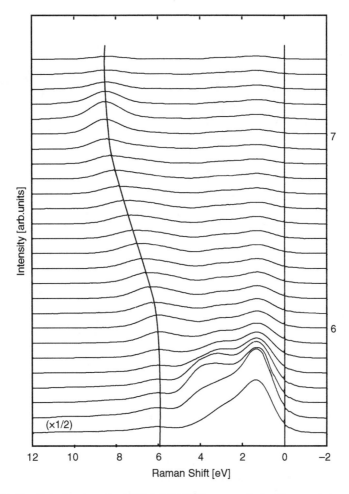

FIGURE 8.41   Calculated results of Ni 2p3d RXES spectra for incident energies around the charge transfer satellite of XAS (6 and 7 in Figure 8.40a). (From Matsubara, M., Uozumi, T., Kotani, A., and Parlebas, J.C., *J. Phys. Soc. Jpn.*, 74, 2052, 2005. With permission.)

The results of theoretical calculations are shown in Figure 8.43 where the calculation was made using the SIAM with the same parameter values as in Figure 8.40, but the spectral broadening is reduced corresponding to the new experimental resolution and the polarization geometry is changed to the vertical (V) polarization (the dashed curves) and the horizontal (H) polarization (the solid curves). For the definition of V and H polarization geometries, see the inset of Figure 8.44. Comparing the RXES spectra of Figure 8.42 with the dashed curves in Figure 8.43, we find that the agreement of the experimental and theoretical results is satisfactory. We see some difference for the experimental and theoretical intensities of the elastic scattering line, but in the theoretical calculations the effect of the reabsorption is not taken into account.

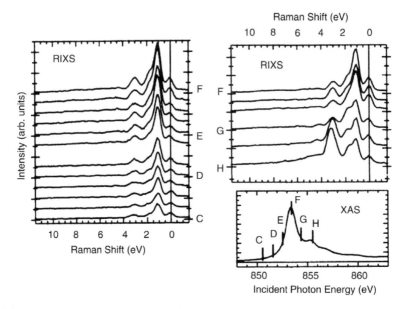

FIGURE 8.42 Experimental results of the Ni L_3 XAS and 2p3d RXES spectra (in the vertical polarization) for NiO. The total resolution is 0.75 eV (FWHM). The incident energies from C to H are indicated on the XAS spectrum. (From Kotani, A., et al., *Rad. Phys. Chem.*, 75, 1670, 2006. Reprinted with permission from IOP Publishing Ltd.)

A more detailed comparison of the experimental and theoretical results is made for the spectra F, G, and H. For this purpose, Ghiringhelli et al. (2005) measured the RXES spectra for both V and H polarization geometries with a better total resolution of 0.55 eV (FWHM). The results are shown in Figure 8.44 with the open (V polarization) and solid (H polarization) circles, and compared with the theoretical results shown with the dashed (V polarization) and solid (H polarization) curves. It can be seen that the experimental and theoretical RXES spectra due to the dd excitation are in good agreement with each other including the polarization dependence.

8.4.3   Ni 2p3d RXES in NiO: Spin-Flip Excitations

Figure 8.45 shows the Ni 2p3d RXES spectra at the 2p x-ray absorption spectrum of NiO. The resonant elastic peak of $^3A_2$ symmetry is visible at 0.0 eV, and the dd transitions to the $^3T_2$, $^3T_1$, and ^1E states are visible at energies of 1.1 eV and at about 2.0 eV energy loss. Additional peaks are visible at low-energy loss. For example, a peak is visible at about 0.25 eV for the spectrum excited at 858 eV. This peak can be assigned to a spin-flip transition in which a Ni^{2+} ion flips its spin state from $S_z = -1.0$ to $S_z = +1.0$. This must be differentiated from the spin state transition of the $^3A_2$ ground state to a ^1E final state. The consequence of this spin-flip is that the antiferromagnetic neighbors now see an apparent ferromagnetic neighbor and the energy involved is the energy loss observed in RXES. The spin-flip peak is highest at an excitation energy of 858 eV, which is at the shoulder of the 2p x-ray absorption

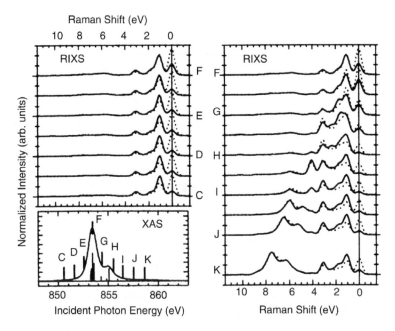

FIGURE 8.43 Calculated results of the Ni L_3 XAS and 2p3d RXES spectra for NiO with the SIAM. The RXES spectra with the dashed and solid curves are obtained in the vertical and horizontal polarizations, respectively. The incident energies from C to K are indicated on the XAS spectrum. (From Ghiringhelli, G., et al., *J. Phys. Cond. Matt.*, 17, 5397, 2005. Reprinted with permission from IOP Publishing Ltd.)

spectral shape. The reason is that this shoulder relates to a state in which the 2p and 3d electrons are essentially antiparallel aligned, in contrast to the parallel alignment of the main peak. This antiparallel alignment facilitates the spin-flip transitions to occur. The measurement of spin-flip transitions is important for the study of the antiferromagnetic coupling in transition-metal oxides. There is, however, an experimental complication, which is that the overall resolution of the experiment should preferably be on the order of 0.2 eV. Recently, high-resolution Ni 3p3d RXES (Chiuzbăian et al., 2005) of NiO have been measured, which seems to confirm this spin flip excitation.

8.4.4 Ni 1s4p RXES of NiO: Pressure Dependence

According to Kao et al. (1996), the Ni 1s4p RXES of NiO also exhibits inelastic peaks close to the positions of the two-charge transfer excitations mentioned previously. Recently, Shukla et al. (2003) measured the pressure dependence of the Ni 1s4p RXES of NiO with the incident energy tuned to the EQ pre-peak of Ni 1s XAS. At ambient pressure, they observed two RXES structures at 5.3 and 8.5 eV, corresponding to the charge-transfer excitations. With increasing pressure up to 100 GPa, the RXES intensity decreases and the 5.3 and 8.5 eV structures are smeared out. The explanation for this is that the width of the charge-transfer excitation band increases with the

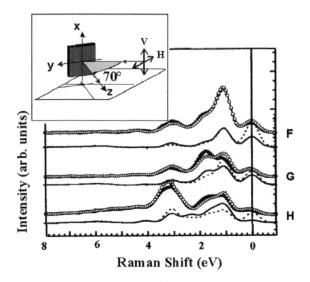

FIGURE 8.44 Comparison of the experimental and theoretical results for dd excitations of the Ni 2p3d RXES of NiO. The open and solid circles are the experimental results in the vertical and horizontal polarizations, respectively, with a total resolution of 0.5 eV (FWHM). The dashed and solid curves are the calculated results for the vertical and horizontal polarizations. The definition of the vertical and horizontal polarizations are given by V and H in the inset. (From Kotani, A., et al., *Rad. Phys. Chem.*, 75, 1670, 2006. Reprinted with permission from IOP Publishing Ltd.)

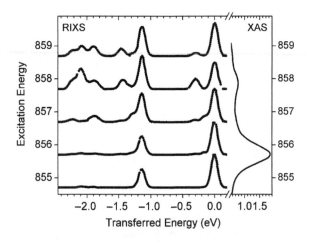

FIGURE 8.45 The 1s2p RXES spectra of NiO are given as a function of their excitation energy, indicated on the left. The corresponding 2p XAS spectrum is given on the right. (Reprinted with permission from de Groot, F.M.F., Kuiper, P., and Sawatzky, G.A., *Phys. Rev. B*, 57, 14584, 1998. Copyright 1998 by the American Physical Society.)

pressure. Since RXES is a photon-in and photon-out process, it is a powerful tool in the study of electronic states under high pressure. Experimental measurements of the pressure-induced change of electronic structure, such as a high-spin and low-spin phase transition, can make use of the 1s3p RXES, as well as 1s3p NXES.

8.4.5   Co 2p3d RXES in CoO and Other Co Compounds

For CoO, Magnuson et al. (2002b) measured the Co 2p3d RXES spectra and performed the analysis of the spectra with the SIAM. Some parameter values used are $\Delta = 4.0$ eV, $V(e_g) = 2.2$ eV, and $10\,Dq = 0.5$ eV. The result is shown in Figure 8.46. They observed both dd excitation (indicated by "$3d^7$" in Figure 8.46) and charge-transfer excitation (indicated by "$3d^8\,L^{-1}$"). The charge transfer excitation looks to occur when the incident x-ray energy resonates with the charge transfer satellite (D and E) of the Co 2p XAS in a similar way to that in NiO. However, the agreement between the calculated and experimental results of the charge-transfer excitation for D is not very good. As a result, it is not very clear whether the charge-transfer excitation behaves like NXES with the change of the incident energy. We note that the CoO spectrum indicates partial oxidation to Co_3O_4, as is evident from comparison with *in situ* soft XAS data on the reduction from Co_3O_4 to CoO (Morales et al., 2004). Furthermore, we see some differences in the calculated and experimental XAS spectra, and a more improved analysis (including finite temperature effects) will probably be necessary, not only for XAS (see Chapter 6) but also for RXES.

Magnuson et al. (2004) also measured the temperature dependence of the Co 2p3d RXES spectra of $LaCoO_3$, which is expected to show a temperature-induced transition of the spin magnetic moment. In this material, the NXES-like spectra are mainly observed, probably reflecting delocalized Co 3d states. The measured spectra suggest a change in the spin state of $LaCoO_3$ as the temperature is raised from 85 to 300 K, while the system remains in the same spin state as the temperature is further increased to 510 K. For CoO, the effect of intra-atomic configuration interaction is investigated both experimentally and theoretically for the Co 2p3s RXES by Braicovich et al. (2001). Similar studies have also been made by Taguchi et al. (2001, 2004) for NiO, $MnFe_2O_4$, and $CoFe_2O_4$.

8.4.6   Co 1s2p RXES of CoO: Effect of Resolution

Figure 8.47 shows the 2D images of Co 1s2p RXES spectra of CoO in the Co 1s3d EQ excitation region, obtained with three different experimental resolutions, compared with a ligand field multiplet (LFM) simulation. The 2D image presentation is made in the same manner as that of the Nd^{3+} system shown in Section 8.2.6, but now the ordinate is taken as $\Omega - \omega$ instead of ω (for more details on the ordinate of the 2D image, see some discussion in Section 8.5.1). The theoretical LFM result simulates the experiments well. This also indicates that the nonlocal dipole peak (as that discussed in Section 8.3.3) is small in the case of CoO because the LFM simulation does not include the nonlocal dipole transition. The LFM simulations also show the importance of interference effects. Excluding interference effects degrades the simulation. There is a significant improvement in the spectral features in going from 1.0 to 0.3 eV resolution. This indicates that if one would like to use the 1s2p RXES

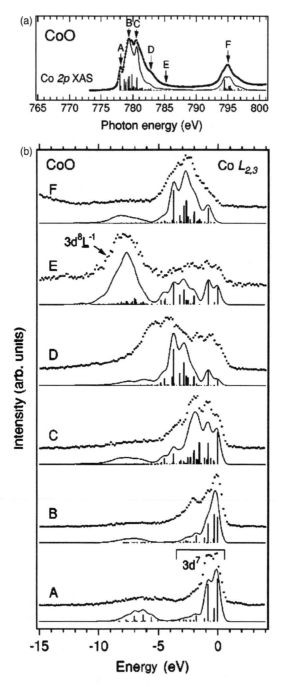

FIGURE 8.46 Upper panel is the experimental (dots) and calculated (solid curve) results of the Co 2p XAS of CoO. Lower panel is the experimental (dots) and calculated (solid curve) results of the Co 2p3d RXES of CoO, where the incident energy is taken at A to F indicated by arrows in the 2p XAS spectrum. (Reprinted with permission from Magnuson, M., et al., *Phys. Rev. B*, 65, 205106, 2002b. Copyright 2002 by the American Physical Society.)

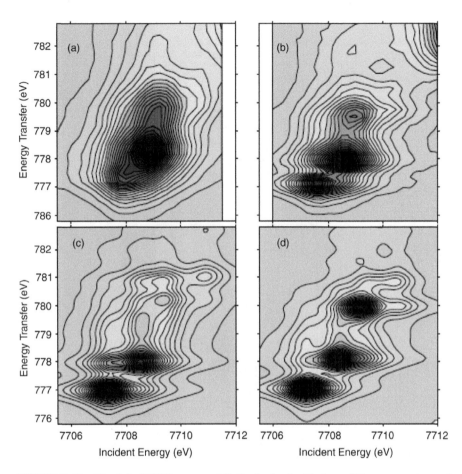

FIGURE 8.47   The 1s2p RXES spectra of CoO, obtained with three different experimental resolutions, respectively 1.0, 0.5, and 0.3 eV, compared with a LFM simulation of the 1s3d quadrupole transition within the 1s2p RXES spectrum. (Figure prepared by Gyorgy Vanko, unpublished.)

spectra for electronic structure determination, obtaining the spectra with 0.3 eV resolution is important.

8.4.7   Co 1s2p RXES: Nonlocal Dipole Transitions

Low-spin Co^{3+} systems have a $3d^6$ ground state with the t_{2g} band occupied and the e_g band empty. Assuming a pre-edge given by EQ transitions only, one expects a K edge with a single pre-edge feature. In Chapter 6, we have seen that the K edge of the low-spin Co^{3+} system $LiCoO_2$ has two pre-edge peaks instead.

Figure 8.48 shows the 2D images of Co 1s2p RXES spectra of $EuCoO_3$ and $LiCoO_2$. The beginning of the edge is visible at E1 = 7716 eV and, before the edge, two separate peaks are visible at 7709 and 7711 eV, respectively. These 1s2p RXES images have also been measured for $Co^{3+}(acac)$, $LaCoO_3$, and $AgCoO_2$, where all

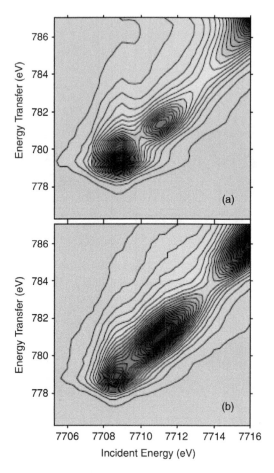

FIGURE 8.48 The 1s2p RXES spectra of the two low-spin Co^{3+} systems $EuCoO_3$ (a) and $LiCoO_2$ (b). (Figure prepared by Gyorgy Vanko, unpublished.)

systems except $Co^{3+}(acac)$ have these two peaks. $Co^{3+}(acac)$ has only the first peak. The 1s2p RXES images show that the first peak has a broadening that runs in the vertical direction, while the second structure is a broad band-like feature that runs in the diagonal direction. This indicates that the first peak is a $3d^6 \rightarrow 1s^13d^7 \rightarrow 2p^53d^7$ local resonance, while the second peak is a $3d^6 \rightarrow 1s^14p^1$ (3d-band) $\rightarrow 2p^54p^1$ (3d-band) transition that occurs at a fixed emission energy, where $4p^1$ (3d-band) means the off-site 3d state that is allowed by the nonlocal ED transition from the 1s core state, as discussed in Section 8.5.3. $Co^{3+}(acac)$ contains no close Co–Co distances hence no Co 3d-band and the second peak is absent. The intensity of the second peak is smallest in $LiCoO_2$, bigger in $AgCoO_2$, and largest in $EuCoO_3$ and $LaCoO_3$. This indicates that there is a direct relation between the relative importance of the nonlocal dipole peak and the coupling between the two Co atoms via the oxygen. The Co–O–Co angle is 90° in $LiCoO_2$ and ~180° in $LaCoO_3$ and $EuCoO_3$, being intermediate in $AgCoO_2$, as described in detail by Vanko et al. (2007). This indicates that $180°$ Co_1–O–Co_2 allows for a much stronger coupling of the 4p states of Co_2 via the oxygen 2p valence band with the empty 3d states of Co_1, where we call this the nonlocal ED transition. These 1s2p RXES results lead to a refined interpretation of the pre-edges of the 3d transition metals, as follows:

1. The pre-edge is caused by EQ transitions from 1s to 3d while the edge is caused by ED from 1s to 4p.
2. There is a nonlocal ED peak that is caused by transition to the 4p states that are hybridized into the 3d-band. The hybridization is via oxygen to the nearest metal neighbor where its strength is determined by the geometry and degree of covalence. Because this nonlocal ED final state is less localized, it is not so much affected by the core hole potential as the localized 3d states, and as such they appear at higher energies than the EQ peaks.
3. If point symmetry is broken, as for example in tetrahedral symmetry, there is local mixing of the 4p and 3d states, yielding local dipole transitions into the localized 3d states. In this case, there are localized, mixed, 3d4p states that, being a single state, must be at the same energy.

We note that low-spin Fe^{2+} systems have the same t_{2g}^6 ground state and Westre et al. (1999) indeed recognized an additional peak between the pre-edge and the edge, which identifies with the nonlocal dipole peak. The nonlocal dipole peak also seems to be dominant in the Fe_2O_3 pre-edge and 1s2p RXES spectrum, which does not correspond well to pure EQ calculations.

8.5   IRON AND MANGANESE COMPOUNDS

8.5.1   Fe 1s2p RXES of Iron Oxides: 2D RXES Images

We will use the 1s2p RXES spectra of iron oxides as examples to explain the advantages of measuring a whole 2D RXES or RXES plane at the 1s3d EQ resonance. The 2D RXES plane is measured as a function of the incident energy Ω and the emitted energy ω. The recorded intensity is proportional to $F(\Omega, \omega)$ and is plotted in a two-dimensional grid. One axis is given by the excitation energy Ω and the other axis is given by the emitted energy ω or the energy transfer $\Omega - \omega$. If the energy transfer is plotted, this implies that the horizontal axis relates to the K pre-edge and the vertical axis to the L edge. There are two different lifetime broadenings that are important for the 2D images, the lifetime broadening of the 1s intermediate state (Γ_{1s}) and the Lorentzian of the final state (Γ_{1s}). Assuming a plot versus energy transfer, Γ_{1s} runs horizontal and Γ_{2p} runs vertical. In addition, an experimental spectrum is broadened by the energy bandwidths of the incident energy monochromator and the crystal analyzer. The incident energy broadening is in the horizontal direction and the analyzer broadening runs diagonal (along ω).

Figure 8.49 shows two 2D images of the 1s2p RXES spectrum of a Fe^{2+} system within the LFM model. It can be observed that the initial state broadenings run horizontal in the left image and diagonal in the right image. This effectively implies

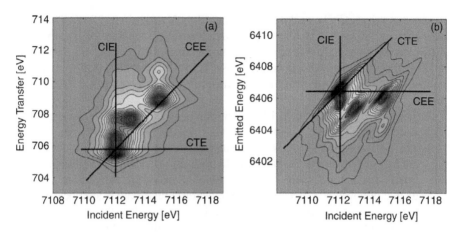

FIGURE 8.49 A theoretical 2D image of the 1s2p RXES spectrum for Fe^{2+} using an experimental broadening of 0.3 eV for both the monochromator and the detector resolution. The contour plots indicate lines of equal intensity where the dark areas have the highest intensity. The difference between the two plots is that in plot (a) the vertical axis gives the energy transfer and in plot (b) it gives the emitted energy. The lines indicate the three different cross-sections, CIE, CEE, and CTE, as discussed in the text. (Reprinted with permission from de Groot, F.M.F., et al., *J. Phys. Chem. B*, 109, 20751, 2005. Copyright 2005 by the American Physical Society.)

that if the 2D image is plotted along the constant emitted energy (CEE) line, the resulting spectrum is a sharpened version of the x-ray absorption spectrum. We now discuss the possible cross sections. They are:

1. Constant incident energy (CIE) spectra, a vertical cross section at fixed excitation energy (7112 eV in Figure 8.49). This relates to resonant x-ray emission spectra.
2. Constant transferred energy (CTE) spectra, a horizontal cross section at a CTE in Figure 8.49a. In Figure 8.49b, the CTE scan is a diagonal cross section. A CTE implies that for all excitation energies, the same final states are probed. This is a similar cross section as the constant final state (CFS) spectra in resonant photoemission spectroscopy (RPES).
3. CEE spectra, a diagonal cross section at a constant emission energy in Figure 8.49a. In Figure 8.49b, the CEE scan is a horizontal cross section at fixed emission energy. This relates to the so-called "lifetime removed" or "lifetime suppressed" spectra. If the emission energy is taken at the peak position of the 1s2p NXES spectrum, the CEE spectrum is the sharpened version of the 1s XAS spectrum, which is called the high-energy resolution fluorescence detected XAS (HERFD-XAS) or the ES with high-energy resolution. Also, this CEE spectrum is sometimes referred to as the PFY.

In addition to these three cross sections, spectra can be measured (or created from a 2D image) with a broader analyzer resolution. In the limit that the complete pre-edge structure is included in such broad line scans, we call them integrated spectra. In principle, this gives the three integrated analogs of the cross sections:

1. Integrated incident energy (IIE) spectra, a vertical band centered on a fixed excitation energy. This yields the total decay intensity of all intermediate states to the 2p final states.
2. Integrated transferred energy (ITE) spectra, a horizontal band centered on a transferred energy in Figure 8.49a. In Figure 8.49b, the ITE scan is a diagonal band. Effectively, this implies that the complete decay at a certain excitation energy is integrated; in other words, this relates to total fluorescence yield.
3. Integrated emission energy (IEE) spectra, a diagonal band centered on constant emission energy in Figure 8.49a. In Figure 8.49b, the CEE scan is a horizontal band. Effectively, this also implies that the complete decay at a certain excitation energy is integrated; in other words, this also relates to total fluorescence yield.

Both an ITE scan and an IEE scan generate the same spectrum as long as the complete pre-edge is integrated. The behavior of ITE scans and IEE scans is different above the edge where IEE scans follow the normal fluorescence and as such relate to the total 1s2p FY.

Figure 8.50 shows the theoretical Fe $1s2p_{3/2}$ RXES spectra of four different iron oxides, respectively (a) the tetrahedral Fe^{2+} oxide $FeAl_2O_4$, (b) the octahedral Fe^{2+} oxide Fe_2SiO_4, (c) the octahedral Fe^{3+} oxide Fe_2O_3, and (d) the tetrahedral Fe^{3+} oxide $FePO_4$. These spectra provide additional information versus the 2p XAS spectral shapes. In addition, it is possible to measure these 1s2p RXES spectra under essentially any experimental condition, inside chemical reactors and at high temperatures and/or high pressures. This figure shows that in order to make full use of the possibilities of 1s2p RXES, an overall experimental resolution of 0.3 eV is needed. For example, with this 0.3 eV resolution, the differences between tetrahedral and octahedral symmetry in Fe^{2+} oxides $FeAl_2O_4$ and Fe_2SiO_4 is clear.

8.5.2 HERFD-XAS OF IRON OXIDES

Figure 8.51 shows the pre-edge region of the Fe K-edge x-ray absorption spectrum of Fe_2SiO_4 (fayalite) measured in two different ways. The bottom portion shows the Fe K pre-edge structure in a conventional absorption spectrum. The subtraction of the modeled main edge contribution from the XAS spectrum over the full energy range yields the isolated pre-edge feature. The pre-edge intensity for the spectra measured by normal XAS strongly depends on the details of the edge subtraction. The top portion of Figure 8.51 shows the Fe_2SiO_4 spectrum measured by 1s2p CEE scans, denoted as 1s2p HERFD-XAS. The small fluorescence energy window selected by the high-resolution detector used in this experiment suppresses the background contribution due to lifetime broadening of the main edge. This results in an almost flat signal before the pre-edge and no background needs to be extracted. The pre-edge integrated intensity and its energy center-of-gravity provide important information regarding the site symmetry and the valence of a metal ion. It is evident

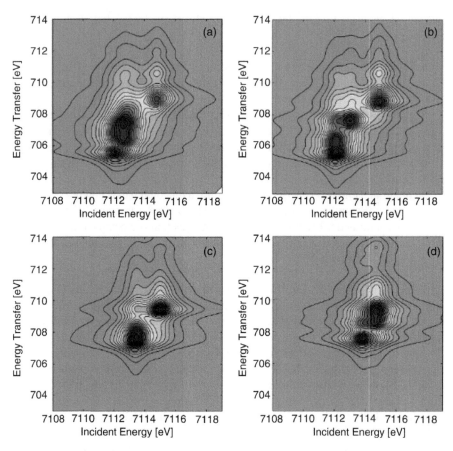

FIGURE 8.50 Theoretical $1s2p_{3/2}$ RXES spectra with an experimental broadening of 0.3 eV for both the monochromator and the detector resolution. The dark area relates to the peak maximum. The LFM parameters have been optimized for (a) $FeAl_2O_4$, (b) Fe_2SiO_4, (c) Fe_2O_3, and (d) $FePO_4$. (Reprinted with permission from de Groot, F.M.F., et al., *J. Phys. Chem. B*, 109, 20751, 2005. Copyright 2005 by the American Chemical Society.)

that HERFD-XAS provides an enormous improvement with respect to normal XAS experiments in this respect.

Figure 8.52 shows the 1s2p HERFD-XAS Fe 1s x-ray absorption spectra of Fe_2SiO_4, $FeAl_2O_4$, Fe_2O_3, and $FePO_4$ normalized to the edge jump. One observes a pre-edge structure around 7115 eV, followed by the absorption edge. The edge energy is 7118 eV for the divalent iron oxides and 7122 eV for the trivalent iron oxides. The onset of the edges is exactly identical for both divalent and trivalent iron oxides. At higher energies, differences occur between Fe_2SiO_4 and $FeAl_2O_4$ as well as between Fe_2O_3 and $FePO_4$ because of differences in their empty DOS. The edge shift of 4 eV between Fe^{2+} and Fe^{4+} oxides is in agreement with the trends found in, for example, bulk Mn oxides (de Vries et al., 2003). These quantum chemical calculations show that it is not so much the valence itself, but the fact that the changes in the Mn 3d occupation and the Madelung potential do not compensate each other.

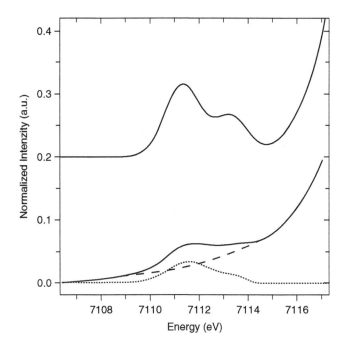

FIGURE 8.51 The Fe K pre-edge feature of Fe_2SiO_4 (fayalite: Fe^{2+}, Oh). Below: conventional fluorescence XAS spectrum (solid), cubic spline function used to model the background (dashed), and the isolated pre-edge (dots). Top: 1s2p-detected HERFD-XAS spectrum, no background subtracted. In both spectra the pre-edge is normalized to the average absorption intensity calculated around 7200 eV. (Reprinted with permission from Heijboer, W.M., et al., *J. Phys. Chem. B*, 108, 10002, 2004. Copyright 2004 by the American Chemical Society.)

8.5.3 Fe 2p XAS SPECTRA MEASURED AT THE Fe K EDGE

The CIE cross sections in Figure 8.49 allow one to measure essentially the metal L edges using K pre-edge excitation and 1s2p XES decay. This effectively allows one to measure soft x-ray absorption spectra with hard x-rays. Figure 8.53 shows three inelastic scattering spectra taken with the incident photon energy tuned to the pre-edge, respectively, the T_{2g} peak (7113.4 eV) and the E_g peak (7114.8 eV). The peak positions of the spectral features in spectra (a) and (b) agree well with the main peaks in the 2p XAS spectrum.

8.5.4 VALENCE SELECTIVE XAS

The 1s3p NXES spectrum can be used to measure valence-selective XAS spectra in mixed-valent compounds. Figure 8.54 shows the 1s3p NXES spectrum of Prussian Blue ($Fe_4[Fe(CN)_6]_3$). The spectra of Fe_2O_3 and $K_4Fe(CN)_6$ are included as models for the high-spin and low-spin Fe sites in Prussian Blue. Their normalized sum is superimposed, and is equivalent to the Prussian Blue spectrum. The valence-selectivity is achieved by tuning the emission analyzer to a particular fluorescence energy.

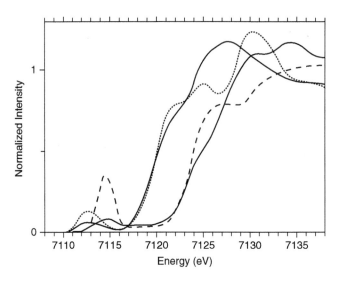

FIGURE 8.52 Experimental Fe 1s2p HERFD-XAS spectra of $FeAl_2O_4$ (dotted), Fe_2SiO_4 (solid), Fe_2O_3 (solid with points), and $FePO_4$ (dashed). (Reprinted with permission from de Groot, F.M.F., et al., *J. Phys. Chem. B*, 109, 20751, 2005. Copyright 2005 by the American Chemical Society.)

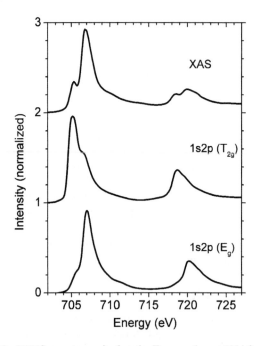

FIGURE 8.53 1s2p RXES spectra excited at the E_g pre-edge at 7114.8 eV (bottom), and at the T_{2g} pre-edge of 7113.4 eV (middle), compared with the 2p XAS spectrum (top). (Reprinted with permission from Caliebe, W.A., et al., *Phys. Rev. B*, 58, 13452, 1998. Copyright 1998 by the American Physical Society.)

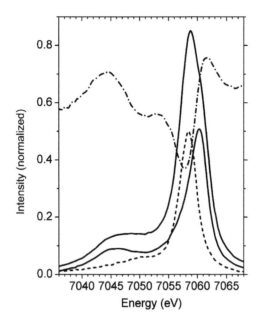

FIGURE 8.54 1s3p XES spectra of Fe_2O_3 (solid), $K_4Fe(CN)_6$ (dashed) and $Fe_4[Fe(CN)_6]_3$ (solid). The $Fe_4[Fe(CN)_6]_3$ is equal to the weighted sum of Fe_2O_3 and $K_4Fe(CN)_6$ spectra. This yields the estimated fraction of signal arising from high-spin Fe^{3+} component (dash-dotted). (Reprinted with permission from Glatzel, P., Jacquamet, L., Bergmann, U., de Groot, F.M.F., and Cramer, S.P., *Inorg. Chem.*, 41, 3121, 2002. Copyright 2002 by the American Chemical Society.)

The fluorescence intensity is then recorded while the incident energy is scanned across the Fe 1s XAS spectrum. The contributions from the different Fe sites in Prussian Blue to such selective absorption scan depend on the chosen emission energy. The resulting 1s XAS do not represent the "pure" spectra from the respective sites, because the 1s3p NXES spectra overlap. It is therefore necessary to apply a mathematical procedure to extract the pure 1s XAS spectra. This procedure provides two independent HERFD-XAS spectra for Fe^{2+} and Fe^{3+}. It is noted that high-spin low-spin systems are particularly suited for this technique because of the distinct change in 1s3p NXES spectra. Separating high spin Fe^{2+} from Fe^{3+}, for example in Fe_3O_4 is more complicated as there is larger spectral overlap of their 1s3p NXES spectra.

8.5.5 Mn 2p3d RXES of MnO

The Mn 2p3d RXES of MnO was measured by Butorin et al. (1996a) and recently by Ghiringhelli et al. (2006) with higher resolution. In Figure 8.55a and 8.55b, we show the result by Butorin et al. (1996a), where (a) shows the Mn 2p XAS and (b) the Mn 2p3d RXES spectra observed in the depolarized geometry are shown with the dots on the lower panel. The solid curves in the lower panel are the results of the atomic calculation, where the effect of the AM coupling is taken into account but the crystal field and charge-transfer effects are disregarded. In MnO, the Mn 3d shell is filled by five parallel-spin electrons to form a stable state, so that the hybridization

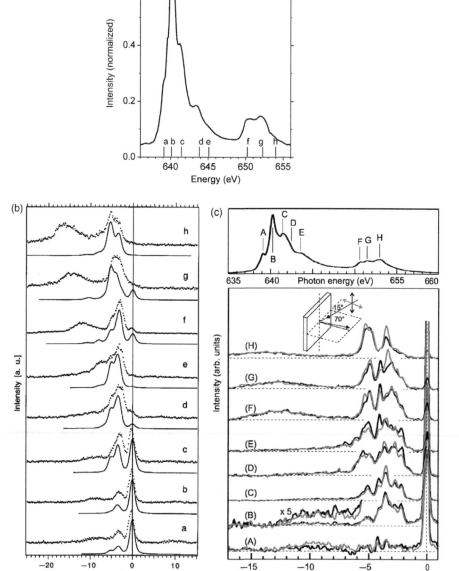

FIGURE 8.55 (a) Experimental results of the Mn 2p XAS; (b) the related Mn 2p3d RXES (dots) of MnO, compared with the RXES spectra by atomic calculations (solid); (c) the same spectra, but the RXES spectra are obtained by high-resolution measurements for two polarization geometries. [Reprinted with permission. Panels (a) and (b) from Butorin, S.M., et al., *Phys. Rev. B*, 54, 4405, 1996a; panel (c) Ghiringhelli, G., et al., *Phys. Rev. B*, 73, 035111, 2006. Copyright 2006 by the American Physical Society.]

effect between the Mn 3d and O 2p states is relatively weak, and the intensity of charge-transfer excitations in RXES is weak. The RXES structures within 7 eV from the elastic scattering (the origin of the abscissa) are due to the dd excitations, and they are in fair agreement with the atomic calculations partly because of the low experimental resolution 1.6 eV (FWHM).

The results of high-resolution experiments by Ghiringhelli et al. (2006) are shown in Figure 8.55c. The resolution is 0.3 eV (FWHM), more than five times better than that by Butorin et al. (1996a). Furthermore, the polarization dependence is also measured with the vertical and horizontal polarizations shown in Figure 8.55c. The incident energies A, B, and C are almost equal to a, b, and c in the experiments by Butorin et al. Comparing the RXES spectra of Figure 8.55 for these incident energies, we see that the dd excitation features given by almost a single peak in Figure 8.55b split into five peaks in Figure 8.55c. Further, the charge-transfer excitations around −10 eV are seen more clearly in Figure 8.55c.

Theoretical analysis is made with the SIAM. The best fit is obtained with the following parameters: $\Delta = 6.5$ eV, $U_{dd} = 7.2$ eV, $U_{dc} = 8.0$ eV, $V(e_g) = -2V(t_{2g}) = 1.8$ eV, $10\,Dq = 0.5$ eV, $R_c = 0.8$ eV, $R_v = 0.9$ eV and the width of the O 2p band $W = 3.0$ eV. The comparison with the experiment is made in Figure 8.56, where we show an expanded view of the dd region with excitation energies from B to F (the spectral structures are labeled s_1 to s_5 as in the panel C). In the L_3 excitation region, the theory reproduces the structures s_1 to s_5 very well. Moreover, the sign of the dichroism (difference of the spectra with vertical and horizontal polarization geometries) is correct in almost all cases. Also in the L_2 region, we find a general agreement between theory and experiment but in a less detailed way especially as far as the dichroism is concerned. This is not surprising because the calculations do not include Coster–Kronig conversion that is dependent on the polarization.

In the comparison of the charge-transfer excitation with the experiment, it is important to avoid superposition with NXES not included in the calculations. This is possible at excitation energies near the L_3 threshold where no sign of features at CEE is found and in the upper L_2 range where most of the NXES is found at transferred energies higher than those of the charge-transfer excitations. The analysis is summarized in Figure 8.56b giving very clear results in spite of the experimental noise. In fact, the features are rather broad but for our purposes, we can dramatically smooth the measured spectra (heavy solid lines). In Figure 8.56b, the theory and the experiment are normalized to the same height of the feature at −3.6 eV (averaged over the polarization) in the dd region. In cases A and B, the comparison with the experiment is excellent: the theory gives the charge-transfer excitations at the correct energies and with the correct intensity. Moreover, the theory predicts the correct dichroism that is much stronger in A than in B. In case C, there is also a general agreement on the tails of the spectra and on the absence of dichroism, but the NXES going across the spectra fills the valley around 6 to 7 eV. At the maximum energy of the present measurements (case H), the theory predicts a sizable charge transfer distribution whose intensity agrees with the measurements. Of course, in the measurements, there is also an NXES contribution extending at higher transferred energies. In conclusion, the SIAM with the above parameters is very satisfactory both in the calculation of the dd and charge-transfer excitations.

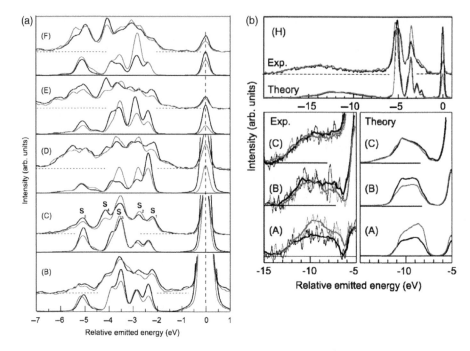

FIGURE 8.56 (a) Experimental and theoretical results of the Mn 2p3d RXES spectra in the dd excitation region of MnO; (b) in the charge transfer region. In (a) and H of (b), the experimental and theoretical results are given by the upper and lower curves, respectively, and gray and black curves correspond to two different polarization geometries, vertical and horizontal geometries, respectively. In A–C of (b), the experimental and theoretical results are given in left and right. (Reprinted with permission from Ghiringhelli, G., et al., *Phys. Rev. B*, 73, 035111, 2006. Copyright 2006 by the American Physical Society.)

8.5.6 Mn 2p3d RXES: Interplay of dd and Charge Transfer Excitations

In the theoretical calculations of dd excitations of Mn 2p3d RXES for MnO, Ghiringhelli et al. (2006) also made the LFM model calculations in addition to the CTM model calculations mentioned previously. They found that the results of LFM calculations could be in good agreement with the experimental ones if the value of 10 Dq is taken as 1.0 eV (twice that used in the CTM calculation) and the Slater integrals are rescaled to 75% (from 80% of the CTM calculation). As an example, we show in Figure 8.57, the result of the LFM calculation (lower panel) for the incident energy (C) and (E) compared with the experiment (upper panel). This suggests that the effect of hybridization (charge-transfer effect) can be approximately included in the LFM model by renormalizing the values of 10 Dq and Slater integrals. This situation is very similar to the calculation of XAS spectra discussed in Chapter 6.

In order to see the condition with which hybridization effect can be renormalized in the LFM model, Matsubara and Kotani (2007) made three different calculations for the 2p3d RXES of Mn^{2+} systems. Figure 8.58(a) shows the result using LFM calculations for various values of 10 Dq. Figure 8.58(b) and (c) show the results of using

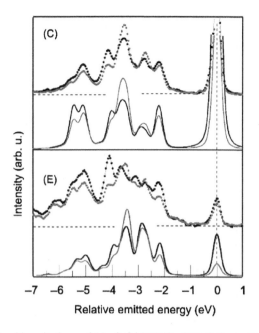

FIGURE 8.57 The dd excitations of Mn 2p3d RXES of MnO for incident energy (C) and (E) calculated with LFM model (solid) and compared with the experimental result (points). The labels (C) and (E) refer to the excitation energies as indicated in Figure 8.55. Gray and black curves correspond respectively to the vertical and horizontal geometries. (Reprinted with permission from Ghiringhelli, G., et al., *Phys. Rev. B*, 73, 035111, 2006. Copyright 2006 by the American Physical Society.)

CTM calculations with the same parameter values as MnO, but Δ and $V(e_g)$ $[= -2V(t_{2g})]$ are, respectively, changed for (b) and (c). The incident energy is fixed at the maximum XAS peak [position (B) of Figure 8.55c], and the polarization geometry is the depolarized one. For simplicity, the rescaling of the Slater integrals is not made for the LFM calculations. The results of $V(e_g) = 1.8$ eV in (b) and $\Delta = 6.5$ eV in (c) correspond to the RXES of MnO (not exactly the same as that in Figure 8.56), and that of $10\,Dq = 1.0$ eV in (a) is a good approximation for this case. Comparing (a) and (b), we see that the increase of $10\,Dq$ in (a) gives the same trend of the spectral change as the increase of $V(e_g)$ in (b). This indicates that the increase of $V(e_g)$ causes the increase in the hybridization effect and the increase of the hybridization effect in the CTM model can be renormalized by an increase of $10\,Dq$ in the LFM model. However, from the comparison of (a) and (c), we find that the increase of the hybridization effect caused by the decrease of Δ in the CTM model cannot be renormalized by an increase of $10\,Dq$ in the LFM model. From these results, we can get the condition of the renormalizability of the hybridization effect in the LFM model, as shown below.

In the case of (b), the charge-transfer energy Δ is large. Thus, the dd excitation and charge-transfer excitation are well separated in their energy range, as shown in the upper panel of Figure 8.59. In this case, the effect of hybridization on the dd excitation can be approximately renormalized in the LFM model by rescaling the value of $10\,Dq$. This can be understood if the separation of dd and charge transfer excitation

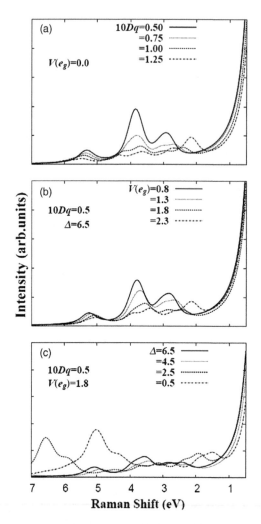

FIGURE 8.58 Three different calculations of Mn 2p3d RXES in the dd excitation region. (a) Result of LFM calculations; (b) and (c) result of CTM calculations.

energies is large enough because the second-order perturbation with respect to the hybridization gives rise to an effective 10 Dq. Even if the second-order perturbation does not converge, an effective 10 Dq can be defined. On the other hand, the value of Δ is decreased in (c), the charge transfer and dd excitation energies overlap each other as shown in the lower panel of Figure 8.59, and then the LFM model breaks down. In this case, the interplay of dd and charge transfer excitations gives rise to complicated excitation modes, which cannot be obtained by the LFM calculations.

Finally, we would like to make two remarks. The first one concerns the comparison of XAS and RXES. Even in XAS, the LFM theory breaks down, in principle, if the charge transfer and dd excitations overlap in the final state. Even though they do not overlap, the LFM theory cannot describe the charge transfer excitation itself.

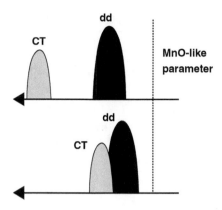

FIGURE 8.59 Schematic representation of two cases where the LFM model can describe the dd excitation (upper panel) and cannot describe it (lower panel). (From Taguchi, M., Kruger, P., Parlebas, J.C., and Kotani, A., *Phys. Rev. B*, 73, 125404, 2006. With permission.)

However, as mentioned in Chapters 3 and 6, the charge-transfer excitation is strongly suppressed in the final state of XAS, and so the validity range of the LFM theory is very wide. In contrast, the charge transfer excitation is not very weak, in general, in RXES, and especially when a weak XAS satellite is tuned as the intermediate state of RXES, the charge-transfer excitation intensity is strongly enhanced in the RXES spectra. Therefore, care should be taken when applying the LFM theory to the calculation of RXES spectra. The second remark concerns the example of the system where the charge transfer and dd excitations overlap. In most insulating TM compounds, the dd excitations seem to be described at least approximately by the LFM theory, but more detailed study will be required on this point. The situation is different for metallic systems and biological systems. Here, we only point out that the systems of a Mn impurity in a Ag host metal and a Mn monolayer on the Ag metal substrate would be examples of overlapping charge transfer and dd excitations as predicted theoretically by Taguchi et al. (2006).

8.5.7 Mn 1s4p RXES of LaMnO$_3$

Spin, charge, and orbital orderings in perovskite-type manganites have attracted much attention due to the affects of strong electron correlation in solids. LaMnO$_3$ is a Mott insulator where the occupied 3d(e$_g$) states exhibit an orbital-ordering below 780 K. Inami et al. (2003) measured the Mn 1s4p RXES of orbital-ordered LaMnO$_3$. An example of observed RXES spectra is shown in Figure 8.60 where RXES is measured by changing the incident photon energy with the momentum transfer fixed at (1.6, 1.6, 0). As indicated by the arrows, they observed three RXES features at 2.5, 8, and 11 eV. The two higher energy features are interpreted as the charge transfer excitations from O 2p bands to the Mn 3d and 4s/4p bands, while the 2.5 eV peak is ascribed to an orbital excitation across the Mott gap. In the ground state, the occupied e$_g$ states of 3d($x^2 - r^2$) and 3d($y^2 - r^2$) orbitals are alternatively ordered corresponding to the LHB states, so that the relevant orbital excitation occurs from these occupied orbitals to the empty e$_g$ states of 3d($y^2 - z^2$) and 3d($z^2 - x^2$) orbitals

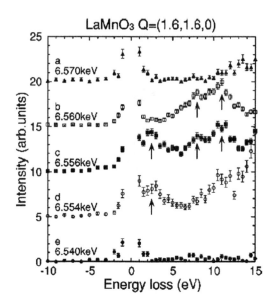

FIGURE 8.60   Experimental results of Mn 1s4p RXES spectra of LaMnO$_3$. Three inelastic excitations indicated with arrows are observed. (Reprinted with permission from Inami, T., et al., *Phys. Rev. B*, 67, 45108, 2003. Copyright 2003 by the American Physical Society.)

to form the double occupancy of e$_g$ states (UHB) and a hole in the LHB. Theoretical calculations were also made for this 2.5 eV peak, and the small momentum dispersion of this peak, as well as the characteristic azimuthal angle dependence, was reproduced considerably well.

The effect of hole-doping in the Mn 1s4p RXES of La$_{1-x}$Sr$_x$MnO$_3$ ($x = 0.2$ and 0.4) has also been studied by Ishii et al. (2004). For $x = 0.4$, the system is in the metallic ferromagnetic state, but the observed RXES spectra show that the inelastic excitation across the Mott gap persists around 2.0 eV. At the same time, the Mott gap is partly filled by low energy excitations, corresponding to the metallic behavior. For $x = 0.2$, the system is in the insulating paramagnetic state at high temperatures, but below 309 K, the metallic ferromagnetic state becomes more stable. Ishii et al. (2004) measured the temperature-dependence of the RXES intensity across the phase transition temperature for the low-energy excitations. They revealed that the temperature-dependence is anisotropic, the intensity increases with decreasing temperature along the $\mathbf{Q} = \langle h00 \rangle$ direction, and there is almost no temperature-dependence along the $\mathbf{Q} = \langle hh0 \rangle$ direction. This anisotropy might be caused by the anisotropy of magnetic interaction, which is strongly correlated with the orbital anisotropy. A similar temperature-dependence of the RXES intensity along the $\mathbf{Q} = \langle h00 \rangle$ direction has also been reported by Grenier et al. (2005).

8.5.8   Mn AND Ni 1s3p XES: CHEMICAL SENSITIVITY

An x-ray emission transition involving core states is, in the first approximation, not expected to be sensitive to the chemical nature of the absorbing atom. A 1s core hole is being replaced with a 3p core hole and as far as the charge is concerned, this will have very little influence on the valence electrons. From the local charge point-of-view, the valence electrons do not notice that a 3p electron is filling a 1s core hole. This implies that the center of gravity of a 1s3p XES spectrum is only determined by the binding energy difference of the 1s and 3p core states, implying it is the same for all Mn systems, independent of valence, covalence, and spin state. Still, there is a large amount of literature concerning the determination of the valence and spin state from the 1s3p XES spectral shapes. The reason for this indirect chemical sensitivity is not the charge of the core hole, but the 3p core hole angular momentum and its very large exchange interaction with the 3d electrons. This makes the 3p core hole in the final state sensitive to the amount of unpaired 3d electrons in the valence band.

Figure 8.61 shows the 1s3p XES spectra of three manganese oxides. There is a linear relationship between the spin state on the Mn atom and the peak position of the main peak. The reason for the shift of the main peak is that when the spin state increases, the exchange splitting increases and the intensity of the satellite moves away further from the center. To compensate for the increased shift of the satellite, the main peak is shifted a little further in the opposite direction, causing its effective shift. In Section 8.6, we will discuss how to obtain the spin state from the experimental spectra in the most reliable manner.

Figure 8.62 might appear surprising as it shows distinct energy shifts of the main peak in a series of Ni^{2+} systems. Ni^{2+} is always high-spin and its spin state is $S = 1$ in all four cases. Still, the main peak shifts and this shift is actually linear with the charge-transfer energy of these four systems, where the same hoppings have been used. As discussed in Chapter 5, a larger charge transfer implies a more pure $3d^8$ system. $NiBr_2$ has more mixing of $3d^9\underline{L}$ and this effectively decreases the number of 3d-holes; hence, the 3p3d exchange interaction decreases, moving the satellite closer to the main peak and shifting the main peak to the center.

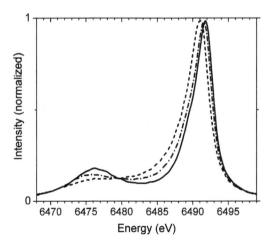

FIGURE 8.61　1s3p XES in the Mn oxides MnO (solid), Mn_2O_3 (dash-dotted), and MnO_2 (dashed). (From Glatzel, P., and Bergmann, U., *Coord. Chem. Rev.*, 249, 65, 2005. With permission from Elsevier Ltd.)

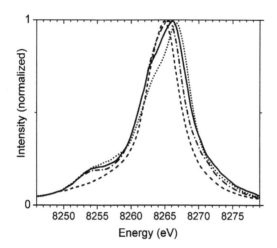

FIGURE 8.62 1s3p XES in the Ni^{2+} systems, respectively NiF_2 (dotted), NiO (solid), $NiCl_2$ (dash-dotted), and $NiBr_2$ (dashed). (From Glatzel, P., and Bergmann, U., *Coord. Chem. Rev.*, 249, 65, 2005. With permission from Elsevier Ltd.)

8.5.9　Mn 1s3p XES: K Capture Versus X-Ray Ionization

The use of K capture isotopes allows the study of the effects of the excitation process. Recently, Bergmann et al. (1999) and Glatzel et al. (2003) compared the resonance x-ray excitation with K-capture excitation for MnO. In the x-ray excitation process, the 1s electron leaves the atom as a continuum electron, and the spectrum is calculated by coupling the 1s3p spectra to the 1s XPS spectrum. In the K-capture process, the 1s electron annihilates a proton and is captured in the core. The calculation is technically the same as that for x-ray excitation as given previously, but the difference is the ordering of the configurations. In the case of K capture, the 1s electron that has been "excited" has become part of the nucleus, implying that the total charge of the nucleus plus the 1s core level is not modified. This implies that the outer valence electrons hardly notice any effect on the positions of the various configurations and the energy difference between $1s^13d^5$ and $1s^13d^6\underline{L}$ is equal to that in the ground state. In contrast, an x-ray excited intermediate state has its $1s^13d^6\underline{L}$ configuration lowered by the core hole potential. Experimentally, the K-capture spectrum is sharper, and the satellite structure is found at a lower energy. This can be explained from additional states in the x-ray excited spectrum (Glatzel et al., 2001, 2003).

Subsequently, we will develop the differences in the calculation between x-ray excitation and K capture in steps. We note that it is not always possible, or desirable, to describe the ground, intermediate, and final states in all possible detail and a number of approximations can be made in order to describe, and understand, the various RXES and NXES experiments. There are a number of useful approximations to calculate the XES experiments:

1. Using AMs.
2. Using LFMs, that is, adding crystal field effects.

3. Using charge transfer multiplets (CTM), that is, adding charge-transfer effects.
4. In the case of NXES experiments, the lowest intermediate state can be used, or the 1s XPS spectrum can be used to generate the full NXES spectral shape.

We will go through these four approximations for a comparison between x-ray excited and K-capture 1s3p XES for the example of a $3d^5$ Mn^{2+} system, where we assume that within the CTM model, the $3d^5$ configuration is mixed with $3d^6\underline{L}$.

8.5.9.1 Atomic Multiplet Calculation

The ground state of Mn^{2+} has its five spin-up electrons filled. The five spin-down states are empty. It is assumed that the core hole creation does not modify the valence electron situation. The x-ray excited intermediate state has a $1s^1 3d^5 \varepsilon$ configuration, where ε is a free electron that is assumed not to affect the decay. After the 1s3p decay, the final state is $3p^5 3d^5$. In the case of K-capture, the initial state is ^{55}Fe that is transferred to ^{55}Mn after K-capture has modified a proton to a neutron. The difference with the x-ray excited transition is the absence of the free electron because the electron has been captured by the nucleus

$$\text{x-ray: } 3d^5 \rightarrow 1s^1 3d^5(\varepsilon) \rightarrow 3p^5 3d^5(\varepsilon) \qquad (8.25)$$

$$\text{K-cap: } 3d^5 \rightarrow 1s^1 3d^5 \rightarrow 3p^5 3d^5. \qquad (8.26)$$

It can be observed from these equations that within the AM calculation, the same calculation is performed for the NXES transition and we neglect any influence from the excited electron (ε). The spectral shape of the NXES calculation consists of two structures separated by the 3p3d exchange interaction. Assuming an Mn^{2+} ion, the atomic ground-state symmetry is ^6S. In the $1s^1 3d^5$ configuration, the total symmetry can be found by multiplying the 1s electron with the ^6S symmetry of the 3d electrons. This either gives a ^5S state for antiparallel alignment of the 1s and 3d electrons or a ^7S for parallel alignment. Both states are split by the 1s3d exchange interaction, which is in the meV energy range. The dipole selection rules imply that transitions are possible from ^5S to ^5P and from ^7S to ^7P, with the energy difference between ^5P and ^7P given by the 3p3d exchange interaction. Writing out all symmetry combinations, three ^5P states and one ^7P state are found.

Figure 8.63 (bottom) shows the atomic multiplet spectrum for which the 3d and 3p spin–orbit couplings have been set to zero. The overall intensities of the states are 7 : 5 for the ^7P against the ^5P states. The main peak is the ^7P state. The shoulder and the satellite are the two main ^5P states. The satellite is related to the antiparallel aligned 3p and 3d electrons, while the shoulder relates to a 3d configuration different from the $^6A_1$ ground state. The three ^5P states are linear combinations of the three LS-like atomic states that are mixed by the strong 3p3d as well as 3d3d multiplet effects. Figure 8.63b shows the effects of the atomic 3p and 3d spin–orbit coupling. Essentially, the ^7P and ^5P states are split into their three substates $^7P_4$, $^7P_3$, and $^7P_2$. The situation for the ^5P states is similar, and the satellite is split into $^5P_3$, $^5P_2$, and $^5P_1$. Comparing spectra a and b of Figure 8.63, it can be seen that the effects of the spin–orbit coupling on the broadened spectral shapes are minimal.

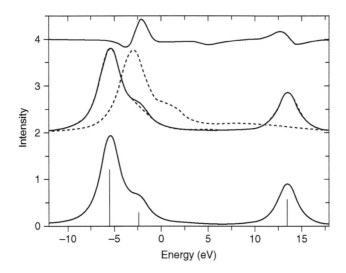

FIGURE 8.63 Theoretical 1s3p XES calculated with the AM (bottom) and LFM (middle) models. Middle: The spin–orbit couplings are set to their atomic values (solid). A cubic crystal field of 2.0 has been added (dotted). A cubic crystal field of 4.0 eV has been added (dashed). Top: Difference between b and c (*10). (Reprinted with permission from de Groot, F., *Chem. Rev.*, 101, 1779, 2001. Copyright 2001 by the American Chemical Society.)

8.5.9.2　LFM Calculation

Figure 8.63c shows the effects of the addition of a cubic crystal field of 2.0 eV. Compared to Figure 8.63b, it can be seen that after broadening the sticks with the experimental and lifetime broadening, few changes are visible. This implies that 1s3p x-ray emission spectra are not very sensitive to the details of the crystal-field effects. The positive aspect of this observation is that one can assume that the 1s3p x-ray emission spectra are virtually the same for all high-spin divalent manganese systems. This finding is important for the use of the 1s3p x-ray emission channel for selective x-ray absorption measurements. Figure 8.63d shows the effect of a cubic crystal field of 4.0 eV where a completely different spectrum is found. The reason is that the ground-state symmetry has changed from $^6A_1$ to $^2T_2$. Thus, only one unpaired 3d electron remains, leading to a very small 3p3d exchange splitting and essentially no satellite. It is evident that 1s3p x-ray emission will immediately show that a system is in its high-spin or low-spin state.

8.5.9.3　Charge Transfer Multiplet Calculation

Describing XES with one configuration, in general, does not result in a very good approximation. The main reason is that, in the 1s excitation step, major screening occurs. Essentially, a core electron is being removed, implying a core charge that has increased by one. This pulls down the valence 3d electrons by the core hole potential, as described in detail in Chapter 5 for XPS. Since the 3d electrons are pulled down in a larger energy than the 4s and 4p electrons, the intermediate-state

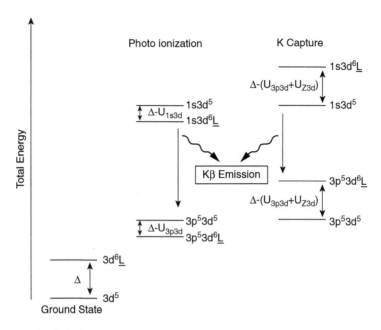

FIGURE 8.64 Ordering in the CTM model for 1s3p XES compared for x-ray ionization and K capture. (Reprinted from Glatzel, PhD thesis, 2001.)

configurations will be different from the ground-state configurations. Essentially, the core hole potential U_{1s3d} has to be subtracted from the ground state energy positions as used in the charge transfer model. This implies that the energy of $1s^1 3d^6 \underline{L}$ is $\Delta - U_{1s3d}$ where, in the case of K-capture, its energy is identical to the ground state at Δ, as indicated in Figure 8.64.

$$\text{X-ray: } 3d^5 + 3d^6 \underline{L} \rightarrow 1s^1 3d^5 + 1s^1 3d^6 \underline{L}(\varepsilon) \rightarrow 3p^5 3d^5 + 3p^5 3d^6 \underline{L}(\varepsilon) \quad (8.27)$$

$$\text{K-cap: } 3d^5 + 3d^6 \underline{L} \rightarrow 1s^1 3d^5 + 1s^1 3d^6 \underline{L} \rightarrow 3p^5 3d^5 + 3p^5 3d^6 \underline{L} \quad (8.28)$$

An approximation that is often used in the calculation of the 1s3p XES spectrum is the assumption that only the lowest energy intermediate state has to be used to calculate the spectral shape. It is implicitly assumed that all intermediate states decay to the "intermediate ground state." This "relaxed" model has been used in the calculations of the 1s3p x-ray emission spectra of divalent nickel compounds (de Groot et al., 1994a).

8.5.9.4 Coherent Calculation of Mn 1s3p NXES Spectra

The "relaxed" charge transfer model is probably not the correct model. It is a better approach to assume that the intermediate states do not relax and, in fact, that NXES is still a coherent second-order optical process. This implies that, in the case of x-ray ionization, the 1s XPS spectral shape is calculated with Equation 8.27, and for all these intermediate states, the 1s3p x-ray emission spectral shape is also calculated.

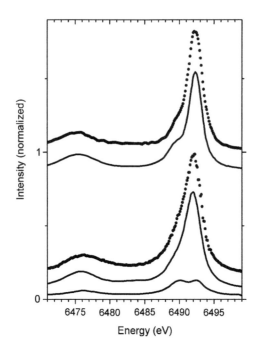

FIGURE 8.65   Comparison between calculated (solid) and experimental (points) 1s3p NXES spectra for (a) the K capture in $^{55}Fe_2O_3$ and (b) for MnO after x-ray ionization. The theoretical curves in (b) show the spectrum resulting from the bonding (top) and antibonding (bottom) combinations in the intermediate states. (Reprinted from Glatzel, PhD thesis, 2001.)

One then combines the 1s3p XPS excitation with the 1s2p XES decay. In the case of K capture, it is a valid assumption to assume that the ordering of the intermediate states is not affected by the capture process of the 1s electron because of the charge neutrality of the core + 1s states. This implies that the 1s core state has the same combination of $1s^13d^5$ and $1s^13d^6\underline{L}$ as the initial state.

Figure 8.65 shows the comparison between the coherent CTM calculations and the experiments for the 1s3p XES spectra for K capture in $^{55}Fe_2O_3$ and x-ray ionization in MnO. The observed additional broadening for the x-ray ionized spectrum is reproduced in the calculations. The cause for this extra broadening is mainly due to the different 1s3p XES spectra of the different intermediate states. Unfortunately, the experimental spectra lack structure, so it is not possible to further fine tune the differences in the electronic structure parameters for both situations. It is noted that 1s XPS would map the intermediate states experimentally, but no data is available as yet. With the recent efforts to measure hard x-ray photoemission spectroscopy (HAXPS) spectra, it should also become possible to measure detailed 1s XPS spectra.

8.6   EARLY TRANSITION METAL COMPOUNDS

We now discuss the RXES of early TM compounds. CaF_2, ScF_3, TiO_2 are nominally d^0 systems. In these d^0 systems, the ground state is the bonding state of the d^0 and $d^1\underline{L}$

configurations (also with a small mixing with the $3d^2\underline{L}^2$ configuration). We introduce some basic aspects of RXES, using a $3d^0$ ground state description of CaF_2 as an example. Then, we show that the 2p3d RXES of d^0 systems exhibits dramatic polarization dependence for the polarized and depolarized geometries. We also discuss the RXES of d^n ($n = 1, 2, 3$) systems.

8.6.1   Ca 2p3s RXES in CaF₂

The 2p3s RXES spectra of CaF_2 is now discussed. CaF_2 is now discussed is an ionic compound that can be well approximated with a pure Ca^{2+} $3d^0$ ground state. Performing a 2p3s RXES experiment first involves a 2p XAS excitation, followed by the detection of the 2p3s XES decay. The 2p XAS spectrum of Ca^{2+} ions follows an analog description to the 2p XAS spectrum of Ti^{4+} ions, as discussed in Chapter 4. The ground state is $3d^0$ $^1A_1$. The 2p XAS final states must have $J = 1$ character and there are three such states for an atom ($^1P_1$, $^3P_1$, and $^3D_1$), where the actual final states are mixtures of these three LS configurations. A cubic crystal field branches a $J = 1$ final state to T_1 symmetry and adds another four states to the $2p^53d^1$ final state matrix elements with T_1 symmetry. In total, this yields a 1×7 transition matrix element.

The 2p3s XES decay yields a final state with a $3s^13d^1$ configuration. This configuration is split by three interactions: (a) the 3s3d exchange interaction, (b) the cubic crystal field and (c) the 3d spin–orbit coupling. The 3d spin–orbit coupling is small and can be neglected. The final state symmetries are found after multiplication of the term symbols of a 3s electron times a 3d electron, that is, $^2S \otimes {}^2D = {}^1D$ and ^3D atomic states. The cubic crystal field splits both states into $^1T_2$ and ^1E, respectively $^3T_2$ and ^3E. These four states are the four peaks that will be visible in the spectral shapes. The actual calculations are performed in double group symmetry because of the large 2p spin–orbit coupling in the intermediate state. Adding the 3d spin–orbit coupling in the $3s^13d^1$ final state splits the ^3E state into T_1 and T_2 and the $^3T_2$ state into A_2, E, T_1, and T_2. Both singlet states are not split.

Table 8.2 shows the configuration, atomic symmetries, and LFM symmetries of the $3d^0$ ground state, $2p^53d^1$ intermediate states, and $3s^13d^1$ final states. It turns out that seven out of the eight $3s^13d^1$ final states have finite intensity in the 2p3s RXES experiment. Only the A_2 symmetry state cannot be reached from the T_1 symmetry intermediate state. The next task is to calculate the energies and transition matrix elements of the 2p XAS excitation and the 2p3s XES decay. These matrix elements are inserted into the Kramers–Heisenberg formula:

$$I(\Omega, \omega) \sim \left| \sum_{2p^53d^1} \frac{\langle 3s^13d^1 \mid \mathbf{e}' \cdot \mathbf{r} \mid 2p^53d^1 \rangle \langle 2p^53d^1 \mid \mathbf{e} \cdot \mathbf{r} \mid 3d^0 \rangle}{E_{2p^53d^1} - E_{3d^0} - \hbar\Omega - i\Gamma_{2p}} \right|^2$$

$$\times \frac{\Gamma_{1s}/\pi}{(E_{3d^0} + \hbar\Omega - E_{3s^13d^1} - \hbar\omega)^2 + \Gamma_{1s}^2} \tag{8.29}$$

This yields seven matrix elements for the 2p3d excitation and a 7×7 matrix with 49 matrix elements for the 2p3s decay. As mentioned previously, the seven $3s^13d^1$ final states group into four quasi-degenerate states.

TABLE 8.2
Atomic Symmetries Are Given for the 3d^0 Ground State, the 2p^{5}3d^1 Intermediate States, and the 3s^{1}3d^1 Final States

Conf.	Atomic Symmetry		Atomic J Values	Allowed Cubic Symmetries
3d^0	^{1}S	^{1}S	0	A$_1$
Dipole	^{1}P	^{1}P	1	T$_1$
2p^{5}3d^1	^{2}P ⊗ ^{2}D	^{1}P, ^{1}D, ^{1}F	1, 2, 3	(T$_1$), (E, T$_2$), (A$_2$, T$_1$, T$_2$)
		^{3}P	0, 1, 2	(A$_1$), (T$_1$), (E, T$_2$)
		^{3}D	1, 2, 3	(T$_1$), (E, T$_2$), (A$_2$, T$_1$, T$_2$)
		^{3}F	2, 3, 4	(E, T$_2$), (A$_2$, T$_1$, T$_2$), (A$_1$, E, T$_1$, T$_2$)
Dipole	^{1}P	^{1}P	1	T$_1$
3s^{1}3d^1	^{2}S ⊗ ^{2}D	^{1}D	2	(E, T$_2$)
		^{3}D	1, 2, 3	(T$_1$), (E, T$_2$), (A$_2$, T$_1$, T$_2$)

Note: The atomic double group *J*-values are given and the last column gives their projections to cubic symmetry in double group notation.

Figure 8.66 shows the stick spectrum of the 2p3s RXES excitation. The stick spectrum is not obtained via the Kramers–Heisenberg formula, but by multiplying the 2p XAS intensities by the 2p3s XES intensities; in other words, by neglecting broadening and interference effects. The seven excitation energies and their decay to the seven 3s^{1}3d^1 final states can be observed, visible for four different energies.

Figure 8.67 shows the same 2p3s RXES spectral shape, after inclusion of the life time broadenings and the experimental resolution, by using Equation 8.29. A landscape with four major peaks is observed. The main peaks are related to the transition highest energy transitions of the L$_3$ edge, respectively the L$_2$ edge to the 3s^{1}3d^1 states where the 3d^1 state is given as an e$_g$ electron. Since the final state is pure in t$_{2g}$ against e$_g$ character, this also implies that the highest energy transitions of the L$_3$ edge and the L$_2$ edge are mainly e$_g$ in character. The four main peaks can all be described, approximately, as 3d^0 → 2p^{5}e$_g$ → 3s^{1}e$_g$. In addition, there are four smaller peaks that can be described as 3d^0 → 2p^{5}t$_{2g}$ → 3s^{1}t$_{2g}$. It is noted that the final state is pure in

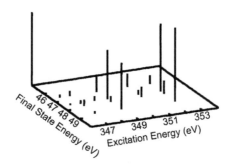

FIGURE 8.66 Intensity line plot of the RXES cross sections, calculated as the x-ray absorption intensity times the x-ray emission intensity. (Reprinted with permission from de Groot, F.M.F., *Phys. Rev. B*, 53, 7099, 1996. Copyright 1996 by the American Physical Society.)

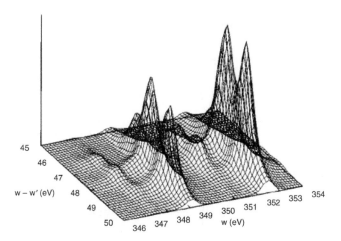

FIGURE 8.67 Intensity plot of the 2p3s RXES cross sections, calculated using Equation 8.29, which includes interference effects. The excitation energy ($\hbar\Omega$) is given on the x axis and the final-state energy ($\hbar\Omega - \hbar\omega$) on the y axis. A Gaussian broadening of 0.5 eV has been used, together with the lifetime broadenings $\Gamma_{2p} = 0.2$ eV and $\Gamma_{3s} = 0.4$ eV. (Reprinted with permission from de Groot, F.M.F., *Phys. Rev. B*, 53, 7099, 1996. Copyright 1996 by the American Physical Society.)

crystal field character and can, in principle, be used as a selective detector for the e_g, respectively t_{2g}, character of the $2p^5 3d^1$ intermediate states (de Groot, 1996).

The 2p3s RXES experiments on CaF_2 have been performed by Rubensson et al. (1994). They measured the 2p3s x-ray emission spectra at several positions within the 2p x-ray absorption edge. Most details of these RXES experiments are reproduced by the LFM calculations for Ca^{2+} as sketched previously (de Groot, 1996). There is a complication with 2p3s RXES experiments that creates a significant difference between the calculated spectra and experiment. This is the fact that a 3s core hole state in calcium will always have a strong interaction with a state that had two 3p holes and an extra 3d valence electron. This mechanism is similar to the situation of 3s XPS as discussed in Chapter 5 (Bagus, 1973). The $\langle 3s3d|1/r|3p3p \rangle$ matrix element is strong because the energy difference between 3d3d and 3p3p is small, which is an example of the near degeneracy effect (NDE) (Bagus et al., 2004, 2006). Dallera et al. (2003) included this 3s3d to 3p3p configuration interaction into the simulation of the 2p3s RXES spectra of CaF_2 and excellent agreement was obtained.

8.6.2 Ti 2p3d RXES of TiO$_2$: Polarization Dependence

The polarization dependence in the Ti 2p3d RXES of TiO_2 was calculated by Matsubara et al. (2000) and measured by Harada et al. (2000). The calculation of RXES spectra is made with the TiO_6 cluster model with O_h symmetry. The used parameter values are $\Delta = 2.9$ eV, $V(e_g) = 3.4$ eV, $U_{dd} = 4.0$ eV, $U_{dc} = 6.0$ eV and $10\,Dq = 1.7$ eV. The results are shown in Figure 8.68, where the Ti 2p XAS and 2p3d2p RXES are shown in Figure 8.68a and 8.68b, respectively. The incident photon energy in RXES is taken at positions a–h, in the XAS spectrum. The spectral structure is

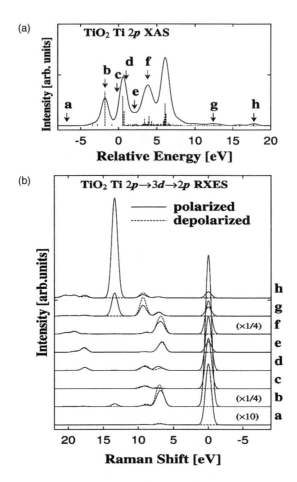

FIGURE 8.68 Calculated results of (a) Ti 2p XAS and (b) Ti 2p3d RXES of TiO_2 with TiO_6 cluster model. (From Matsubara, M., et al., *J. Phys. Soc. Jpn.*, 69, 1558, 2000. With permission.)

divided into three categories: (*i*) elastic line at 0 eV, (*ii*) inelastic spectra at 7 and 9 eV, and (*iii*) inelastic line at 14 eV. The mechanism of these spectra is explained by the energy level scheme shown in Figure 8.69.

The ground state of TiO_2 is the bonding state between the $3d^0$ and $3d^1\underline{L}$ configurations, and the antibonding state is located about 14 eV above the ground state. Both bonding and antibonding states are specified by irreducible representation A_{1g} of the O_h symmetry group. In addition to these states, there are nonbonding $3d^1\underline{L}$ states with T_{1g}, T_{2g}, and E_g symmetries about 7–9 eV above the ground state. When a Ti 2p electron is excited to the 3d state by the incident photon, we have $2p^53d^1$ and $2p^53d^2\underline{L}$ configurations, which are mixed strongly by the covalence hybridization. The main peak of the Ti 2p XAS corresponds to the bonding state between the $2p^53d^1$ and $2p^53d^2\underline{L}$ configurations, while the satellite corresponds to the antibonding state between

Intermediate

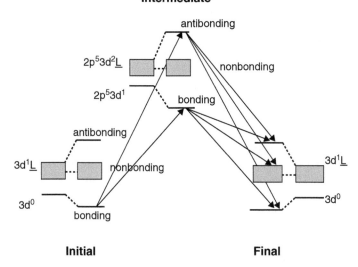

FIGURE 8.69 Schematic representation of the Ti 2p3d RXES transition of TiO_2. (From Matsubara, M., et al., *J. Phys. Soc. Jpn.*, 69, 1558, 2000. With permission.)

them. The intensity of the satellite is very weak because of the phase cancellation between the wave functions of the ground and photo-excited states. Also, the x-ray absorption is almost forbidden to the nonbonding $2p^53d^2\underline{L}$ states. In Figure 8.69, we disregard the effects of the spin–orbit splitting of the 2p states and the crystal field splitting of the 3d states, for simplicity. If we take into account these effects, the main peak (and also the satellite) splits into four peaks, as seen in Figure 8.68.

The resonantly excited intermediate states, which correspond to the main peak and the satellite of the XAS, decay radiatively to each final state of RXES (i.e. the bonding, nonbonding, and antibonding states). The categories (*i*), (*ii*), and (*iii*) of the calculated spectra correspond to the bonding, nonbonding, and antibonding final states, respectively. Spectrum (*ii*) has two peaks, which correspond to the crystal field splitting of the nonbonding $3d^1\underline{L}$ configuration [$3d^1(t_{2g})\underline{L}$ and $3d^1(e_g)\underline{L}$]. Spectrum (*iii*) occurs for the incident photon energy tuned to the satellite of the XAS spectrum. This is because the XAS satellite corresponds to the antibonding intermediate state of RXES, so that the intensity of the antibonding final state is dramatically enhanced. The spectra (*i*) and (*iii*) are allowed only for the polarized geometry, while spectrum (*ii*) is allowed for both polarized and depolarized geometries.

The mechanism of the polarization dependence of the RXES spectra is as follows. The spin–orbit interaction is neglected for simplicity although it is included in the calculation in Figure 8.68. Since the ground state symmetry of TiO_2 is A_{1g} and the electric dipole transition operator (both for x-ray absorption and emission processes) is represented by T_{1u}, the irreducible representations in the final state are given by reducing the product representation $A_{1g} \otimes T_{1u} \otimes T_{1u}$. In order to obtain the selection rule for the polarized and depolarized geometries, we take the scattering plane as the zx plane and the component of the dipole excitation operator as $T_{1u}(y)$ for the polarized geometry and as $T_{1u}(z)$ for the depolarized geometry. Then, the irreducible representations allowed in the final state of RXES are given by

$$\sum_{\gamma=x,y} A_{1g} \otimes T_{1u}(y) \otimes T_{1u}(\gamma) = A_{1g}, E_g, T_{1g}, T_{2g}, \text{ (polarized)} \qquad (8.30)$$

$$\sum_{\gamma=x,y} A_{1g} \otimes T_{1u}(z) \otimes T_{1u}(\gamma) = T_{1g}, T_{2g}, \text{ (depolarized)} \qquad (8.31)$$

Therefore, the elastic peak (bonding state) and the 14 eV inelastic peak (antibonding state) are allowed for the polarized geometry, but they are forbidden for the depolarized geometry. The nonbonding states are allowed both for the polarized and depolarized geometries. The results of Equations 8.30 and 8.31 are easily found from the Clebsch–Gordan coefficients $\langle \Gamma'\gamma\Gamma'' \gamma'' | \Gamma\gamma \rangle$ listed in Table 8.3, where both Γ' and Γ' are the T_{1u} representation. The irreducible representations in Equation 8.30 are Γ which makes $\langle \Gamma'\gamma' \Gamma'' \gamma'' | \Gamma\gamma \rangle$ nonzero for $\Gamma'(\gamma') = T_{1u}(y)$ and $\Gamma''(\gamma'') = T_{1u}(x)$ or $T_{1u}(y)$, and those in Equation 8.31 are for $\Gamma'(\gamma') = T_{1u}(z)$ and $\Gamma''(\gamma'') = T_{1u}(x)$ or $T_{1u}(y)$. Experimental results of RXES for TiO_2 is shown in Figure 8.70. The three categories of RXES spectra (*i*) to (*iii*) are clearly seen, in addition to the spectra indicated by the vertical bars, which are absent in the calculated results (Figure 8.68). The elastic scattering peak at 0 eV (category *i*) and the inelastic one at 14 eV (category *iii*) are allowed only for the polarized geometry, and the intensity of the 14 eV peak is dramatically enhanced when the incident photon energy is tuned to the satellite of the XAS spectrum. Near the middle of the elastic (0 eV) and inelastic (14 eV) scattering peaks, there are inelastic scattering spectra (category *ii*), which are allowed both for the polarized and depolarized geometries. These results are in good agreement with the calculated ones. The spectral width of (*ii*) is much larger than that of the calculated result, and this broadening comes mainly from the energy bandwidth of the O 2p states, which is disregarded in the cluster model.

   It is to be mentioned that some differences between the calculated and experimental results is seen for the Ti 2p XAS. The second peak of the main structure is split into two in the experiment, but only one peak is seen in the calculation. This

TABLE 8.3
Clebsch–Gordon Coefficients $\langle \Gamma'\gamma' \Gamma'' \gamma'' | \Gamma\gamma \rangle$ for $T_1 \times T_1$

$\Gamma'(\gamma')$	$\Gamma''(\gamma'')$				$\Gamma(\gamma)$					
		A_1	$E(3z^2 - r^2)$	$E(x^2 - y^2)$	$T_1(x)$	$T_1(y)$	$T_1(z)$	$T_2(yz)$	$T_2(zx)$	$T_2(xy)$
$T_1(y)$	$T_1(x)$	0	0	0	0	0	$\frac{1}{\sqrt{2}}$	0	0	$-\frac{1}{\sqrt{2}}$
$T_1(y)$	$T_1(y)$	$\frac{-1}{\sqrt{3}}$	$\frac{1}{\sqrt{6}}$	$\frac{1}{\sqrt{2}}$	0	0	0	0	0	0
$T_1(z)$	$T_1(x)$	0	0	0	0	$-\frac{1}{\sqrt{2}}$	0	0	$-\frac{1}{\sqrt{2}}$	0
$T_1(z)$	$T_1(y)$	0	0	0	$\frac{1}{\sqrt{2}}$	0	0	$-\frac{1}{\sqrt{2}}$	0	0

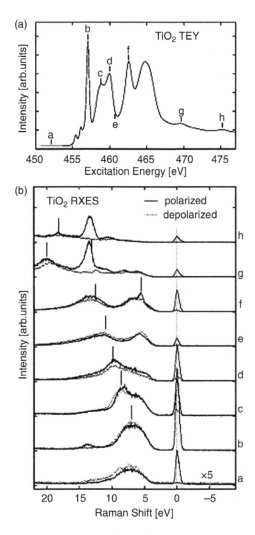

FIGURE 8.70 Experimental results of (a) Ti 2p XAS and (b) Ti 2p3d RXES of TiO_2. (From Matsubara, M., et al., *J. Phys. Soc. Jpn.*, 69, 1558, 2000. With permission.)

discrepancy is due to the approximation that the local symmetry around Ti is treated as O_h, but actually it is D_{2h}. The lower symmetry calculation reproduces this splitting correctly, as has also been discussed in Chapter 6.

Now, we discuss the origin of the spectra, indicated by the vertical bars in Figure 8.70. The energy position of these bars changes almost proportionally to the change of the incident photon energy, so that the emitted photon energy is almost independent of the incident photon energy, similar to the NXES spectrum, which is usually observed for the incident photon energy well above the XAS threshold. In the experimental result in Figure 8.70, however, these spectra are observed near the XAS threshold (so-called NXES-like spectra). Idé and Kotani (1998, 2000) calculated RXES with a one-dimensional d-p model (a simplified version of the nominally

$3d^0$ system) with multi-TM sites. They showed that NXES-like spectra are absent in the cluster with a single transition metal site but, for larger clusters, NXES-like spectra can occur near the XAS threshold because of the existence of spatially extended XAS final states (RXES intermediate states) due to the effect of multi-transition-metal sites. Therefore, in order to reproduce the NXES-like spectra of TiO_2, it is necessary to extend the cluster size to be larger than the TiO_6 cluster. This mechanism of the NXES-like spectra is similar to that observed in NiO, but now the intermediate states with continuous excitation energy occur due to the multi-metal-sites effect, whereas they occur in NiO as energy continuum of the charge-transfer satellite in XAS.

8.6.3   Sc 2p3d RXES of the ScF_3, $ScCl_3$, and $ScBr_3$

The Sc 2p3d RXES of ScF_3, $ScCl_3$, and $ScBr_3$ were calculated by Matsubara et al. (2000, 2004) with the ScX_6 (X = F, Cl, and Br) cluster model. These calculations are similar to the calculations for TiO_2. The main difference in the electronic states of Sc halides and TiO_2 is that the charge-transfer energy Δ of Sc halides is much larger than that of TiO_2. The value of Δ is 11.5, 6.0, 5.5, and 2.9 eV for ScF_3, $ScCl_3$, $ScBr_3$, and TiO_2, respectively. As a result, the nonbonding excitation energy in RXES of Sc halides is very close to the antibonding excitation energy. For instance, the calculated Sc 2p3d RXES spectra of ScF_3 are shown in Figure 8.71 (right), where the nonbonding excitation energies are about 12 and 14 eV [for $3d^1\underline{L}$ states with $3d(t_{2g})$ and $3d(e_g)$, respectively] and the antibonding excitation energy is about 16 eV. The selection rule for the polarized and depolarized geometries is the same as that in TiO_2, but the experimental measurements of RXES for Sc halides have so far been made only in the depolarized geometry.

An interesting point found from the theoretical calculation is that the resonance behavior of the nonbonding $3d^1(e_g)\underline{L}$ and $3d^1(t_{2g})\underline{L}$ final states is somewhat different for ScF_3 and $ScCl_3$. Figure 8.71 shows that for ScF_3 the intensity of $3d^1(t_{2g})\underline{L}$ [indicated by $I(t_{2g})$] is enhanced for the incident energy tuned to the $2p^53d^1(t_{2g})$ peaks a and c, and that of $3d^1(e_g)\underline{L}$ [indicated by $I(e_g)$] is enhanced for the incident energy tuned to $2p^53d^1(e_g)$ peaks b and d. This is a normal behavior. On the other hand, for $ScCl_3$ as shown in Figure 8.71, the resonance behavior for the incident energies a, b, and c is similar to the case of ScF_3, but for d the intensity of the $3d^1(t_{2g})\underline{L}$ peak is more enhanced than the $3d^1(e_g)\underline{L}$ peak. Namely, the behavior in $ScCl_3$ is anomalous for the $2p^53d^1(e_g)$ intermediate state of the $2p_{1/2}$ core hole so that the intensity of the $3d^1(t_{2g})\underline{L}$ final state is enhanced. A similar anomaly is also found for $ScBr_3$. This anomaly in the resonant enhancement was also confirmed by experimental observations and explained theoretically by the 3d state in the XAS peak d being almost a pure e_g state for ScF_3, but a strong mixture between e_g and t_{2g} states for $ScCl_3$ and $ScBr_3$, because of the stronger hybridization effect as well as the multiplet coupling effect.

8.6.4   TM 2p3d RXES of d^n ($n = 1, 2, 3$) Systems

Extending the study for the d^0 systems TiO_2 and Sc halides, the polarization dependence in 2p3d RXES of TM compounds are studied for d^n systems with $n = 1, 2,$ and 3 (TiF_3, VF_3, and Cr_2O_3), both theoretically and experimentally, by Matsubara et al. (2002). From group theoretical consideration, the selection rule in RXES is easily

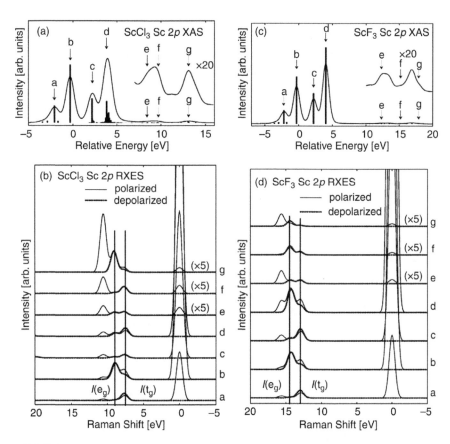

FIGURE 8.71 Calculated results of ScF_3 (right) and $ScCl_3$ (left). The upper panel is the Sc 2p XAS and the lower panel is the Sc 2p3d RXES with ScX_6 cluster model (X = F and Cl). (From Matsubara, M., et al., *J. Phys. Soc. Jpn.*, 73, 711, 2004. With permission.)

obtained for each system. In Table 8.4, we show the irreducible representations of the ground state and the allowed final states in polarized and depolarized geometries. From this table, it is seen that the bonding and antibonding final states are allowed for TiF_3 (with T_{2g} ground state) and VF_3 (with T_{1g} ground state) in both polarized and depolarized geometries, whereas they are forbidden for Cr_2O_3 (with A_{2g} ground state) in the depolarized geometry (as in the case of d^0 system).

In Figure 8.72, the calculated and experimental results of XAS and RXES spectra for TiF_3 are shown. It is seen that the elastic peak is allowed for both polarized and depolarized geometries, but the intensity of antibonding and nonbonding charge transfer excitations is too weak to be clearly seen. The strong inelastic peak about 2 eV from the elastic peak is due to the crystal field excitation (dd excitation), which is absent in TiO_2. Generally, with increasing 3d electron number n, the spectral structure of RXES becomes more complicated because of the crystal field and multiplet coupling effects. On the other hand, the polarization dependence of RXES, as well as the strength of the antibonding resonance, is shown to become weaker with increasing n due to the decreasing hybridization effect and the increasing energy spread of

TABLE 8.4
Ground State Symmetries and the Allowed Final State
Symmetries of Each Compound

| | | Final States | |
		Polarized	Depolarized
Compounds	Ground State	Polarized	Depolarized
TiO_2 (d^0)	A	A_1, E, T_1, T_2	T_1, T_2
TiF_3 (d^1)	T_2	A_1, A_2, E, T_1, T_2	A_1, A_2, E, T_1, T_2
VF_3 (d^2)	T_1	A_1, A_2, E, T_1, T_2	A_1, E, T_1, T_2
Cr_2O_3 (d^3)	A_2	A_2, E, T_1, T_2	T_1, T_2

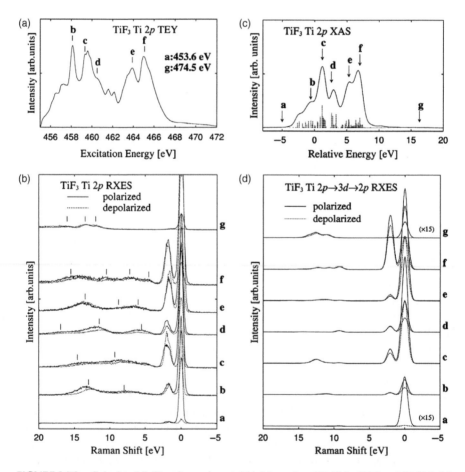

FIGURE 8.72 Calculated (left) and experimental (right) results of (a) Ti 2p XAS and (b) Ti 2p3d RXES for TiF_3. (From Matsubara, M., Uozumi, T., Kotani, A., Harada, Y., and Shin, S., *J. Phys. Soc. Jpn.*, 71, 347, 2002. With permission.)

the multiplet structures. As a result, the polarization dependence in d^n ($n = 1, 2, 3$) systems is less dramatic than that in d^0 systems, even for Cr_2O_3 with essentially the same selection rule as TiO_2.

8.6.5   V 2p3d RXES OF VANADIUM OXIDES

The V 2p3d RXES of V compounds, V_2O_3, VO_2, NaV_2O_6, and V_6O_{13}, were measured by Schmitt et al. (2002, 2004a,b) and some of the results were theoretically analyzed by cluster model calculations and energy band calculations of the density functional theory. For a Mott insulator, NaV_2O_6, for instance, a sharp RXES peak at about 1.7 eV and a broad RXES peak around 6.5 eV are observed and assigned to the dd excitation and the charge-transfer excitation, respectively (see Figure 8.73). More recently, very similar RXES spectra with a sharp peak at 1.6 eV and a broad one around 7 eV have been observed for the mixed valence compound V_6O_{13}, and naturally assigned to the dd excitation and the charge transfer excitation, respectively, quite consistently with the cluster model calculations. On the other hand, essentially the same RXES spectra were measured for NaV_2O_6 by Zhang et al. (2002), but the lower energy peak (they denoted it as "−1.56 eV energy loss feature") is interpreted in a different way. Zhang et al. (2002) calculated the RXES spectrum with a ladder model consisting of eight V atoms, where only the V $d(xy)$ orbital is taken into account with on-site and inter-site correlations. They interpreted, based on their calculation, that the relevant RXES peak is due to the excitation between the LHB and UHB. For comment on the controversy of this issue, see Duda et al. (2004), van Veenendaal and Fedro (2004), and the reply by Zhang et al. (2004).

8.7   ELECTRON SPIN STATES DETECTED BY RXES AND NXES

8.7.1   LOCAL SPIN-SELECTIVE EXCITATION SPECTRA

As an interesting application of RXES, the spin-dependent excitation spectra in Mn compounds is now discussed. Let us consider the Mn 1s3p RXES ($K\beta$ RXES) in MnF_2, where a Mn 1s electron is excited to a Mn 4p conduction band by the incident x-ray, and then a Mn 3p electron makes a transition to the 1s state by emitting an x-ray photon. If we consider the case where the excited Mn 1s electron has a down-spin (here down-spin means that the spin direction is antiparallel to the 3d spin direction in the same atomic site), then the net spin of the 3p final state is parallel to the 3d spin, and the RXES spectrum (not shown here) exhibits a high energy main peak ^7P because of the gain in the 3p3d exchange interaction. On the other hand, if the excited Mn 1s electron has an up-spin, then the net Mn 3p spin is antiparallel to the 3d spin and we have a low energy satellite with ^5P in the RXES. Therefore, if the emitted photon energy is fixed at the main peak (satellite) and the change of the RXES intensity measured by changing the incident photon energy near the threshold of the Mn 1s XAS, then the excited Mn 1s electron has necessarily the down-spin (up-spin), and the observed excitation spectrum is expected to reflect the partial DOS of the conduction band with down-spin (up-spin).

Such experiments have been done by Hämäläinen et al. (1992) for MnF_2 and MnO, and the results for MnF_2 are shown in the inset of Figure 8.74. The spin-down

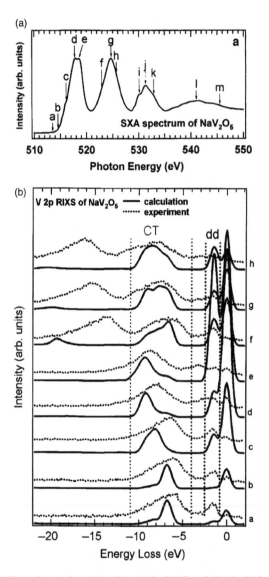

FIGURE 8.73 (a) Experimental results of the V 2p XAS and (b) the V 2p3d RXES (points) of NaV$_2$O$_5$. The RXES spectra calculated with the cluster model are shown with solid curves in the lower panel. (From Schmitt, T., et al., *J. Alloy Comp.*, 362, 143, 2004b. With permission from Elsevier Ltd.)

and spin-up curves correspond to the excitation spectra at the main peak and the satellite, respectively. These spectra were found not to agree well with the spin dependent DOS of the Mn 4p conduction band of MnF$_2$, calculated by the spin-dependent linearized augmented plane-wave method (Dufek et al., 1993, 1994). In order to explain the mechanism of these spin-dependent excitation spectra, Taguchi et al. (1997) calculated the excitation spectra with an MnF$_6$ cluster model and using the second-order optical formula, where they used a model of the spin-dependent

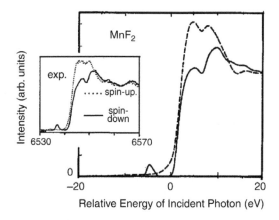

FIGURE 8.74 Calculated results of spin-dependent excitation spectra for MnF_2 compared with experimental ones in the inset. (Reprinted with permission from Taguchi, M., Uozumi, T., and Kotani, A., *J. Phys. Soc. Jpn.*, 66, 247, 1997. Copyright 1997 by the American Physical Society.)

Mn 4p DOS by simulating the result of the energy band calculation. In the calculation of the excitation spectra, they took into account the following effects: (*i*) corehole potential acting on the conduction band states, (*ii*) term-dependent lifetime broadening of the Mn 3p core hole, (*iii*) spin-flip effect by the 3d3d exchange interaction, and (*iv*) covalency hybridization between Mn 3d and F 2p states. They obtained the excitation spectra shown in Figure 8.74, which are in good agreement with the experimental data. From this analysis, it was concluded that the effects of the core-hole potential and the term-dependent lifetime broadening were very important in explaining the difference between the excitation spectra and the DOS.

In Figure 8.75, a weak pre-edge peak is found below the absorption threshold of the spin-down spectrum, but no pre-edge peak for the spin-up spectrum. This preedge peak originates from the Mn 1s3d EQ excitation. For Mn^{2+}, the 3d majority spin states are completely filled, so that the EQ transition is forbidden for the spin-up spectrum. The Mn 1s3p RXES of MnF_2 was also measured by tuning the incident photon energy to the pre-edge peak excitation, and the results were analyzed theoretically with the cluster model by Taguchi et al. (2000). The results are shown in Figure 8.75. The structures a and b of the pre-edge peak of XAS experiments, which are shown in the upper panel of Figure 8.75, correspond to the crystal field splitting of the Mn 3d state. The experimental Mn 1s3p RXES spectra with the incident photon energy tuned to a and b are shown with the dotted curves a and b, respectively. The dotted curve of the NXES is observed for the incident photon energy well above the absorption edge. The solid curves are the calculated results and we find good agreement between the theoretical and experimental results.

8.7.2 Spin-Dependent TM 1s3p NXES Spectra

Figure 8.76 shows the separation into spin-up and spin-down using ligand-field multiplet calculations. In the case of the 1s3p NXES spectra, a pure "spin-down" peak is seen for the satellite, while the main peak is mostly "spin-up." This is particularly clear in the middle of the series where the exchange interaction is largest. This large

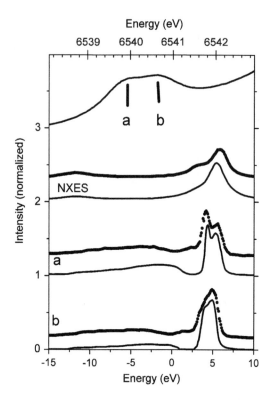

FIGURE 8.75 Experimental data of a pre-edge peak of Mn 1s XAS (top) and Mn 1s3p RXES (dots) in MnF_2. The calculated results of RXES are shown with the solid curves. (Reprinted with permission from Taguchi, M., Parlebas, J.C., Uozumi, T., Kotani, A., and Kao, C.C., *Phys. Rev. B*, 61, 2553, 2000. Copyright 2000 by the American Chemical Society.)

exchange interaction makes 1s3p NXES spectra ideal for local-spin selective XAS experiments (Peng et al., 1994a,b; Wang et al., 1997).

8.7.3 TM 1s3p NXES AND SPIN-TRANSITIONS

The 1s3p NXES spectra essentially probe the number of unpaired 3d spins via the spin-polarized 3p core hole in the final state. This makes it an ideal probe for spin transitions. Figure 8.77 presents the 1s3p NXES spectra of Fe^{3+}, Fe^{2+}, and Co^{2+} spin-transition complexes. A clear difference is visible between high-spin and low-spin spectra with the high-spin spectra showing a clear satellite. The spectra calculated with LFM theory (performed with different crystal field splittings to obtain different spin states), are in good qualitative agreement with the measured spectra and reproduce the nature of the observed changes. (The satellite peak heights are exaggerated in the calculations, which is a consequence of the term-dependent lifetime broadening that is not accounted for in these LFM calculations.) The calculations confirm that the satellite region is essentially absent for low-spin systems, making it a clear marking for high-spin ground states.

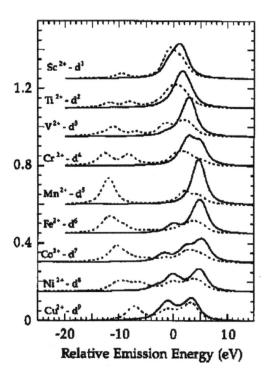

FIGURE 8.76 Theoretical spin-polarized 1s3p NXES spectra of the divalent 3d transition-metal ions using the ligand-field multiplet calculation. Plotted are "spin-down" (dashed) and "spin-up" spectra (solid). (Reprinted with permission from Wang, X., de Groot, F.M.F., and Cramer, S.P., *Phys. Rev. B*, 56, 4553, 1997. Copyright 1997 by the American Physical Society.)

The 1s3p NXES spectra are often measured for systems under extreme conditions, for example, under high-pressure in diamond anvil cells. The goal of the experiments is to determine at which pressure a spin transition occurs for the iron materials that make up the inner core of the earth (Badro et al., 2003, 2004). An important issue in such studies is to determine the "spin moment," or put another way, the percentage of high-spin and low-spin spectra that simulate the measured spectrum. Vanko et al. (2006) have made a detailed study of the various routes to determine this "spin moment" from the measured spectral shapes. An experimental fact is that because it is not always possible to maintain high reliability in the absolute XES energies, the measured spectra for high-spin and low-spin must be aligned and normalized. Both alignment and normalization can be performed in two ways:

1. Aligning peak positions (p). This is experimentally the most straightforward approach.
2. Aligning the center of gravity of the spectra (c). This is, theoretically, the correct procedure. Because the 1s2p NXES excitation only involves core states, the energy of the transition is a property of the element and not affected by the nature of the valence electrons, including its spin state and valence.

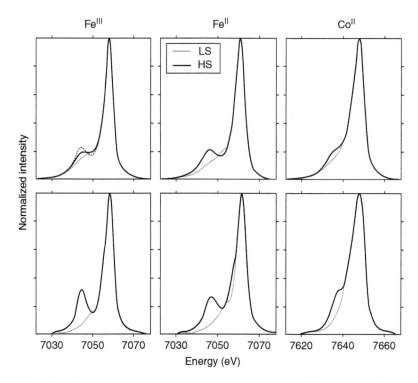

FIGURE 8.77 Measured (top) and calculated (bottom) 1s3p NXES spectra of Fe^{3+}, Fe^{2+}, and Co^{2+} spin-transition complexes. The spectra are normalized to the maximum for better visibility of the satellite region. (Reprinted with permission from Vanko, G., et al., *J. Phys. Chem. B*, 110, 11647, 2006. Copyright 2006 by the American Chemical Society.)

3. Normalizing the maximum (m), the most direct approach to the experiment.
4. Normalizing the integrated area (a), the theoretically correct approach because the 1s2p NXES overall intensity must be constant.

There are three aspects of the spectral shape that can be quantified:

1. The energy difference between characteristic aspects of the spectral shape (E).
2. The intensity of the satellite (I).
3. The integrated difference between the spectra (S).

E-methods include the energy difference between main peak and center-of-gravity and also the energy difference between peak and satellite. The E-method is not dependent on alignment and normalization. It is, however, highly dependent on the specific spectral shapes and the experimental resolution. It turns out that all E-methods are highly nonlinear in their determination of the spin state. I-methods are critically dependent on normalization. They also depend on the alignment because the intensity must be compared in the same energy interval. The choice of this interval gives another important factor. It turns out that if one uses c-alignment and a-normalization, the I-method is reliable. All other combinations of normalization and alignment fail. S-methods take the integrated difference between two spectra; that is, at each energy the absolute difference is measured and integrated over the whole spectral shape. This implies that S-methods also depend on alignment and (a little) on normalization. Using the c-alignment and a-normalization, the S-method is essentially exact. Because the S-method is also not reliable on the choice of the energy interval, it is a general approach for all 3d systems. The conclusion is that it is crucial to align spectra at their center-of-gravity and normalize them to their integrated area. This seems a straightforward approach but it is noted that most experimental studies use another alignment or normalization procedure. Both the S-method (integrated difference) and the I-method (satellite intensity) are linear in spin state, where the S-method is not dependent on additional factors, such as the energy interval of the satellite region. In conclusion, we would strongly suggest all experimental spectra to align spectra at their center-of-gravity, normalize them to their integrated area, and measure their integrated difference (Vanko et al., 2006).

8.7.4 LOCAL-SPIN SELECTIVE XAS AND XMCD

Local-spin selective XAS is determined by the local spin-orientation of the valence band with respect to the core hole. This implies that only the local order is important and not the interactions with the neighbors. In this sense, spin-selective XES as well as local-spin selective XAS are not magnetic probes; that is, they do not probe magnetic order.

In the case of a ferromagnetic system, the local-spin selective XAS can be measured however, and the difference between spin-up and spin-down with the difference between left and right x-ray excitation can be compared. The local-spin-selective x-ray absorption spectrum ($\Delta\mu_{LSS}$) identifies with the spin-polarized DOS.

$$\Delta\mu_{LSS} \equiv \mu^{up} - \mu^{down} = R^+\rho^+ - R^-\rho^- \tag{8.32}$$

K-edge MCD measures the spin-polarized DOS times the Fano factor.

$$\Delta\mu_{MCD} \equiv \mu_+(\mathbf{B}) - \mu_-(\mathbf{B}) = P(R^+\rho^+ - R^-\rho^-) \tag{8.33}$$

By measuring both the MCD and the spin-selective x-ray absorption spectra of the same system, it is possible to directly determine the magnitude of the Fano factor, including its potential energy dependence. Measurements on a ferromagnetic MnP sample show a large energy dependence of the Fano factor, reaching −4% at the edge and decreasing to a much smaller value above the edge (de Groot et al., 1995).

8.8 MCD IN RXES OF FERROMAGNETIC SYSTEMS

8.8.1 LONGITUDINAL AND TRANSVERSE GEOMETRIES IN MCD-RXES

Let us consider the situation shown in Figure 8.78. The magnetization of a ferromagnetic thin-film sample is parallel to the sample surface and its direction is

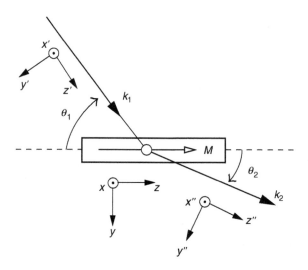

FIGURE 8.78 Geometrical alignment of MCD-RXES. (From Fukui, K., et al., *J. Phys. Soc. Jpn.*, 70, 1230, 2001a. With permission.)

taken in the z axis. The z axis is also taken to be the quantization axis of the angular momentum of the system. The angle between the incident x-ray (emitted x-ray) and the z axis is denoted as θ_1 (θ_2), where the scattering plane includes the sample surface normal and the magnetization direction. The MCD in RXES (denoted by the MCD-RXES) spectrum is defined by the difference of RXES spectra for incident photons with − and + helicities, where the helicity of the emitted photon is not detected.

Before giving some expressions of MCD-RXES, we decompose the expression of RXES spectrum as follows:

$$F(\Omega, \omega) = \left\{ \sum_{j,i} \left| \frac{\langle j | T_2^{\lambda_2} | i \rangle \langle i | T_1^{\lambda_1} | g \rangle}{E_g + \hbar\Omega - E_i + i\Gamma_i} \right|^2 \right.$$
$$\left. + \sum_{j,i,i'} \left[\frac{\langle j | T_2^{\lambda_2} | i \rangle \langle i | T_1^{\lambda_1} | g \rangle (\langle j | T_2^{\lambda_2} | i' \rangle \langle i' | T_1^{\lambda_1} | g \rangle)^*}{(E_g + \hbar\Omega - E_i + i\Gamma_i)(E_g + \hbar\Omega - E_{i'} - i\Gamma_{i'})} + c.c. \right] \right\}$$
$$\times \frac{\Gamma_j / \pi}{(E_j + \hbar\omega - E_g - \hbar\Omega)^2 + \Gamma_j^2}, \qquad (8.34)$$

where the helicities of the incident and emitted photons λ_1 and λ_2 are written explicitly, as well as the spectral broadening Γ_j in the final state. The first and second terms in the curly brackets are the diagonal and cross terms in the expansion of $|\ldots|^2$. The diagonal term represents the two-step process of RXES, where the whole process is described by two successive processes: x-ray absorption and x-ray emission. On the other hand, the cross term describes the RXES process from the ground state to the final state via different intermediate states that interfere each other. Therefore, the cross term is denoted by the interference term.

By the definition mentioned previously, the spectrum of MCD-RXES $\Delta F(\Omega, \omega)$ is given by

$$\Delta F(\Omega, \omega) \equiv \sum_{\lambda_2} \left\{ [F(\Omega, \omega)]_{(\lambda_1=-)} - [F(\Omega, \omega)]_{(\lambda_1=+)} \right\}. \qquad (8.35)$$

With the atomic model and the ED transition both for excitation and de-excitation processes, we can calculate analytically the dependence of MCD-RXES on θ_1 and θ_2. To this end, we introduce three Cartesian coordinates (x, y, z), (x', y', z'), and (x'', y'', z'') where all x, x' and x'' axes are perpendicular to the scattering plane, and z' and z'' axes, respectively, are parallel to the incident and emitted x-ray directions. Then, the electric dipole transition operators $T_1^{\lambda_1}$ $(\lambda_1 = \pm)$ for the circular polarized incident x-rays are written as

$$T_1^{(\pm)} = r\left(C_{\pm 1}^{(1)}\right)' = \pm r\left[\frac{1}{2}(\cos\theta_1 \mp 1)C_{-1}^{(1)} + \frac{1}{2}(\cos\theta_1 \pm 1)C_{+1}^{(1)} + \frac{i}{\sqrt{2}}\sin\theta_1 C_0^{(1)}\right], \qquad (8.36)$$

where $C_q^{(1)}$ and $(C_q^{(1)})'$ are the spherical tensor operators in the (x, y, z) and (x', y', z') systems, respectively. For $T_2^{\lambda_2}$, we can take either of the circular polarizations or the linear polarizations, since we take the summation over the polarization λ_2 in Equation 8.35. Here, we take linear polarizations in the x'' and y'' directions, and express $T_2^{\lambda_2}$ in the (x, y, z) system as follows:

$$T_2^{(x'')} = T_2^{(x)} = r\left[\frac{1}{\sqrt{2}} C_{-1}^{(1)} - \frac{1}{\sqrt{2}} C_{+1}^{(1)}\right], \qquad (8.37)$$

$$T_2^{(y'')} = r\left[\frac{i}{\sqrt{2}} \left(C_{-1}^{(1)}\right)'' + \frac{i}{\sqrt{2}} \left(C_{+1}^{(1)}\right)''\right]$$

$$= r\left[\frac{i}{\sqrt{2}} \cos\theta_2 C_{-1}^{(1)} + \frac{i}{\sqrt{2}} \cos\theta_2 C_{+1}^{(1)} - \sin\theta_2 C_0^{(1)}\right]. \qquad (8.38)$$

We insert Equations 8.36–8.38 into Equation 8.35 to obtain the angular dependence of Equation 8.35 explicitly. As we use the atomic model, all the eigenstates of the material system are expressed as the standard form of atomic wavefunctions $|\alpha J M\rangle$, where J is the total angular momentum, M is its z component, and α represents the electronic configuration; we describe $|g\rangle$, $|i\rangle$ and $|j\rangle$ as $|\alpha_g, J_g, M_g\rangle$, $|\alpha_i, J_i, M_i\rangle$ and $|\alpha_j, J_j, M_j\rangle$, respectively. Then, the angular integrals for the transition matrix elements in Equation 8.34 can be performed explicitly, and by keeping only nonzero matrix elements, we obtain the following explicit expression of the angle dependence of MCD-RXES (Fukui et al., 2001a,b):

$$\Delta F(\Omega, \omega) = \sum_j \left\{ -\frac{1}{2}\cos\theta_1 \left[(1+\cos^2\theta_2)F_1 + 2\sin^2\theta_2 F_2\right] - \frac{1}{4}(\sin\theta_1 \sin 2\theta_2)F_3 \right\}$$
$$\times \frac{\Gamma_j/\pi}{(E_g + \hbar\Omega - E_g - \hbar\omega)^2 + \Gamma_j^2}, \tag{8.39}$$

where F_1, F_2, and F_3 are factors independent of θ_1 and θ_2. The explicit expressions of F_1, F_2, and F_3 are given by

$$F_1 = |f_{+1,+1}|^2 - |f_{+1,-1}|^2 + |f_{-1,+1}|^2 - |f_{-1,+1}|^2 \tag{8.40}$$

$$F_2 = |f_{0,+1}|^2 - |f_{0,-1}|^2 \tag*{•(8.41)}$$

$$F_3 = (f_{-1,0})^* f_{0,-1} + (f_{0,-1})^* f_{-1,0} - (f_{+1,0})^* f_{0,+1} - (f_{0,+1})^* f_{+1,0}$$
$$+ (f_{-1,+1})^* f_{0,0} + (f_{0,0})^* f_{-1,+1} - (f_{+1,-1})^* f_{0,0} - (f_{0,0})^* f_{+1,-1} \tag{8.42}$$

with f_{q_1,q_2} defined by

$$f_{q_1,q_2} \equiv \sum_i \frac{\langle j | rC_{q_2}^{(1)} | i \rangle \langle i | rC_{q_1}^{(1)} | g \rangle}{E_i - E_g - \hbar\Omega - i\Gamma_i}$$
$$= \sum_i \frac{\langle \alpha_j J_j M_j | rC_{q_2}^{(1)} | \alpha_i J_i M_i \rangle \langle \alpha_i J_i M_i | rC_{q_1}^{(1)} | \alpha_g J_g M_g \rangle}{E_j - E_g - \hbar\Omega - i\Gamma_i} \tag{8.43}$$

The present calculation is valid not only in the atomic model, but also in more general models (for instance, in the SIAM) with SO_3 symmetry.

This result indicates a remarkable fact. In the curly bracket of Equation 8.39, the first and second terms, which are proportional to $\cos\theta_1$ and $\sin\theta_1$, originate from the diagonal term and interference term contributions, respectively. Therefore, when the incident x-ray direction is parallel to the magnetization [denoted by the longitudinal geometry (LG)], we only have the diagonal term contribution, while for the incident x-ray perpendicular to the magnetization [denoted by the transverse geometry (TG)], we only have the interference term contribution. For an arbitrary value of θ_1 between $0°$ and $90°$, MCD-RXES is given by a superposition of the diagonal and interference terms, but it has been checked that the diagonal term contribution is dominant except for θ_1 very close to $90°$. Another remarkable fact is that in TG, the θ_2 dependence of MCD-RXES is given simply by $\sin 2\theta_2$. On the other hand, the θ_2 dependence of MCD-RXES in LG is much smaller. Similar calculations can also be made for the case where the x-ray excitation is due to the EQ transition and the x-ray de-excitation due to the ED transition (Fukui et al., 2004).

Here, we have considered the simple geometrical alignment shown in Figure 8.78. For more general cases, where the directions of the incident and emitted x-rays are described by (θ_1, ϕ_1) and (θ_2, ϕ_2), the angle dependence of MCD-RXES has been given by Ogasawara et al. (2004) and Ferriani et al. (2004).

8.8.2 MCD-RXES in LG of CeFe$_2$

In this section, we give a theoretical prediction to apply MCD-RXES in LG for detecting hidden structures in MCD-XAS. As an example, we discuss the hidden 4f^2 contribution in the Ce L$_3$ MCD-XAS of mixed valence ferromagnetic compound CeFe$_2$.

The calculations of XAS, MCD-XAS, RXES, and MCD-RXES were made by Asakura et al. (2004a, 2005). In Figure 8.79, we show the calculated results (solid curves) of the Ce L$_3$ XAS and its MCD, compared with the experimental results (cross marks) by Giorgetti et al. (1993). This calculation is essentially the same as that presented in Figure 7.23 of Chapter 7, but the calculation in Chapter 7 was made with the Ce$_{17}$Fe$_{12}$ cluster for the Ce 5d and Fe 3d LCAO states, but the present result is obtained with a Ce$_{35}$Fe$_{76}$ cluster in order to calculate the RXES and MCD-RXES spectra that require more precise calculations.

Both XAS and MCD-XAS spectra exhibit the double-peak structure: the higher energy peak of XAS (and MCD-XAS) corresponds mainly to the Ce 4f^0 configuration and the lower energy one to the Ce 4f^1 and 4f^2 configurations. The contributions from the Ce 4f^1 and 4f^2 configurations cannot be observed separately because the spectral broadening by Γ_L is very large and the intensity of the 4f^2 contribution is smaller than that of the 4f^1. It can be shown that the Ce Lα_1 RXES (especially MCD-RXES in LD) is a powerful tool to detect the hidden 4f^2 contribution.

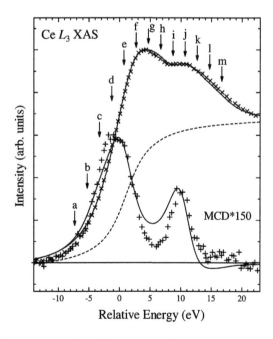

FIGURE 8.79 Calculated results (solid curves) and the experimental ones (crosses) of XAS and MCD-XAS at the L$_3$ edge of CeFe$_2$. The dashed curve represents the background contribution taken in the calculation. (From Asakura, K., Kotani, A., and Harada, I., *J. Phys. Soc. Jpn.*, 74, 1328, 2005. With permission.)

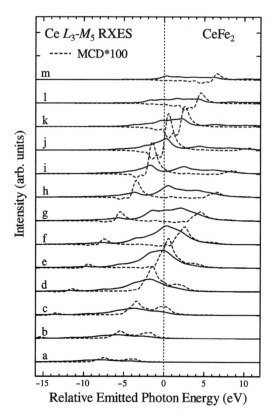

FIGURE 8.80 Calculated results of the Ce L_3M_5 RXES (solid curves) and MCD-RXES (dashed curves) for $CeFe_2$. The incident x-ray energies are taken at positions a to m shown in Figure 8.79. (From Asakura, K., Kotani, A., and Harada, I., *J. Phys. Soc. Jpn.*, 74, 1328, 2005. With permission.)

In Figure 8.80, we show the calculated Ce $L\alpha_1$RXES (solid curves) and its MCD-RXES (dashed curves) as a function of the emitted photon energy whose origin is taken at the energy difference of the Ce $3d_{5/2}$ and $2p_{3/2}$ core levels. The angles θ_1 and θ_2 are taken to be $0°$ and $54.7°$, respectively. The incident photon energy is tuned at positions a to m shown in Figure 8.79, and the effect of the background contribution in XAS is disregarded for simplicity. The spectral broadening Γ_M is taken as 0.7 eV, which corresponds to the lifetime broadening of the Ce 3d core hole. In the region a to m, the RXES (especially MCD-RXES) spectra exhibit a double peak structure, consisting of the Ce $4f^2$ (higher energy peak) and $4f^1$ (lower energy one) contributions, which can be separated due to the spectral broadening Γ_M smaller than Γ_L.

In order to get a less broadened version of the MCD-XAS (XAS), we should take the ES of MCD-RXES (RXES). In Figure 8.81, we show the excitation spectra of the Ce $L\alpha_1$ RXES and its MCD-RXES with the emitted photon energy fixed at the energy difference of the Ce $3d_{5/2}$ and $2p_{3/2}$ core levels. Namely, the amplitudes of

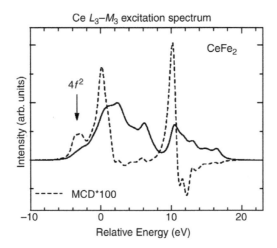

FIGURE 8.81 Calculated results of excitation spectra of the Ce $L_3 M_5$ RXES (solid curves) and its MCD (dashed curves) for $CeFe_2$. (From Asakura, K., Kotani, A., and Harada, I., *J. Phys. Soc. Jpn.*, 74, 1328, 2005. With permission.)

the RXES and its MCD-RXES just along the dotted line in Figure 8.80 are shown in Figure 8.81. Here, we can clearly see the $4f^2$ contribution (indicated by an arrow) especially in the ES of MCD-RXES (dashed curve). It can be shown, furthermore, that if we reduce the value of Γ_L in the calculation of Figure 8.79 from 3.0 eV to 0.7 eV, the calculated results agree almost perfectly with Figure 8.81. In Sections 8.2.5 and 8.2.6, we have mentioned that the ES of the RE $L\alpha_1$ RXES is different from the less broadened version of L_3 XAS mainly because of the 4f3d interaction in the final state of RXE. In $CeFe_2$, however, the Ce 4f state is almost in the spin singlet and orbital singlet state, so that the 4f3d interaction can be disregarded.

We have shown theoretically that the technique of the ES of MCD-RXES is very useful to observe fine structures of MCD-XAS beyond the lifetime broadening of the L_3 core hole. This technique is an extension of the technique of HERFD-XAS by Hämäläinen et al. (1991) to MCD-RXES. It is highly desirable that the present theoretical prediction of observing the Ce $4f^2$ signal by the ES of MCD-RXES will be confirmed by experimental observations, and that the present technique will be used more generally in order to obtain the high-resolution MCD spectra.

8.8.3 EXPERIMENTS AND THEORY OF MCD-RXES IN TG

MCD-RXES in TG was first observed experimentally by Braicovich et al. (1999) for 2p3d excitation and 3s2p de-excitation in $NiFe_2O_4$ and Co metal. They observed nonvanishing MCD-RXES and explained this as being an effect of the polarization of the core hole (Thole et al., 1995). They did not report the details of the spectral analysis. After that, Fukui et al. (2001a,b) measured nonvanishing MCD-RXES for the Gd L_3 excitation and Gd $L\alpha_1$RXES of $Gd_{33}Co_{67}$ amorphous alloy. They made a theoretical analysis with an atomic model for Gd^{3+}, and showed that MCD-RXES in

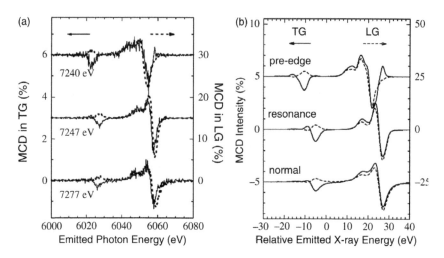

FIGURE 8.82 (a) Experimental and (b) theoretical results of the Gd $L_3M_{4,5}$ MCD-RXES for incident x-ray energies at the pre-edge (7240 eV), resonance (7247 eV) and normal (7277 eV) excitations. The results TG and LG are shown with the solid and dashed curves, respectively. (From Fukui, K., et al., *J. Phys. Soc. Jpn.*, 70, 3457, 2001b. With permission.)

TG is caused by the interference process in the coherent second-order optical process, while that in LG consists of two successive real processes.

The results of measured Gd $L\alpha_1$ MCD-RXES spectra are shown in Figure 8.82a, where the incident x-ray energy is taken at three different values, 7240 eV (pre-threshold region), 7247 eV (main peak resonance), and 7277 eV (high energy continuum for normal fluorescence), and the MCD-RXES spectra are shown with the solid curves for TG and the dashed curves for LG. Here, the angle θ_2 is fixed at 45°. The strong MCD-RXES in the higher emitted x-ray energy corresponds to the contribution from the $3d_{5/2}$ state, while the weak one in the lower energy corresponds to the contribution from the $3d_{3/2}$ state. The spectral shape in TG is similar to that in LG for the $3d_{5/2}$ contribution, while the sign of TG is opposite to that of LG for the $3d_{3/2}$ contribution. It is to be noted that the scale of the MCD-RXES intensity in Figure 8.82a is different for TG and LG, and it is found that the amplitude of the MCD-RXES in TG is about one fifth of that in the LG.

The calculated results corresponding to the experimental data are shown in Figure 8.82b. The calculated and experimental results are in good agreement with each other both in the spectral shape and the spectral intensity ratio in TG and LG. Furthermore, the θ_2 dependence of MCD-RXES spectra in the transverse geometry has also been measured with the incident x-ray energy at 7247 eV, and it is confirmed that the dependence is well described by $\sin 2\theta_2$ as given by the theoretical calculation. The result is shown in Figure 8.83, where A (●), B (+), and C (○) are intensities of an MCD-RXES peak by the $3d_{3/2}2p_{3/2}$ transition, and of two MCD-RXES peaks by the $3d_{5/2}2p_{3/2}$ transition, respectively, and they are displayed as a function of the angle θ_2 (actually, as a function of the scattering angle $\theta_2 + 90°$). The theoretical result is shown with the solid curves and is found to be in reasonable agreement with

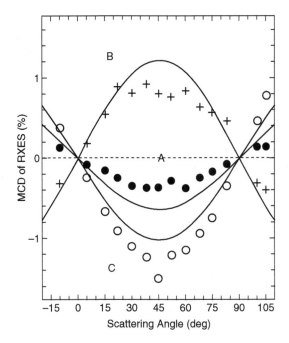

FIGURE 8.83 Experimental result of $2\theta_2$ dependence of the Gd $L_3M_{4,5}$ MCD-RXES in the transverse geometry for three different emitted photon energies A (closed circle), B (cross), and C (open circle). The solid curves are the calculated ones. (From Fukui, K., et al., *J. Phys. Soc. Jpn.*, 70, 3457, 2001b. With permission.)

the experimental result. This is the evidence that almost pure interference contribution has been observed in MCD-RXES experiments in TG.

Finally, we would like to discuss the sum rule of MCD-RXES, which is an important application of MCD-RXES in TG. Since the RXES is the second-order optical process, it is necessary, for the derivation of the sum rule, to approximately take out the resonance denominator $(E_g + \Omega - E_i + i\Gamma_i)$ from the summation over i. This approximation is called "fast collision approximation," which implies that Γ_i is much larger than the energy separation of different multiplet terms that interfere in the RXES process. Then the MCD-RXES intensity integrated over both Ω and ω (denoted by integrated MCD-RXES) is related with the multipole moments of physical quantities, charge, angular momentum and so on, in the ground state. Therefore, from the angular dependence of the integrated MCD-RXES, one can estimate the ground state multipole moments.

First, experimental observations of the integrated MCD-RXES in TG were recently made by Braicovich et al. (2003) for Co 2p3d excitation and 3s2p de-excitation of $CoFe_2O_4$. The angle between the incident and emitted x-rays is fixed at a finite value, but the emitted x-ray direction is changed with the incident direction as an axis of rotation. Here, both excitation and de-excitation processes are due to the ED transition, and then the integrated MCD-RXES is related with the first-order (dipole) and third-order (octupole) moments of ground state quantities. The angular dependence of the integrated MCD-RXES is given, essentially in the same manner as that of the MCD in resonant photoemission (van der Laan, 1995), in terms of ground state multipole moments. Braicovich et al. (2003) obtained the ground state moments of Co 3d holes up to fourth-order by combining the measurements of MCD-RXES with the sum of + and − helicities in RXES, and linear and circular dichroism in XAS. The obtained moments are compared with those calculated with the AM and the CTM model.

Van der Laan et al. (2004) made a similar analysis of ground state moments for $CoFe_2O_4$ and $NiFe_2O_4$. The obtained dipole, quadrupole, and octupole moments of the 3d hole charge in Co and Ni were compared with those calculated with atomic and cluster models. The difference in the quenching of these moments for Co and Ni ferrites was recognized, and the trend is rather consistent with the cluster model calculation.

It should be noted that the above-mentioned approach toward the sum rule of MCD-RXES is still being tested, and further development is expected in the near future. The applicability of the fast collision approximation should be examined more carefully in order to obtain more quantitative and accurate results. However, the consistency of the obtained results with the cluster model calculation is encouraging. Furthermore, beyond the sum rule of MCD-RXES, the information on the electronic and magnetic properties of photo-excited states will be the topics expected in future investigations of MCD-RXES.

Appendix A

Precise Derivation of XPS Formula

With the derivation of $F(E_B)$ in Section 3.2.1 of Chapter 3, some implicit simplifying assumptions have been made. Here, a more precise derivation of $F(E_B)$ is given with which the assumptions then become clear. These assumptions are mostly satisfied in the usual XPS experiments.

Let us consider the XPS process where a photon with a wavevector $\mathbf{q}$ and polarization λ is incident on a material sample and a photoelectron emitted from the sample is observed at a point $\mathbf{R}$ well apart from the sample. We write the Hamiltonian of the material system as H_m and denote its ground state as $|g\rangle$ (with energy E_g). Then the initial state of the photoemission is expressed as

$$|\Psi_0\rangle = |\mathbf{q}\lambda\rangle|g\rangle. \tag{A.1}$$

After switching on the electron–photon interaction, by which the incident photon is absorbed and a core electron c is excited to a photoelectron state, the eigenstate of the total system is expressed by using the scattering theory as

$$|\Psi\rangle = |\Psi_0\rangle + \frac{1}{\hbar\Omega + E_g - H_m + i\eta} M_c^+ |g\rangle, \quad \eta \to +0, \tag{A.2}$$

where $\hbar\Omega$ is the incident photon energy and M_c^+ represents the photoexcitation operator of the core electron to the photoelectron state.

Now, the material system is divided into two subsystems: the excited photoelectron (described by the Hamiltonian h) and the remaining material system (described by the Hamiltonian H_m'). Therefore, H_m is written as

$$H_m = h + H_m' + V, \tag{A.3}$$

where V is the interaction between the two subsystems. The Hamiltonian h is further divided as follows:

$$h = h_0 + v, \tag{A.4}$$

where h_0 is the Hamiltonian of a free electron and v is the periodic potential of the crystalline lattice (v vanishes outside the sample). We write the eigenstates of h_0 and H'_m as $|\mathbf{k}\rangle$ and $|m'\rangle$, respectively, so that

$$\varepsilon_\mathbf{k} = \hbar^2 k^2 / 2m, \tag{A.5}$$

$$H'_m |m'\rangle = E'_m |m'\rangle. \tag{A.6}$$

Then, the overlap integral between $|\Psi\rangle$ and $|\mathbf{k}m'\rangle (= |\mathbf{k}\rangle|m'\rangle)$ is given by

$$\langle \mathbf{k}m'|\Psi\rangle = \frac{1}{\hbar\Omega + E_g - E'_m - \varepsilon_\mathbf{k} + i\eta} \langle \mathbf{k}m'|(1 + TG)M_c^+|g\rangle, \tag{A.7}$$

where we have used the following resolvent expansion formula (with respect to the interaction $v + V$):

$$\frac{1}{\hbar\Omega + E_g - H_m + i\eta} = G(1 + TG) \tag{A.8}$$

with

$$G = \frac{1}{\hbar\Omega + E_g - H'_m - h_0 + i\eta}, \tag{A.9}$$

$$T = (v + V)(1 + GT). \tag{A.10}$$

In order to obtain the probability amplitude that the photoelectron is detected at $\mathbf{R}$, we use the real-space representation of the photoelectron state,

$$|\mathbf{r}\rangle = \sum_\mathbf{k} \frac{1}{(2\pi)^{3/2}} e^{i\mathbf{k}\cdot\mathbf{r}} |\mathbf{k}\rangle. \tag{A.11}$$

The asymptotic form of $\langle \mathbf{R}m'|\Psi\rangle$ for large $\mathbf{R}$ is obtained as

$$\langle \mathbf{R}m'|\Psi\rangle = -\frac{2m}{\hbar^2} \frac{\exp(ik_f R)}{4\pi R} \langle \mathbf{k}_f m'|(1 + TG)M_c^+|g\rangle, \tag{A.12}$$

where

$$k_f = \sqrt{2m(\hbar\Omega + E_g - E'_m)}/\hbar \tag{A.13}$$

and the direction of $\mathbf{k}_f$ is parallel to $\mathbf{R}$. Thus, we obtain the photocurrent per unit solid angle at $\mathbf{R}$ as follows:

$$I = \sum_m \frac{mk_f}{4\pi^2\hbar^3} |\langle \mathbf{k}_f m'|(1 + TG)|g\rangle|^2 \delta(\varepsilon_{\mathbf{k}_f} + E'_m - E_g - \hbar\Omega). \tag{A.14}$$

Equation A.14 is the general expression of the XPS spectrum, but now we impose some simplifying assumptions. The effect of V is disregarded and then the photocurrent is expressed as

$$I = \sum_m A(\mathbf{k}_f) |\langle m'|a_c|g\rangle|^2 \delta(\varepsilon_{\mathbf{k}_f} + E'_m - E_g - \hbar\Omega), \tag{A.15}$$

where

$$A(\mathbf{k}_f) = \frac{e^2 k_f}{2\pi\hbar^2 m\Omega} |\langle \phi_{\mathbf{k}_f}|(1 + tg)\mathbf{p}\cdot\mathbf{e}_{q\lambda}|\phi_c\rangle|^2, \tag{A.16}$$

$$g = 1/(\varepsilon_{\mathbf{k}_f} - h_0 + i\eta), \tag{A.17}$$

$$t = v(1 + gt). \tag{A.18}$$

The effect of V can be disregarded when the kinetic energy of the photoelectron $\varepsilon_{\mathbf{k}_f}$ is much larger than the energy scale of V. It is further assumed that $A(\mathbf{k}_f)$ can be regarded as a constant after averaging over the angle of $\mathbf{k}_f$, which is acceptable for the sufficiently large $\varepsilon_{\mathbf{k}_f}$, again. Then, the photoemission spectrum F is defined, which is normalized so as to be unity when integrated over the photoelectron kinetic energy $\varepsilon (= \varepsilon_{\mathbf{k}_f})$:

$$F(\varepsilon, \Omega) = \sum_{m'} |\langle m'|a_c|g\rangle|^2 \delta(\varepsilon + E'_m - E_g - \hbar\Omega). \tag{A.19}$$

Finally, the system is decomposed into the core electrons and the valence electron states (VES), and the operator a_c and the core states from the expression of F are eliminated. In the initial state of the photoemission, the Hamiltonian of the VES is expressed as H_0, and in the final state of the photoemission, it is written as $H = H_0 + U$. Then Equation A.19 reduces to Equation 3.9 of Chapter 3.

We would like to say a few words on the sudden approximation. When the kinetic energy of the photoelectron is large enough, the photoelectron moves away suddenly from the core hole site. Then, the core hole potential (say U_{fc}) is suddenly applied to VES. As an extreme opposite, if the kinetic energy of the photoelectron is very small, the Coulomb interaction V cancels with U_{fc} just after the core electron excitation. Then, the potential U_{fc} is applied adiabatically corresponding to the slow removal of the photoelectron from the core hole site. This situation is called the adiabatic limit. The transition from the adiabatic to the sudden limit with the increase in the photoelectron kinetic energy has already been discussed [see, e.g. Hedin et al. (1998) and Lee et al. (1999)].

The photoemission process described here is called the "one-step model," where the processes from the photo-excitation of the core electron to the detection of the photocurrent are treated by a single scattering event. However, the whole process can be divided approximately into three successive processes:

1. The absorption of the x-ray inside the solid, creating a photoelectron with an energy equal to the photon energy minus its binding energy.
2. The "transport" of the excited photoelectron to the surface.

3. The escape of the photoelectron from the surface. This subtracts energy of
   the work function from the kinetic energy of the electron, that is, the energy
   difference between the vacuum level and the Fermi level.

This approximate model is denoted by the three-step model. Of these three steps,
the second step includes the effect of the mean free path of the photoelectron, which
is caused by the interaction V between the photoelectron and the remaining electron
systems. The finite mean free path, which strongly depends on the kinetic energy of
the photoelectron, determines the surface-sensitivity of XPS, as described in
Chapters 3 and 5.

Appendix B
Derivation of Equation 3.88 in Chapter 3

The XPS spectrum is rewritten as

$$F(E_B) = -\frac{1}{\pi} \mathrm{Im} \langle 0 | \frac{1}{z - H} | 0 \rangle, \tag{B.1}$$

where

$$z = E_B + E_0 + i\eta, \quad \eta \to +0. \tag{B.2}$$

In the final state of type (A), the 4f state is occupied. Therefore, it is convenient to introduce a new state

$$|f\rangle = a_f^+ |0\rangle, \tag{B.3}$$

and rewrite $\langle 0 | 1/(z - H) | 0 \rangle$ as follows:

$$
\begin{aligned}
\langle 0 | \frac{1}{z - H} | 0 \rangle &= \langle f | a_f^+ \frac{1}{z - H_{0f} - H'} a_f | f \rangle \\
&= G_f^0 + (V G_f^0)^2 G,
\end{aligned}
\tag{B.4}
$$

where

$$H_{0f} = H_0 + \varepsilon_f a_f^+ a_f, \tag{B.5}$$

$$H' = V \sum_{\mathbf{k}} (a_{\mathbf{k}}^+ a_f + a_f^+ a_{\mathbf{k}}), \tag{B.6}$$

$$G_f^0 = \langle f | a_f^+ \frac{1}{z - H_{0f}} a_f | f \rangle = \frac{1}{E_B + i\eta}, \tag{B.7}$$

$$G = \sum_{k < k_F} \sum_{k' < k_F} \langle f | a_{\mathbf{k}}^+ \frac{1}{z - H} a_{\mathbf{k}'} | f \rangle. \tag{B.8}$$

When we define a "generating function" $\hat{g}(t)$ by

$$-\frac{1}{\pi}\operatorname{Im}G = \frac{1}{2\pi\hbar}\int_{-\infty}^{\infty}dt\,\exp\left(i\frac{E_B}{\hbar}t\right)\hat{g}(t),\qquad(\mathrm{B}.9)$$

$\hat{g}(t)$ can be written by the linked-cluster theorem as

$$\hat{g}(t) = \sum_{k<k_F}\sum_{k'<k_F}\langle f|a_\mathbf{k}^+\exp\left[-\frac{i}{\hbar}(H-E_0)t\right]a_{\mathbf{k}'}\,|\,f\rangle = L_2(t)\exp\left[L_1(t)\right],\qquad(\mathrm{B}.10)$$

where

$$\exp\left[L_1(t)\right] = \langle f|S(t)|f\rangle,\qquad(\mathrm{B}.11)$$

$$L_2(t) = \sum_{k<k_F}\sum_{k'<k_F}\exp\left[-\frac{i}{\hbar}(\varepsilon_f-\varepsilon_\mathbf{k})t\right]\left[\langle f\,|\,a_\mathbf{k}^+S(t)a_{\mathbf{k}'}\,|\,f\rangle\right]_c.\qquad(\mathrm{B}.12)$$

Here, $S(t)$ is the S-matrix defined by

$$S(t) = \exp\left(i\frac{H_{0f}}{\hbar}t\right)\exp\left(-i\frac{H}{\hbar}t\right),\qquad(\mathrm{B}.13)$$

and $[\ ]_c$ means the contribution from connected diagrams. By the most divergent term approximation, $L_1(t)$ and $L_2(t)$ are obtained as

$$L_1(t) \cong \int_0^t d\tau_1\int_0^{\tau_1}d\tau_2\int_0^{\tau_2}d\tau_3\int_0^{\tau_3}d\tau_4\langle f|H'(\tau_1)H'(\tau_2)H'(\tau_3)H'(\tau_4)|f\rangle$$

$$\cong -i\frac{\Delta_f}{\hbar}t - (\rho v_{\mathrm{eff}})^2\log\left(\frac{iDt}{\hbar}\right),\qquad(\mathrm{B}.14)$$

$$L_2(t) \cong \frac{\rho}{it}\exp\left[\frac{i}{\hbar}(\varepsilon_F-\varepsilon_f)t\right]\left(\frac{iDt}{\hbar}\right)^{-2\rho v_{\mathrm{eff}}},\qquad(\mathrm{B}.15)$$

where Δ_f represents an appropriate energy shift and v_{eff} is defined by

$$v_{\mathrm{eff}} \equiv -\frac{V^2}{\varepsilon_F-\varepsilon_f}.\qquad(\mathrm{B}.16)$$

The diagramatic representation of $L_1(t)$ and $L_2(t)$ is given in Figure B.1 where the solid and wavy lines with arrows represent the propagators of the conduction electron and the 4f electron, respectively.

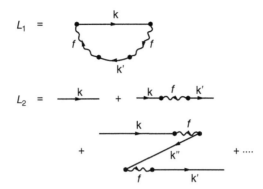

FIGURE B.1 Diagramatic representation of $L_1(t)$ and $L_2(t)$. (From Kotani, A., and Toyozawa, Y., *J. Phys. Soc. Jpn.*, 35, 1073, 1973a. With permission.)

Substituting Equations B.10, B.14, and B.15 into Equation B.9, we obtain

$$-\frac{1}{\pi}\mathrm{Im}\,G = \begin{cases} \dfrac{\rho}{\Gamma(1-2g_{\mathrm{eff}}+g_{\mathrm{eff}}^2)\left(\dfrac{E_B+\varepsilon_F-\tilde{\varepsilon}_f}{D}\right)^{2g_{\mathrm{eff}}-g_{\mathrm{eff}}^2}} & \text{for } E_B \geq -(\varepsilon_F-\tilde{\varepsilon}_f), \\[20pt] 0 & \text{for } E_B < -(\varepsilon_F-\tilde{\varepsilon}_f), \end{cases} \tag{B.17}$$

where

$$g_{\mathrm{eff}} = -\rho v_{\mathrm{eff}}, \tag{B.18}$$

$$\tilde{\varepsilon}_f = \varepsilon_f + \Delta_f. \tag{B.19}$$

The XPS spectrum near the threshold $E_B = -(\varepsilon_F - \tilde{\varepsilon}_f)$ is expressed as

$$F(E_B) = \frac{\rho v^2}{(\varepsilon_F - \tilde{\varepsilon}_f)^2}\left(-\frac{1}{\pi\rho}\mathrm{Im}\,G\right). \tag{B.20}$$

The substitution of Equation B.17 into Equation B.20 gives Equation 3.88 of Chapter 3.

Appendix C
Fundamental Tensor Theory

The fundamental theory of photoemission using a tensor description has been developed by Thole and van der Laan in a series of papers on "Spin polarization and magnetic dichroism in photoemission from core and valence states in localized magnetic systems." In their first paper (Thole and van der Laan, 1991), group theory was used to derive a general model for spin polarization and magnetic dichroism in photoemission in the presence of atomic interactions between the hole created and the valence holes. Eight fundamental spectra were introduced: the isotropic, spin, orbit (circular dichroism), spin–orbit, spin-magnetic-quadrupole, anisotropic (linear dichroism), spin magnetic, and spin magnetic-octupole spectrum (see Table C.1).

If the sum rules were applied to core level photoemission, all integrals would be zero because in the ground state, all core states are filled and hence carry no spin and orbital momentum. The spectral shapes, however, are not zero. Thole and van der Laan (1991) showed that all measurable spectra as a function of electron spin and that x-ray polarization can be calculated from eight fundamental spectra. The isotropic spectrum (00) does not involve any x-ray or electron polarization. The magnetic circular dichroism (MCD) spectrum relates to an x-ray polarization of 1 (10), the linear dichroism to an x-ray polarization of 2 (20), and the spin-polarized spectrum has an electron polarization of 1 (01). In addition, there are four combinations of the x-ray and electron polarization.

Figure C.1 shows the six different spectra for copper 2p photoemission using three different polarizations and a spin detector. The six fundamental spectra $I(\sigma\varepsilon)$ are derived from the six primitive spectra by linear combinations. $I(00)$ is the normalized addition of all spectra, $I(10)$ the difference between $\sigma = 1$ and $\sigma = -1$, as indicated in Table C.1. In their second paper, "Emission from open shells" (van der Laan and Thole, 1993), the various sum rules for x-ray absorption and photoemission spectra were derived on the basis of the tensor description. In their third paper, "Angular distributions" (Thole and van der Laan, 1994), a general analysis for angular dependent spectra was presented. It is shown that the eight fundamental spectra are able to describe all angular dependence. The angular dependence is indicated with a function U. U is written as a function of the magnetization M, the angular dependence (α), and the x-ray polarization (σ).

Table C.2 shows that the magnetic moment can be measured with MCD and also with magnetic linear dichroism in angular distribution (MLDAD). The angular distribution of the emitted electrons can be measured. The precise angular distribution is dependent on the combination of a number of vectors and their mutual angles.

TABLE C.1
Core Spectra of p-Electrons and Their Various
Properties with Respect to the X-Ray Polarization (σ),
Electron Spin (ε), and Magnetization (M)

Fundamental Spectra	Tensor $I(\sigma\varepsilon\text{-}M)$	Primitive Spectra
Isotropic	00–0	$\Sigma(\text{all})$
Circular dichroism	10–1	$\Sigma(1) - \Sigma(-1)$
Linear dichroism	20–2	$\Sigma(1) + \Sigma(-1) - 2\Sigma(0)$
Spin	01–1	$\Sigma(\uparrow) - \Sigma(\downarrow)$
Spin–orbit	11–0	$\Sigma(1\uparrow, -1\downarrow)$
Spin–orbit quadrupole	11–2	$-\Sigma(1\downarrow, -1\uparrow)$
Spin magnetic	21–1	$\Sigma(1\uparrow, -1\uparrow) + 2\Sigma(0\downarrow)$
Spin-magnetic octupole	21–3	$-\Sigma(1\downarrow, -1\downarrow) + 2\Sigma(0\uparrow)$

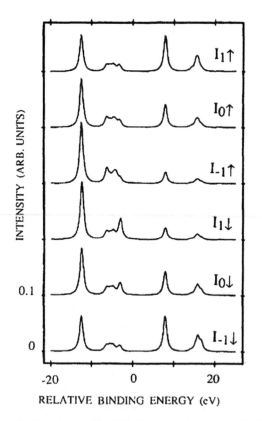

FIGURE C.1 The primitive copper 2p photoemission spectra for all six combinations of x-ray polarization (−1, 0, or 1) and electron spin ($\uparrow$ or $\downarrow$). (Reprinted with permission from Thole, B.T., and van der Laan, G., *Phys. Rev. Lett.*, 44, 12424, 1991a. Copyright 1991 by the American Physical Society.)

Table C.2
Circular (MCD) and Linear (MLD) Dichroic Spectra
Compared with Their Angular Distribution Spectra

Spectrum	$U(M\alpha\sigma)$	Moment
MCD	101	$\langle M \rangle$
Magnetic circular dichroism in angular distribution (MCDAD)	221	$\langle M^2 \rangle$
MLD	202	$\langle M^2 \rangle$
MLDAD	122	$\langle M \rangle$

Note: Third column gives the sensitivity to the moment and the squared moment.

The vectors include the x-ray polarization (σ), the magnetization (M), the photoelectron direction (ε), plus the crystal field effects and the resulting point group symmetry. Without crystal field effects, the angular dependent functions $U(\sigma, \varepsilon, M)$ are given in van der Laan (1995), where instead of σ, P is used for the x-ray polarization.

Figure C.2 shows an experimental measurement of the magnetic linear dichroism (MLD) and MLDAD spectra of the Fe 3p XPS peak in iron metal. The MLD

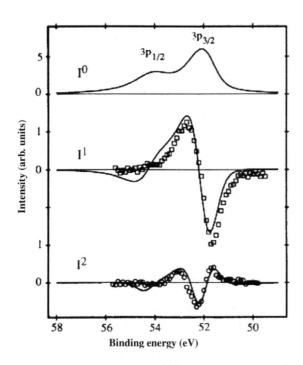

FIGURE C.2 The Fe 3p XPS spectrum (top) with its MLDAD spectrum (middle) and MLD spectrum (bottom). Experimental measurements are shown with symbols and theory is represented with solid lines. (Reprinted with permission from van der Laan, G., *Phys. Rev. B*, 51, 240, 1995. Copyright 1995 by the American Physical Society.)

spectrum (bottom) is the difference in the 3p XPS spectra measured with the magnetic field parallel to the electric field of the x-ray beam ($M \parallel \sigma$) and perpendicular ($M \perp \sigma$). Experimentally, this can be measured by looking at a surface in the x, y plane, enter with the x-ray grazing in the x-direction ($\Omega \parallel x$) with the polarization in the y-direction. The magnetic field must be aligned in the x-direction ($[M \parallel x] \perp [\sigma \parallel y]$) and the y-direction ($[M \parallel y] \parallel [\sigma \parallel y]$). For the MLDAD experiment, the x-ray polarization is changed to the z-direction. The detection angle is important now and the electrons are counted in the normal emission, that is, along the z-axis. If the magnetic field is in the x-direction, this yields ($[\varepsilon \parallel z] \parallel [\sigma \parallel z] \perp [M \parallel x] \parallel [\Omega \parallel x]$). This geometry measures the same spectrum independent of the direction of the magnetic field. The MLDAD spectrum is measured if the magnetic field is rotated to the y-direction. This yields ($[\varepsilon \parallel z] \parallel [\sigma \parallel z] \perp [M \parallel y] \perp [\Omega \parallel x]$) and this chiral geometry, where the magnetic field M, x-ray propagation direction Ω and x-ray polarization σ (+ detection direction ε) are all perpendicular to each other, yields the MLDAD effect if the magnetic field direction is inverted from y to $-y$ (Roth et al., 1993). MLDAD spectra have been measured for the Fe and Cr 2p XPS of Cr adsorbates on Fe (Bethe et al., 2005) and Fe/Co/Fe multilayers (Bruno et al., 2001).

Panacione et al. (2001) used an elegant way to derive the MCD, CDAD, and MLDAD spectra from four basic experiments. They first realized that the measurement of a true MCD spectrum is difficult because one is never sure of a perfect nonchiral geometry of the (σ, ε, M) axis, and because of the finite opening angle of the electron detector. They showed that reversing the polarization σ gives a circular dichroism spectrum as a linear combination of MCD and CDAD. Reversing the magnetic field M gives a spectrum that is a linear combination of MCD and MLDAD. Reversing both M and σ gives four combinations (σ^+, M^+), (σ^+, M^-), (σ^-, M^+), and (σ^-, M^-). From the formulas describing the angular dependence, three expressions can be derived to obtain the MCD, CDAD, and MLDAD from σ and M only.

$$\text{MCD} \propto (\sigma^+, M^+) + (\sigma^-, M^-) - (\sigma^+, M^-) - (\sigma^-, M^+) \tag{C.1}$$

This equation seems evident in that it is a combination of both parallel situations of σ and M minus the two antiparallel cases. Its importance is that, by adding both combinations, any spurious effects from nonchiral geometries are effectively removed (i.e. the CDAD and MLDAD effects).

$$\text{CDAD} \propto (\sigma^+, M^+) + (\sigma^+, M^-) - (\sigma^-, M^-) - (\sigma^-, M^+) \tag{C.2}$$

$$\text{MLDAD} \propto (\sigma^+, M^+) + (\sigma^-, M^+) - (\sigma^-, M^-) - (\sigma^+, M^-) \tag{C.3}$$

Equations C.2 and C.3 show the determination of the CDAD and MLDAD effects. The CDAD effect is obtained by reversing the polarization and the MLDAD effect by reversing the magnetization. Both effects are zero in a perfect nonchiral geometry. For example, an x-ray absorption experiment is by definition nonchiral and using these equations will always yield zero CDAD and MLDAD spectra. In an x-ray photoemission experiment, the electron detector has to be put somewhere, yielding potentially nonzero CDAD and MLDAD spectra.

Appendix D

Derivation of the Orbital Moment Sum Rule

We derive the orbital moment sum rule after Altarelli (1993). For simplicity, we confine ourselves to the electric dipole transition of 2p(or 3p)-3d in transition metal (TM) systems or 3d(or 4d)-4f in rare earth (RE) systems, and the ground state configuration is written as $3d^n$ or $4f^n$. With a single electron orbital angular momentum $\ell = 2$ or 3. The ground state $|0\rangle$ is expanded in a set of Slater determinants $\{n_{m\sigma}\}$ labeled by the quantum numbers of the $h = 2(2\ell + 1) - n$ missing electrons in the ℓ shell:

$$|0\rangle = \sum_{\{n_{m\sigma}\}} c^{(0)}(\{n_{m\sigma}\})|\{n_{m\sigma}\}\rangle, \tag{D.1}$$

where

$$\{n_{m\sigma}\} = \{n_{-\ell\uparrow}, n_{-\ell\downarrow}, \dots, n_{m\sigma}, \dots, n_{\ell\uparrow}, n_{\ell\downarrow}\} \tag{D.2}$$

with each of the $n_{m\sigma}$ either 0 or 1, and they add up to h. The expansion coefficients $c^{(0)}(\{n_{m\sigma}\})$ are general and the z component of the orbital angular momentum is given, in terms of these coefficients, by

$$\langle 0|L_z|0\rangle = -\sum_{\{n_{m\sigma}\}} |c^{(0)}(\{n_{m\sigma}\})|^2 \sum_{m,\sigma} m n_{m\sigma}. \tag{D.3}$$

The integrated intensity of x-ray absorption spectroscopy (XAS) by the electric dipole transitions D_q ($q = -1, 0, 1$) is written as

$$I_q = \sum_f \omega_{0f} |\langle 0|D_q|f\rangle|^2, \tag{D.4}$$

where each final state $|f\rangle$ can also be expanded on a basis of Slater determinants $\{c_z, \sigma_c; n'_{m\sigma}\}$ specifying the quantum numbers c_z, σ_z of the core hole (in the shell $c = \ell - 1$), as well as those of the $h - 1$ holes of the ℓ shell, $n'_{m\sigma}$. By assuming that ω_{0f} is replaced by an average ϖ, I_q is rewritten as

$$I_q = \varpi \sum_f |\langle 0|D_q|f\rangle|^2 = \varpi \langle 0|D_q \sum_f |f\rangle\langle f|D_{-q}|0\rangle. \tag{D.5}$$

Since the projection operator $\sum_f |f\rangle\langle f|$ can be replaced by $\sum_{f'} |f'\rangle\langle f'|$ in terms of any complete basis $|f'\rangle$ for the same excited state configuration, we choose the Slater determinants $|\{c_z, \sigma_c; n'_{m\sigma}\}\rangle$ and obtain

$$I_q = \varpi \sum_{\{n_{m\sigma}\}} \sum_{\{c_z, \sigma_c; n'_{m\sigma}\}} |c^{(0)}(\{n_{m\sigma}\})|^2 \, |\langle\{n_{m\sigma}\}|D_q|\{c_z, \sigma_c; n'_{m\sigma}\}\rangle|^2, \qquad (D.6)$$

Then the matrix element of D_q is a simple one-hole dipole transition from $(m, \sigma) = (c_z + q, \sigma_c)$ to (c_z, σ_c), so that its square is proportional to

$$n_{c_z+q,\,\sigma_c} \begin{bmatrix} c & 1 & \ell \\ -c_z & -q & c_z+q \end{bmatrix}^2. \qquad (D.7)$$

Substituting the expressions for $3j$ symbols, we obtain

$$I_1 + I_0 + I_{-1} = C\sum_{m=-\ell}^{\ell}\sum_{\sigma} n_{m,\sigma}/(2\ell+1) = C[2(2\ell+1)-n]/(2\ell+1) \qquad (D.8)$$

and

$$I_1 - I_{-1} = C\sum_{\{n_{m\sigma}\}} |C^0\{n_{m\sigma}\}|^2 \sum_{m=-\ell}^{\ell}\sum_{\sigma} n_{m,\sigma}m/(2\ell+1)l = -C\langle 0|L_z|0\rangle/(2\ell+1)\ell, \qquad (D.9)$$

where C is a constant factor including the radial matrix element of D_q. It is assumed that the radial matrix element does not depend on the spin of the excited electron, the spin–orbit split core states and the incident x-ray energy. By dividing Equation D.9 by Equation D.8, we obtain

$$\langle 0|L_z|0\rangle = -\frac{I_1 - I_{-1}}{I_1 + I_0 + I_{-1}}\,\ell\langle N_h\rangle, \qquad (D.10)$$

where the hole number $h \equiv 2(2\ell+1) - n$ is written as $\langle N_h\rangle$.

Appendix E
Theoretical Test of the Spin Sum Rule

The expectation values $\langle L_z \rangle$, $\langle S_z \rangle$, $\langle T_z \rangle$, and $\langle S_{\text{eff}} \rangle$ are calculated for divalent 3d and 4d TM ions in an octahedral crystal field. The high-spin 3d systems have been calculated in a cubic field of 1.0 eV and the low-spin 4d systems in a field of 3.5 eV. The results are listed in Table E.1. For the high-spin 3d systems, all holes are paired and $\langle S_z \rangle$ is equal to –0.5 times the number of holes. It can be checked that the spin–orbit coupling slightly decreases the spin-moment of all ground states with a partly-filled t_{2g} shell ($3d^6$, $3d^7$, and $4d^5$) and has no effect on all other systems. The orbital moment $\langle L_z \rangle$ is zero if the spin–orbit coupling is off. Large orbital moments are found for all ground states with a partly-filled t_{2g} shell ($3d^6$, $3d^7$, and $4d^5$) and smaller orbital moments are found for the other ground states.

If the orbital moment sum rule is applied, the exact orbital moment will be obtained as calculated explicitly for the ground state, with all numerical errors smaller than 10^{-4}. The situation is completely different for the spin sum rule where large deviations are found between the actual spin-moment, corrected for the spin-quadrupole coupling $\langle T_z \rangle$, and the values as determined by the sum rules. Table 7.5 of Chapter 7 shows a theoretical test of the effective spin sum rule if applied to the 3d metals $3d^5$ Mn^{2+}, $3d^6$ Fe^{2+}, $3d^7$ Co^{2+}, $3d^8$ Ni^{2+}, and $3d^9$ Cu^{2+}. Two sets of calculations have been performed, one with the 3d spin–orbit coupling at its atomic value and the other with zero 3d spin–orbit coupling. The calculations with spin–orbit coupling reproduce the results obtained by Teramura et al. (1996) where they use the symbols X_E for the theoretical value of S_{eff} and X_I for the experimental sum rule value. Without spin–orbit coupling, there are no orbital moments and spin-quadrupole coupling $\langle T_z \rangle$. This implies that without the 3d spin–orbit coupling, the spin sum rule should yield the spin moment. We have calculated the x-ray magnetic circular dichroism (XMCD) spectral shapes and integrated their L_3 and L_2 edges. In the case of the d^9 systems Cu^{2+} and Ag^{2+}, there are no multiplet effects and the spin sum rule is exact. Note, however, that with 3d spin–orbit coupling, the value of $\langle T_z \rangle$ is very large.

In the case of $3d^8$, the sum rules work relatively well. Spin–orbit coupling has little effect on the effective spin moment in the case of a $^3A_2$ ground state. Both with and without 3d spin–orbit coupling, the sum rule predicts a spin moment of ~0.9, which is 0.1 too small. In the case of $3d^7$ Co^{2+}, without spin–orbit coupling, the sum rule gives 1.25 instead of 1.5, an error of –15%. Interestingly, with 3d spin–orbit

TABLE E.1
Spin and Orbital Moments for the Late 3d and 4d TM Ions

	−LS $\langle S_z \rangle$	+LS $\langle S_z \rangle$	+LS $\langle S_{eff} \rangle$	+LS $\langle T_z \rangle$	+LS $\langle L_z \rangle$
3d⁵ [⁶A₁]	−2.50	−2.50	−2.50	<−0.01	<−0.01
3d⁶ [⁵T₂]	−2.00	−1.97	−1.69	0.08	−0.99
3d⁷ [⁴T₁]	−1.50	−1.44	−1.59	−0.04	−1.57
3d⁸ [³A₂]	−1.00	−0.99	−0.98	<0.01	−0.32
3d⁹ [²E]	−0.50	−0.50	−1.38	<−0.78	−0.37
4d⁵ [²T₁]	−0.50	−0.19	0.01	0.06	−0.81
4d⁶ [¹A₁]	0	<−0.01	<−0.01	<−0.01	<−0.01
4d⁷ [²E]	−0.50	−0.50	−0.50	<0.01	−0.08
4d⁸ [³A₂]	−1.00	−1.00	−0.99	<0.01	−0.21
4d⁹ [²E]	−0.50	−0.50	−1.41	−0.91	−0.37

Note: The high-spin 3d systems have been calculated for a cubic crystal field of 1.0 eV; the low-
spin 4d systems for a crystal field of 3.5 eV. The respective ground state symmetries are
indicated in square brackets.

coupling, the effective spin moment increases to 1.6, while the sum rule decreases
to 1.2. It is noted that the experimental application of the sum rule in the case of
Co impurities on platinum yields a $\langle S_z \rangle$ value of ~0.9 against the density functional
theory (DFT) calculated value of ~1.1 (i.e. an error ~−20%) (Gambardella et al.,
2003). In the case of 3d⁶ Fe^{2+}, things become worse and without 3d spin–orbit
coupling, a completely wrong moment is obtained out of the sum rule (i.e. 0.6 instead
of 2.0). In the case of Mn^{2+}, the 3d spin–orbit coupling has no influence, but the sum
rule results are dramatically off again, yielding 1.7 instead of 2.5. In conclusion,
it can be stated that it is only in the case of 3d⁸ systems that the effective spin sum
rule can be reliably used with errors of the order of 10%. In the case of 3d⁵ and 3d⁷
systems, 20–40% values were obtained that were too low, and in the case of 3d⁶
systems, the result depends strongly on the 3d spin–orbit coupling being quenched
(which is typical in metals but not in oxides), and the sum rule gives values that
can be either too low or too high.

In Table E.2, calculations are shown for the low-spin 4d transition metal (TM)
ions 4d⁵ Tc^{2+}, 4d⁶ Ru^{2+}, 4d⁷ Rh^{2+}, 4d⁸ Pd^{2+}, and 4d⁹ Ag^{2+}, where much closer agreement is found between the sum rule and the actual calculated spin-moment value for
the ground state. In the case of the 4d-systems, we use a crystal field splitting of
3.5 eV, which creates low-spin ground states in all cases. A major difference between
the 3d and 4d systems is that the 2p spin–orbit coupling is much larger so that there
will be no issue to separate the L_2 and L_3 edges. Without 4d spin–orbit coupling,
perfect agreement of the effective spin sum rule and the actual spin moments is
obtained. The reason is that $\langle T_z \rangle$ is zero and there are no multiplet effects and no 4d
spin–orbit coupling effects. The 4d spin–orbit coupling has essentially no effect on

TABLE E.2
Effective Spin Moment Sum Rule is Checked for 4d Systems

	$\langle S_z \rangle$	$\langle S_{eff} \rangle$	$\langle S_{eff} \rangle$	$\langle S_{eff} \rangle$
−LS	**Theory**	**Theory**	**Sum Rule**	**Error**
$4d^5$	−0.50	−0.50	−0.50	<0.1%
$4d^6$	0	0	0	<0.1%
$4d^7$	−0.50	−0.50	−0.50	<0.1%
$4d^8$	−1.00	−1.00	−1.00	<0.1%
$4d^9$	−0.50	−0.50	−0.50	0.0%
+LS				
$4d^5$	−0.19	0.01	−0.42	>100%
$4d^6$	<−0.01	<−0.01	<−0.01	<0.1%
$4d^7$	−0.50	−0.50	−0.50	<0.1%
$4d^8$	−0.99	−0.99	−1.00	<0.1%
$4d^9$	−0.50	−1.41	−1.41	0.0%

Note: The theoretical values for $\langle S_z \rangle$ and $\langle S_{eff} \rangle$ are compared with the results of the sum rule without the inclusion of the 3d spin–orbit coupling (−LS, top) and with 3d spin–orbit coupling (+LS, bottom). The simulations have been performed for an octahedral system with a crystal field value of 3.5 eV. The cases are given where the sum rule error is large.

the $4d^6$, $4d^7$, and $4d^8$ ground state, mainly because the t_{2g} shell is filled. The $4d^9$ ground state has strong selection rules similar to Cu^{2+}. The $4d^5$ ground state has a partly-filled t_{2g} shell and now $\langle T_z \rangle$ is significantly different from zero and the ground state neither gives the spin moment nor the effective spin moment. Note that the 4d spin–orbit coupling is larger than the 3d spin–orbit coupling and is usually not quenched.

Appendix F
Calculations of XAS Spectra with Single Electron Excitation Models

Single electron excitation models assume that the x-ray absorption spectroscopy (XAS) spectrum can be calculated from the transition of a core state to an empty state, using the following simple equation for the Fermi Golden Rule: $I_{XAS} \sim M^2\rho$. The XAS spectrum can be determined from the empty density of states (DOS) (ρ) multiplied by the one-electron transition matrix element (M). The actual calculation of ρ is, however, far from trivial. In principle, all electronic structure calculations can be used to determine the DOS related to this ground state. The dominating basic method for electronic structure determination is the density functional theory the (DFT). Within the DFT, a range of methods and codes are applied to the determination of the empty DOS and the XAS spectral shape. These methods can be grouped into:

1. Real space multiple scattering methods.
2. Real space wave function methods.
3. Reciprocal space band structure methods.

From an electronic structure point-of-view, DFT is just the starting point and a large range of more precise methods has been developed. Some of them are also applied to x-ray absorption.

F.1 DENSITY FUNCTIONAL THEORY WITH LOCAL DENSITY APPROXIMATION

DFT states that the ground state energy can be expressed as a function of the electron density. The practical implementation of this theorem in the local (spin) density approximation (LSDA) has allowed solid-state calculations based on this formalism. In principle, DFT only provides the total energy. However, the electronic structure in terms of the DOS, which, in turn, is used to analyze XAS spectra (Rehr and Albers, 2000). In LSDA of DFT, the kinetic, nuclear, and Hartree potentials are taken into account with the exchange and correlation effects described by a combined potential. All potentials are local functions of the electron density and all complications are collected in the exchange-correlation potential. For this, it is assumed that in a solid, its value is equal to that of a homogeneous electron gas for a particular density n. A range of alternative formulations are used for the exchange-correlation potential (Becke, 1993; Perdew et al., 1996). The electronic structure and properties of solids are described with a number of alternative realizations of LSDA. These methods vary with the use of plane waves [pseudo-potentials and augmented plane waves (APW)] or spherical waves [augmented spherical waves (ASW)]; the use of fixed basis sets such as the linear combination of atomic orbitals (LCAO); the use of the electron-scattering formulation such as real space multiple scattering (MS) and the Koringa–Kohn–Rostoker (KKR) method; linearized versions such as linearized muffin-tin orbitals (LMTO) and linearized APW (LAPW). For each of these methods, in general, several computer codes exist. Some methods exist in versions for spin-polarized calculations, the inclusion of spin–orbit coupling, fully relativistic codes, full potentials, spin-moment, or direction-restricted calculations (Jones and Gunnarsson, 1990; Zeller 1992; Fulde, 1995).

Crucial aspects with respect to the simulation of XAS spectral shapes within DFT are:

1. The accuracy of the potential (as defined within LSDA).
2. The use of real space or reciprocal space calculations.
3. The (partial) inclusion of the core hole potential.
4. The calculation over a longer energy range covering the empty states.
5. The inclusion of spin-polarization, spin–orbit coupling, and relativistic effects.

Beyond the LSDA approximation, the following aspects can be noted (they are further discussed below):

1. The inclusion of orbital polarization.
2. The inclusion of local correlations in LSDA + U.
3. The inclusion of the self-interaction correction.
4. The use of the GW approximation.
5. The use of the Bethe–Salpeter equation (BSE).

We will now, in short, describe a number of popular methods with respect to the issues mentioned above. For a code to become popular, in addition to the aspects mentioned previously, it is as important that the method is relatively easy to use. Three codes dominate the single electron excitation simulation of x-ray absorption spectra: FEFF, Wien2K, and STOBE. FEFF is a real space multiple scattering code, Wien2K is a pseudo-potential band structure code and STOBE is a real space quantum chemistry code. These codes are described below, and some similar codes as well as some many-body codes that are being developed are also mentioned.

F.2   FEFF AND OTHER MULTIPLE SCATTERING CODES

We describe FEFF as an example of a real space multiple scattering code. The FEFF programs are being developed in Seattle by John Rehr et al. (Rehr et al., 1991;

Rehr et al., 1992; Zabinsky et al., 1995, Ankudinov et al., 1998). For a detailed description of the codes, we refer the reader to their review papers. FEFF makes use of the standard quasiparticle model, using the (screened) core hole potential within the final state rule and a parameterized energy-dependent self-energy.

FEFF is a very flexible and easy-to-use code. The cluster around the absorbing atom is determined and the XAS spectrum calculated with a certain choice for the potential. The standard potential used is the Hedin–Lundqvist self-energy within LSDA, where the space within the cluster is divided using a muffin-tin approach (i.e. touching spheres with a constant potential in the interstitial regions). Obviously, this is less accurate than full-potentials, as used in Wien2K. FEFF has proven to be accurate for essentially all XAS features starting at ~10 eV above the edge, where we note that most comparisons with experiment use XAS spectra with lifetime (and/or experimental) broadenings above 1 eV. In other words, FEFF is precise to this limit of 1 eV broadening.

Due to the large multiplet effects, FEFF is not effective for the 3d metal L and M edges and the rare earth (RE) M and N edges. Due to large charge-transfer effects, the L edges of mixed-valence cerium systems also cannot be simulated from FEFF alone. An interesting set of spectra is provided by the K edges of the 3d transition metals. In general, these spectra can be simulated well with FEFF, starting from the edge energy and upwards. The pre-edge region presents more complications, first due to the quadrupole nature of the 1s3d transitions, but recent high-resolution data has revealed the presence of two different sets of peaks due to the 3d-band. This needs some additional simulation tools, in addition to normal FEFF (or other DFT) calculations.

An alternative multiple scattering code is the family of programs developed by Natoli et al. (Benfatto et al., 1987; Filipponi et al., 1991, 1995). This full multiple scattering code has been developed for over more than 20 years. As far as near edge XAS simulations are concerned, the code is now a full multiple scattering code similar to FEFF. Important additions include the applications to x-ray magnetic circular dichroism (XMCD) and the attempt to create a multi-channel multiple scattering (Natoli et al., 1990). These multiple scattering codes are included in GNXAS, a multiple scattering program for extended x-ray absorption fine structure (EXAFS), and in MXAN, a full multiple scattering code that, in addition, is able to optimize the geometric structure from XAS analysis (Benfatto et al., 2003).

F.3   WIEN2K AND OTHER BAND STRUCTURE CODES

Band structure calculations are performed in reciprocal space. Note that both multiple scattering and band structure calculations are performed within the DFT, implying that they should yield the same answer if all approximations used are valid. The physical picture of band structure calculations is rather different from multiple scattering. The XAS features in multiple scattering are seen as arising from the scattering of electrons in the potentials of the neighbors. In band structure calculations, they are seen as the product of the electronic structure due to long-range interactions into the electron waves that are pictured in the band structure (i.e. the energy of the waves in a particular point in reciprocal space). The conservation of momentum effectively implies that the XAS spectrum identifies with the integrated band structure, or in other words, the DOS of the system.

Wien2K is used as an example of a modern band structure code to describe XAS. Wien2K is mainly developed with respect to electron energy loss spectroscopy (EELS) instead of XAS, but the differences between EELS and XAS are minor and essentially both relate to the empty DOS, as has been discussed in Chapter 2. Wien2K is based on the linearized-augmented-plane-wave (LAPW) description of the potential, which is one of the most accurate methods. Calculations of this type can be done for systems containing about 100 atoms per unit cell. The use of reciprocal space implies periodic boundary conditions, but a size of 100 atoms allows for the calculation of surfaces, interfaces, and so on.

A large range of other band structure codes has been used or is being used to interpret XAS spectral shapes. From the 1960s, DFT methods based on APW, localized muffin-tin orbitals (LMTO), ASW, linearized-spherical-waves (LSW), and pseudo-potentials have been used. Historically, the TM oxides were calculated using APW by Mattheiss (1972a,b) and spin-polarized ASW by Terakura et al. (1984). Presently, Wien2K, as well as all modern spin-polarized band structure calculations, all yield good results for the oxygen K edges of TM oxides. This will be further discussed in Chapter 4. We note that the lifetime broadening of the oxygen K edge is 0.3 eV, so any code that is accurate to within 0.3 eV will be sufficient to simulate the oxygen K edges. It is expected that for most TM oxides, the many-body effects on the oxygen K edge will be smaller than 0.3 eV, and hence not visible in the experiment. Band structure calculations are mainly performed to determine the geometric structure of systems. The calculation of the empty states is trivial close to the edge, but depending on the method used, at higher energies less accurate results are found. An elegant solution to this problem has been developed by Taillefumier et al. (2002), who have used a recursion method to determine the DOS over more than 100 eV above the edge.

F.4   STOBE AND OTHER MOLECULAR DFT CODES

STOBE and other molecular DFT codes use real space geometric structure input like multiple scattering codes, but their treatment of the electronic structure within DFT is more like band structure codes. Traditionally, these codes have been developed for molecules and they are excellent in the interpretation of molecular XAS spectra. For example, the analysis of a series of C_6-ring compounds with STOBE yields very good agreement (Kolczewski et al., 2006). We note that molecules are usually less well described with multiple scattering codes due to the directional bonds.

An example of a molecular DFT code is STOBE, which stands for STOckholm-BErlin. With respect to XAS analysis, the advantage of STOBE over other molecular DFT codes is that STOBE contains all the routines needed for XAS analysis. STOBE is based on the linear combination of Gaussian type orbitals (GTOs), which is essentially a linear combination of atomic orbitals (LCAO) method. STOBE can be applied to molecules and adsorbates, but also to solids, surfaces, and adsorbates. STOBE has been used, for example, to calculate the oxygen K edges of TM oxides (Kolczewski and Hermann, 2005). Alternative molecular DFT codes include Amsterdam Density Functional (ADF), Gaussian and TurboMole. ADF is, for example, used in coordination compounds of 3d metals (DeBeer-George et al., 2005). Since these molecular DFT codes use very accurate potentials, essentially any molecular DFT program will yield a DOS that can be compared with XAS experiments, with the exception of edges that have charge transfer and/or multiplet effects such as the L edges of transition metals. The finite difference methods, as applied to XAS by Joly (2001), can also be considered as an accurate real space DFT approach. Though the calculation procedure is different, finite difference methods bear a close resemblance to the molecular DFT codes.

F.5   CALCULATIONS BEYOND LSDA

The calculation tools within LSDA (multiple scattering, band structure, and molecular DFT) provide, in general, good or excellent agreement with XAS experiments. This already indicates that, within the energy resolution of the experiments, effects that go beyond LSDA will be small. The exceptions are, of course, strong correlations (charge transfer) and multiplet effects.

F.6   GW APPROACH

A well-defined model to describe electronic excitations is the GW approach in which the Green function (G) is calculated with a screened Coulomb interaction (W) and is used to calculate the excitation energies. The calculation requires a nonlocal, energy-dependent self-energy operator. A GW calculation gives the density of quasi-particle states for the occupied states of the $N-1$ system and the empty states of the $N+1$ system. Hence, it gives a direct description of (inverse) photoemission spectra. This is in contrast to LSDA calculations, which can be considered as GW calculations with a local, energy-independent self-energy operator, and thereby yielding an "N-particle DOS." A GW calculation of NiO yields much improvement compared with LSDA, but photoemission satellites due to strong correlation effects are not found (Aryasetiawan and Gunnarsson, 1998). As far as the XAS spectra is concerned, GW is expected to yield similar results to that of LSDA.

F.7   BETHE–SALPETER EQUATION APPROACH

Neutral excitations, including x-ray absorption spectra, can be described through the solution of the Bethe–Salpeter equation (BSE), which includes the interacting electron-hole pairs. In the case of all but strongly correlated systems, good agreement between theory and experiment is achieved by using the BSE approach. Shirley (2004, 2005) used the BSE approach to couple a core state to a valence state and as such to calculate the core level spectrum of $SrTiO_3$. In principle, the BSE calculation could include all local two-electron integrals (in other words, all multiplet effects). In practice, the BSE approach is not feasible yet for systems with 3d electrons in the ground state. However, in the future it could provide an *ab initio* route that includes multiplet effects. Charge-transfer effects would not be included and they would probably need an approximation such as that used in LSDA + U calculations.

F.8   LSDA + U APPROACH

In LSDA + U calculations, the total energy functional is expressed as the LSDA energy corrected for U and the exchange parameter J (Anisimov et al., 1991; Anisimov and Gunnarsson, 1991). The potential is corrected for the deviations from the average occupation. LSDA + U calculations improve the electronic structure and band gap description, but when applied to the analysis of XAS spectra, there does not seem to be much difference from normal LSDA calculations (McComb et al., 2003). A related approach is the inclusion of self-interaction corrections into LSDA calculations.

References

Abbate, M., de Groot, F.M.F., Fuggle, J.C., Fujimori, A., Tokura, Y., Fujishima, Y., Strebel, O., Domke, M., Kaindl, G., van Elp, J., Thole, B.T., Sawatzky, G.A., Sacchi, M., and Tsuda, N. (1991a) *Phys. Rev. B*, 44, 5419.

Alders, D., Tjeng, L.H., Voogt, F.C., Hibma, T., Sawatzky, G.A., Chen, C.T., Vogel, J., Sacchi, M., and Iacobucci, S. (1998) *Phys. Rev. B*, 57, 11623.

Allen, J.W. (1985) *J. Magn. Magn. Mat.*, 52, 135.

Altarelli, M. (1993) *Phys. Rev. B*, 47, 597.

Anderson, P.W. (1961) *Phys. Rev.*, 124, 41.

Anderson, P.W. (1967) *Phys. Rev. Lett.*, 18, 1049.

Anisimov, V.I., and Gunnarsson, O. (1991) *Phys. Rev. B*, 43, 7570.

Anisimov, V.I., Zaanen, J., and Andersen, O.K. (1991) *Phys. Rev. B*, 44, 943.

Ankudinov, A.L., Ravel, B., Rehr, J.J., and Conradson, S.D. (1998) *Phys. Rev. B*, 58, 7565.

Aono, M., Chiang, T.C., Knapp, J.A., Tanaka, T., and Eastman, D.E. (1980) *Phys. Rev. B*, 21, 2661.

Arenholz, E., van der Laan, G., Chopdekar, R.V., and Suzuki, Y. (2006) *Phys. Rev. B*, 74, 094407.

Arenholz, E., van der Laan, G., Chopdekar, R.V., and Suzuki, Y. (2007) *Phys. Rev. Lett.*, 98, 197201.

Arrio, M.A., Sainctavit, P., Brouder, C., and Deudon, C. (1995) *Physica B*, 209, 27.

Arrio, M.A., Sainctavit, P., Moulin, C.C.D., Brouder, C., de Groot, F.M.F., Mallah, T., and Verdaguer, M. (1996a) *J. Phys. Chem.*, 100, 4679.

Arrio, M.A., Sainctavit, P., Moulin, C.C.D., Mallah, T., Verdaguer, M., Pellegrin, E., and Chen, C.T. (1996b) *J. Am. Chem. Soc.*, 118, 6422.

Aryasetiawan, F., and Gunnarsson, O. (1998) *Rep. Prog. Phys.*, 61, 237.

Asakura, K., Nakahara, J., Harada, I., Ogasawara, H., Fukui, K., and Kotani, A. (2002) *J. Phys. Soc. Jpn.*, 71, 2771.

Asakura, K., Fukui, K., Ogasawara, H., Harada, I., Parlebas, J.C., and Kotani, A. (2004a) *J. Phys. Soc. Jpn.*, 73, 2008.

Asakura, K., Fukui, K., Ogasawara, H., Harada, I., and Kotani, A. (2004b) *Physica B*, 345, 205.

Asakura, K., Kotani, A., and Harada, I. (2005) *J. Phys. Soc. Jpn.*, 74, 1328.

Badro, J., Fiquet, G., Guyot, F., Rueff, J.P., Struzhkin, V.V., Vanko, G., and Monaco, G. (2003) *Science*, 300, 789.

Badro, J., Rueff, J.P., Vanko, G., Monaco, G., Fiquet, G., and Guyot, F. (2004) *Science*, 305, 383.

Bagus, P., Freeman, A.J., and Sasaki, F. (1973) *Phys. Rev. Lett.*, 30, 850.

Bagus, P.S., Broer, R., and Ilton, E.S. (2004) *Chem. Phys. Lett.*, 394, 150.

Bagus, P.S., Broer, R., and Parmigiani, F. (2006) *Chem. Phys. Lett.*, 421, 148.

Bagus, P.S., and Ilton, E.S. (2006) *Phys. Rev. B*, 73, 155110.

Bair, R.A., and Goddard III, W.A. (1980) *Phys. Rev. B*, 22, 2767.

Bartolomé, F., Tonnerre, J.M., Séve, L., Raoux, D., Chaboy, J., Garcia, L.M., Krisch, M., and Kao, C.C. (1997) *Phys. Rev. Lett.*, 79, 3775.

Baudelet, F., Giorgetti, Ch., Pizzini, S., Brouder, Ch., Dartyge, E., Fontaine, A., Kappler, J.P., and Krill, G. (1993) *J. Electron Spectrosc. Relat. Phenom.*, 62, 153.

Beaurepaire, E., Kappler, J.P., Lewonczuk, S., Ringeissen, J., Khan, M.A., Parlebas, J.C., Iwamoto, Y., and Kotani, A. (1993a) *J. Phys. Cond. Matt.*, 5, 5841.

Beaurepaire, E., Lewonczuk, S., Ringeissen, J., Parlebas, J.C., Uozumi, T., Okada, K., and Kotani, A. (1993b) *Europhys. Lett.*, 22, 463.

Becke, A.D. (1993) *J. Chem. Phys.*, 98, 1372.

Belliere, V., Joorst, G., Stephan, O., de Groot, F.M.F., and Weckhuysen, B.M. (2006) *J. Phys. Chem. B*, 110, 9984.

Benfatto, M., and Natoli, C.R. (1987) *J. Non-Cryst. Sol.*, 95–6, 319.

Benfatto, M., Della Longa, S., and Natoli, C.R. (2003) *J. Synchrot. Rad.*, 10, 51.

Bergmann, U., Horne, C.R., Collins, T.J., Workman, J.M., and Cramer, S.P. (1999) *Chem. Phys. Lett.*, 302, 119.

Bergmann, U., Mullins, O.C., and Cramer, S.P. (2000) *Anal. Chem.*, 72, 2609.

Bergmann, U., Wernet, P., Glatzel, P., Cavalleri, M., Pettersson, L.G.M., Nilsson, A., and Cramer, S.P. (2002) *Phys. Rev. B*, 66, 92107.

Bergmann, U., Groenzin, H., Mullins, O.C., Glatzel, P., Fetzer, J., and Cramer, S.P. (2003) *Chem. Phys. Lett.*, 369, 184.

Bergmann, U., Groenzin, H., Mullins, O.C., Glatzel, P., Fetzer, J., and Cramer, S.P. (2004) *Petrol. Sci. Tech.*, 22, 863.

Berlasso, R., Dallera, C., Borgatti, F., Vozzi, C., Sansone, G., Stagira, S., Nisoli, M., Ghiringhelli, G., Villoresi, P., Poletto, L., Pascolini, M., Nannarone, S., De Silvestri, S., and Braicovich, L. (2006) *Phys. Rev. B*, 73, 115101.

Bethke, C., Kisker, E., Weber, N.B., and Hillebrecht, F.U. (2005) *Phys. Rev. B*, 71, 24413.

Bianconi, A., Marcelli, A., Davoli, I., Stizza, S., and Campagna, M. (1984) *Solid State Commun.*, 49, 409.

Bianconi, A., Marcelli, A., Dexpert, H., Karnatak, R., Kotani, A., Jo, T., and Petiau, J. (1987) *Phys. Rev. B*, 35, 806.

Bianconi, A., Kotani, A., Okada, K., Giorgi, R., Gargano, A., Marcelli, A., and Miyahara, T. (1988) *Phys. Rev. B*, 38, 3433.

Billas, I.M.L., Chatelain, A., and Deheer, W.A. (1994) *Science*, 265, 1682.

Blugel, S. (1992) *Phys. Rev. Lett.*, 68, 851.

Blühm, H., Hävecker, M., Knop-Gericke, A., Kleimenov, E., Schlögl, R., Teschner, D., Bukhtiyarov, V.I., Ogletree, D.F., and Salmeron, M. (2004) *J. Phys. Chem. B*, 108, 14340.

Bocquet, A.E., Mizokawa, T., Saitoh, T., Namatame, H., and Fujimori, A. (1992a) *Phys. Rev. B*, 46, 3771.

Bocquet, A.E., Saitoh, T., Mizokawa, T., and Fujimori, A. (1992b) *Solid State Commun.*, 83, 11.

Bocquet, A.E., Mizokawa, T., Morikawa, K., Fujimori, A., Barman, S.R., Maiti, K., Sarma, D.D., Tokura, Y., and Onoda, M. (1996) *Phys. Rev. B*, 53, 1161.

Bonfim, M., Ghiringhelli, G., Montaigne, F., Pizzini, S., Brookes, N.B., Petroff, F., Vogel, J., Camarero, J., and Fontaine, A. (2001) *Phys. Rev. Lett.*, 86, 3646.

Bonnenberg, D., Hempel, K.A., and Wijn, H.P.J. (1986) *Magnetic Properties of 3d, 4d, and 5d Elements, Alloys and Compounds*, edited by K.-H. Hellwege and O. Madelung, Landolt-Bornstein, New Series (Springer-Verlag, Berlin), Vol. III/19a, p. 178.

Braicovich, L., Brookes, N.B., Dallera, C., Salvietti, M., and Olcese, G.L. (1997) *Phys. Rev. B*, 56, 15047.

Braicovich, L., van der Laan, G., Ghiringhelli, G., Tagliaferri, A., van Veenendaal, M.A., Brookes, N.B., Chervinskii, M.M., Dallera, C., De Michelis, B., and Dürr, H.A. (1999) *Phys. Rev. Lett.*, 82, 1566.

Braicovich, L., Taguchi, M., Borgatti, F., Ghiringhelli, G., Tagliaferri, A., Brookes, N.B., Uozumi, T., and Kotani, A. (2001) *Phys. Rev. B*, 63, 245115.

Braicovich, L., Tagliaferri, A., van der Laan, G., Ghiringhelli, G., and Brookes, N.B. (2003) *Phys. Rev. Lett.*, 90, 117401.

Bressler, C., and Chergui, M. (2004) *Chem. Rev.*, 104, 1781.

Brice-Profeta, S., Arrio, M.A., Tronc, E., Menguy, N., Letard, I., Moulin, C.C.D., Nogues, M., Chaneac, C., Jolivet, J.P., and Sainctavit, P. (2005) *J. Magn. Magn. Mat.*, 288, 354.

Brik, M.G., Ogasawara, K., Ikeno, H., and Tanaka, I. (2006) *Eur. Phys. J. B*, 51, 345.

Brookes, N.B., Ghiringhelli, G., Tjernberg, O., Tjeng, L.H., Mizokawa, T., Li, T.W., and Menovsky, A.A. (2001) *Phys. Rev. Lett.*, 87, 237003.

Brouder, C. (1990) *J. Phys. Cond. Matt.*, 2, 701.

Brouder, C., and Hikam, M. (1991) *Phys. Rev. B*, 43, 3809.

Bruno, F., Cvetko, D., Floreano, L., Gotter, R., Morgante, A., Verdini, A., Panaccione, G., Sirotti, F., Sacchi, M., Torelli, P., and Rossi, G. (2001) *J. Magn. Magn. Mat.*, 233, 123.

Brydson, R., Garvie, L.A.J., Craven, A.J., Sauer, H., Hofer, F., and Cressey, G. (1993) *J. Phys. Cond. Matt.*, 5, 9379.

Burroughs, P., Hamnett, A., Orchard, A.F., and Thornton, G. (1976) *J. Chem. Soc. Dalton Trans.*, 17, 1686.

Butler, P.H. (1981) *Point Group Symmetry, Applications, Methods and Tables* (Plenum, New York).

Butorin, S.M., Guo, J.H., Magnuson, M., Kuiper, P., and Nordgren, J. (1996a) *Phys. Rev. B*, 54, 4405.

Butorin, S.M., Mancini, D.C., Guo, J.H., Wassdahl, N., Nordgren, J., Nakazawa, M., Tanaka, S., Uozumi, T., Kotani, A., Ma, Y., Myano, K.E., Karlin, B.A., and Shuh, D.K. (1996b) *Phys. Rev. Lett.*, 77, 574.

Butorin, S.M., Duda, L.C., Guo, J.H., Wassdahl, N., Nordgren, J., Nakazawa, M., and Kotani, A. (1997a) *J. Phys. Cond. Matt.*, 9, 8155.

Butorin, S.M., Guo, J.H., Magnuson, M., and Nordgren, J. (1997b) *Phys. Rev. B*, 55, 4242.

Butorin, S.M. (2000) *J. Electron Spectrosc. Relat. Phenom.*, 110, 213.

Caliebe, W.A., Kao, C.C., Hastings, J.B., Taguchi, M., Uozumi, T., and de Groot, F.M.F. (1998) *Phys. Rev. B*, 58, 13452.

Carra, P., Harmon, B.N., Thole, B.T., Altarelli, M., and Sawatzky, G.A. (1991) *Phys. Rev. Lett.*, 66, 2495.

Carra, P., Fabrizio, M., and Thole, B.T. (1995) *Phys. Rev. Lett.*, 74, 3700.

Cavalleri, A., Chong, H.H.W., Fourmaux, S., Glover, T.E., Heimann, P.A., Kieffer, J.C., Mun, B.S., Padmore, H.A., and Schoenlein, R.W. (2004) *Phys. Rev. B*, 69, 153106.

Chakarian, V., Idzerda, Y.U., Meigs, G., Chaban, E.E., Park, J.H., and Chen, C.T. (1995) *Appl. Phys. Lett.*, 66, 3368.

Chen, C.T., and Sette, F. (1989) *Rev. Sci. Instrum.*, 60, 1616.

Chen, C.T., Sette, F., Ma, Y., and Modesti, S. (1990) *Phys. Rev. B*, 42, 7262.

Chen, C.T., Smith, N.V., and Sette, F. (1991a) *Phys. Rev. B*, 43, 6785.

Chen, C.T., Sette, F., Ma, Y., Hybertsen, M.S., Stechel, E.B., Foulkes, W.M.C., Schluter, M., Cheong, S.W., Cooper, A.S., Rupp, L.W., Batlogg, B., Soo, Y.L., Ming, Z.H., Krol, A., and Kao, Y.H. (1991b) *Phys. Rev. Lett.*, 66, 104.

Chen, C.T., Tjeng, L.H., Kwo, J., Kao, H.L., Rudolf, P., Sette, F., and Fleming, R.M. (1992) *Phys. Rev. Lett.*, 68, 2543.

Chen, C.T., Idzerda, Y.U., Lin, H.J., Meigs, G., Chaiken, A., Prinz, G.A., and Ho, G.H. (1993) *Phys. Rev. B*, 48, 642.

Chen, C.T., Idzerda, Y.U., Lin, H.J., Smith, N.V., Meigs, G., Chaban, E., Ho, G.H., Pellegrin, E., and Sette, F. (1995) *Phys. Rev. Lett.*, 75, 152.

Chiuzbăian, S.G., Ghiringhelli, G., Dallera, C., Grioni, M., Amann, P., Wang, X., Braicovich, L., and Patthey, L. (2005) *Phys. Rev. Lett.*, 95, 197402.

Cini, M. (1977) *Solid State Commun.*, 24, 681.

Cini, M. (1978) *Phys. Rev. B.*, 17, 2788.

Citrin, P.H., Wertheim, G.K., and Baer, Y. (1977) *Phys. Rev. B*, 16, 4256.

Collart, E., Shukla, A., Rueff, J.P., Leininger, P., Ishii, H., Jarrige, I., Cai, Y.Q., Cheong, S.W., and Dhalenne, G. (2006) *Phys. Rev. Lett.*, 96, 157004.

Colliex, C., Manoubi, T., and Krivanek, O.L. (1986) *J. Elec. Microsc.*, 35, 307.

Colliex, C. (1991) *Micr. Micr. Micr.*, 2, 403.

Colliex, C., Tence, M., Lefevre, E., Mory, C., Gu, H., Bouchet, D., and Jeanguillaume, C. (1994) *Mikrochim. Acta*, 114, 71.

Coulthard, I., Antel, W.J., Frigo, S.P., Freeland, J.W., Moore, J., Calaway, W.S., Pellin, M.J., Mendelsohn, M., Sham, T.K., Naftel, S.J., and Stampfl, A.P.J. (2000) *J. Vac. Sci. Tech. A*, 18, 1955.

Cowan, R.D. (1981) *The Theory of Atomic Structure and Spectra* (University of California Press, Berkeley).

Cox, P.A., Lang, J.K., and Baer, Y. (1981a) *J. Phys. F*, 11, 113.

Cox, P.A., Lang, J.K., and Baer, Y. (1981b) *J. Phys. F*, 11, 121.

Cramer, S.P., de Groot, F.M.F., Ma, Y., Chen, C.T., Sette, F., Kipke, C.A., Eichhorn, D.M., Chan, M.K., Armstrong, W.H., Libby, E., Christou, G., Brooker, S., McKee, V., Mullins, O.C., and Fuggle, J.C. (1991) *J. Am. Chem. Soc.*, 113, 7937.

Cramer, S.P., Peng, G., Christiansen, J., Chen, J., van Elp, J., George, S.J., and Young, A.T. (1996) *J. Electron Spectrosc. Relat. Phenom.*, 78, 225.

Crocombette, J.P., and Jollet, F. (1994) *J. Phys. Cond. Matt.*, 6, 10811.

Crocombette, J.P., Pollak, M., Jollet, F., Thromat, N., and Gautiersoyer, M. (1995) *Phys. Rev. B*, 52, 3143.

Crocombette, J.P., and Jollet, F. (1996) *J. Phys. Cond. Matt.*, 8, 5253.

Cros, V., Petroff, F., Vogel, J., Fontaine, A., Kappler, J.P., Krill, G., Rogalev, A., and Goulon, J. (1997) *J. Appl. Phys.*, 81, 3774.

Czekaj, S., Nolting, F., Heyderman, L.J., Willmott, P.R., and van der Laan, G. (2006) *Phys. Rev. B*, 73, 020401.

Dagotto, E. (1994) *Rev. Mod. Phys.*, 66, 763.

Dallera, C., Grioni, M., Shukla, A., Vanko, G., Sarrao, J.L., Rueff, J.P., and Cox, D.L. (2002) *Phys. Rev. Lett.*, 88, 196403.

Dallera, C., Taguchi, M., Hague, C.F., Journel, L., Mariot, J.M., Rueff, J.P., Sacchi, M., Braicovich, L., Ghiringhelli, G., Tagliaferri, A., Palenzona, A., and Brookes, N.B. (2003a) *Phys. Rev. B*, 67, 113104.

Dallera, C., Annese, E., Rueff, J.P., Palenzona, A., Vanko, G., Braicovich, L., Shukla, A., and Grioni, M. (2003b) *Phys. Rev. B*, 68, 245114.

Damascelli, A., Hussain, Z., and Shen, Z.X. (2003) *Rev. Mod. Phys.*, 75, 473.

Danger, J., Le Fèvre, P., Magnan, H., Chandesris, D., Bourgeois, S., Jupille, J., Eickhoff, T., and Drube, W. (2002) *Phys. Rev. Lett.*, 88, 243001.

Dartyge, E., Fontaine, A., Giorgetti, C., Pizzini, S., Baudelet, F., Krill, G., Brouder, C., and Kappler, J.P. (1992) *Phys. Rev. B*, 46, 3155.

de Groot, F.M.F., Grioni, M., Fuggle, J.C., Ghijsen, J., Sawatzky, G.A., and Petersen, H. (1989) *Phys. Rev. B*, 40, 5715.

de Groot, F.M.F., Fuggle, J.C., Thole, B.T., and Sawatzky, G.A. (1990a) *Phys. Rev. B*, 41, 928.

de Groot, F.M.F., Fuggle, J.C., Thole, B.T., and Sawatzky, G.A. (1990b) *Phys. Rev. B*, 42, 5459.

de Groot, F.M.F. (1991) *X-Ray Absorption of Transition Metal Oxides* (PhD thesis, Nijmegen University).

de Groot, F.M.F., Figueiredo, M.O., Basto, M.J., Abbate, M., Petersen, H., and Fuggle, J.C. (1992) *Phys. Chem. Miner.*, 19, 140.

de Groot, F.M.F., Faber, J., Michiels, J.J.M., Czyzyk, M.T., Abbate, M., and Fuggle, J.C. (1993a) *Phys. Rev. B*, 48, 2074.

de Groot, F.M.F., Abbate, M., van Elp, J., Sawatzky, G.A., Ma, Y.J., Chen, C.T., and Sette, F. (1993b) *J. Phys. Cond. Matt.*, 5, 2277.

de Groot, F.M.F. (1994) *J. Electron Spectrosc. Relat. Phenom.*, 67, 529.

de Groot, F.M.F., Fontaine, A., Kao, C.C., and Krisch, M. (1994a) *J. Phys. Cond. Matt.*, 6, 6875.

de Groot, F.M.F., Arrio, M.A., Sainctavit, P., Cartier, C., and Chen, C.T. (1994b) *Solid State Commun.*, 92, 991.

de Groot, F.M.F., Hu, Z.W., Lopez, M.F., Kaindl, G., Guillot, F., and Tronc, M. (1994c) *J. Chem. Phys.*, 101, 6570.

de Groot, F.M.F., Pizzini, S., Fontaine, A., Hämäläinen, K., Kao, C.C., and Hastings, J.B. (1995) *Phys. Rev. B*, 51, 1045.

de Groot, F.M.F. (1996) *Phys. Rev. B*, 53, 7099.

de Groot, F.M.F., Kuiper, P., and Sawatzky, G.A. (1998) *Phys. Rev. B*, 57, 14584.

de Groot, F. (2001) *Chem. Rev.*, 101, 1779.

de Groot, F.M.F., Krisch, M.H., and Vogel, J. (2002) *Phys. Rev. B*, 66.

de Groot, F. (2005) *Coord. Chem. Rev.*, 249, 31.

de Groot, F.M.F., Glatzel, P., Bergmann, U., van Aken, P.A., Barrea, R.A., Klemme, S., Hävecker, M., Knop-Gericke, A., Heijboer, W.M., and Weckhuysen, B.M. (2005) *J. Phys. Chem. B*, 109, 20751.

de Groot, F., and Vogel, J. (2006) *Neutron and X-Ray Spectroscopy* (Grenoble, Springer, 2006), 3.

de Vries, A.H., Hozoi, L., and Broer, R. (2003) *Int. J. Quantum Chem.*, 91, 57.

de Vries, C.P., den Herder, J.W., Kaastra, J.S., Paerels, F.B., den Boggende, A.J., and Rasmussen, A.P. (2003) *Astron. Astrophys.*, 404, 959.

DeBeer-George, S., Brant, P., and Solomon, E.I. (2005) *J. Am. Chem. Soc.*, 127, 667.

Demeter, M., Neumann, M., and Reichelt, W. (2000) *Surface Science*, 454, 41.

Dicicco, A., and Filipponi, A. (1994) *Phys. Rev. B*, 49, 12564.

Doniach, S., and Šunjić, M.J. (1970) *J. Phys. C*, 3, 285.

Döring, G., Sternemann, C., Kaprolat, A., Mattila, A., Hämäläinen, K., and Schulke, W. (2004) *Phys. Rev. B*, 70, 85115.

Drube, W., Lessmann, A., and Materlik, G. (1993) *Jpn. J. Appl. Phys.*, 32, 173.

Drube, W., Treusch, R., Sham, T.K., Bzowski, A., and Soldatov, A.V. (1998) *Phys. Rev. B*, 58, 6871.

Drube, W., Sham, T.K., Kravtsova, A., and Soldatov, A.V. (2003) *Phys. Rev. B*, 67, 35122.

Duda, L.C. (1996) PhD thesis (Uppsala University).

Duda, L.C., Dräger, G., Tanaka, S., Kotani, A., Guo, J.H., Heumann, D., Bocharov, S., Wassdahl, N., and Nordgren, J. (1998) *J. Phys. Soc. Jpn.*, 67, 416.

Duda, L.C., Downes, J., Mc Guinness, C., Schmitt, T., Augustsson, A., Smith, K.E., Dhalenne, G., and Revcolevschi, A. (2000) *Phys. Rev. B*, 61, 4186.

Duda, L.C., Schmitt, T., Nordgren, J., Kuiper, P., Dhalenne, G., and Revcolevschi, A. (2004) *Phys. Rev. Lett.*, 93, 196701.

Dufek, P., Schwarz, K., and Blaha, P. (1993) *Phys. Rev. B*, 48, 12672.

Dufek, P., Blaha, P., Sliwko, V., and Schwarz, K. (1994) *Phys. Rev. B*, 49, 10170.

Dürr, H.A., van der Laan, G., and Thole, B.T. (1996) *Phys. Rev. Lett.*, 76, 3464.

Eisenberger, P., Platzman, P.M., and Winick, H. (1976a) *Phys. Rev. Lett.*, 36, 623.

Eisenberger, P., Platzman, P.M., and Winick, H. (1976b) *Phys. Rev. B*, 13, 2377.

Erskine, J.L., and Stern, E.A. (1975) *Phys. Rev. B*, 12, 5016.

Esteva, J.M., Karnatak, R.C., Fuggle, J.C., and Sawatzky, G.A. (1983) *Phys. Rev. Lett.*, 50, 910.

Fadley, C.S., Shirley, D.A., Freeman, A.J., Bagus, P.S., and Mallow, J.V. (1969) *Phys. Rev. Lett.*, 23, 1397.

Fadley, C.S., and Shirley, D.A. (1970) *Phys. Rev. A*, 2, 1109.

Fano, U. (1961) *Phys. Rev.*, 124, 1866.

Fano, U. (1969) *Phys. Rev.*, 178, 131.

Farges, F., Brown, G.E., and Rehr, J.J. (1997) *Phys. Rev. B*, 56, 1809.

Ferriani, P., Bertoni, C.M., and Ferrari, G. (2004) *Phys. Rev. B*, 69, 104433.

Filipponi, A., Dicicco, A., Tyson, T.A., and Natoli, C.R. (1991) *Solid State Commun.*, 78, 265.

Filipponi, A., DiCicco, A., and Natoli, C.R. (1995) *Phys. Rev. B*, 52, 15122.

Finazzi, M., Brookes, N.B., and de Groot, F.M.F. (1999) *Phys. Rev. B*, 59, 9933.

Fink, J., Muller-Heinzerling, T., Scheerer, B., Speier, W., Hillebrecht, F.U., Fuggle, J.C., Zaanen, J., and Sawatzky, G.A. (1985) *Phys. Rev. B*, 32, 4899.

Fink, J. (1992) *Top. Appl. Phys.*, 69, 203.

Flipse, C.F.J., Rouwelaar, C.B., and de Groot, F.M.F. (1999) *Europ. Phys. J. D*, 9, 479.

Fomichev, V.A., Zimkina, T.M., Gribovskii, S.A., and Zhukova, I.I. (1967) *Fizika Tverdogo Tela* (Sankt, Peterburg), 9, 1490.

Fromme, B., Koch, C., Deussen, R., and Kisker, E. (1995) *Phys. Rev. Lett.*, 75, 693.

Fromme, B., Bocatius, V., and Kisker, E. (2001) *Phys. Rev. B*, 64, 125114.

Froud, C.A., Rogers, E.T.F., Hanna, D.C., Brocklesby, W.S., Praeger, M., de Paula, A.M., Baumberg, J.J., and Frey, J.G. (2006) *Opt. Lett.*, 31, 374.

Fuggle, J.C., Hillebrecht, F.U., Zolnierik, Z., Lasser, R., Freiburg, C., Gunnarsson, O., and Schönhammer, K. (1983) *Phys. Rev. B*, 27, 7330.

Fujii, T., de Groot, F.M.F., Sawatzky, G.A., Voogt, F.C., Hibma, T., and Okada, K. (1999) *Phys. Rev. B*, 59, 3195.

Fujimori, A., and Minami, F. (1984) *Phys. Rev. B*, 30, 957.

Fujimori, A., Minami, F., and Sugano, S. (1984) *Phys. Rev. B*, 29, 5225.

Fujimori, A., Bocquet, A.E., Saitoh, T., and Mizokawa, T. (1993) *J. Electron Spectrosc. Relat. Phenom.*, 62, 141.

Fukui, K., Ogasawara, H., Kotani, A., Iwazumi, T., Shoji, H., and Nakamura, T. (2001a) *J. Phys. Soc. Jpn.*, 70, 1230.

Fukui, K., Ogasawara, H., Kotani, A., Iwazumi, T., Shoji, H., and Nakamura, T. (2001b) *J. Phys. Soc. Jpn.*, 70, 3457.

Fukui, K., Ogasawara, H., Kotani, A., Harada, I., Maruyama, H., Kawamura, N., Kobayashi, K., Chaboy, J., and Marcelli, A. (2001c) *Phys. Rev. B*, 64, 104405.

Fukui, K., and Kotani, A. (2004) *J. Phys. Soc. Jpn.*, 73, 1059.

Fulde, P. (1995) *Electron Correlations in Molecules and Solids* (Springer, Berlin).

Funk, T., Friedrich, S., Young, A.T., Arenholz, E., Delano, R., and Cramer, S.P. (2004) *Rev. Sci. Instrum.*, 75, 756.

Funk, T., Deb, A., George, S.J., Wang, H.X., and Cramer, S.P. (2005) *Coord. Chem. Rev.*, 249, 3.

Gambardella, P., Dhesi, S.S., Gardonio, S., Grazioli, C., Ohresser, P., and Carbone, C. (2002) *Phys. Rev. Lett.*, 88.

Gambardella, P., Rusponi, S., Veronese, M., Dhesi, S.S., Grazioli, C., Dallmeyer, A., Cabria, I., Zeller, R., Dederichs, P.H., Kern, K., Carbone, C., and Brune, H. (2003) *Science*, 300, 1130.

Garvie, L.A.J., and Buseck, P.R. (2004) *Amer. Miner.*, 89, 485.

Gawelda, W., Johnson, M., de Groot, F.M.F., Abela, R., Bressler, C., and Chergui, M. (2006) *J. Am. Chem. Soc.*, 128, 5001.

Gehring, G.A. (2002) *J. Phys. Cond. Matt.*, 14, V5.

Gel'mukhanov, F., and Ågren, H. (1994) *Phys. Rev. A*, 49, 4378.

George, S.J., van Elp, J., Chen, J., Ma, Y., Chen, C.T., Park, J.B., Adams, M.W.W., Searle, B.G., de Groot, F.M.F., Fuggle, J.C., and Cramer, S.P. (1992) *J. Am. Chem. Soc.*, 114, 4426.

Ghiringhelli, G., Tagliaferri, A., Braicovich, L., and Brookes, N.B. (1998) *Rev. Sci. Instrum.*, 69, 1610.

Ghiringhelli, G., Brookes, N.B., Annese, E., Berger, H., Dallera, C., Grioni, M., Perfetti, L., Tagliaferri, A., and Braicovich, L. (2004) *Phys. Rev. Lett.*, 92, 117406.

Ghiringhelli, G., Matsubara, M., Dallera, C., Fracassi, F., Gusmeroli, R., Piazzalunga, A., Tagliaferri, A., Brookes, N.B., Kotani, A., and Braicovich, L. (2005) *J. Phys. Cond. Matt.*, 17, 5397.

Ghiringhelli, G., Matsubara, M., Dallera, C., Fracassi, F., Tagliaferri, A., Brookes, N.B., Kotani, A., and Braicovich, L. (2006) *Phys. Rev. B*, 73, 035111.

Gilbert, B., Frazer, B.H., Belz, A., Conrad, P.G., Nealson, K.H., Haskel, D., Lang, J.C., Srajer, G., and De Stasio, G. (2003) *J. Phys. Chem. A*, 107, 2839.

Giorgetti, C., Pizzini, S., Dartyge, E., Fontaine, A., Baudelet, F., Brouder, C., Bauer, P., Krill, G., Miraglia, S., Fruchart, D., and Kappler, J.P. (1993) *Phys. Rev. B*, 48, 12732.

Giorgetti, C., Dartyge, E., Baudelet, F., and Brouder, C. (2001) *Appl. Phys. A*, 73, 703.

Glatzel, P. (2001) (PhD thesis, Hamburg University), http://www.physnet.uni-hamburg.de/services/fachinfo/dissfb12_2001.html

Glatzel, P., Bergmann, U., de Groot, F.M.F., and Cramer, S.P. (2001) *Phys. Rev. B*, 6404.

Glatzel, P., Jacquamet, L., Bergmann, U., de Groot, F.M.F., and Cramer, S.P. (2002) *Inorg. Chem.*, 41, 3121.

Glatzel, P., Bergmann, U., Yano, J., Visser, H., Robblee, J.H., Gu, W.W., de Groot, F.M.F., Christou, G., Pecoraro, V.L., Cramer, S.P., and Yachandra, V.K. (2004) *J. Am. Chem. Soc.*, 126, 9946.

Glatzel, P., and Bergmann, U. (2005) *Coord. Chem. Rev.*, 249, 65.

Goedkoop, J.B., Thole, B.T., van der Laan, G., Sawatzky, G.A., de Groot, F.M., and Fuggle, J.C. (1988a) *Phys. Rev. B*, 37, 2086.

Goedkoop, J.B., Fuggle, J.C., Thole, B.T., van der Laan, G., and Sawatzky, G.A. (1988b) *J. Appl. Phys.*, 64, 5595.

Gorschluter, A., and Merz, H. (1994) *Phys. Rev. B*, 49, 17293.

Gorschluter, A., and Merz, H. (1998) *J. Electron Spectrosc. Relat. Phenom.*, 87, 211.

Goulon, J., Rogalev, A., Wilhelm, F., Goulon-Ginet, C., Carra, P., Marri, I., and Brouder, C. (2003) *J. Exp. Theo. Phys.*, 97, 402.

Grenier, S., Hill, J.P., Kiryukhin, V., Ku, W., Kim, Y.J., Thomas, K.J., Cheong, S.W., Tokura, Y., Tomioka, Y., Casa, D., and Gog, T. (2005) *Phys. Rev. Lett.*, 94, 47203.

Grioni, M., Weibel, P., Malterre, D., F. Jeanneret, F., Baer, Y., and Olcese, G. (1995) *Physica B*, 206–207, 71.

Grioni, M., Weibel, P., Malterre, D., Baer, Y., and Duò, L. (1997) *Phys. Rev. B*, 55, 2056.

Grunes, L.A. (1983) *Phys. Rev. B*, 27, 2111.

Guevara, J., Llois, A.M., and Weissmann, M. (1998) *Phys. Rev. Lett.*, 81, 5306.

Guillot, C., Ballu, Y., Paigné, J., Lecante, J., Jain, K.P., Thiry, P., Pinchaux, R., Petroff, Y., and Falicov, L.M. (1977) *Phys. Rev. Lett.*, 39, 1632.

Gunnarsson, O., and Schönhammer, K. (1983a) *Phys. Rev. Lett.*, 50, 604.

Gunnarsson, O., and Schönhammer, K. (1983b) *Phys. Rev. B*, 28, 4315.

Gunnarsson, O., and Jepsen, O. (1988) *Phys. Rev. B*, 38, 3568.

Gweon, G.H., Park, J.G., and Oh, S.J. (1993) *Phys. Rev. B*, 48, 7825.

Haak, H.W., Sawatzky, G.A., and Thomas, T.D. (1978) *Phys. Rev. Lett.*, 41, 1825.

Hague, C.F., Underwood, J.H., Avila, A., Delaunay, R., Ringuenet, H., Marsi, M., and Sacchi, M. (2005) *Rev. Sci. Instrum.*, 76.

Hämäläinen, K., Siddons, D.P., Hastings, J.B., and Berman, L.E. (1991) *Phys. Rev. Lett.*, 67, 2850.

Hämäläinen, K., Kao, C.C., Hastings, J.B., Siddons, D.P., Berman, L.E., Stojanoff, V., and Cramer, S.P. (1992) *Phys. Rev. B*, 46, 14274.

Harada, Y., Kinugasa, T., Eguchi, R., Matsubara, M., Kotani, A., Watanabe, M., Yagishita, A., and Shin, S. (2000) *Phys. Rev. B*, 61, 12854.

Harada, Y., Okada, K., Eguchi, R., Kotani, A., Takagi, H., Takeuchi, T., and Shin, S. (2002) *Phys. Rev. B*, 66, 165104.

Harrison, W.A. (1989) *Electronic Structure and the Properties of Solids* (Dover, New York).

Hasan, M.Z., Isaacs, E.D., Shen, Z.X., Miller, L.L., Tsutsui, K., Tohyama, T., and Maekawa, S. (2000) *Science*, 288, 1811.

Hasan, M.Z., Montano, P.A., Isaacs, E.D., Shen, Z.X., Eisaki, H., Sinha, S.K., Islam, Z., Motoyama, N., and Uchida, S. (2002) *Phys. Rev. Lett.*, 88, 177403.

Hasselström, J., Föhlisch, A., Denecke, R., Nilsson, A., and de Groot, F.M.F. (2000) *Phys. Rev. B*, 62, 11192.

Haverkort, M.W., Hu, Z., Tanaka, A., Ghiringhelli, G., Roth, H., Cwik, M., Lorenz, T., Schussler-Langeheine, C., Streltsov, S.V., Mylnikova, A.S., Anisimov, V.I., de Nadai, C., Brookes, N.B., Hsieh, H.H., Lin, H.J., Chen, C.T., Mizokawa, T., Taguchi, Y., Tokura, Y., Khomskii, D.I., and Tjeng, L.H. (2005a) *Phys. Rev. Lett.*, 94.

Haverkort, M.W., Hu, Z., Tanaka, A., Reichelt, W., Streltsov, S.V., Korotin, M.A., Anisimov, V.I., Hsieh, H.H., Lin, H.J., Chen, C.T., Khomskii, D.I., and Tjeng, L.H. (2005b) *Phys. Rev. Lett.*, 95.

Hayashi, H., Udagawa, Y., Caliebe, W.A., and Kao, C.C. (2002) *Phys. Rev. B*, 66, 33105.

Hayashi, H., Takeda, R., Kawata, M., Udagawa, Y., Watanabe, Y., Takano, T., Nanao, S., Kawamura, N., Uefuji, T., and Yamada, K. (2004) *J. Electron Spectrosc. Relat. Phenom.*, 136, 199.

Hedin, L., Michiels, J., and Inglesfield, J. (1998) *Phys. Rev. B*, 58, 15565.

Heijboer, W.M., Battiston, A.A., Knop-Gericke, A., Hävecker, M., Mayer, R., Blühm, H., Schlögl, R., Weckhuysen, B.M., Koningsberger, D.C., and de Groot, F.M.F. (2003) *J. Phys. Chem. B*, 107, 13069.

Heijboer, W.M., Battiston, A.A., Knop-Gericke, A., Hävecker, M., Mayer, R., Blühm, H., Schlögl, R., Weckhuysen, B.M., Koningsberger, D.C., and de Groot, F.M.F. (2004) *J. Phys. Chem. B*, 108, 10002.

Heitler, W. (1944) *The Quantum Theory of Radiation* (Oxford Univ. Press).

Herbst, J.F. (1991) *Rev. Mod. Phys.*, 63, 819.

Hill, J.P., Kao, C.C., and McMorrow, D.F. (1997) *Phys. Rev. B*, 55, R8662.

Hill, J.P., Kao, C.C., Caliebe, W.A.L., Matsubara, M., Kotani, A., Peng, J.L., and Greene, R.L. (1998) *Phys. Rev. Lett.*, 80, 4967.

Himpsel, F.J., Karlsson, U.O., McLean, A.B., Terminello, L.J., de Groot, F.M.F., Abbate, M., Fuggle, J.C., Yarmoff, J.A., Thole, B.T., and Sawatzky, G.A. (1991) *Phys. Rev. B*, 43, 6899.

Hinkers, H., Stiller, R., and Merz, H. (1989) *Phys. Rev. B*, 40, 10594.

Hocking, R.K., Wasinger, E.C., de Groot, F.M.F., Hodgson, K.O., Hedman, B., and Solomon, E.I. (2006) *J. Am. Chem. Soc.*, 128, 10442.

Hopfield, J.J. (1969) *Comm. Solid State Phys.*, 2, 40.

Horiba, K., Taguchi, M., Chainani, A., Takata, Y., Ikenaga, E., Miwa, D., Nishino, Y., Tamasaku, K., Awaji, M., Takeuchi, A., Yabashi, M., Namatame, H., Taniguchi, M., Kumigashira, H., Oshima, M., Lippmaa, M., Kawasaki, M., Koinuma, H., Kobayashi, K., Ishikawa, T., and Shin, S. (2004) *Phys. Rev. Lett.*, 93, 236401.

Hu, Z., Kaindl, G., Warda, S.A., Reinen, D., de Groot, F.M.F., and Muller, B.G. (1998a) *Chem. Phys.*, 232, 63.

Hu, Z., Mazumdar, C., Kaindl, G., de Groot, F.M.F., Warda, S.A., and Reinen, D. (1998b) *Chem. Phys. Lett.*, 297, 321.

Hu, Z., Golden, M.S., Ebbinghaus, S.G., Knupfer, M., Fink, J., de Groot, F.M.F., and Kaindl, G. (2002) *Chem. Phys.*, 282, 451.

Hu, Z., Wu, H., Haverkort, M.W., Hsieh, H.H., Lin, H.J., Lorenz, T., Baier, J., Reichl, A., Bonn, I., Felser, C., Tanaka, A., Chen, C.T., and Tjeng, L.H. (2004) *Phys. Rev. Lett.*, 92.

Huang, D.J., Wu, W.B., Guo, G.Y., Lin, H.J., Hou, T.Y., Chang, C.F., Chen, C.T., Fujimori, A., Kimura, T., Huang, H.B., Tanaka, A., and Jo, T. (2004) *Phys. Rev. Lett.*, 92.

Huang, H.B., and Jo, T. (2004) *Physica B*, 351, 313.

Hüfner, S., Steiner, P., Sander, I., Neumann, M., and Witzel, S. (1991) *Z. Phys. B*, 83, 185.

Hüfner, S. (1995) *Photoelectron Spectroscopy* (Springer, 1995).

Hund, F. (1925) *Z. Phys.*, 33, 855.

Hund, F. (1927) *Linienspektren und periodisches System der Elemente* (Springer, Berlin).

Idé, T., and Kotani, A. (1998) *J. Phys. Soc. Jpn.*, 67, 3621.

Idé, T., and Kotani, A. (1999) *J. Phys. Soc. Jpn.*, 68, 3100.

Idé, T., and Kotani, A. (2000) *J. Phys. Soc. Jpn.*, 69, 1895.

Ikeda, T., Okada, K., Ogasawara, H., and Kotani, A. (1990) *J. Phys. Soc. Jpn.*, 59, 622.

Ikeno, H., Tanaka, I., Miyamae, L., Mishima, T., Adachi, H., and Ogasawara, K. (2004) *Materials Transactions*, 45, 1414.

Ikeno, H., Tanaka, I., Koyama, Y., Mizoguchi, T., and Ogasawara, K. (2005) *Phys. Rev. B*, 72, 75123.

Ikeno, H., Mizoguchi, T., Koyama, Y., Kumagai, Y., and Tanaka, I. (2006) *Ultramicr.*, 106, 970.

Inami, T., Fukuda, T., Mizuki, J., Ishihara, S., Kondo, H., Nakao, H., Matsumura, T., Hirota, K., Murakami, Y., Maekawa, S., and Endoh, Y. (2003) *Phys. Rev. B*, 67, 45108.

Ishii, H., Ishiwata, Y., Eguchi, R., Harada, Y., Watanabe, M., Chainani, A., and Shin, S. (2001) *J. Phys. Soc. Jpn.*, 70, 1813.

Ishii, K., Inami, T., Ohwada, K., Kuzushita, K., Mizuki, J., Murakami, Y., Ishihara, S., Endoh, Y., Maekawa, S., Hirota, K., and Moritomo, Y. (2004) *Phys. Rev. B*, 70, 224437.

Ishii, K., Tsutsui, K., Endoh, Y., Tohyama, T., Kuzushita, K., Inami, T., Ohwada, K., Maekawa, S., Masui, T., Tajima, S., Murakami, Y., and Mizuki, J. (2005a) *Phys. Rev. Lett.*, 94, 187002.

Ishii, K., Tsutsui, K., Endoh, Y., Tohyama, T., Maekawa, S., Hoesch, M., Kuzushita, K., Tsubota, M., Inami, T., Mizuki, J., Murakami, Y., and Yamada, K. (2005b) *Phys. Rev. Lett.*, 94, 207003.

Iwamoto, Y., Nakazawa, M., Kotani, A., and Parlebas, J.C. (1995) *J. Phys. Cond. Matt.*, 7, 1149.

Iwashita, K., Oguchi, T., and Jo, T. (1996) *Phys. Rev. B*, 54, 1159.

Jensen, E., Bartynski, R.A., Hulbert, S.L., Johnson, E.D., and Garrett, R. (1989) *Phys. Rev. Lett.*, 62, 71.

Jo, T., and Kotani, A. (1985) *Solid State Commun.*, 54, 451.

Jo, T., and Kotani, A. (1988a) *Phys. Rev. B*, 38, 830.

Jo, T., and Kotani, A. (1988b) *J. Phys. Soc. Jpn.*, 57, 2288.

Jo, T., and Sawatzky, G.A. (1991) *Phys. Rev. B*, 43, 8771.

Jo, T., and Imada, S. (1993) *J. Phys. Soc. Jpn.*, 62, 3721.

Jo, T., and Imada, S. (1998) *J. Phys. Soc. Jpn.*, 67, 3617.

Jo, T., Imada, S., and van Elp, J. (1999) *Physica B*, 261, 1151.

Johnson, P.D. (1997) *Rep. Prog. Phys.*, 60, 1217.

Jollet, F., Ortiz, V., and Crocombette, J.P. (1997) *J. Electron Spectrosc. Relat. Phenom.*, 86, 83.

Joly, Y., Cabaret, D., Renevier, H., and Natoli, C.R. (1999) *Phys. Rev. Lett.*, 82, 2398.

Joly, Y. (2001) *Phys. Rev. B*, 63, 125120.

Jones, R.O., and Gunnarsson, O. (1989) *Rev. Mod. Phys.*, 61, 689.

Jones, P., Inglesfield, J.E., Michiels, J.J.M., Noble, C.J., Burke, V.M., and Burke, P.G. (2000) *Phys. Rev. B*, 62, 13508.

Journel, L., Mariot, J.M., Rueff, J.P., Hague, C.F., Krill, G., Nakazawa, M., Kotani, A., Rogalev, A., Wilhelm, F., Kappler, J.P., and Schmerber, G. (2002) *Phys. Rev. B*, 66, 45106.

Juett, A.M., Schulz, N.S., and Chakrabarty, D. (2004) *Astrophys. J.* 612, 308.

Kaga, H., Kotani, A., and Toyozawa, Y. (1976) *J. Phys. Soc. Jpn.*, 41, 1851.

Kaindl, G., Wertheim, G.K., Schmiester, G., and Sampathkumaran, E.V. (1987) *Phys. Rev. Lett.*, 58, 606.

Kalkowski, G., Kaindl, G., Brewer, W.D., and Krone, W. (1987) *Phys. Rev. B*, 35, 2667.

Kanai, K., Tezuka, Y., Fujisawa, M., Harada, Y., Shin, S., Schmerber, G., Kappler, J.P., Parlebas, J.C., and Kotani, A. (1997) *Phys. Rev. B*, 55, 2623.

Kanai, K., Tezuka, Y., Terashima, T., Muro, Y., Ishikawa, M., Uozumi, T., Kotani, A., Schmerber, G., Kappler, J.P., Parlebas, J.C., and Shin, S. (1999) *Phys. Rev. B*, 60, 5244.

Kanai, K., Terashima, T., Kotani, A., Uozumi, T., Schmerber, G., Kappler, J.P., Parlebas, J.C., and Shin, S. (2001) *Phys. Rev. B*, 63, 033106.

Kao, C.C., Chen, C.T., Johnson, E.D., Hastings, J.B., Lin, H.J., Ho, G.H., Meigs, G., Brot, J. M., Hulbert, S.L., Idzerda, Y.U., and Vettier, C. (1994) *Phys. Rev. B*, 50, 9599.

Kao, C.C., Caliebe, W.A.L., Hastings, J.B., and Gillet, J.M. (1996) *Phys. Rev. B*, 54, 16361.

Keast, V.J., Scott, A.J., Brydson, R., Williams, D.B., and Bruley, J. (2001) *J. Microsc.*, 203, 135.

Kim, C., Matsuura, A.Y., Shen, Z.X., Motoyama, N., Eisaki, H., Uchida, S., Tohyama, T., and Maekawa, S. (1996) *Phys. Rev. Lett.*, 77, 4054.

Kim, Y.J., Hill, J.P., Burns, C.A., Wakimoto, S., Birgeneau, R.J., Casa, D., Gog, T., and Venkataraman, C.T. (2002) *Phys. Rev. Lett.*, 89.

Kim, Y.J., Hill, J.P., Benthien, H., Essler, F.H.L., Jeckelmann, E., Choi, H.S., Noh, T.W., Motoyama, N., Kojima, K.M., Uchida, S., Casa, D., and Gog, T. (2004a) *Phys. Rev. Lett.*, 92, 137402.

Kim, Y.J., Hill, J.P., Chou, F.C., Casa, D., Gog, T., and Venkataraman, C.T. (2004b) *Phys. Rev. B*, 69, 155105.

Kim, Y.J., Hill, J.P., Komiya, S., Ando, Y., Casa, D., Gog, T., and Venkataraman, C.T. (2004c) *Phys. Rev. B*, 70, 94524.

Kim, Y.J., Hill, J.P., Gu, G.D., Chou, F.C., Wakimoto, S., Birgeneau, R.J., Komiya, S., Ando, Y., Motoyama, N., Kojima, K.M., Uchida, S., Casa, D., and Gog, T. (2004d) *Phys. Rev. B*, 70, 205128.

Kimura, S., Ufuktepe, Y., Nath, K.G., Kinoshita, T., Kumigashira, H., Takahashi, T., Matsumura, T., Suzuki, T., Ogasawara, H., and Kotani, A. (1998) *J. Magn. Magn. Mat.*, 177, 349.

Kinoshita, T., Ufuktepe, Y., Nath, K.G., Kimura, S., Kumigashira, H., Takahashi, T., Matsumura, T., Suzuki, T., Ogasawara, H., and Kotani, A. (1998) *J. Electron Spectrosc. Relat. Phenom.*, 88, 377.

Kinoshita, T., Gunasekara, H., Takata, Kimura, S., Okuno, M., Haruyama, Y., Kosugi, N., Nath, K.G., Wada, H., Mitsuda, A., Shiga, M., Okuda, T., Harasawa, A., Ogasawara, H., and Kotani, A. (2002) *J. Phys. Soc. Jpn.*, 71, 148.

Kobayashi, K., Mizokawa, T., Ino, A., Matsuno, J., Fujimori, A., Samata, H., Mishiro, A., Nagata, Y., and de Groot, F.M.F. (1999) *Phys. Rev. B*, 59, 15100.

Koide, T. (1994) *Oyo Butsuri* (in Japanese), 63, 1210.

Koide, T., Miyauchi, H., Okamoto, J., Shidara, T., Fujimori, A., Fukutani, H., Amemiya, K., Takeshita, H., Yuasa, S., Katayama, T., and Suzuki, Y. (2001) *Phys. Rev. Lett.*, 87, 257201.

Koide, T., Miyauchi, H., Okamoto, J., Shidara, T., Fujimori, A., Fukutani, H., Amemiya, K., Takeshita, H., Yuasa, S., Katayama, T., and Suzuki, Y. (2004) *J. Electron Spectrosc. Relat. Phenom.*, 136, 107.

Kolczewski, C., and Hermann, K. (2005) *Theo. Chem. Acc.*, 114, 60.

Kolczewski, C., Puttner, R., Martins, M., Schlachter, A.S., Snell, G., Sant'Anna, M.M., Hermann, K., and Kaindl, G. (2006) *J. Chem. Phys.*, 124.

Kotani, A., and Toyozawa, Y. (1973a) *J. Phys. Soc. Jpn.*, 35, 1073.

Kotani, A., and Toyozawa, Y. (1973b) *J. Phys. Soc. Jpn.*, 35, 1082.

Kotani, A., and Toyozawa, Y. (1974) *J. Phys. Soc. Jpn.*, 37, 912.

Kotani, A., and Toyozawa, Y. (1979) *Synchrotron Radiation*, edited by C. Kunz (Springer-Verlag, Berlin), page 169.

Kotani, A., Mizuta, H., Jo, T., and Parlebas, J.C. (1985) *Solid State Commun.*, 53, 805.

Kotani, A., and Jo, T. (1986) *J. Physique*, C8/915.

Kotani, A. (1987) *Handbook on Synchrotron Radiation, Vol. 2*, edited by G.V. Marr (North-Holland, Amsterdam).

Kotani, A., Jo, T., Okada, K., Nakano, T., Okada, M., Bianconi, A., Marcelli, A., and Parlebas, J.C. (1987a) *J. Magn. Magn. Mat.*, 70, 28.

Kotani, A., Okada, K., and Okada, M. (1987b) *Solid State Commun.*, 64, 1479.

Kotani, A., Okada, M., Jo, T., Bianconi, A., Marcelli, A., and Parlebas, J.C. (1987c) *J. Phys. Soc. Jpn.*, 56, 798.

Kotani, A., Jo, T., and Parlebas, J.C. (1988) *Adv. Phys.*, 37, 37.

Kotani, A., Ogasawara, H., Okada, K., Thole, B.T., and Sawatzky, G.A. (1989) *Phys. Rev. B*, 40, 65.

Kotani, A., and Okada, K. (1990) *Progress of Theoretical Physics Supplement*, 101, 329.

Kotani, A., and Ogasawara, H. (1992) *J. Electron Spectrosc. Relat. Phenom.*, 60, 257.

Kotani, A. (1993) *Jpn. J. Appl. Phys.*, 32, 3.

Kotani, A., and Okada, K. (1993) *Recent Advances in Magnetism of Transition Metal Compounds*, edited by A. Kotani and N. Suzuki (World Scientific, Singapore).

Kotani, A., and Shin, S. (2001) *Rev. Mod. Phys.*, 73, 203.

Kotani, A. (2004) *J. Electron Spectrosc. Relat. Phenom.*, 137–140, 669.

Kotani, A. (2005) *Eur. Phys. J. B*, 47, 3.

Kotani, A., Parlebas, J.C., Le Fèvre, P., Chandesris, D., Magnan, H., Asakura, K., Harada, I., and Ogasawara, H. (2005) *Nucl. Instrum. Meth. A*, 547, 124.

Kotani, A. (2006) *J. Molecular Structure: Theochem*, 777, 17.

Kotani, A., Matsubara, M., Uozumi, T., Ghiringhelli, G., Fracassi, F., Dallera, C., Tagliaferri, A., Brookes, N.B., and Braicovich, L. (2006) *Rad. Phys. Chem.*, 75, 1670.

Kotani, A., and Okada, K. (2006) *Second International Workshop on Hard X-Ray Photoelectron Spectroscopy.*

Kotani, A., Okada, K., Calandra, M., and Shukla, A. (2007) *AIP Conference Series (XAFS13).*

Kowalczyk, S.P., Ley, L., McFeely, F.R., and Shirley, D.A. (1975) *Phys. Rev. B*, 11, 1721.

Kramers, H.A., and Heisenberg, W. (1925) *Z. Phys.*, 31, 681.

Krisch, M.H., Kao, C.C., Sette, F., Caliebe, W.A., Hämäläinen, K., and Hastings, J.B. (1995) *Phys. Rev. Lett.*, 74, 4931.

Krisch, M.H., Sette, F., Masciovecchio, C., and Verbeni, R. (1997) *Phys. Rev. Lett.*, 78, 2843.

Krivanek, O.L., Manoubi, T., and Colliex, C. (1985) *Ultramicr.*, 18, 155.

Krivanek, O.L., and Paterson, J.H. (1990) *Ultramicr.*, 32, 313.

Kuepper, K., Bondino, F., Prince, K.C., Zangrando, M., Zacchigna, M., Takacs, A.F., Crainic, T., Matteucei, M., Parmigiani, F., Winiarski, A., Galakhov, V.R., Mukovskii, Y.M., and Neumann, M. (2005) *J. Phys. Chem. B*, 109, 15667.

Kuiper, P., Kruizinga, G., Ghijsen, J., Grioni, M., Weijs, P.J.W., de Groot, F.M.F., Sawatzky, G.A., Verweij, H., Feiner, L.F., and Petersen, H. (1988) *Phys. Rev. B*, 38, 6483.

Kuiper, P., Guo, J.H., Såthe, C., Duda, L.C., Nordgren, J., Pothuizen, J.J.M., de Groot, F.M.F., and Sawatzky, G.A. (1998) *Phys. Rev. Lett.*, 80, 5204.

Kurata, H., Lefevre, E., Colliex, C., and Brydson, R. (1993) *Phys. Rev. B*, 47, 13763.

Lagarde, P., Flank, A.M., Ogasawara, H., and Kotani, A. (2003) *J. Electron Spectrosc. Relat. Phenom.*, 128, 193.

Laubschat, C., Weschke, E., Holtz, C., Domke, M., Strebel, O., and Kaindl, G. (1990) *Phys. Rev. Lett.*, 65, 1639.

Le Fèvre, P., Magnan, H., Chandesris, D., Vogel, J., Formoso, V., and Comin, F. (1998) *Phys. Rev. B*, 58, 1080.

Le Fèvre, P., Magnan, H., Chandesris, D., Vogel, J., Formoso, V., and Comin, F. (2004) *J. Electron Spectrosc. Relat. Phenom.*, 136, 37.

Le Fèvre, P., Magnan, H., Chandesris, D., Jupille, J., Bourgeois, S., Barbier, A., Drube, W., Uozumi, T., and Kotani, A. (2005) *Nucl. Instrum. Meth. A*, 547, 176.

Lee, G., and Oh, S.J. (1991) *Phys. Rev. B*, 43, 14674.

Lee, J.D., Gunnarsson, O., and Hedin, L. (1999) *Phys. Rev. B*, 60, 8034.

Lopez, M.F., Gutierrez, A., Laubschat, C., and Kaindl, G. (1995) *Solid State Commun.*, 94, 673.

Lund, C.P., Thurgate, S.M., and Wedding, A.B. (1997) *Phys. Rev. B*, 55, 5455.

Magnuson, M., Nilsson, A., Weinelt, M., and Martensson, N. (1999) *Phys. Rev. B*, 60, 2436.

Magnuson, M., Butorin, S.M., Guo, J.H., Agui, A., Nordgren, J., Ogasawara, H., Kotani, A., Takahashi, T., and Kunii, S. (2001) *Phys. Rev. B*, 63, 75101.

Magnuson, M., Butorin, S.M., Agui, A., and Nordgren, J. (2002a) *J. Phys. Cond. Matt.*, 14, 3669.

Magnuson, M., Butorin, S.M., Guo, J.H., and Nordgren, J. (2002b) *Phys. Rev. B*, 65, 205106.

Magnuson, M., Butorin, S.M., Såthe, C., Nordgren, J., and Ravindran, P. (2004) *Europhys. Lett.*, 68, 289.

Mahan, G.D. (1967) *Phys. Rev.*, 163, 612.

Martensson, N., Weinelt, M., Karis, O., Magnuson, M., Wassdahl, N., Nilsson, A., Stöhr, J., and Samant, M. (1997) *Appl. Phys. A*, 65, 159.

Martensson, N., Karis, O., and Nilsson, A. (1999) *J. Electron Spectrosc. Relat. Phenom.*, 100, 379.

Martins, M., Godehusen, K., Richter, T., Wernet, P., and Zimmermann, P. (2006) *J. Phys. B*, 39, R79.

Matsubara, M., Uozumi, T., Kotani, A., Harada, Y., and Shin, S. (2000) *J. Phys. Soc. Jpn.*, 69, 1558.

Matsubara, M., Uozumi, T., Kotani, A., Harada, Y., and Shin, S. (2002) *J. Phys. Soc. Jpn.*, 71, 347.

Matsubara, M., Harada, Y., Shin, S., Uozumi, T., and Kotani, A. (2004a) *J. Phys. Soc. Jpn.*, 73, 711.

Matsubara, M., Uozumi, T., Kotani, A., and Parlebas, J.C. (2005) *J. Phys. Soc. Jpn.*, 74, 2052.

Matsubara, M., and Kotani, A. (2007) (unpublished).

Matsuyama, H., Harada, I., and Kotani, A. (1997) *J. Phys. Soc. Jpn.*, 66, 337.

Matsuyama, H., Fukui, K., Maruyama, H., Harada, I., and Kotani, A. (1998) *J. Magn. Magn. Mat.*, 177, 1029.

Mattheiss, L.F. (1972a) *Phys. Rev. B*, 5, 290.

Mattheiss, L.F. (1972b) *Phys. Rev. B*, 5, 306.

McComb, D.W., Craven, A.J., Chioncel, L., Lichtenstein, A.I., and Docherty, F.T. (2003) *Phys. Rev. B*, 68.

McMahan, A.K., Martin, R.M., and Satpathy, S. (1988) *Phys. Rev. B*, 38, 6650.

Mirone, A., Dhesi, S.S., and van der Laan, G. (2006) *Eur. Phys. J. B*, 53, 23.

Mitterbauer, C., Kothleitner, G., Grogger, W., Zandbergen, H., Freitag, B., Tiemeijer, P., and Hofer, F. (2003) *Ultramicr.*, 96, 469.

Morales, F., de Groot, F.M.F., Glatzel, P., Kleimenov, E., Blühm, H., Havecker, M., Knop-Gericke, A., and Weckhuysen, B.M. (2004) *J. Phys. Chem. B*, 108, 16201.

Moroni, R., Moulin, C.C.D., Champion, G., Arrio, M.A., Sainctavit, P., Verdaguer, M., and Gatteschi, D. (2003) *Phys. Rev. B*, 68.

Mosser, A., Romeo, M.A., Parlebas, J.C., Okada, K., and Kotani, A. (1991) *Solid State Commun.*, 79, 641.

Nagel, M., Biswas, I., Nagel, P., Pellegrin, E., Schuppler, S., Peisert, H., and Chasse, T. (2007) *Phys. Rev. B*, 75, 195426.

Nakai, S., Nakamori, H., Tomita, A., Tsutsumi, K., Nakamura, H., and Sugiura, C. (1974) *Phys. Rev. B*, 9, 1870.

Nakajima, R., Stöhr, J., and Idzerda, Y.U. (1999) *Phys. Rev. B*, 59, 6421.

Nakamura, M., Takata, Y., and Kosugi, N. (1996) *J. Electron Spectrosc. Relat. Phenom.*, 78, 115.

Nakamura, T., Shoji, H., Hirai, E., Nanao, S., Fukui, K., Ogasawara, H., Kotani, A., Iwazumi, T., Harada, I., Katano, R., and Isozumi, Y. (2003) *Phys. Rev. B*, 67.

Nakano, T., Okada, K., and Kotani, A. (1988) *J. Phys. Soc. Jpn.*, 57, 655.

Nakazawa, M., Tanaka, S., Uozumi, T., and Kotani, A. (1996) *J. Phys. Soc. Jpn.*, 65, 2303.

Nakazawa, M. (1998) PhD thesis (University of Tokyo).

Nakazawa, M., Ogasawara, H., Kotani, A., and Lagarde, P. (1998) *J. Phys. Soc. Jpn.*, 67, 323.

Nakazawa, M., Ogasawara, H., and Kotani, A. (2000) *J. Phys. Soc. Jpn.*, 69, 4071.

Nakazawa, M., Fukui, K., Ogasawara, H., Kotani, A., and Hague, C.F. (2002) *Phys. Rev. B*, 66, 113104.

Nakazawa, M., Fukui, K., and Kotani, A. (2003) *J. Sol. State Chem.*, 171, 295.

Namatame, H., Fujimori, A., Takagi, H., Uchida, S., de Groot, F.M.F., and Fuggle, J.C. (1993) *Phys. Rev. B*, 48, 16917.

Naslund, L.A., Luning, J., Ufuktepe, Y., Ogasawara, H., Wernet, P., Bergmann, U., Pettersson, L.G.M., and Nilsson, A. (2005) *J. Phys. Chem. B*, 109, 13835.

Nath, K.G., Ufuktepe, Y., Kimura, S., Kinoshita, T., Kumigashira, H., Takahashi, T., Matsumura, T., Suzuki, T., Ogasawara, H., and Kotani, A. (1998) *J. Electron Spectrosc. Relat. Phenom.*, 88, 369.

Nath, K.G., Ufuktepe, Y., Kimura, S., Haruyama, Y., Kinoshita, T., Matsumura, T., Suzuki, T., Ogasawara, H., and Kotani, A. (2003) *J. Phys. Soc. Jpn.*, 72, 1792.

Natoli, C.R., Benfatto, M., Brouder, C., Lopez, M.F.R., and Foulis, D.L. (1990) *Phys. Rev. B*, 42, 1944.

Niemantsverdriet, J.W. (1993) *Spectroscopy in Catalysis* (VCH, 1993).

Nolting, F., Scholl, A., Stöhr, J., Seo, J.W., Fompeyrine, J., Siegwart, H., Locquet, J.P., Anders, S., Luning, J., Fullerton, E.E., Toney, M.F., Scheinfein, M.R., and Padmore, H.A. (2000) *Nature*, 405, 767.

Nordgren, J., Bray., G., Gamn, S., Nyholm, R., Rubensson, J., and Wassdahl, N. (1989) *Rev. Sci. Instrum.*, 60, 1960.

Nordgren, J., and Kurmaev, E.Z. (2000) *J. Electron Spectrosc. Relat. Phenom.*, 110–111, 1.

Nozières, P., and De Dominicis, C.T. (1969) *Phys. Rev.*, 178, 1097.

Nücker, N., Fink, J., Renker, B., Ewert, D., Politis, C., Weijs, P.J.W., and Fuggle, J.C. (1987) *Z. Phys. B*, 67, 9.

Nücker, N., Fink, J., Fuggle, J.C., Durham, P.J., and Temmerman, W.M. (1988) *Phys. Rev. B*, 37, 5158.

Ogasawara, H., Kotani, A., Okada, K., and Thole, B.T. (1991a) *Phys. Rev. B*, 43, 854.

Ogasawara, H., Kotani, A., Potze, R., Sawatzky, G.A., and Thole, B.T. (1991b) *Phys. Rev. B*, 44, 5465.

Ogasawara, H., Kotani, A., Thole, B.T., Ichikawa, K., Aita, O., and Kamada, M. (1992) *Solid State Commun.*, 81, 645.

Ogasawara, H., Kotani, A., and Thole, B.T. (1994) *Phys. Rev. B*, 50, 12332.

Ogasawara, H., and Kotani, A. (1996) *J. Electron Spectrosc. Relat. Phenom.*, 78, 119.

Ogasawara, H., and Kotani, A. (1998) *J. Electron Spectrosc. Relat. Phenom.*, 88, 261.

Ogasawara, H., Kotani, A., Le Fèvre, P., Chandesris, D., and Magnan, H. (2000) *Phys. Rev. B*, 62, 7970.

Ogasawara, H., Fukui, K., and Matsubara, M. (2004) *J. Electron Spectrosc. Relat. Phenom.*, 136, 161.

Oh, S.J., Gweon, G.H., and Park, J.G. (1992) *Phys. Rev. Lett.*, 68, 2850.

Ohldag, H., Scholl, A., Nolting, F., Anders, S., Hillebrecht, F.U., and Stöhr, J. (2001) *Phys. Rev. Lett.*, 86, 2878.

Okada, K., and Kotani, A. (1989a) *J. Phys. Soc. Jpn.*, 58, 1095.

Okada, K., and Kotani, A. (1989b) *J. Phys. Soc. Jpn.*, 58, 2578.

Okada, K., Kotani, A., and Thole, B.T. (1992) *J. Electron Spectrosc. Relat. Phenom.*, 58, 325.

Okada, K., and Kotani, A. (1992a) *J. Phys. Soc. Jpn.*, 61, 4619.

Okada, K., and Kotani, A. (1992b) *J. Phys. Soc. Jpn.*, 61, 449.

Okada, K., and Kotani, A. (1993) *J. Electron Spectrosc. Relat. Phenom.*, 62, 131.

Okada, K., and Kotani, A. (1995a) *Phys. Rev. B*, 52, 4794.

Okada, K., and Kotani, A. (1996) *J. Electron Spectrosc. Relat. Phenom.*, 78, 53.

Okada, K., Kotani, A., Maiti, K., and Sarma, D.D. (1996) *J. Phys. Soc. Jpn.*, 65, 1844.

Okada, K., and Kotani, A. (1997a) *J. Electron Spectrosc. Relat. Phenom.*, 86, 119.

Okada, K., and Kotani, A. (1997b) *J. Phys. Soc. Jpn.*, 66, 341.

Okada, K., and Kotani, A. (1999a) *J. Electron Spectrosc. Relat. Phenom.*, 103, 757.

Okada, K., and Kotani, A. (1999b) *J. Phys. Soc. Jpn.*, 68, 666.

Okada, K., and Kotani, A. (2002) *Phys. Rev. B*, 65, 144530.

Okada, K., and Kotani, A. (2003) *J. Phys. Soc. Jpn.*, 72, 797.

Okada, K., and Kotani, A. (2005) *J. Phys. Soc. Jpn.*, 74, 653.

Okada, K., and Kotani, A. (2006) *J. Phys. Soc. Jpn.*, 75, 44702.

Orchard, A.F., and Thornton, G. (1978) *J. Electron Spectrosc. Relat. Phenom.*, 13, 27.

Panaccione, G., van der Laan, G., Dürr, H.A., Vogel, J., and Brookes, N.B. (2001) *Eur. Phys. J. B*, 19, 281.

Panaccione, G., Altarelli, M., Fondacaro, A., Georges, A., Huotari, S., Lacovig, P., Lichtenstein, A., Metcalf, P., Monaco, G., Offi, F., Paolasini, L., Poteryaev, A., Tjernberg, O., and Sacchi, M. (2006) *Phys. Rev. Lett.*, 97, 116401.

Park, J., Ryu, S., Han, M.S., and Oh, S.J. (1988) *Phys. Rev. B*, 37, 10867.

Park, J.H., Chen, C.T., Cheong, S.W., Bao, W., Meigs, G., Chakarian, V., and Idzerda, Y.U. (1996) *Phys. Rev. Lett.*, 76, 4215.

Park, J.H., Tjeng, L.H., Tanaka, A., Allen, J.W., Chen, C.T., Metcalf, P., Honig, J.M., de Groot, F.M.F., and Sawatzky, G.A. (2000) *Phys. Rev. B*, 61, 11506.

Parlebas, J.C., Khan, M.A., Uozumi, T., Okada, K., and Kotani, A. (1995) *J. Electron Spectrosc. Relat. Phenom.*, 71, 117.

Parlebas, J.C., Asakura, K., Fujiwara, A., Harada, I., and Kotani, A. (2006) *Phys. Rep.*, 431, 1.

Paterson, J.H., and Krivanek, O.L. (1990) *Ultramicr.*, 32, 319.

Pattrick, R.A.D., van der Laan, G., Henderson, C.M.B., Kuiper, P., Dudzik, E., and Vaughan, D.J. (2002) *Eur. J. Miner.*, 14, 1095.

Pecher, K., McCubbery, D., Kneedler, E., Rothe, J., Bargar, J., Meigs, G., Cox, L., Nealson, K., and Tonner, B. (2003) *Geochim. Cosmochim. Acta*, 67, 1089.

Pedio, M., Fuggle, J.C., Somers, J., Umbach, E., Haase, J., Lindner, T., Hofer, U., Grioni, M., de Groot, F.M.F., Hillert, B., Becker, L., and Robinson, A. (1989) *Phys. Rev. B*, 40, 7924.

Pellegrin, E., Tjeng, L.H., de Groot, F.M.F., Hesper, R., Sawatzky, G.A., Moritomo, Y., and Tokura, Y. (1997) *J. Electron Spectrosc. Relat. Phenom.*, 86, 115.

Pen, H.F., Tjeng, L.H., Pellegrin, E., de Groot, F.M.F., Sawatzky, G.A., van Veenendaal, M. A., and Chen, C.T. (1997) *Phys. Rev. B*, 55, 15500.

Peng, G., de Groot, F.M.F., Hämäläinen, K., Moore, J.A., Wang, X., Grush, M.M., Hastings, J.B., Siddons, D.P., Armstrong, W.H., Mullins, O.C., and Cramer, S.P. (1994a) *J. Am. Chem. Soc.*, 116, 2914.

Peng, G., Wang, X., Randall, C.R., Moore, J.A., and Cramer, S.P. (1994b) *Appl. Phys. Lett.*, 65, 2527.

Perdew, J.P., Burke, K., and Ernzerhof, M. (1996) *Phys. Rev. Lett.*, 77, 3865.

Petersen, H. (1982) *Opt. Comm.*, 40, 402.

Pettifor, D.G. (1977) *J. Phys. F*, 7, 613.

Piamonteze, C., de Groot, F.M.F., Tolentino, H.C.N., Ramos, A.Y., Massa, N.E., Alonso, J.A., and Martinez-Lope, M.J. (2005) *Phys. Rev. B*, 71, 020406.

Pichler, T., Hu, Z., Grazioli, C., Legner, S., Knupfer, M., Golden, M.S., Fink, J., de Groot, F.M.F., Hunt, M.R.C., Rudolf, P., Follath, R., Jung, C., Kjeldgaard, L., Bruhwiler, P., Inakuma, M., and Shinohara, H. (2000) *Phys. Rev. B*, 62, 13196.

Pompa, M., Flank, A.M., Lagarde, P., Rife, J.C., Stekhin, I., Nakazawa, M., Ogasawara, H., and Kotani, A. (1997) *Phys. Rev. B*, 56, 2267.

Potze, R.H., Sawatzky, G.A., and Abbate, M. (1995) *Phys. Rev. B*, 51, 11501.

Poulopoulos, P. (2005) *Int. J. Mod. Phys. B*, 19, 4517.

Poumellec, B., Cortes, R., Tourillon, G., and Berthon, J. (1991) *Phys. Stat. Sol. B*, 164, 319.

Prince, K.C., Matteucci, M., Kuepper, K., Chiuzbǎian, S.G., Bartkowski, S., and Neumann, M. (2005) *Phys. Rev. B*, 71, 85102.

Qian, Q., Tyson, T.A., Kao, C.C., Rueff, J.P., de Groot, F.M.F., Croft, M., Cheong, S.W., Greenblatt, M., and Subramanian, M.A. (2000) *J. Phys. Chem. Sol.*, 61, 457.

Regan, T.J., Ohldag, H., Stamm, C., Nolting, F., Luning, J., Stöhr, J., and White, R.L. (2001) *Phys. Rev. B*, 6421, art. no.

Rehr, J.J., Deleon, J.M., Zabinsky, S.I., and Albers, R.C. (1991) *J. Am. Chem. Soc.*, 113, 5135.

Rehr, J.J., Albers, R.C., and Zabinsky, S.I. (1992) *Phys. Rev. Lett.*, 69, 3397.

Rehr, J.J., and Albers, R.C. (2000) *Rev. Mod. Phys.*, 72, 621.

Röntgen, W.C. (1896) *Nature* (translated), 53, 274.

Rosencwaig, A., Wertheim, G.K., and Guggenheim, H.J. (1971) *Phys. Rev. Lett.*, 27, 479.

Roth, C., Hillebrecht, F.U., Rose, H.B., and Kisker, E. (1993) *Phys. Rev. Lett.*, 70, 3479.

Rubensson, J., Eisebitt, S., Nicodemus, M., Boske, T., and Eberhardt, W. (1994) *Phys. Rev. B*, 49, 1507.

Rubensson, J.E. (2000) *J. Electron Spectrosc. Relat. Phenom.*, 110, 135.

Rudolf, P., Sette, F., Tjeng, L.H., Meigs, G., and Chen, C.T. (1992) *J. Magn. Magn. Mat.*, 109, 109.

Rueff, J.P., Kao, C.C., Struzhkin, V.V., Badro, J., Shu, J., Hemley, R.J., and Hao, H.K. (1999) *Phys. Rev. Lett.*, 82, 3284.

Rueff, J.P., Hague, C.F., Mariot, J.M., Journel, L., Delaunay, R., Kappler, J.P., Schmerber, G., Derory, A., Jaouen, N., and Krill, G. (2004) *Phys. Rev. Lett.*, 93, 067402.

Safonova, O.V., Tromp, M., van Bokhoven, J.A., de Groot, F.M.F., Evans, J., and Glatzel, P. (2006) *J. Phys. Chem. B*, 110, 16162.

Sakurai, J.J. (1967) *Advanced Quantum Mechanics* (Addison-Wesley, Reading, MA), Chap. 2.

Sato, H., Shimada, K., Arita, M., Hiraoka, K., Kojima, K., Takeda, Y., Yoshikawa, K., Sawada, M., Nakatake, M., Namatame, H., Taniguchi, M., Takata, Y., Ikenaga, E., Shin, S., Kobayashi, K., Tamasaku, K., Nishino, Y., Miwa, D., Yabashi, M., and Ishikawa, T. (2004) *Phys. Rev. Lett.*, 93, 246404.

Sawatzky, G.A. (1977) *Phys. Rev. Lett.*, 39, 504.

Schattschneider, P., Hebert, C., Franco, H., and Jouffrey, B. (2005) *Phys. Rev. B*, 72, 45142.

Schmitt, T., Duda, L.C., Augustsson, A., Guo, J.H., Nordgren, J., Downes, J.E., McGuinness, C., Smith, K.E., Dhalenne, G., Revcolevshci, A., Klemm, M., and Horn, S. (2002) *Surf. Rev. Lett.*, 9, 1369.

Schmitt, T., Duda, L.C., Matsubara, M., Mattesini, M., Klemm, M., Augustsson, A., Guo, J. H., Uozumi, T., Horn, S., Ahuja, R., Kotani, A., and Nordgren, J. (2004a) *Phys. Rev. B*, 69, 125103.

Schmitt, T., Duda, L.C., Matsubara, M., Augustsson, A., Trif, F., Guo, J.H., Gridneva, L., Uozumi, T., Kotani, A., and Nordgren, J. (2004b) *J. Alloy Comp.*, 362, 143.

Schneider, W.D., Delley, B., Wuilloud, E., Imer, J.M., and Baer, Y. (1985) *Phys. Rev. B*, 32, 6819.

Scholl, A., Stöhr, J., Luning, J., Seo, J.W., Fompeyrine, J., Siegwart, H., Locquet, J.P., Nolting, F., Anders, S., Fullerton, E.E., Scheinfein, M.R., and Padmore, H.A. (2000) *Science*, 287, 1014.

Schütz, G., Wagner, W., Wilhelm, W., Kienle, P., Zeller, R., Frahm, R., and Materlik, G. (1987) *Phys. Rev. Lett.*, 58, 737.

Schütz, G., Knülle, M., Wienke, R., Wilhelm, W., Wagner, W., Kienle, P., and Frahm, R. (1988) *Z. Phys. B*, 73, 67.

Schütz, G., Wienke, R., Wilhelm, W., Wagner, W., Kienle, P., Zeller, R., and Frahm, R. (1989) *Z. Phys. B*, 75, 495.

Schütz, G., Fischer, P., Attenkofer, K., Knulle, M., Ahlers, D., Stahler, S., Detlefs, C., Ebert, H., and de Groot, F.M.F. (1994) *J. Appl. Phys.*, 76, 6453.

Seah, M.P., and Dench, W.A. (1979) *Surf. Interf. Anal.*, 1, 2.

Sekiyama, A., Iwasaki, T., Matsuda, K., Saitoh, Y., Onukl, Y., and Suga, S. (2000) *Nature*, 403, 396.

Sen, S.K., Riga, J., and Verbist, J. (1976) *Chem. Phys. Lett.*, 39, 560.

Sham, T.K. (1983) *J. Am. Chem. Soc.*, 105, 2269.

Sham, T.K. (1985) *Phys. Rev. B*, 31, 1888.

Shin, S., Suga, S., Kanzaki, H., Shibuya, S., and Yamaguchi, T. (1981) *Solid State Commun.*, 38, 1281.

Shirley, E.L. (2004) *J. Electron Spectrosc. Relat. Phenom.*, 136, 77.

Shirley, E.L. (2005) *J. Electron Spectrosc. Relat. Phenom.*, 144, 1187.

Shukla, A., Rueff, J.P., Badro, J., Vanko, G., Mattila, A., de Groot, F.M.F., and Sette, F. (2003) *Phys. Rev. B*, 67.

Shukla, A., Calandra, M., Taguchi, M., Kotani, A., Vanko, G., and Cheong, S.W. (2006) *Phys. Rev. Lett.*, 96, 77006.

Siegbahn, K., Nordling, C., Fahlman, A., Nordberg, R., Hamrin, K., Hedman, J., Johansson, G., Bergmark, T., Karlsson, S., Lindgren, I., and Lindberg, B. (1967) *ESCA—Atomic, Molecular and Solid State Structure Studied by Means of Electron Spectroscopy*, Nova Acta Regiae Soc. Sci. Ups., Ser IV, 20.

Sinkovic, B., Tjeng, L.H., Brookes, N.B., Goedkoop, J.B., Hesper, R., Pellegrin, E., de Groot, F.M.F., Altieri, S., Hulbert, S.L., Shekel, E., and Sawatzky, G.A. (1997) *Phys. Rev. Lett.*, 79, 3510.

Soderholm, L., Liu, G.K., Antonio, M.R., and Lytle, F.W. (1998) *J. Chem. Phys.*, 109, 6745.

Soldatov, A.V., Ivanchenko, T.S., Dellalonga, S., Kotani, A., Iwamoto, Y., and Bianconi, A. (1994) *Phys. Rev. B*, 50, 5074.

Sparks Jr., C.J. (1974) *Phys. Rev. Lett.*, 33, 262.

Steiner, P., Zimmermann, R., Reinert, F., Engel, T., and Hüfner, S. (1996) *Z. Phys. B*, 99, 479.

Stewart, B., Peacock, R.D., Alagna, L., Prosperi, T., Turchini, S., Goulon, J., Rogalev, A., and Goulon-Ginet, C. (1999) *J. Am. Chem. Soc.*, 121, 10233.

Stöhr, J. (1992) *XANES Spectroscopy* (Springer, Berlin).

Stöhr, J., and König, H. (1995) *Phys. Rev. Lett.*, 75, 3748.

Stöhr, J., and Nakajima, R. (1998) *IBM J. Res. Dev.*, 42, 73.

Suga, S., Imada, S., Higashiya, A., Shigemoto, A., Kasai, S., Sing, M., Fujiwara, H., Sekiyama, A., Yamasaki, A., Kim, C., Nomura, T., Igarashi, J., Yabashi, M., and Ishikawa, T. (2005a) *Phys. Rev. B*, 72, 81101.

Suga, S., Sekiyama, A., Imada, S., Shigemoto, A., Yamasaki, A., Tsunekawa, M., Dallera, C., Braicovich, L., Lee, T.L., Sakai, O., Ebihara, T., and Onukis, Y. (2005b) *J. Phys. Soc. Jpn.*, 74, 2880.

Sugano, S., Tanabe, Y., and Kamimura, H. (1970) *Multiplets of Transition Metal Ions* (Academic Press, New York).

Suzuki, S., Nagakura, I., Ishii, T., Satoh, T., and Sagawa, T. (1972) *Phys. Lett.*, 41A, 95.

Suzuki, S., Ishii, T., and Sagawa, T. (1974) *J. Phys. Soc. Jpn.*, 37, 1334.

Taguchi, M., Uozumi, T., and Kotani, A. (1997) *J. Phys. Soc. Jpn.*, 66, 247.

Taguchi, M., Parlebas, J.C., Uozumi, T., Kotani, A., and Kao, C.C. (2000) *Phys. Rev. B*, 61, 2553.

Taguchi, M., Braicovich, L., Borgatti, F., Ghiringhelli, G., Tagliaferri, A., Brookes, N.B., Uozumi, T., and Kotani, A. (2001) *Phys. Rev. B*, 63, 245114.

Taguchi, M., Braicovich, L., Annese, E., Dallera, C., Ghiringhelli, G., Tagliaferri, A., and Brookes, N.B. (2004) *Phys. Rev. B*, 69, 212414.

Taguchi, M., Chainani, A., Horiba, K., Takata, Y., Yabashi, M., Tamasaku, K., Nishino, Y., Miwa, D., Ishikawa, T., Takeuchi, T., Yamamoto, K., Matsunami, M., Shin, S., Yokoya, T., Ikenaga, E., Kobayashi, K., Mochiku, T., Hirata, K., Motoya, K., Hori, J., Ishii, K., Nakamura, F., and Suzuki, T. (2005a) *Phys. Rev. Lett.*, 95, 177002.

Taguchi, M., Chainani, A., Kamakura, N., Horiba, K., Takata, Y., Yabashi, M., Tamasaku, K., Nishino, Y., Miwa, D., Ishikawa, T., Shin, S., Ikenaga, E., Yokoya, T., Kobayashi, K., Mochiku, T., Hirata, K., and Motoya, K. (2005b) *Phys. Rev. B*, 71, 155102.

Taguchi, M., Kruger, P., Parlebas, J.C., and Kotani, A. (2006) *Phys. Rev. B*, 73, 125404.

Taillefumier, M., Cabaret, D., Flank, A.M., and Mauri, F. (2002) *Phys. Rev. B*, 66.

Tanaka, S., Ogasawara, H., Kayanuma, Y., and Kotani, A. (1989a) *J. Phys. Soc. Jpn.*, 58, 1087.

Tanaka, S., Okada, K., and Kotani, A. (1989b) *J. Phys. Soc. Jpn.*, 58, 813.

Tanaka, S., Kayanuma, Y., and Kotani, A. (1990) *J. Phys. Soc. Jpn.*, 59, 1488.

Tanaka, S., and Kotani, A. (1993) *J. Phys. Soc. Jpn.*, 62, 464.

Tanaka, S., Ogasawara, H., Okada, K., and Kotani, A. (1993) *Jpn. J. Appl. Phys.*, 32, 101.

Tanaka, S., Okada, K., and Kotani, A. (1994) *J. Phys. Soc. Jpn.*, 63, 2780.

Tanaka, A., and Jo, T. (1994) *J. Phys. Soc. Jpn.*, 63, 2788.

Tanaka, A., and Jo, T. (1996) *J. Phys. Soc. Jpn.*, 65, 615.

Tanaka, A., and Jo, T. (1997) *J. Phys. Soc. Jpn.*, 66, 1591.

Tanaka, I., Mizoguchi, T., and Yamamoto, T. (2005) *J. Am. Ceram. Soc.*, 88, 2013.

Terakura, K., Oguchi, T., Williams, A.R., and Kübler, J. (1984) *Phys. Rev. B*, 30, 4734.

Teramura, Y., Tanaka, A., and Jo, T. (1996) *J. Phys. Soc. Jpn.*, 65, 1053.

Tezuka, Y., and Shin, S. (2004) *J. Electron Spectrosc. Relat. Phenom.*, 136, 151.

Theil, C., van Elp, J., and Folkmann, F. (1999) *Phys. Rev. B*, 59, 7931.

Thole, B.T., Cowan, R.D., Sawatzky, G.A., Fink, J., and Fuggle, J.C. (1985a) *Phys. Rev. B*, 31, 6856.

Thole, B.T., van der Laan, G., and Sawatzky, G.A. (1985b) *Phys. Rev. Lett.*, 55, 2086.

Thole, B.T., van der Laan, G., Fuggle, J.C., Sawatzky, G.A., Karnatak, R.C., and Esteva, J. (1985c) *Phys. Rev. B*, 32, 5107.

Thole, B.T., and van der Laan, G. (1988a) *Phys. Rev. B*, 38, 3158.

Thole, B.T., and van der Laan, G. (1988b) *Phys. Rev. A*, 38, 1943.

Thole, B.T., and van der Laan, G. (1988c) *Phys. Rev. B*, 38, 3158.

Thole, B.T., and van der Laan, G. (1991a) *Phys. Rev. Lett.*, 67, 3306.

Thole, B.T., and van der Laan, G. (1991b) *Phys. Rev. B*, 44, 12424.

Thole, B.T., Carra, P., Sette, F., and van der Laan, G. (1992) *Phys. Rev. Lett.*, 68, 1943.

Thole, B.T., and van der Laan, G. (1993) *Phys. Rev. Lett.*, 70, 2499.

Thole, B.T., and van der Laan, G. (1994a) *Phys. Rev. B*, 49, 9613.

Thole, B.T., and van der Laan, G. (1994b) *Phys. Rev. B*, 50, 11474.

Thole, B.T., van der Laan, G., and Fabrizio, M. (1994) *Phys. Rev. B*, 50, 11466.

Thole, B.T., and van der Laan, G. (1994c) *Phys. Rev. B*, 50, 11474.

Thole, B.T., Dürr, H.A., and van der Laan, G. (1995) *Phys. Rev. Lett.*, 74, 2371.

Thurgate, S.M. (1996) *J. Electron Spectrosc. Relat. Phenom.*, 81, 1.

Tjeng, L.H., Chen, C.T., Rudolf, P., Meigs, G., van der Laan, G., and Thole, B.T. (1993) *Phys. Rev. B*, 48, 13378.

Tjeng, L.H., Sinkovic, B., Brookes, N.B., Goedkoop, J.B., Hesper, R., Pellegrin, E., de Groot, F.M.F., Altieri, S., Hulbert, S.L., Shekel, E., and Sawatzky, G.A. (1997) *Phys. Rev. Lett.*, 78, 1126.

Tjeng, L.H., de Groot, F.M.F., Sawatzky, G.A., Sinkovic, B., Brookes, N.B., Goedkoop, J.B., Hesper, R., Altieri, S., Pellegrin, E., and Hulbert, S.L. (1998) *Phys. Rev. Lett.*, 81, 734.

Tjeng, L.H., Brookes, N.B., Goedkoop, J.B., Sinkovic, B., de Groot, F.M.F., Hesper, R., Altieri, S., Pellegrin, E., Tanaka, A., Hulbert, S.L., Shekel, E., Hien, N.T., Menovsky, A.A., and Sawatzky, G.A. (1999) *Jpn. J. Appl. Phys.*, 38, 344.

Tjernberg, O., Soderholm, S., Karlsson, U.O., Chiaia, G., Qvarford, M., Nylen, H., and Lindau, I. (1996) *Phys. Rev. B*, 53, 10372.

Tjernberg, O., Finazzi, M., Duo, L., Ghiringhelli, G., Ohresser, P., and Brookes, N.B. (2000) *Physica B*, 281, 723.

Tjernberg, O., Tjeng, L.H., Steeneken, P.G., Ghiringhelli, G., Nugroho, A.A., Menovsky, A.A., and Brookes, N.B. (2003) *Phys. Rev. B*, 67.

Tolentino, H., Medarde, M., Fontaine, A., Baudelet, F., Dartyge, E., Guay, D., and Tourillon, G. (1992) *Phys. Rev. B*, 45, 8091.

Tsutsui, K., Tohyama, T., and Maekawa, S. (1999) *Phys. Rev. Lett.*, 83, 3705.

Tsutsui, K., Tohyama, T., and Maekawa, S. (2000) *Phys. Rev. B*, 61, 7180.

Tulkki, J., and Åberg, T. (1980) *J. Phys. B*, 13, 3341.

Ufuktepe, Y., Kimura, S., Kinoshita, T., Nath, K.G., Kumigashira, H., Takahashi, T., Matsumura, T., Suzuki, T., Ogasawara, H., and Kotani, A. (1998) *J. Phys. Soc. Jpn.*, 67, 2018.

Uhlenbrock, S., Scharfschwerdt, C., Neumann, M., Illing, G., and Freund, H.J. (1992) *J. Phys. Cond. Matt.*, 4, 7973.

Uozumi, T., Okada, K., Kotani, A., Durmeyer, O., Kappler, J.P., Beaurepaire, E., and Parlebas, J.C. (1992) *Europhys. Lett.*, 18, 85.

Uozumi, T., Okada, K., Kotani, A., Zimmermann, R., Steiner, P., Hüfner, S., Tezuka, Y., and Shin, S. (1997) *J. Electron Spectrosc. Relat. Phenom.*, 83, 9.

Uozumi, T., Kanai, K., Shin, S., Kotani, A., Schmerber, G., Kappler, J.P., and Parlebas, J.C. (2002) *Phys. Rev. B*, 65, 45105.

Uozumi, T., Kotani, A., and Parlebas, J.C. (2004) *J. Electron Spectrosc. Relat. Phenom.*, 137–40, 623.

van Acker, J.F., Stadnik, Z.M., Fuggle, J.C., Hoekstra, H., Buschow, K.H.J., and Stroink, G. (1988) *Phys. Rev. B*, 37, 6827.

van Aken, P.A., Styrsa, V.J., Liebscher, B., Woodland, A.B., and Redhammer, G.J. (1999) *Phys. Chem. Miner.*, 26, 584.

van Aken, P.A., and Liebscher, B. (2002) *Phys. Chem. Miner.*, 29, 188.

van Bokhoven, J.A., Louis, C., T Miller, J., Tromp, M., Safonova, O.V., and Glatzel, P. (2006) *Angew. Chem.*, 45, 4651.

van der Laan, G., Westra, C., Haas, C., and Sawatzky, G.A. (1981) *Phys. Rev. B*, 23, 4369.

van der Laan, G., Thole, B.T., Sawatzky, G.A., Goedkoop, J.B., Fuggle, J.C., Esteva, J.M., Karnatak, R., Remeika, J.P., and Dabkowska, H.A. (1986) *Phys. Rev. B*, 34, 6529.

van der Laan, G., Thole, B.T., Sawatzky, G.A., and Verdaguer, M. (1988) *Phys. Rev. B*, 37, 6587.

van der Laan, G. (1990) *Phys. Rev. B*, 41, 12366.

van der Laan, G. (1991) *J. Phys. Cond. Matt.*, 3, 7443.

van der Laan, G. (1995) *Phys. Rev. B*, 51, 240.

van der Laan, G., Ghiringhelli, G., Tagliaferri, A., Brookes, N.B., and Braicovich, L. (2004) *Phys. Rev. B*, 69, 104427.

van der Laan, G. (2006) *Curr. Op. Sol. St. Mat. Sci.*, 10, 120.

van Dorssen, G.E., Koningsberger, D.C., and Ramaker, D.E. (2002) *J. Phys. Cond. Matt.*, 14, 13529.

van Elp, J., Wieland, J.L., Eskes, H., Kuiper, P., Sawatzky, G.A., de Groot, F.M.F., and Turner, T.S. (1991) *Phys. Rev. B*, 44, 6090.

van Elp, J., and Searle, B.G. (1997) *J. Electron Spectrosc. Relat. Phenom.*, 86, 93.

van Elp, J., and Tanaka, A. (1999) *Phys. Rev. B*, 60, 5331.

van Veenendaal, M.A., Eskes, H., and Sawatzky, G.A. (1993) *Phys. Rev. B*, 47, 11462.

van Veenendaal, M.A., and Sawatzky, G.A. (1993) *Phys. Rev. Lett.*, 70, 2459.

van Veenendaal, M.A., and Sawatzky, G.A. (1994) *Phys. Rev. B*, 49, 3473.

van Veenendaal, M.A., Alders, D., and Sawatzky, G.A. (1995) *Phys. Rev. B*, 51, 13966.

van Veenendaal, M., Goedkoop, J.B., and Thole, B.T. (1997) *Phys. Rev. Lett.*, 78, 1162.

van Veenendaal, M., and Fedro, A.J. (2004) *Phys. Rev. Lett.*, 92, 219701.

van Veenendaal, M. (2006) *Phys. Rev. B*, 74, 085118.

Vanko, G., Neisius, T., Molnar, G., Renz, F., Karpati, S., Shukla, A., and de Groot, F.M.F. (2006) *J. Phys. Chem. B*, 110, 11647.

Vanko, G. (2008) (unpublished).

Veal, B.W., and Paulikas, A.P. (1985) *Phys. Rev. B*, 31, 5399.

Vogel, J., and Sacchi, M. (1994) *J. Electron Spectrosc. Relat. Phenom.*, 67, 181.

Vogel, J., Kuch, W., Bonfim, M., Camarero, J., Pennec, Y., Offi, F., Fukumoto, K., Kirschner, J., Fontaine, A., and Pizzini, S. (2003) *Appl. Phys. Lett.*, 82, 2299.

Wang, X., Randall, C.R., Peng, G., and Cramer, S.P. (1995) *Chem. Phys. Lett.*, 243, 469.

Wang, X., de Groot, F.M.F., and Cramer, S.P. (1997) *Phys. Rev. B*, 56, 4553.

Wasinger, E.C., de Groot, F.M.F., Hedman, B., Hodgson, K.O., and Solomon, E.I. (2003) *J. Am. Chem. Soc.*, 125, 12894.

Watanabe, M., Harada, Y., Nakazawa, M., Ishiwata, Y., Eguchi, R., Takeuchi, T., Kotani, A., and Shin, S. (2002) *Surf. Rev. Lett.*, 9, 983.

Weibel, P., Grioni, M., Malterre, D., Dardel, B., and Baer, Y. (1994) *Phys. Rev. Lett.*, 72, 1252.

Weinelt, M., Nilsson, A., Magnuson, M., Wiell, T., Wassdahl, N., Karis, O., Fohlisch, A., Martensson, N., Stöhr, J., and Samant, M. (1997) *Phys. Rev. Lett.*, 78, 967.

Weissbluth, M. (1978) *Atoms and Molecules* (Academic Press, New York).

Weller, D., Stöhr, J., Nakajima, R., Carl, A., Samant, M.G., Chappert, C., Megy, R., Beauvillain, P., Veillet, P., and Held, G.A. (1995) *Phys. Rev. Lett.*, 75, 3752.

Wende, H. (2004) *Rep. Prog. Phys.*, 67, 2105.

Wernet, P., Nordlund, D., Bergmann, U., Cavalleri, M., Odelius, M., Ogasawara, H., Naslund, L.A., Hirsch, T.K., Ojamae, L., Glatzel, P., Pettersson, L.G.M., and Nilsson, A. (2004) *Science*, 304, 995.

Wernet, P., Testemale, D., Hazemann, J.L., Argoud, R., Glatzel, P., Pettersson, L.G.M., Nilsson, A., and Bergmann, U. (2005) *J. Chem. Phys.*, 123.

Westre, T.E., Kennepohl, P., DeWitt, J.G., Hedman, B., Hodgson, K.O., and Solomon, E.I. (1997) *J. Am. Chem. Soc.*, 119, 6297.

Wilke, M., Farges, F., Petit, P.E., Brown, G.E., and Martin, F. (2001) *Amer. Miner.*, 86, 714.

Wu, R.Q., Wang, D.S., and Freeman, A.J. (1993) *Phys. Rev. Lett.*, 71, 3581.

Wu, R.Q., and Freeman, A.J. (1994) *Phys. Rev. Lett.*, 73, 1994.

Wu, Z.Y., Xian, D.C., Hu, T.D., Xie, Y.N., Tao, Y., Natoli, C.R., Paris, E., and Marcelli, A. (2004) *Phys. Rev. B*, 70, 33104.

Wuilloud, E., Delley, B., Schneider, W.D., and Baer, Y. (1984) *Phys. Rev. Lett.*, 53, 202.

X-Ray-Data-Booklet (2001) (LBNL, Berkeley).

Zaanen, J., Sawatzky, G.A., and Allen, J.W. (1985a) *Phys. Rev. Lett.*, 55, 418.

Zaanen, J., Sawatzky, G.A., Fink, J., Speier, W., and Fuggle, J.C. (1985b) *Phys. Rev. B*, 32, 4905.

Zaanen, J., Westra, C., and Sawatzky, G.A. (1986) *Phys. Rev. B*, 33, 8060.

Zaanen, J., and Sawatzky, G.A. (1987) *Can. J. Phys.*, 65, 1262.

Zabinsky, S.I., Rehr, J.J., Ankudinov, A., Albers, R.C., and Eller, M.J. (1995) *Phys. Rev. B*, 52, 2995.

Zeller, R. (1992) in *Unoccupied Electronic States* (Springer, Berlin), p. 25.

Zhang, F.C., and Rice, T.M. (1988) *Phys. Rev. B*, 37, 3759.

Zhang, Z.M., Jeng, S.P., and Henrich, V.E. (1991) *Phys. Rev. B*, 43, 12004.

Zhang, G.P., Callcott, T.A., Woods, G.T., Lini, L., Sales, B., Mandrus, D., and He, J. (2002) *Phys. Rev. Lett.*, 88, 77401.

Zhang, G.P., Callcott, T.A., Woods, G.T., Lin, L., Sales, B., Mandrus, D., and He, J. (2004) *Phys. Rev. Lett.*, 93, 169702.

Zhang, G.P., and Callcott, T.A. (2006) *Phys. Rev. B*, 73, 125102.

Zimkina, T.M., Shulakov, A.S., Fomichev, V.A., Braiko, A.P., and Stepanov, A.P. (1984) *Int. Conf. X-Ray Inn.-Shell Processes At., Mol. Solids, Conf. Proc.*, 263.

Zimmermann, R., Claessen, R., Reinert, F., Steiner, P., and Hüfner, S. (1998) *J. Phys. Cond. Matt.*, 10, 5697.

Index

An environmentally friendly book printed and bound in England by www.printondemand-worldwide.com

PEFC Certified

This product is
from sustainably
managed forests
and controlled
sources

www.pefc.org

PEFC/16-33-415

This book is made of chain-of-custody materials; FSC materials for the cover and PEFC materials for the text pages.

#0005 - 210416 - C0 - 234/156/27 [29] - CB - 9780849390715